AN INTRODUCTION TO MATHEMATICS
FOR STUDENTS OF ECONOMICS

By the same author

*

BUILDING CYCLES AND BRITAIN'S GROWTH
URBAN DECAY
(*with D. F. Medhurst*)

AN INTRODUCTION
TO MATHEMATICS

for Students of Economics

J. PARRY LEWIS

SECOND EDITION

MACMILLAN
ST MARTIN'S PRESS

First edition 1959
Reprinted 1961, 1962, 1964, 1965, 1967, 1968
Second edition 1969
Reprinted 1970, 1971

Published by
THE MACMILLAN PRESS LTD
London and Basingstoke
Associated companies in New York Toronto
Dublin Melbourne Johannesburg and Madras

Library of Congress catalog card no. 77–76382

SBN 333 07302 9 (hard cover)

Printed in Great Britain by
ROBERT MACLEHOSE AND CO LTD
The University Press, Glasgow

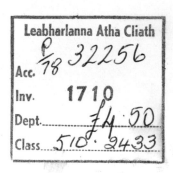

PREFACE TO THE SECOND EDITION

This new edition has benefited from ten years of further teaching experience and comment from students and teachers. It remains a book about mathematics, aiming at clarity and an emphasis on fundamentals, and written specifically for students of economics, even though others may find it useful.

It differs from the first edition in several ways. There has been substantial rewriting, usually to achieve greater clarity. The main mathematical additions are an expanded section on probability, some elementary statistical ideas, additional proofs, a section on reduction formulae, a chapter on multiple, linear and surface integration, six chapters on sets and linear algebra (including linear programming) and a few new appendices, including one on Beta and Gamma Functions. There are also many more examples drawn from economics.

Two cautionary notes must be sounded. It has become fashionable to praise " new mathematics " and to look upon calculus as " old-fashioned ". The economist can certainly use set theory and linear algebra to solve important problems, and he needs to have some knowledge of them: but he also needs a much greater facility in calculus than is sometimes recognised, and if it comes to choice between " old " and " new " – which it should not – it is my view that for the student of economics the " old " is the more useful. This new edition can serve as an introduction to both.

The second note of caution stems from the development of mathematical economics. Mathematics is a tool which can enable its user to make the most of his raw materials: but the quality of the output will still depend on that of the input. If mathematics is used to explore the consequences of some artificial or simplified assumptions then the emerging results are necessarily correct only when those assumptions hold. There comes a point when the artificiality of the assumptions turns mathematical economics into an intellectual exercise of little, if any, relevance to real problems. Mathematical technique outstrips understanding of economic realities, and the economist is transformed into a mathematician. But the student should remember that if he is to deal with real and complicated economic problems then he may need powerful techniques. The more elementary techniques taught in this book are a necessary prelude to them; and the very simplified economic assumptions are made in order to help in the teaching of these techniques, and to demonstrate the area of application. They are no substitute for a study of economic realities.

I am greatly indebted to my Research Assistant, Mr. Adrian Lovett, for help with the manuscript and proofs. Other valuable assistance came from Mrs. Margaret Barrow, Miss Nola Bowden and Mrs. Dorothy Evans. My greatest debt is to all of those readers who have given me their comments on the first edition.

UNIVERSITY OF MANCHESTER   J. PARRY LEWIS
    January 1969

PREFACE TO THE FIRST EDITION

THIS book is based on a non-examined course of lectures given to those students of economics at Cardiff University College who care to attend. It is not a book about economics; nor is it written for anybody but the student of economics. It is essentially what its title claims, and attempts to provide the student who has done little, if any, mathematics with an introduction which starts at a very simple level but goes quite a long way.

Although the reader of modern economic theory needs a considerable insight into mathematical principles and techniques, his requirements are highly specialist. Some branches of mathematics are quite unnecessary to him; and although he needs to know something about homogeneous functions and Euler's Thereom, a proof of that —beautiful though it may be—is quite unnecessary. Because of this the whole book is both selective and practical. Proofs are often omitted ; but basic principles are firmly stressed, and there are carefully graded exercises.

The book is divided into sections. The first revises elementary algebra, starting from the very beginning, and then leading to more complicated ideas. The second introduces some trigonometry and co-ordinate geometry, once again without assuming any previous knowledge of these subjects.

The third section is concerned with the calculus, and is illustrated by references to a number of economic problems. In the next section these ideas are extended to cases where many variables are involved. The fifth section develops certain ideas which are of use in cycle theory.

Appendices deal with a number of matters, the principle-one being determinants. This particular appendix may be read quite separately from the rest of the book. There is also a bibliography.

In writing this book I have been most fortunate with my helpers. Some of the earlier chapters benefited from comments made by Mr. Gordon Hargreaves, Mr. A. Hedley Pope and Mr. Alastair White. Dr. F. J. Jones acted as a most useful guinea-pig, using my drafts in an attempt to learn some mathematics. My colleague Mr. J. D. Pole most kindly read the whole of the first draft and made many valuable suggestions. Other suggestions and much kind encouragement came from Professor Brinley Thomas, to whom I am also in-

ix

debted for the way in which he took me under his wing at a time when I knew no economics, and Professor R. G. D. Allen whose interest and encouragement have been most generous, and whose suggestions most valuable. The final typescript was read by Mrs. Kathleen Urwin, and the proofs by Mrs. Hilda Hughes, both of whom have saved me from a number of errors. If any remain it is likely that they are due to my own final alterations, rather than to any oversight on the part of these many and generous assistants. Mrs. Urwin also provided about half of the answers to the exercises, and others were provided by my former teacher and colleague, Mr. J. Maldwyn Howells.

While my debt to all these people will be properly appreciated only by anybody who has written a mathematics book, and cannot be overstressed, I feel that if this book has merit as a teaching book then most of it must nevertheless be attributed to four people to whose teaching I myself am so indebted : Mr. J. M. Howells and Mr. W. S. Osborne, who helped me both as a pupil and as a young teacher at Caerphilly Grammar School, Dr. D. G. Taylor whose lectures had an influence even greater than his infectious enthusiasm, and the Rev. Dr. A. D. Amos who has the rare gift of being able to explain simple things.

To my publisher and printers I must also express my thanks and admiration for the ways in which they have performed their duties and saved me from many pitfalls.

J. P. L.

UNIVERSITY COLLEGE
    CARDIFF
    July 1958

CONTENTS

APPENDICES:

SECTION I

ALGEBRA

Chapter I.   Revision of Elementary Algebra.

Chapter II.   The Summation and Convergence of Some Simple Series.

Chapter III. Permutations, Combinations, the Binomial Theorem, and Probability.

Most readers will already be familiar with most of the first chapter. Time spent on revision, and on the exercises, will bring considerable benefit later on.

Although a thorough knowledge of the contents of Chapters II and III is desirable, it will be possible to read most of this book even if the more difficult parts of these chapters are omitted from a first reading.

CHAPTER I

REVISION OF ELEMENTARY ALGEBRA

I think the devil put it in his mind.
In all arithmétic you couldn't find
Until today so tricky an equation.
CHAUCER : *The Summoner's Tale*

1. The nature of mathematics

A large part of economics is conerned with relationships between quantities. Some of the relationships are simply a consequence of definition; some are the result of the ways in which people behave. Some have been observed in the real world, while others are the result of a theory which may be quite untested.

Mathematics is largely concerned with analysing the logical consequences of these relationships, taken singly or together. If, as economists, we know (or believe that we know) a relationship between income and consumer expenditure, and another between consumer expenditure and investment, then we may ask whether these two relationships enable us to establish a third, as a logical consequence of the two we have, between income and investment. Whether they do enable us to do this will, of course, depend on the actual relationships with which we begin. It will also depend on our ability to manipulate the relationships in a way which will reveal their logical consequences.

This emphasis on the logical consequences of the assumed, observed, or defined relationships with which we begin our analysis is important. Whether the final conclusion is right or wrong depends only on the validity of our initial relationships—unless we make a mistake in our mathematics. When used correctly, mathematics will help us to see what our relationships imply: but it will not tell us anything else.

2. Elementary ideas

At the risk of seeming to be too concerned with pure mathematics, we begin by looking afresh at some elementary matters.

We all have some ideas about numbers and about the operations of addition, subtraction, multiplication and division. It is important for us to be a little more precise about some of these. There are, in

3

fact, very many different kinds of numbers, and many different kinds of addition and multiplication. Even in this very elementary book we shall be concerned with some of these different kinds of numbers and operations.

To begin, we are concerned with *real numbers* of the kind with which we are all familiar. Without giving a precise definition or an exhaustive list of their properties, we may note that these numbers can be represented by points on a straight line. If we begin with the

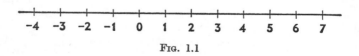

FIG. 1.1

positive *integers* 1, 2, 3, 4, ... Then we may denote them by equally spaced successive marks proceeding from the *zero*, 0.

We can produce a set of *negative integers*, denoted by equally spaced marks to the left of zero.

There will be a negative integer corresponding to every positive integer, and denoted by the same number prefaced with a minus sign. Zero will be the only integer equal to its negative.

Definition:

We may now *define* the operation of addition *so far as it is concerned with real integers* in the following way:

(a) To add the *positive integer a* to a positive or negative integer b, locate b on Figure 1.1 and move a spaces to the right.

(b) To add the *negative integer –a* to a positive or negative integer b, locate b on the scale in Figure 1.1 and move a places to the left.

EXERCISE 1.1

Show that, when addition is defined as above,

7 added to	3	results in	10				$7+3$	$=10$
6	,,	,,	-2	,,	,,	4	$6+(-2)$	$=4$
6	,,	,,	-9	,,	,,	-3	$6+(-9)$	$=-3$
6	,,	,,	-6	,,	,,	0	$6+(-6)$	$=0$
6	,,	,,	0	,,	,,	6	$6+0$	$=6$
-4	,,	,,	7	,,	,,	3	$-4+7$	$=3$
-4	,,	,,	4	,,	,,	0	$-4+4$	$=0$
-4	,,	,,	-5	,,	,,	-9	$-4+(-5)$	$=-9$
-4	,,	,,	0	,,	,,	-4	$-4+0$	$=-4$

Definition:

We define *subtraction* to be negative addition, in the sense that we move along the scale in Figure 1.1 in the direction opposite to that required for addition. *If we subtract a positive integer,* we move to the left (instead of to the right as we do when we add). *If we subtract a negative integer,* we move to the right (instead of to the left as we do when we add).

EXERCISE 1.2

(i) Show that, according to the above definition,

4 subtracted from			7 results in			3	$7-4$	$=3$
4	,,	,,	4	,,	,,	0	$4-4$	$=0$
4	,,	,,	-5	,,	,,	-9	$-5-4$	$=-9$
4	,,	,,	0	,,	,,	-4	$0-4$	$=-4$
-4	,,	,,	7	,,	,,	11	$7-(-4)=11$	
-4	,,	,,	4	,,	,,	8	$4-(-4)=8$	
-4	,,	,,	-4	,,	,,	0	$-4-(-4)=0$	
-4	,,	,,	-6	,,	,,	-2	$-6-(-4)=-2$	

(ii) Compare the first four results of (i) with those of Exercise 1.1. Note the equivalence of adding -4 and subtracting 4.

(iii) Note that when two integers are added or subtracted the result is another integer.

Definition:

We define *multiplication* of the integer a by the integer b to be the process of adding a to zero b times, if b is positive, and of subtracting a from zero b times if b is negative.

Example:

3 multiplied by			4 is $0+3+3+3+3$	$=$	12
-3	,,	,,	4 is $0+(-3)+(-3)+(-3)+(-3)$	$=$	-12
3	,,	,,	-4 is $0-3-3-3-3$	$=$	-12
-3	,,	,,	-4 is $0-(-3)-(-3)-(-3)-(-3)$	$=$	12

Notice that

a *positive* integer multiplied by a *positive* integer = *positive* integer.

a *negative*   ,,   ,,   ,, ,, *positive*   ,,   = *negative*   ,,

a *positive*   ,,   ,,   ,, ,, *negative*   ,,   = *negative*   ,,

a *negative*   ,,   ,,   ,, ,, *negative*   ,,   = *positive*   ,,

EXERCISE 1.3

Use the above definitions to show that,

$$
\begin{array}{ccccc}
7 & \text{multiplied by} & -6 & \text{yields} & -42 \\
0 & ,, & ,, & -3 & ,, & 0 \\
3 & ,, & ,, & 0 & ,, & 0 \\
-4 & ,, & ,, & -2 & ,, & 8 \\
-2 & ,, & ,, & 3 & ,, & -6 \\
-0 & ,, & ,, & 4 & ,, & 0 \\
\end{array}
$$

We now consider division. To do so we begin by acknowledging the existence of *fractional* numbers whose values lie between consecutive integers. We agree to define addition, subtraction and multiplication for these fractional numbers in a way similar to that already used for integers, but while we define multiplication of a fraction by an integer, we do not yet define multiplication of anything by a fraction.

Definition:

We define the *inverse* (or *reciprocal*) of an integer to be that fractional number which is such that when it is multiplied by the integer it yields unity. (1).

We denote the inverse of a by $\dfrac{1}{a}$.

Example:

The inverse of 3 is denoted by $\frac{1}{3}$. Its value is such that

$\frac{1}{3}$ multiplied by 3 yields **1**.

i.e. $0 + \frac{1}{3} + \frac{1}{3} + \frac{1}{3} = 1$

We can now define *division*.

Definition:

We define *division* of an integer a by an integer b to be the multiplication of the inverse of b by a. It is denoted by $\dfrac{a}{b}$.

Example:

4 divided by 3 is $\frac{1}{3}$ multiplied by 4

$$= 0 + \tfrac{1}{3} + \tfrac{1}{3} + \tfrac{1}{3} + \tfrac{1}{3}$$

which we denote by $\frac{4}{3}$.

3. Linear equations

The simplest kind of mathematical relationship is one such as

$$x + 4 = 6$$

which tells us something about the quantity x. Consideration of this statement will show that it is true for only one value of x, namely 2. By this statement, x is equated to 2; and the statement is called an *equation*. The quantity denoted by x, whose value has been found by considering this equation, is originally unknown; and because of this we refer to x as *the unknown.* Since only one value of the unknown will satisfy this equation, we say that the equation is *linear*, or of *the first degree*, in x.

Definition:

We define a linear equation in x (or an equation of the first degree in x) to be an equation of the form

$$ax + b = 0$$

where a and b are constants, a being positive or negative, and b being positive, negative or zero.

For example, $2x + 3 = 0$, $x - 6 = 0$ and $14x - \frac{1}{2} = 0$ are all linear equations.

We also define the *solution* of an equation. It is the value of the unknown which makes the equation a correct statement.

Examples of linear equations, with their solutions, are:

$x - 6 = 0$	Solution: $x = 6$
$14x - 3 = 0$	Solution: $x = \frac{3}{14}$
$2x + 3 = 6$	Solution: $x = \frac{3}{2}$

There are certain methods for the derivation of the solution. They are based on the fact that as the two sides of an equation are (by definition) equal, we may multiply, divide, add to, or subtract from, one side, provided that we do the same to the other side. For example,

(i) if $x - 6 \quad = 0$
     then $x - 6 + 6 = 0 + 6$
     and so $x = 6$

(ii) if $14x - 3 \quad = 0$
     then $14x - 3 + 3 = 3$
     and so $14x = 3$
     whence $x = 3 \div 14 = \frac{3}{14}$

(iii) if $2x + 3 \quad = 6$
      then $2x + 3 - 3 = 6 - 3$
      therefore $2x = 3$
      and so $x = \frac{3}{2}$

(iv) if $ax + b \quad = 0$
     then $ax + b - b = -b$
                          $ax = -b$
                          $x = -\dfrac{b}{a}$

(v) Sometimes we may have an equation such as

$$\frac{3x + 2}{4} = \frac{2x + 6}{5}$$

The easiest way of solving this is to multiply both sides by the lowest number into which both 4 and 5 can be divided, i.e., by 20. Remembering that multiplication signs can be replaced by appropriately located brackets, we have

$$20 \times \frac{3x + 2}{4} = 20 \times \frac{2x + 6}{5}$$

i.e. $\cancel{20}\dfrac{(3x + 2)}{\cancel{4}} = \cancel{20}\dfrac{(2x + 6)}{\cancel{5}}$

$5(3x + 2) = 4(2x + 6)$ (by cancelling)

Giving $15x + 10 = 8x + 24$

We now subtract all terms containing x from the right hand side and subtract the *same terms* from the other side. This means subtracting $8x$ from both sides, to get

$15x - 8x + 10 = 24$

i.e. $7x + 10 = 24$

Next we remove all terms *not* containing x from the left hand side, and subtract the same amount from the right hand side. This gives

$$7x = 24 - 10$$

$$= 14$$

whence $\qquad\qquad x = 2$

We may note that, starting with

$$\frac{3x+2}{4} = \frac{2x+6}{5}$$

we obtained the solution by a process which at one stage yielded

$$5(3x+2) = 4(2x+6)$$

We could, in fact, have reached this stage more quickly by the process of *cross-multiplication*, in which the top of one side is multiplied by the bottom of the other, and vice versa. For example, to solve

$$\frac{7x+8}{3x+1} = \frac{5}{3}$$

we can begin by writing it as

$$3(7x+8) = 5(3x+1)$$

whence $\qquad\qquad 21x + 24 = 15x + 5$

$$6x + 19 = 0$$

$$x = -\tfrac{19}{6}$$

The validity of this process depends on the fact that

if $\qquad\qquad \dfrac{a}{b} = \dfrac{c}{d}$

then $\qquad\qquad ad = bc$

We may prove this as follows:

if $\qquad\qquad \dfrac{a}{b} = \dfrac{c}{d}$

then multiplying both sides by b gives

$$a = \frac{bc}{d}$$

Further multiplication of both sides by d gives

$$ad = bc$$

which is the required result.

EXERCISE 1.4

Solve the following equations:

1. $3x - 17 = 0$ 2. $3x + 17 = 0$
3. $3x + 17 = 20$ 4. $3x + 17 = x + 20$
5. $3x + 17 = 2x + 20$ 6. $3(x + 6) = 21$
7. $3(x + 6) = x + 20$ 8. $3(x + 6) = 2(x + 3)$
9. $3(x + 6) = 2(x + 3) + 4$ 10. $3(x + 6) = 2(x + 3) - 4$
11. $3(x - 2) = 2(3x - 1)$ 12. $4(x - 7) = 3(2x + 1) - 5$

Note: These exercises will be more easily solved if they are tackled in the order set. The general principle is to rearrange each equation in the standard form $ax + b = 0$ and then to solve as above. Finally the answer may be checked by substitution. For example,

$$3(2x + 3) - 2 = 4x - 17$$

gives
$$6x + 9 - 2 = 4x - 17$$

$$6x + 7 = 4x - 17$$

whence
$$6x - 4x = -17 - 7$$

yielding
$$2x = -24$$

$$x = -12$$

If we put $x = -12$ in the original equation we obtain

L.H.S: $3(-24 + 3) - 2$ R.H.S: $-48 - 17$

$= 3(-21) - 2 = -63 - 2$ $= -65$

$= -65 = $ R.H.S.

which checks the correctness of the answer.

EXERCISE 1.5

If the demand (D) for a good is given by

$$D = 40 - 3p$$

where p is its price, and the supply (S) is given by

$$S = 10 + 2p$$

at what price are demand and supply equal? *Hint:* put $D = S$ and so obtain a linear equation in p. Then solve it.

4. Quadratic equations

The statement

$$3x^2 - 15x + 18 = 0$$

is true when $x = 2$. This can be shown by putting 2 into the left-hand side of the equation wherever we have x.

We have

$$(3 \times 2^2) - (15 \times 2) + 18$$

$$= (3 \times 4) \ - (15 \times 2) + 18$$

$$= \quad 12 \ \ - \quad 30 \quad + 18$$

$$= \quad 30 \ \ - \quad 30$$

which is identical with the right-hand side. However, it is also true when $x = 3$, the left-hand side becoming

$$27 - 45 + 18$$

which is again zero. It can be shown that there is no other value of x which satisfies the equation. This equation, in fact, equates x to 2 or 3. We say " to 2 or 3 " because we do not want to suggest that it is two things at once. The equation is true if $x = 2$. It is also true if $x = 3$. These are alternatives. If $x = 2$ or if $x = 3$ the equation is true. Because there are two and only two values of x to satisfy this equation, we call it an equation of the *second degree* in x. Equations of the second degree are often called *quadratic* equations. We call the values of x that satisfy such an equation its *solutions*; and we see that a quadratic equation, of the kind just considered, has two solutions. Later we shall meet quadratic equations which at first sight appear to have only one solution, or even none; but this is a difficulty that we shall consider when we come to it.

Definition:

A quadratic equation (or equation of the second degree) in x is an equation of the form

$$ax^2 + bx + c = 0$$

where a, b and c are constants, a being positive or negative, and b and c being positive, negative or zero.

For example, we have

$$x^2 - 3x - 1 = 0, \quad 7x^2 - 2x = 0 \quad \text{and} \quad 2x^2 - \tfrac{1}{2} = 0.$$

There are several methods for solving quadratic equations. We shall now consider three of them. It is the third method which the student is likely to use most frequently, but the other methods are useful.

First we shall consider solution by factorising, which is the more basic method and can often be used to great advantage. Then we shall consider solution by use of a formula, which is a severely practical method.

Consider the equation

$$(x-3)(2x+5)=0$$

We may clearly rewrite this as

$$x(2x+5)-3(2x+5)=0$$

i.e.,
$$2x^2+5x-6x-15=0$$

whence
$$2x^2-x-15=0$$

which we recognise as a quadratic equation in the standard form.

Now we know that a quadratic equation has two solutions, and that these solutions are the values of x which make the left-hand side zero. If, therefore, we can find two values of x which have this property they will be the solutions to this quadratic equation.

These values are easily found when we consider this equation in its earlier form

$$(x-3)(2x+5)=0$$

Here we have two brackets such that their product is zero. We could think of the first bracket as being replaced by u and the second by v so that we have

$$uv=0$$

where $u=x-3$ and $v=2x+5$. In this form we see all the more easily that the solution to the quadratic equation is immediately obtainable for if $uv=0$, then clearly either $u=0$ or $v=0$. If the product of two items is zero, then one of those items must be zero. Thus we have that the equation can be true only if either $u=0$ or $v=0$ which means that either $x-3=0$ or $2x+5=0$. It follows that the quadratic equation is true only if

either $x-3=0$ *or* $2x+5=0$

i.e., $x=3$ *or* $x=-\frac{5}{2}$

and these two values of x provide the solutions to that equation.

One method of solving a quadratic equation is therefore to write it in the standard form and to attempt to factorise the left-hand side. For example,

$$3x^2-7x+2=0$$

may be expressed as

$$(3x-1)(x-2)=0$$

which is true only if *either* $\qquad 3x-1=0 \quad or \quad x-2=0$

i.e., $\qquad\qquad x=\frac{1}{3} \quad or \qquad x=2$

These solutions may be checked by substitution in the original form.

A method of obtaining factors of expressions of this kind is given in Appendix 1, which also contains some exercises on this subject. Students who feel that they should already know this method may attempt the following exercise, turning to the appendix if they have much difficulty.

We may note that the equation

$$(x-3)(2x+5)=0$$

can also be written as

$$(x-3)(x+\tfrac{5}{2})=0$$

(for we have simply divided something that is equal to zero by 2, which has no effect on its value).

In this form the solutions are immediately apparent, for

$$either \quad x-3=0 \quad or \quad x+\tfrac{5}{2}=0$$

and so the solutions are $x=3$ and $x=-\frac{5}{2}$.

More generally, we have that if the solutions to a quadratic equation are given by x_1 and x_2 (where the suffixes 1 and 2 simply denote different values of x), then the equation can be written as

$$(x-x_1)(x-x_2)=0$$

For example, it will be found from the following exercise that the equation

$$4x^2+4x-3=0$$

has the solutions $x_1=\frac{1}{2}$ and $x_2=-\frac{3}{2}$. The equation can therefore be written as

$$(x-\tfrac{1}{2})(x+\tfrac{3}{2})=0$$

If this is multiplied out, we obtain

$$x^2-\frac{x}{2}+\frac{3x}{2}-\frac{1}{2}\cdot\frac{3}{2}=0$$

i.e. $\qquad\qquad x^2+x-\tfrac{3}{4}=0$

Multiplication of this by 4 yields

$$4x^2+4x-3=0$$

which is the original equation.

EXERCISE 1.6

Solve the following quadratic equations by the method given above, and rewrite the equations in the form $(x - x_1)(x - x_2) = 0$:

1. $3x^2 + 7x + 2 = 0$ 2. $3x^2 - 5x - 2 = 0$
3. $3x^2 + 5x - 2 = 0$ 4. $4x^2 - 4x - 3 = 0$
5. $4x^2 + 8x + 3 = 0$ 6. $4x^2 + 4x - 3 = 0$
7. $4x^2 - 8x + 3 = 0$ 8. $4x^2 - 13x + 3 = 0$
9. $4x^2 + 13x + 3 = 0$ 10. $4x^2 - 7x + 3 = 0$
11. $4x^2 + 7x + 3 = 0$ 12. $4x^2 - x - 3 = 0$
13. $4x^2 - 10x + 6 = 0$ 14. $4x^2 - 14x + 6 = 0$
15. $4x^2 - 11x + 6 = 0$ 16. $4x^2 + 5x - 6 = 0$
17. $4x^2 - 12x + 9 = 0$ 18. $4x^2 - 37x + 9 = 0$

A second method is known as " completing the square ". This is the method which justifies our third method, which is simply to use a formula which gives the answer. Factorising may not always be possible—and even when it is the answers may be far from obvious. The methods we now consider rest on the following facts:

For all values of x and of a

(i) $(x + a)^2 = (x + a)(x + a)$

$$= x^2 + 2ax + a^2$$

(ii) $(x - a)^2 = (x - a)(x - a)$

$$= x^2 - 2ax + a^2$$

For example,

$$(x + 4)^2 = (x + 4)(x + 4)$$

$$= x^2 + 8x + 16$$

Suppose now that we have an equation such as

$$3x^2 + 7x + 2 = 0.$$

We know that we can add to the left-hand side, subtract from it, multiply it or divide it, provided that we treat the right-hand side in the same way.

Let us try to convert the left-hand side into

$$x^2 + 2ax + a^2$$

where a is some number yet to be determined. This will probably mean altering the right-hand side.

We have

$$3x^2 + 7x + 2 = 0$$

First, divide by 3, to get

$$x^2 + \tfrac{7}{3}x + \tfrac{2}{3} = 0$$

Compare the L.H.S. of this with
$$x^2 + 2ax + a^2$$
If we match the coefficients of x by writing
$$2a = \tfrac{7}{3}$$

then $\qquad\qquad\qquad\qquad a = \tfrac{7}{6}$

and $\qquad\qquad\qquad\qquad a^2 = \tfrac{49}{36}$

How can we obtain from the L.H.S. the expression

$$x^2 + \tfrac{7}{3}x + \tfrac{49}{36}$$

which is exactly of the form we want?

Take the calculation a step further. We have

$$3x^2 + 7x + 2 = 0$$

$$x^2 + \tfrac{7}{3}x + \tfrac{2}{3} = 0 \quad \text{(by dividing)}$$

$$x^2 + \tfrac{7}{3}x \quad\;\; = -\tfrac{2}{3}$$

To make the L.H.S. as we want it we add $\tfrac{49}{36}$. And so we also add this to the R.H.S.

$$x^2 + \tfrac{7}{3}x + \tfrac{49}{36} = -\tfrac{2}{3} + \tfrac{49}{36}$$

$$= -\tfrac{24}{36} + \tfrac{49}{36}$$

$$= \tfrac{25}{36}$$

Now using the result that

$$x^2 + 2ax + a^2 = (x+a)^2$$

we have $\qquad\qquad x^2 + \tfrac{7}{3}x + \tfrac{49}{36} = (x + \tfrac{7}{6})^2$

∴ our equation
$$x^2 + \tfrac{7}{3}x + \tfrac{49}{36} = \tfrac{25}{36}$$

gives us $\qquad\qquad (x + \tfrac{7}{6})^2 \quad\;\; = \tfrac{25}{36}$

and so $\qquad\qquad x + \tfrac{7}{6} \quad\;\; = +\sqrt{\tfrac{25}{36}}$

Here we must be careful. There is a positive square root *and* a negative square root. Just as

$$(+5) \times (+5) = +25$$

$$\text{so} \quad (-5) \times (-5) = +25$$

Thus the equation

$$x + \tfrac{7}{6} = \sqrt{\tfrac{25}{36}} = \frac{\sqrt{25}}{\sqrt{36}}$$

may mean that

$$x + \tfrac{7}{6} = \tfrac{5}{6} \quad (\text{since } \tfrac{5}{6} = \sqrt{\tfrac{25}{36}})$$

or it may mean that

$$x + \tfrac{7}{6} = -\tfrac{5}{6}$$

In the former case

$$x + \tfrac{7}{6} = \tfrac{5}{6}$$

i.e.

$$x = -\tfrac{2}{6} = -\tfrac{1}{3}$$

while in the latter case

$$x + \tfrac{7}{6} = -\tfrac{5}{6}$$

$$x = -\tfrac{12}{6} = -2$$

Thus the two solutions are $x = -2$ and $x = -\tfrac{1}{3}$.

While the explanation of this method is tedious, its application is quicker. We give a few examples.

Examples

(i) $4x^2 + 8x + 3 = 0$

Divide by the coefficient of x^2 (i.e. by 4)

$$x^2 + 2x + \tfrac{3}{4} = 0$$

Shift the constant term to the other side, changing sign,

$$x^2 + 2x = -\tfrac{3}{4}$$

Add the square of half of the coefficient of x (i.e., the square of half of 2) to both sides

$$x^2 + 2x + (1)^2 = -\tfrac{3}{4} + (1)^2$$

$$x^2 + 2x + 1 = -\tfrac{3}{4} + 1 = \tfrac{1}{4}$$

i.e. $(x+1)^2 = (\tfrac{1}{2})^2$

∴ $x + 1 = \sqrt{(\tfrac{1}{2})^2}$

$$x + 1 = +\tfrac{1}{2}$$

or $x + 1 = -\tfrac{1}{2}$

whence $x = +\tfrac{1}{2} - 1 = -\tfrac{1}{2}$

or $x = -\tfrac{1}{2} - 1 = -\tfrac{3}{2}$

(ii) $$4x^2 - 8x + 3 = 0$$

Divide by the coefficient of x^2

$$x^2 - 2x + \tfrac{3}{4} = 0$$

Shift the constant (changing sign)

$$x^2 - 2x \quad = -\tfrac{3}{4}$$

Add the square of half of the coefficient of x.

$$x^2 - 2x + (-1)^2 = -\tfrac{3}{4} + (-1)^2$$

$$x^2 - 2x + \quad 1 \quad = -\tfrac{3}{4} + 1$$

whence
$$(x-1)^2 = \tfrac{1}{4}$$

$$(x-1) \ = \sqrt{\tfrac{1}{4}}$$

and so either $$x - 1 = +\tfrac{1}{2}$$

or $$x - 1 = -\tfrac{1}{2}$$

i.e. either $$x = 1 + \tfrac{1}{2} = \tfrac{3}{2}$$

or $$x = 1 - \tfrac{1}{2} = \tfrac{1}{2}$$

(iii) $$3x^2 + 5x - 2 = 0$$

$$x^2 + \tfrac{5}{3}x - \tfrac{2}{3} = 0$$

$$x^2 + \tfrac{5}{3}x \quad = \tfrac{2}{3}$$

$$x^2 + \tfrac{5}{3}x + (\tfrac{5}{6})^2 = \tfrac{2}{3} + (\tfrac{5}{6})^2$$

$$(x + \tfrac{5}{6})^2 = \tfrac{24}{36} + \tfrac{25}{36} = \tfrac{49}{36}$$

$$x + \tfrac{5}{6} = \sqrt{\tfrac{49}{36}}$$

∴ either $$x + \tfrac{5}{6} = +\tfrac{7}{6}$$

or $$x + \tfrac{5}{6} = -\tfrac{7}{6}$$

whence
$$x = \tfrac{1}{3} \text{ or } -2$$

B

Sometimes, of course, the numbers work out less early, as in

$$3x^2 + 5x - 3 = 0$$
$$x^2 + \tfrac{5}{3}x - 1 = 0$$
$$x^2 + \tfrac{5}{3}x = 1$$
$$x^2 + \tfrac{5}{3}x + (\tfrac{5}{6})^2 = 1 + (\tfrac{5}{6})^2$$
$$(x + \tfrac{5}{6})^2 = \tfrac{36}{36} + \tfrac{25}{36} = \tfrac{61}{36}$$
$$(x + \tfrac{5}{6}) = \sqrt{\tfrac{61}{36}} = +1\cdot302 \text{ or } -1\cdot302$$

(which is written $\pm 1\cdot302$)

and so $$x = -\tfrac{5}{6} \pm 1\cdot302$$

EXERCISE 1.7

Rework the questions of Exercise 1.6 using the method just described.

The third method is derived as follows.
Consider

$$ax^2 + bx + c = 0$$

where a, b and c are any numbers, positive, negative or zero, provided that a is not zero. Let us solve this, finding x in terms of a, b and c, by the method of completing the square. We have

$$ax^2 + bx + c = 0$$

Divide by a
$$x^2 + \frac{b}{a}x + \frac{c}{a} = 0$$

Shift the constant term

$$x^2 + \frac{b}{a}x = -\frac{c}{a}$$

Complete the square

$$x^2 + \frac{b}{a}x + \left(\frac{b}{2a}\right)^2 = -\frac{c}{a} + \left(\frac{b}{2a}\right)^2$$

$$\left(x + \frac{b}{2a}\right)^2 = -\frac{c}{a} + \frac{b^2}{4a^2}$$

$$= \frac{-4ac}{4a^2} + \frac{b^2}{4a^2}$$

$$= \frac{b^2 - 4ac}{4a^2}$$

$$x + \frac{b}{2a} = \sqrt{\frac{b^2 - 4ac}{4a^2}}$$

$$= \frac{\pm\sqrt{b^2 - 4ac}}{2a}$$

whence

$$(3) \qquad x = -\frac{b}{2a} \pm \frac{\sqrt{b^2 - 4ac}}{2a}$$

$$= \frac{-b \pm \sqrt{b^2 - 4ac}}{2a}$$

We can now use this result, which should be memorised, to solve a quadratic equation.

For example, to solve

$$4x^2 + 12x = 7$$

we write it as

$$4x^2 + 12x - 7 = 0$$

and compare this with

$$ax^2 + bx + c = 0$$

which shows that

$$a = 4; \quad b = 12; \quad c = -7$$

Substituting these values of a, b and c in the formula above, we have that the solutions are given by

$$x = \frac{-12 \pm \sqrt{12^2 - 4(4)(-7)}}{2(4)}$$

$$= \frac{-12 \pm \sqrt{144 + 112}}{8}$$

$$= \frac{-12 \pm \sqrt{256}}{8} = \frac{-12 \pm 16}{8}$$

This means that one solution is given by $x = \dfrac{-12 + 16}{8} = \dfrac{4}{8} = \dfrac{1}{2}$, and the other by $x = \dfrac{-12 - 16}{8} = \dfrac{-28}{8} = -3\frac{1}{2}$. In other words, the equation is satisfied by both $x = 0 \cdot 5$ and $x = -3 \cdot 5$ which may be checked either by substitution or by solving the equation by factorisation.

EXERCISE 1.8

1. Solve the exercises of 1.6 by using the above formula.

2. Also solve the following by this method, noting that in this case the figures involved are rather more awkward, and lead to solutions which could not be found by any attempt to factorise:

(i) $3x^2 + 4x - 5 = 0$ (ii) $x^2 - 4x = 8$

(iii) $2x^2 = 3x + 7$ (iv) $2x^2 - 8x + 3 = 0$

Before leaving the subject of quadratic equations there are two points for us to notice.

The first is that if we examine the solutions to the equations we have solved here, and to those in the excercises, we will find that in every case the sum of the two solutions is equal to $-b/a$ and their product is equal to c/a. For example, in solving $4x^2 + 12x - 7 = 0$ we found that the two solutions (or, as they are more commonly called, the two " roots ") were given by $\frac{1}{2}$ and $-3\frac{1}{2}$. The sum of these is -3, which is equal to $-\frac{12}{4}$; and their product is $-1\frac{3}{4}$, which is equal to $-\frac{7}{4}$. This is because if x_1 and x_2 are the two values of x which satisfy a quadratic equation, then the equation must be capable of being written as

$$(x - x_1)(x - x_2) = 0$$

which yields

$$x^2 - (x_1 + x_2)x + x_1 x_2 = 0$$

which may be compared with the standard form

$$ax^2 + bx + c = 0$$

This may be written

$$x^2 + \frac{bx}{a} + \frac{c}{a} = 0$$

Comparison with

$$x^2 - (x_1 + x_2)x + x_1 x_2 = 0$$

shows that

$$x_1 + x_2 = -\frac{b}{a} \quad \text{and} \quad x_1 x_2 = \frac{c}{a}$$

This is a highly important result, as we shall see later. Meanwhile it may be noted that it provides a method of solution. For example, if we wish to solve

$$2x^2 + 7x - 15 = 0$$

we may note that if x_1 and x_2 are the roots, then

$$x_1 + x_2 = -\tfrac{7}{2} \quad \text{and} \quad x_1 x_2 = -\tfrac{15}{2}.$$

A moment's inspection shows that if $x_1 = -5$ and $x_2 = \frac{3}{2}$ then these requirements are fulfilled, and so these values of x_1 and x_2 are the roots of the quadratic equation. Basically, this is the method of factorising.

The second point is illustrated if we attempt to solve the quadratic equation

$$2x^2 - 2x + 1 = 0$$

Here $a=2$, $b=-2$ and $c=1$. Substitution in the formula gives the roots

$$x = \frac{-(-2) \pm \sqrt{(-2)^2 - 4(2)(1)}}{4}$$

$$= \frac{2 \pm \sqrt{4-8}}{4}$$

$$= \frac{2 \pm \sqrt{-4}}{4}$$

A difficulty now appears. We have to find the square root of -4, i.e., we have to find a number such that its square is equal to -4. If we square $+2$ we obtain $+4$. If we square -2 we again obtain $+4$. In fact, as far as elementary algebra and arithmetic go, we cannot find the square root of a negative quantity. All the numbers we have met have been either positive or negative (or zero). The square of any positive number is bound to be positive ; and so is the square of any negative number. For reasons which we shall see later, we call a root of this kind, involving the square root of a negative quantity, a *complex root*. We shall have a great deal more to say about them later on. Meanwhile we may note that

   (i) *a quadratic equation has complex roots if, and only if,* $b^2 < 4ac$ (i.e., b^2 is less than $4ac$), for this is the condition that the quantity inside the square-root sign shall be negative ;
   (ii) *complex roots always occur in pairs*, being of the form $p + \sqrt{-q}$ and $p - \sqrt{-q}$ where q is positive.

Complex roots are particularly important in the study of cyclical fluctuations. Until we come to prepare for our study of these we shall ignore them and consider only *real* roots, not involving $\sqrt{-1}$.

5. Equations of Third and Higher Degree

An equation which is satisfied by three (and only three) values of x is called an equation of the *third degree*, or a *cubic* equation. An example is

$$2x^3 - 5x^2 + x + 2 = 0$$

which has three solutions $x=2$, $x=1$ and $x=-\frac{1}{2}$.

Definition :

   We define a cubic equation (or an equation of the third degree) to be of the form

$$ax^3 + bx^2 + cx + d = 0$$

where a, b, c and d are constants, a being positive or negative, the others being positive, negative or zero.

Examination of the above equations and definitions shows that a linear equation contains x but no higher powers of it; that a quadratic equation contains x^2 but no higher powers; and that a cubic equation contains x^3 but no higher powers. These observations suggest an extension of our definition to cover equations of the nth degree.

Definition :

> More generally we define an *equation of the nth degree*, possessing n solutions (i.e., satisfied by n values of the unknown), to be an equation of the form
>
> $$ax^n + bx^{n-1} + cx^{n-2} + \ldots + px + q = 0$$
>
> where a is a non-zero constant, and b, c, $\ldots$, p, q are constants which may be positive, negative or zero.

For example, we have

$$3x^4 + 7x^3 - 2x^2 + 4x - \tfrac{7}{8} = 0 \qquad \text{Degree 4}$$
$$3x^5 + 7x^3 + 4x - \tfrac{7}{8} = 0 \qquad \text{Degree 4}$$
$$8x^6 - 17x^3 = 0 \qquad \text{Degree 6}$$
$$-17x^5 = 0 \qquad \text{Degree 5}$$

The important points to notice are:

(i) that it is the degree of the highest power of x *when the equation is written out in the form indicated in the definition* that determines the degree of the equation; and

(ii) that if some solutions are difficult to find, or even seem not to exist, it in no way alters the degree of the equation.

For example, $3x^2 - 2y = 1/2x$ is of degree 3, since when written out in the proper (or, as we shall say, the " standard ") form, it becomes $6x^3 - 4x^2 - 1 = 0$ (as can be seen by multiplying both sides by $2x$). We also have that $x^8 = 1$ is an equation of the eighth degree, although it may seem difficult to find eight solutions.

6. The Remainder Theorem

Equations of the third and higher degree are usually difficult to solve, and some of them cannot be solved exactly. We shall say little about them except to illustrate the use of the remainder theorem. First, however, we may note that Appendices 2–4 discuss the solution of certain equations in more detail and that Appendix 5 describes a method of reaching a solution by successive numerical approximations.

The *Remainder Theorem* states that if

$$ax^n + bx^{n-1} + \dots + px + q$$

is divided by $x - x_1$, then the remainder after division may be obtained by putting $x = x_1$ in this expression.

For example, if we divide $x^2 - 2x + 3$ by $x - 2$, we should get a remainder of $2^2 - 2(2) + 3 = 3$.

Division shows this to be the case:

$$
\begin{array}{r}
x \\
\hline
x - 2 \ \big|\ x^2 - 2x + 3 \\
x^2 - 2x \\
\hline
+ 3 \ \text{remainder}
\end{array}
$$

It can be shown that, as a consequence, the following statements are true about the roots of the equation

$$ax^n + bx^{n-1} + \dots + px + q = 0$$

   (i) If r is a root of this equation then the left-hand side may be divided by $(x - r)$ without any remainder.
  (ii) If substituting $x = x_1$ gives the left-hand side a positive value, while subsituting $x = x_2$ gives it a negative value, then there is an odd number of roots lying between x_1 and x_2.
 (iii) If substituting $x = x_1$ gives the same sign as substituting $x = x_2$, then there is either no root between these values of x or there is an even number of roots between them.

An example will make this clear. Suppose that we wish to find the roots of

$$2x^3 + 2x^2 - 15x + 6 = 0$$

We notice first of all that this is a cubic equation, and therefore has three roots. Since complex roots always occur in pairs, either one or three of the roots must be real. We begin by considering a few extreme values.

Suppose $x = 10$. Then clearly the left-hand side is very large, and positive. On the other hand, if $x = -10$ the left-hand side is very large and negative. This means that there is an odd number of roots between $x = -10$ and $x = +10$, from result (ii) above.

Now let us consider

$x = 0$ 　　　　　L.H.S. $= 0 + 0 - 0 + 6$ (positive)

Since the left-hand side is also positive for $x = +10$, this means that there is either no solution between 0 and 10 or an even number of solutions in this range. Also, since the left-hand side is negative for $x = -10$ and positive for $x = 0$, there is an odd number of solutions between $x = 0$ and $x = -10$.

Now consider

$x = 1$ L.H.S. $= 2 + 2 - 15 + 6 = -5$ (negative)

However, we have just seen that for $x = 0$ the left-hand side is positive. There is thus an odd number of roots between $x = 0$ and $x = 1$. Furthermore, the left-hand side is positive for $x = 10$, showing that there is an odd number of roots between $x = 1$ and $x = 10$. We have also seen that there is an odd number of roots between $x = 0$ and $x = -10$. Since there are only three roots in all, they must lie in the ranges -10 to 0; 0 to 1 and 1 to 10.

The next step is to narrow down these ranges, attempting to locate the roots more exactly. We will begin with the range -10 to 0, remembering that for $x = -10$ the expression was negative, and that for $x = 0$ it was positive. Now let us try

$x = -5$ L.H.S. $= -250 + 50 + 75 + 6 = -119$

This shows that the single root lying between -10 and 0 must lie between -5 and 0, since it is in this narrower range that the left-hand side changes sign. To locate it more exactly, try

$x = -2$ L.H.S. $= -16 + 8 + 30 + 6 = 28$

This shows that the change of sign, and therefore the root, occurs somewhere between -5 and -2.

Proceeding in this way, we will soon find that in fact the root lies between -3 and -4.

We now have to decide whether to try to find this root more exactly, or to see whether there is a simpler root which we can find first. Substituting integers in the above expression is quite easy: but if we are to find this particular root by this method we will have to substitute fractions, which become complicated. Because of this we see if there is some more attractive root, which will later help us to simplify things.

There are only 3 roots and we have already found that there is an " awkward " one between -10 and 0. The root between 0 and 1 cannot be an integer. Therefore if there is an integral root it must lie between 1 and 10.

Let us try $x = 2$. We have

$x = 2$ L.H.S. $= 16 + 8 - 30 + 6 = 0$

This shows that $x = 2$ is a root, and that therefore the left-hand side is divisible by $x - 2$. If we perform this division we will be left with an equation of one degree less (i.e., with a quadratic equation) which we may solve for the remaining roots.

The division proceeds:

$$\begin{array}{r} 2x^2 + 6x \quad -3 \\ x-2 \enclose{longdiv}{2x^3 + 2x^2 - 15x + 6} \\ 2x^3 - 4x^2 \\ \hline 6x^2 - 15x \\ 6x^2 - 12x \\ \hline -3x + 6 \\ -3x + 6 \\ \hline 0 + 0 \end{array}$$

and this enables us to write the equation as

$$(x-2)(2x^2 + 6x - 3) = 0$$

Here we have a product of two factors equal to zero. Therefore

either $$x - 2 = 0$$

or $$2x^2 + 6x - 3 = 0$$

Solution of the second alternative may now proceed in the usual way, by using the formula for a quadratic equation, yielding the remaining roots 0·44 and −3·44 (approximately).

It will be noticed that we were wise not to proceed further with our attempt to locate the root between −3 and −4 since this would have involved the labour of cubing −3·44 (without any guarantee that it would give us the answer). Often it is a good practice in such cases to locate the root approximately, as we did, and then to use some other method, such as Newton's method (given in Appendix 5) to locate it more accurately, and with less trouble than is involved in the repeated application of the Remainder Theorem.

EXERCISE 1.9

Solve the following equations:

1. $x^3 - 4x^2 - 4x + 16 = 0$
2. $x^3 - 5x^2 - 29x + 105 = 0$
3. $2x^4 - 6x^3 - 14x^2 - 54x - 36 = 0$

Before leaving this subject we must utter a word of caution. Let us consider, first of all, the equation

$$2x^2 - 2x + 1 = 0$$

We may try as many values of x as we like (provided we do not introduce complex numbers) and will always find that the left-hand

side is positive. Now if two values of x give the same sign to the
left-hand side there is either no root between them, or there is an
even number of roots. If there is an even number of roots we expect
the left-hand side to be zero for some value in between, coinciding
with one of the roots. But in this case we fail to find any such value.
The reason is that this is an equation with no real roots.

Now consider

$$9x^2 - 30x + 25 = 0$$

and examine its values for different values of x. We have:

$x < 0$	L.H.S. > 0	positive
$x = 0$	L.H.S. $= 25$	positive
$x = 1$	L.H.S. $= 4$	positive
$x = 2$	L.H.S. $= 1$	positive
$x = 3$	L.H.S. $= 16$	positive
$x = 4$	L.H.S. $= 100$	positive
$x > 4$	L.H.S. increasingly positive	

Once again we have to conclude either that there are no real roots
or that there is an even number of roots hiding between some pair of
values. Here there is a useful clue. If we examine the values of the
left-hand side listed above it seems very likely that, if the left-hand
side is zero for any real value of x, then that value lies between 1
and 3, since the L.H.S. appears to approach zero more closely in
this range than elsewhere. Noticing that the coefficients in the
expression itself have 3 and 5 as factors, we might be tempted to
try some such root as $\frac{5}{3}$. If we do this we find that the expression
vanishes, showing that this is a root, and that therefore $(3x - 5)$ is a
factor. Furthermore, the expression is positive for *any other* value
of x. If we consider a value a trifle larger than $\frac{5}{3}$ we obtain a positive
left-hand side. If we consider a value a trifle smaller we again
obtain a positive left-hand side. Now two positive signs must mean
that for some value in between there is either no root or an even
number of roots. Since we have found one root, there must be an
even number of them. But we cannot find the second root. The
reason for this is that the expression has two coincident roots. The
expression is, in fact, a perfect square, having two identical roots.
It may be written as

$$(3x - 5)(3x - 5) = 0$$

The significance of this kind of result will appear in Chapter V
where we consider a geometrical interpretation of the Remainder
Theorem.

7. Identities

There is a totally different kind of mathematical relationship, which is often confused with an equation, but which is, in fact, something very different. We have already stated that, for all values of x and a

$$(x+a)^2 = x^2 + 2ax + a^2$$

Let us consider an example of this more carefully

Consider

$$(x+3)^2 = x^2 + 6x + 9$$

Substitution will show that this is true for *all* values of x. Put $x = 1$ on both sides of this equation. We have

$$(1+3)^2 = 1 + 6 + 9$$

i.e., $$(4)^2 = 16$$

which is true. We may try some negative number, such as -8, and have

$$(-8+3)^2 = (-8)^2 + 6(-8) + 9$$

Remembering that the square of a negative quantity is positive, we can write

$$(-5)^2 = 64 + (-48) + 9$$

$$25 = 64 - 48 + 9$$

which is correct. Fractions yield the same result. For example, putting $x = \frac{1}{2}$ we have

$$(\tfrac{1}{2}+3)^2 = (\tfrac{1}{2})^2 + 6(\tfrac{1}{2}) + 9$$

$$(\tfrac{7}{2})^2 = \tfrac{1}{4} + 3 + 9$$

$$\tfrac{49}{4} = 12\tfrac{1}{4}$$

which checks again.

A statement of this kind, which is true for *all* values of the unknown, rather than for only a limited number of them, is called an *identity*. Usually the equality sign ($=$) is replaced by the identically equals sign ($\equiv$).

Other identities are

$$x + 2x \equiv 3x$$

$$x^2 - 9 \equiv (x+3)(x-3)$$

$$\frac{2x^2 + 3x + 1}{x+1} \equiv 2x + 1$$

There is an important economic parallel. If we divide output (Y) into two and only two kinds of goods, one called consumer goods (C) other investment goods (I), then, *by definition*

$$Y \equiv C + I.$$

But if we observe, or suppose, that the output of consumer goods is always a fixed fraction c of total output, then

$$C = cY$$

is true, not by definition, but because of an observed or an assumed pattern of human behaviour.

We distinguish between *economic identities*, such as

$$Y \equiv C + I$$

and *economic equations*, such as

$$C = cY$$

The former must be true as a matter of definition or pure accountancy. The latter may or may not be true, and summarises real or hypothetical behaviour.

The balance of payments (B) is, by definition, the difference between exports (E) and imports (M)

$$B \equiv E - M$$

Imports (M) turn out to be a fraction m of our total spending (Y')

$$M = mY'$$

and if we spend exactly what we receive, and receive the exact value of our output (Y) then, as a result of this assumption about behaviour

$$Y' = Y$$

and so

$$M = mY$$

It follows that

$$B \equiv E - M$$

(which is an identity) leads to

$$B = E - mY$$

(which is an equation, since it is no longer necessarily true simply as a result of definition or accountancy).

In any economic theory, distinguish carefully between identities and equations. Only the equations have empirical content. The identities are truisms which can help us to examine relationships, but can have empirical meaning only when one or more of the quantities has been replaced by another quantity obtained from an equation (as mY replaced by M above).

8. Solution of simultaneous equations

We have just considered methods of solving certain equations. These methods have all had the object of finding the value of the unknown (usually denoted by x) which satisfies the equation. Sometimes, however, we are given several equations involving, between them, several unknowns, and we have to find the set of values which enables these unknowns to satisfy all of the equations at the same time. Such equations are called " simultaneous equations ". In this section we shall consider some of the simpler methods of solution.

(a) *Simultaneous linear equations in two unknowns :*

The method of solution is probably best explained by considering a specific example.

Suppose we have the following information:

   3 lb. of butter and 2 lb. of sugar cost £1·06
   2 lb. of butter and 4 lb. of sugar cost £0·92

and that we wish to find the prices of a pound of sugar and of a pound of butter.

Here we have two unknowns—the two prices. We denote them by x and y so that a pound of butter costs x pence and a pound of sugar y pence. We also have certain information which we may summarise thus:

$$3x + 2y = 106 \tag{1}$$

and
$$2x + 4y = 92 \tag{2}$$

In this form we see that the information yields two equations. The prices x and y are such that these two equations are satisfied by them at the same time.

The solution may be found in several ways. We shall now consider two of them, deferring others until a later chapter. Basically the methods are identical, but the detailed procedure varies a little.

The basic method is to write equation (1) in the form

$$3x = 106 - 2y$$

which gives

$$x = \frac{106 - 2y}{3}$$

We now put this value of x in equation (2), which becomes

$$\frac{2(106 - 2y)}{3} + 4y = 92$$

i.e.,
$$\frac{212 - 4y}{3} + 4y = 92$$

whence, multiplying by 3,

$$212 - 4y + 12y = 276$$

$$-4y + 12y = 276 - 212$$

$$8y = 64$$

$$y = 8$$

i.e., sugar costs 8p per pound, and therefore from equation (1) we have, by putting $y = 8$

$$3x + 16 = 106$$

$$3x = 90$$

$$x = 30$$

so that butter costs £0·30 per pound.

It will be noticed that the correctness of this solution may be checked by substituting $x = 30$ and $y = 8$ in the left-hand sides of the *two* equations, and seeing if the results are the same as the right-hand sides.

A second method is to *eliminate* x by multiplying (1) by 2 and (2) by 3, and then subtracting, thus,

$$(1) \times 2 \quad \text{gives} \quad 6x + 4y = 212$$
$$(2) \times 3 \quad \text{gives} \quad 6x + 12y = 276$$

Subtracting the first of these from the second we have

$$8y = 64$$

$$y = 8$$

and then we proceed as above.

As we can see, the essence of this method is to multiply the two equations by amounts which will result in the coefficients of x (or of y) becoming the same in the two equations so that the process of subtraction eliminates that unknown completely.

For reasons which will later become apparent, it is as well to consider here the application of this method to the simultaneous equations

$$a_1x + b_1y = c_1$$

and
$$a_2x + b_2y = c_2$$

where x and y are the unknowns (as before) and a_1, a_2, b_1, b_2, c_1 and c_2 are numerical constants, having any values we may care to give them, but which are written in this way for a special reason connected with some later work.

If we multiply the first of these equations by a_2 and the second by a_1 we will have the new equations

$$a_2 a_1 x + a_2 b_1 y = a_2 c_1$$

and

$$a_1 a_2 x + a_1 b_2 y = a_1 c_2$$

Subtraction now gives

$$a_2 b_1 y - a_1 b_2 y = a_2 c_1 - a_1 c_2$$

whence

$$y = \frac{a_2 c_1 - a_1 c_2}{a_2 b_1 - a_1 b_2}$$

and substitution of this result (or starting afresh but multiplying by b_2 and b_1 instead of by a_2 and a_1) will yield the other solution

$$x = -\frac{b_2 c_1 - b_1 c_2}{a_2 b_1 - a_1 b_2}$$

We shall refer to this result again.

EXERCISE 1.10

Solve the following pairs of simultaneous equations, doing the first few by several methods, including that of substituting in the general result just derived. Check the solutions by substitution in the original equations.

1. $4x + 3y = 7$; $3x - 2y = 9$
2. $3x + 2y = 14$; $2x + 3y = 14$
3. $3x - 4y + 30 = 0$; $3x - 3y = 20$
4. $2x + 3(y - 6) = 0$; $3x - 4(2 - y) = 19$
5. $3x - 4(y - 2) = 2$; $2x + 3(y - 3) = 4$

EXERCISE 1.11

1. Let petrol cost £p_1 per gallon and beer £p_2 per gallon. One week I buy 4 gallons of petrol and 2 gallons of beer and spend £3 in doing so. The next week I buy 6 gallons of petrol and 6 gallons of beer. I also get fined £50 for failing the breathalyser test, and my total outlay that week is £57·5. What are the unit prices of the two liquids?

2. In 1964 I bought n gramophone records, each at a price of £p, and so spent £42. In 1965 each record bore a tax of £0·35, but by reducing my purchases to $n - 10$ I still spent £42. In 1966 the tax rose to £0·63 per record, and so in order still to spend exactly £42 on them I bought only $n - 15$. How many records did I buy in 1964, and at what price?

*(b) Simultaneous equations in two unknowns, one equation being linear
    and the other quadratic :*

Suppose that we have the following information. Two pictures
are to be framed. One is square and the other is twice as long as it
is broad. Framing them requires 14 feet of wood and 6 square feet
of glass.

We want to find the size of each picture.

Let the side of the square picture be of length x feet. Then, in
order to frame it, we need $4x$ feet of framing and x^2 square feet of
glass.

Let the other picture measure y feet by $2y$ feet, so that it requires
$6y$ feet of framing and $2y^2$ square feet of glass.

$$\text{Then the total length of framing} = 4x + 6y = 14 \text{ feet}$$
$$\text{and the total area of glass} \quad = x^2 + 2y^2 = 6 \text{ square feet}$$

To solve these equations, of which the first is linear and the second
quadratic, we proceed as in the first method of the last section.
From the first equation we find that

$$x = \frac{14 - 6y}{4}$$

and we can substitute this into the second equation, obtaining

$$\left(\frac{14 - 6y}{4}\right)^2 + 2y^2 = 6$$

whence

$$\left(\frac{7 - 3y}{2}\right)^2 + 2y^2 = 6$$

yielding

$$\frac{49 - 42y + 9y^2}{4} + 2y^2 = 6$$

i.e., on multiplication throughout by 4 and collection of terms,

$$17y^2 - 42y + 25 = 0$$

This quadratic equation in y has two solutions, namely

$$y = 1 \text{ and } y = \tfrac{25}{17}$$

Let us consider the first of these. If $y = 1$ then our information
about the length of wood tells us that $x = 2$. This means that two
pictures, a square one of side 2 feet, and another measuring
1 foot × 2 feet will require 14 feet of framing and 6 square feet of
glass. The alternative solution, in which $y = \tfrac{25}{17}$, leads to an x of $\tfrac{22}{17}$,

which means that a square picture of side $\frac{22}{17}$ feet and a rectangular one measuring $\frac{25}{17}$ feet by $\frac{50}{17}$ feet will satisfy the same requirements.

Just as a single quadratic equation has two solutions, so the combination of a quadratic and a linear equation will lead to two solutions.

(c) *Other simultaneous equations :*

If the simultaneous equations involve only two unknowns the basic method of solution is as above, but if both of the equations are quadratic, or possess some other complicating feature, the elimination of one variable may not be so straightforward. Reference should be made to some standard work on algebra, or to one of the other books listed in the Bibliography.

If more than two unknowns are involved, then probably the equations will be most easily solved by the use of determinants, which are considered in Chapter XXIX. (See also Appendix 6)

EXERCISE 1.12

Solve the following pairs of simultaneous equations :

1. $x^2 + 3y^2 = 57$; $3x - y = 5$
2. $x^2 + 3y^2 = 57$; $3x - 2y = 1$
3. $xy = 90$; $2x + y = 36$
4. $3x^2 + 2xy + 8 = 0$; $4x - y = 13$
5. $3x^2 + 2xy + y^2 = 11$; $3x - 4y = 19$
6. $\dfrac{x + y}{4} = \dfrac{4}{x - y}$; $2x + 3y = 19$

9. Irrational equations

During the course of this book we shall have to solve some equations involving square roots. We shall now consider a few examples, in order to illustrate the methods of solution.

We may begin with a simple example of the kind

$$\sqrt{x + 3} = 12$$

To solve this we notice that since the left-hand side is given as equal to the right-hand side, then the square of the left-hand side must equal the square of the right-hand side. Upon squaring we have

$$x + 3 = 144$$

yielding

$$x = 141$$

It will be noted that this solution has been obtained quite easily simply because on the one side we had nothing but the square root. The importance of this will be seen when we consider

$$\sqrt{x+3} + 2x = 15$$

If we attempt to square both sides we have difficulty. Remembering that $(a+b)^2 = a^2 + 2ab + b^2$, and writing $\sqrt{x+3} = a$, and $2x = b$, we obtain

$$(x+3) + 4x\sqrt{x+3} + 4x^2 = 225$$

which is even more complicated. An easier way is to rewrite the equation so that the " irrational " part (i.e., the part involving the square (or other) root) is alone on one side of the equation, while everything else is on the other side. This gives us the equation in the form

$$\sqrt{x+3} = 15 - 2x$$

Squaring both sides we now obtain

$$x+3 = 225 - 60x + 4x^2$$

which may more conveniently be written as

$$4x^2 - 61x + 222 = 0$$

This is a quadratic equation in the standard form, and use of the formula will yield solutions

$$x = 6 \text{ and } x = \tfrac{37}{4}$$

We shall shortly see that only one of these solutions is correct, but it is better to leave this point for a moment.

Sometimes we may come across a somewhat more complicated equation, such as

$$\sqrt{x+3} + \sqrt{3x+7} = 8$$

Here we have two square root signs on one side and everything else is conveniently on the other side. We commence our solution as before, squaring both sides. Remembering that $(a+b)^2 = a^2 + 2ab + b^2$ and that therefore $(\sqrt{a} + \sqrt{b})^2 = a + 2\sqrt{ab} + b$, we have that this process of squaring yields

$$(x+3) + 2\sqrt{(x+3)(3x+7)} + (3x+7) = 64$$

This still looks rather complicated, but by keeping only the irrational part on the one side we may rewrite it as

$$2\sqrt{(x+3)(3x+7)} = 64 - (x+3) - (3x+7)$$
$$= 64 - 10 - 4x$$

whence

$$\sqrt{(x+3)(3x+7)} = 27 - 2x$$

The final solution is now obtained by squaring once again, obtaining

$$(x+3)(3x+7) = 729 - 108x + 4x^2$$

$$3x^2 + 16x + 21 = 729 - 108x + 4x^2$$

i.e., $$x^2 - 124x + 708 = 0$$

which has solutions $x = 6$ and $x = 118$. Once again, only one of these is correct.

We must now consider why only cne of the solutions obtained by this method is correct. Let us consider the second example,

$$\sqrt{x+3} + 2x = 15$$

for which we obtained the solutions $x = 6$ and $x = \frac{37}{4}$. If we substitute $x = 6$ in the left-hand side we obtain

$$\sqrt{9} + 12$$

which gives us 15, which is equal to the right-hand side. If, however, we substitute the solution $x = \frac{37}{4}$ we have for the left-hand side

$$\sqrt{\tfrac{49}{4}} + \tfrac{37}{2}$$

At first sight this appears to be

$$\tfrac{7}{2} + \tfrac{37}{2}$$

which comes to 22, which is not equal to the right-hand side. This is because we are forgetting that every quantity has two roots, one positive and one negative. The square root of 4 may be $+2$ or -2, and the same is true in the examples we have just considered. When we take the solution $x = 6$ we are able to obtain the right answer by using the positive root of 9; but when we use the solution $x = \frac{37}{4}$ we obtain the correct answer only if we take the negative square root of $\frac{49}{4}$ giving us a left-hand side of

$$-\tfrac{7}{2} + \tfrac{37}{2} = 15$$

A similar argument holds in the third example. It must be emphasised that unless a square root is prefaced by a negative sign we normally think of the positive root; but in the process of squaring, this sign may disappear. This is not a case of merely distorting numbers to suit a method that has gone wrong. Every positive quantity has two square roots, and one of these is always negative.   The fact that we more readily think of the positive one is our fault.   Mathematically the one root is as important as the other, and as we become better mathematicians we will begin to think of them in these terms.   When the equation arises from some practical problem, it may happen that either the negative or the positive root has no sensible meaning within the terms of the problem ; but the root still exists.   We shall later see examples of this.

EXERCISE 1.13

Solve :

1. $\sqrt{x+3} + 2x = 30$ 2. $\sqrt{x+3} - 2x = 0$

3. $\sqrt{4x+5} - 2x = -5$ 4. $\sqrt{4x+24} - x = 7$

5. $\sqrt{x+3} + \sqrt{2x+4} = 7$ 6. $\sqrt{3x+22} + \sqrt{x+7} = 3$

10. Inequalities

The beginner in mathematics often finds difficulty in handling inequalities.   He should not be disheartened by this, because many practised mathematicians feel much the same.   This short section shows how some of the difficulties may be avoided.

First of all we must introduce our symbols.   " a is greater than b " will be written as " $a > b$ "; " a is less than b " will be written as " $a < b$ ";   " a is not greater than b " may be written as

either " $a \not> b$ "
or   " $a \leqslant b$ "

where this last sign really means " a is less than ($<$) or equal to ($=$) b ".

" a is not less than b " may be written as

either " $a \not< b$ "
or   " $a \geqslant b$ "

where the last sign really means " a is greater than ($>$) or equal to ($=$) b ".

There are a great many rules about the manipulation of inequalities.   For the sake of reference we append some of these, along with examples of their use, but there is a great deal to be said for turning inequalities into equations.   For example, if we are told that $a > b$ then we may write

$$a = b + p \quad \text{where } p \text{ is positive} \quad (\text{i.e., } p > 0).$$

If we are also told that $c > d$ then we may write this as

$$c = d + q \quad \text{where } q > 0$$

We now see that

$$ac = (b + p)(d + q)$$
$$= bd + pd + bq + pq$$

Now p and q are positive. If, in addition, b and d are positive, then clearly every term on the right-hand side is positive and so

$$ac > bd$$

This may appear to be a lengthy process for obtaining a very obvious result, but it is worthwhile being careful. It is very easy to make mistakes in this matter. It is quite obvious that since $4 > 2$ and $7 > 3$ then $28 > 6$: but let us not forget that, although $4 > -2$ and $7 > -50$, it is not true that $28 > 100$. The above procedure, introducing p and q, shows us where to be careful. Let us suppose, for example, that we are told that b is negative (i.e., $b < 0$). To emphasise this, let us write it as $-B$ where B is positive. Then the right-hand side of

$$ac = bd + pd + bq + pq$$

becomes $\qquad -Bd + pd - Bq + pq$

and this is greater than bd (or $-Bd$) only if

$$pd - Bq + pq > 0$$

which may or may not be the case.

Enough has been said to illustrate the reason for care. A complete analysis of the above case would be a fine exercise for a student who really wishes to familiarise himself with these matters, but there is little point in labouring it here.

The following rules may be noted :

(1) *An inequality will still hold after each side has been increased, diminished, multiplied or divided by the same* positive *quantity.* For example, if $a > b$ then, provided $c > 0$,

$$a + c > b + c$$

$$a - c > b - c$$

$$ac > bc$$

$$\frac{a}{c} > \frac{b}{c}$$

(2) *Any term in an inequality may be moved from one side to the other provided that its sign is changed.* For example,

if $\qquad\qquad a - c > b$

then $\qquad\qquad a > b + c$

(3) *If the sides of an inequality be interchanged then the sign must also be changed.* Thus,

if
$$a > b$$

then
$$b < a$$

(4) *If both sides of an inequality be multiplied or divided by the same negative quantity then the sign of the inequality must be reversed,* i.e.,

if
$$a > b \quad \text{and} \quad p < 0$$

then
$$pa < pb$$

and
$$\frac{a}{p} < \frac{b}{p}$$

(5) *If the two sides of an inequality, each having the same sign, be inverted (i.e., turned upside down) then the sign must be reversed.*

If
$$\frac{a}{c} > \frac{b}{d}$$

then
$$\frac{c}{a} < \frac{d}{b}$$

and in particular if
$$a > b$$

then
$$\frac{1}{a} < \frac{1}{b}$$

It will be noted that the first two of these are rules which have become familiar enough in handling equations. ✕

EXERCISE 1.14

1. Prove the above rules by turning each inequality into an equation.

2. Prove that if $a > b$ then $\dfrac{a}{b} > \dfrac{a+x}{b+x}$ if $x > 0$ but that the reverse is true if $x < 0$, provided that $b > 0$ and $x > -b$.

Some standard results

To end this pure revision of basic algebra we list some important results with which the reader is assumed to be familiar.

(a) *Factors and expansions*

$$(a+b)^2 \equiv a^2 + 2ab + b^2 \, ; \qquad (a-b)^2 \equiv a^2 - 2ab + b^2$$
$$a^2 + b^2 \text{ has no (real) factors} \, ; \qquad a^2 - b^2 \equiv (a-b)(a+b)$$
$$(a+b)^3 \equiv a^3 + 3a^2b + 3ab^2 + b^3 \, ; \qquad (a-b)^3 \equiv a^3 - 3a^2b + 3ab^2 - b^3$$
$$a^3 + b^3 \equiv (a+b)(a^2 - ab + b^2) \, ; \qquad a^3 - b^3 \equiv (a-b)(a^2 + ab + b^2)$$

(b) Indices

$$x^m \times x^n \equiv x^{m+n}$$

e.g., $4^3 \times 4^5 = 4^8$

$$(x^m)^n \equiv x^{mn}$$

$$(4^3)^5 = 4^{15}$$

$$x^m \div x^n \equiv x^{m-n}$$

$$4^5 \div 4^3 = 4^{5-3} = 4^2$$

$$x^{-m} \equiv \frac{1}{x^m}$$

$$4^{-3} = \frac{1}{4^3}$$

$$x^{1/n} \equiv \sqrt[n]{x}$$

$$4^{1/2} = \sqrt[2]{4}$$

$$x^{m/n} \equiv \sqrt[n]{x^m} \equiv (\sqrt[n]{x})^m$$

$$8^{2/3} = \sqrt[3]{8^2} = \sqrt[3]{64}$$

$$= 4 = (\sqrt[3]{8})^2$$

$$x^{-1/n} \equiv \frac{1}{x^{1/n}} \equiv \frac{1}{\sqrt[n]{x}}$$

$$4^{-1/2} = \frac{1}{4^{1/2}} = \frac{1}{2}$$

$$x^0 \equiv 1$$

$$4^0 = 15^0 = (-3)^0 = 1$$

(c) Ratios

If $\dfrac{a}{b} = \dfrac{x}{y}$ then $\dfrac{a}{b} = \dfrac{a+x}{b+y}$

$$= \frac{a-x}{b-y}$$

$$= \frac{ma+nx}{mb+ny} \qquad m \text{ and } n \text{ being any real}$$

quantities

$$= \frac{ma-nx}{mb-ny}$$

$$= \frac{32a-5x}{32b-5y}, \text{ etc.}$$

CHAPTER II

THE SUMMATION AND CONVERGENCE OF SOME SIMPLE SERIES

So he matured by a progression
CHRISTOPHER FRY : *The Lady's Not for Burning*

1. The arithmetic progression

One of the most common examples of an Arithmetic Progression is afforded by a salary scale in which there is a constant increment. This constant increment is, in fact, the distinguishing feature of an Arithmetic Progression, which is defined as follows :

Definition :

A series of quantities taken in order form an Arithmetic Progression if they increase or decrease by a constant amount, which is called the *common difference.*

We have, for example,

$$600, 650, 700, 750, 800, \ldots$$
$$8, \quad 11, \quad 14, \quad 17, \quad 20, \ldots$$
$$14, \quad 10, \quad 6, \quad 2, \quad -2, \ldots$$
$$4, \quad 2\tfrac{1}{2}, \quad 1, \quad -\tfrac{1}{2}, \quad -2, \ldots$$

as particular instances of the A.P., while more generally we may write the A.P. in the form

$$a, a+d, a+2d, a+3d, \ldots$$

where a is the first term and d is the common difference, both a and d being allowed to be positive or negative. Clearly the common difference is found by subtracting one term from the following term. In the above examples it is 50, 3, -4 and $-1\tfrac{1}{2}$ respectively.

When we examine the series in the form

$$a, a+d, a+2d, a+3d, \ldots$$

we see that the 4th term is $a+3d$

7th term is $a+6d$

10th term is $a+9d$

and, more generally,

the pth term is $a + (p-1)d$

Clearly if we consider n terms of the series, and denote the nth term (or the last term) by l we shall have that

$$l = a + (n-1)d$$

It can be shown that the sum of n terms of an A.P. is given by

$$S = \frac{n(a+l)}{2}$$

$$= \frac{n}{2}\left\{2a + (n-1)d\right\}$$

Proofs of these results are given in Appendix 8. It follows from the first of these that for the *positive integers up to* n

$$S = 1 + 2 + 3 + 4 + \ldots + n$$

$$= \frac{n(1+n)}{2}$$

so that the sum of the positive integers from 1 to 100 is immediately obtained as $\dfrac{100 \times 101}{2} = 5050$.

Use of the second formula shows that the sum of the *first n odd integers* is given by

$$S = 1 + 3 + 5 + 7 + \ldots, \text{ for } n \text{ terms}$$

$$= \frac{n}{2}\left\{2 + (n-1)2\right\} \qquad \text{since } a = 1 \\ \text{and } d = 2$$

$$= n^2$$

This may be easily checked in special cases. The sum of the first two odd integers, $1 + 3$, is equal to 2^2; the sum of the first three, $1 + 3 + 5$, is 9, which is 3^2.

Exercise 2.1

Sum the following series :

1. $7 + 14 + 21 + 28 + \ldots$, to 20 terms
2. $-4 - 1 + 2 + 5 + \ldots$, to 20 terms
3. The first n even integers, checking your result for simple values of n.
4. $17 + 13 + 9 + \ldots$, to 6 terms, and to 12 terms
5. $80 + 79{\cdot}5 + 79 + 78{\cdot}5 + \ldots$, to 160 terms and to 321 terms

2. The geometric progression

As Malthus pointed out, a population tends to grow in a geometric progression. If two people beget four children, so that two
married couples beget a total of eight children, then (provided the
sexes work out nicely and that there are no untimely deaths), there
will soon be four new pairs of married couples. If this generation
shows the same fertility then these four couples will give rise to
sixteen children, who, in turn, will give rise to a new generation of
thirty-two children. The successive generations will thus be of the
sizes

$$2, 4, 8, 16, 32, \ldots$$

In fact, each generation is twice the size of the previous one.

This is an example of a series which has the property of increasing
(or decreasing) by a constant ratio (as opposed to a constant increment). It is this constant ratio that is the distinguishing feature of
the G.P.

Definition :

A series of quantities taken in order form a G.P. if they increase
or decrease by a constant ratio, which is called the *common ratio*.

We have, for example,

2,	6,	18,	54,	162,	486, ...;	common ratio	3
5,	10,	20,	40,	80,	160, ...;	common ratio	2
5,	-10,	20,	-40,	80,	-160, ...;	common ratio	-2
64,	32,	16,	8,	4,	2, ...;	common ratio	$\frac{1}{2}$
81,	-27,	9,	-3,	1,	$-\frac{1}{3}$, ...;	common ratio	$-\frac{1}{3}$

as particular examples. The general form is given by

$$a, ar, ar^2, ar^3, \ldots$$

where a is the first term and r is the common ratio. This common
ratio is obtained by dividing any term by the preceding one. We
see that the

$$p\text{th term is } ar^{p-1}$$

It is shown in Appendix 8 that the sum of n terms of a G.P. is
given by

$$S = \frac{a(r^n - 1)}{r - 1}$$

$$= \frac{a(1 - r^n)}{1 - r}$$

The former of these is the more convenient when $r>1$, but the latter when $r<1$. /|

As examples of the use of these formulae we have that

$$1 + 3 + 9 + 27 + \ldots, \text{ to 10 terms}$$

has a sum given by

$$\frac{1 \times (3^{10} - 1)}{3 - 1}$$

$$= \frac{1 \times (59{,}049 - 1)}{2}$$

$$= 29{,}524$$

The series

$$4 + 2 + 1 + \tfrac{1}{2} + \tfrac{1}{4} + \ldots$$

summed to 10 terms has a sum

$$\frac{4(1 - (\tfrac{1}{2})^{10})}{1 - \tfrac{1}{2}}$$

$$= \frac{4(1 - \tfrac{1}{1024})}{\tfrac{1}{2}}$$

$$= 8 \times \tfrac{1023}{1024}$$

$$= 7\tfrac{127}{128}$$

It is often necessary to approximate slightly by evaluating r^n to an adequate number of places with logarithms.

Example :

I am entitled to an annual income of £1,000 now and in each of the next nine years. Because the promise of money in the future is less useful to me than having that sum of money now, I would be prepared to exchange the promise of £1,000 next year for £900 now. I would also be prepared to exchange the promise of £1,000 in the following year for £810 now. And so on. For how much would I sell my complete entitlement? How different would my selling price be if the annuity lasted for (*a*) 20 years? (*b*) 40 years?

The present value of £1,000 now is   £1,000
,,   ,,   ,,   ,, £1,000 next year is £900 $= £1.000 \times 0.9$
,,   ,,   ,,   ,, £1,000 in 2 years' time is £810 $= £1{,}000 \times 0.9^2$
etc.

Total present value is

$$£1{,}000 + £1{,}000\,(0.9) + £1{,}000\,(0.9)^2 + \ldots$$

If the annuity has a life of 10 years then its present value is

$$S_{10} = \frac{\pounds1,000 \ (1 - 0 \cdot 9^{10})}{1 - 0 \cdot 9}$$

$$\log 0 \cdot 9 = \bar{1} \cdot 9542$$

$$10 \times \log 0 \cdot 9 = \overline{10} + 9 \cdot 542 = \bar{1} \cdot 5420$$

$$\text{antilog } \bar{1} \cdot 5420 = 0 \cdot 3483$$

Thus,
$$1 - 0 \cdot 9^{10} = 1 - 0 \cdot 3483 = 0 \cdot 6517$$

$$S_{10} = \frac{\pounds1,000 \times 0 \cdot 6517}{0 \cdot 1} = \pounds6,517$$

If the annuity has a life of 20 years then we need to evaluate

$$\frac{\pounds1,000 \ (1 - 0 \cdot 9^{20})}{1 - 0 \cdot 9}$$

which comes to £8,787.

For 40 years the present value is £9,853.

This example illustrates that if we *discount* a future income by a rate r^n where n is the number of time periods which we have to wait, then income in the remote future has very little value now. This means that the present value of an annuity for 20 years is not much increased if the annuity is extended beyond 20 years.

Example :

In an underemployed economy the national output Y_t in month t consists of consumer goods valued at C_t and investment goods valued I_t. This gives an identity

$$Y_t \equiv C_t + I_t$$

We are told that in each month consumer spending just equals 90% of the previous month's national output, so that we have the behavioural equation

$$C_t = 0 \cdot 90 \, Y_{t-1}$$

For several months I_t, C_t and Y_t have been constant at the levels of £1m., £9m. and £10m. respectively. The publishers of the *General Theory* decide to spend some of the profits which they have made from it by digging large holes at a cost of £0·1m. each month. What happens to the national output?

The idea here, of course, is that the £0·1m. is spent employing people who would otherwise have idle time. Their income rises and so does their spending on consumer goods—and this means more work producing these goods, and so even more people are being employed, and being paid, and so add to the demand for yet more goods. Let us spell it out mathematically.

£m

t Month	Y_{t-1} Income of Previous Month	C_t Consumer Spending $=0{\cdot}9\,Y_{t-1}$	I_t Investment	Y_t Total Output (equals Income)	Y_t-Y_{t-1} Increase in Output
1	10	9	1	10	0
2	10	9	1	10	0
3	10	9	1·1	10·1	0·1
4	10·1	9·09	1·1	10·19	0·09
5	10·19	9·171	1·1	10·271	0·081
6	10·271	9·2439	1·1	10·3439	0·0729
7	10·3439	.	.	—	.
.	.	.	.	.	.
.	.	.	.	.	.
.	.	.	.	.	.

For the first two periods we have the static original situation. In period 3 the level of investment is permanently raised by £0·1m. This adds to incomes in period 3, and so more is available for spending in period 4. The increased spending (of £9·09m.) added to the new level of investment (of £1·1m.) produces a new national income of £10·19m. Thus even more can be spent in the next month, and to make the consumer goods valued at £9·171m. extra labour is needed, and paid. The total income generated that month is £10·271m., which (being higher than before) allows a further increase in spending. And so on . . . but for how long?

The answer is that this process will go on for ever, provided that there is always some underemployed labour available, and unless, of course, some other event intervenes. But the fact that it will go on for ever does not mean that income will grow to indefinitely high levels. We may notice that the annual increases in output are

$$0{\cdot}1,\ 0{\cdot}09,\ 0{\cdot}081,\ 0{\cdot}0729,\ \ldots$$

and these are successive terms in a G.P. whose first term is 0·1 and whose common ratio is 0·9.

By the sixth month, four of these terms have arisen, each representing an increase over the previous month. *The total increase* is the sum of these terms which is

$$a\,\frac{1-r^n}{1-r}$$

$$=0{\cdot}1\,\frac{1-0{\cdot}9^4}{1-0{\cdot}9}$$

$$=\frac{0{\cdot}1\,(1-0{\cdot}6561)}{0{\cdot}1}$$

$$=0{\cdot}3439$$

which checks with the entry in our table of 10·3439.

By the tenth month there will have been 8 increases, and their sum will be

$$\frac{0{\cdot}1\,(1-0{\cdot}9^8)}{1-0{\cdot}9}$$

which yields $(1-0{\cdot}4293)=0{\cdot}5707$ and so the level of national output in month 10 is

$$Y_{10}=10{\cdot}5707$$

By the eighteenth month, there will have been 16 increases, and the total of these will be

$$(1-0{\cdot}1843)=0{\cdot}8157$$

yielding

$$Y_{18}=10{\cdot}8157$$

If we go as far as the thirty-fourth month, by when there will have been growth for 32 months, we obtain a total increase of

$$(1-0{\cdot}0340)=0{\cdot}9660$$

yielding

$$Y_{34}=10{\cdot}9660$$

If we look at these results we see that the total increase is always

$$(1-x)$$

where x gets progressively smaller as time increases—but it is always a positive power of 0·9. It gets smaller and smaller but it never can become negative. Thus the total increase

$$(1-x)$$

can never become bigger than unity—and so, no matter how long we go on, the value of Y_t can never get above

$$Y_t=10+1$$

$$=11$$

In fact, since x will always be positive (even though very small), $(1-x)$ will never quite become zero, and so Y_t will never quite reach 11.

We return to this idea in Chapter VII, where we discuss limits. Meanwhile let us look at it algebraically.

After n periods of growth the total increase in income is

$$A \frac{1-c^n}{1-c}$$

where A is the "autonomous" increase in spending (due to the decision to dig holes) *which is sustained throughout the future*, and c is the "average propensity to consume" (being the proportion of income devoted to consumption, and assumed constant).

Now just as 0.9^n became smaller and smaller as time went on, so will c^n become smaller (provided $c < 1$), and so, to all intents and purposes, when n is very large

$$1 - c^n$$

can be replaced by unity. At that stage the total increase in income is

$$A \frac{1}{1-c}$$

Thus, a sustained autonomous increase of £0·1m. will eventually lead to an increase levelling out at

$$£0.1 \frac{1}{1-0.9} \text{ m.}$$
$$= £1\text{m.}$$

above the original level if $c = 0.9$. But if $c = 0.8$ then the increase would be

$$£0.1 \frac{1}{1-0.8} \text{ m.}$$
$$= £0.5\text{m.}$$

The expression $\dfrac{1}{1-c}$ is called "the multiplier", since it multiplies A to yield the change in income.

Notice that there is another method for getting this result. Consider the original equilibrium situation

$$Y_1 \equiv C_1 + I_1 \qquad \text{(identity)}$$
$$C_2 = cY_1 \qquad \text{(behaviour)}$$
$$Y_2 = Y_1 \qquad \text{(no growth)}$$
$$Y_2 \equiv C_2 + I_2$$

which yield

$$Y_2 = cY_1 + I_2$$

$$= cY_2 + I_2$$

whence

$$Y_2 - cY_2 = I_2$$

$$Y_2(1-c) = I_2$$

$$Y_2 = \frac{I_2}{1-c}$$

Now let I increase by an amount A, and suppose that after many periods we are again in a (virtually) steady state, with everything (virtually) constant. Then

$$Y_n \equiv C_n + I_n$$

where

$$I_n = I_2 + A$$

and

$$C_n = cY_{n-1} = cY_n$$

These equations yield

$$Y_n = \frac{I_2 + A}{1-c}$$

and so the increase in income is

$$Y_n - Y_2 = \frac{I_2 + A}{1-c} - \frac{I_2}{1-c}$$

$$= \frac{A}{1-c}$$

as before.

EXERCISE 2.2

Sum the following G.P.'s :

1. $1 + 4 + 16 + 64 + \ldots$, to 6 terms
2. $1 - 4 + 16 - 64 + \ldots$, to 6 terms, to 7 terms and to 8 terms
3. $64 + 32 + 16 + 8 + \ldots$, to 6 terms, to 7 terms and to 8 terms
4. $64 - 32 + 16 - 8 + \ldots$, to 6 terms, to 7 terms and to 8 terms

Consider whether it is possible for the sum of the series 3 above ever to reach 128, and whether the sum of series 4 can be lower than 42·6 or higher than 42·7 if we take as many as fifty terms (or more).

3. The natural numbers

We have already seen that the sum of the first n positive integers (or, as they are sometimes called, the first n natural numbers) is given by

$$S_1 = \frac{n}{2}(n+1)$$

It is shown in Appendix 8 that the *Sum of the Squares* of the first n natural numbers is given by

$$S_2 = 1^2 + 2^2 + 3^2 + 4^2 + \ldots + n^2$$

$$= \frac{n(n+1)(2n+1)}{6}$$

This series, of course, is neither an A.P. nor a G.P. It can also be shown that the *Sum of the Cubes* of the first n natural numbers is given by

$$S_3 = 1^3 + 2^3 + 3^3 + 4^3 + \ldots + n^3$$

$$= \frac{n^2(n+1)^2}{4}$$

$$= \left\{ \frac{n(n+1)}{2} \right\}^2$$

This last result shows that the sum of the cubes of the first n natural numbers is equal to the square of the sum of the first n natural numbers. For example, the sum of the first 20 natural numbers is

$$S_1 = \frac{n(n+1)}{2}$$

$$= \frac{20 \times 21}{2}$$

$$= 210$$

The sum of the squares of these numbers is

$$S_2 = \frac{n(n+1)(2n+1)}{6}$$

$$= \frac{20 \times 21 \times 41}{6}$$

$$= 2870$$

c

The sum of the cubes is

$$S_2 = \left\{ \frac{n(n+1)}{2} \right\}^2$$
$$= 210^2 = S_1{}^2$$
$$= 44100$$

In order to avoid confusion it is customary to indicate the sum of a series by some symbol other than S which has already been overworked. The usual method has the merit of being extremely precise. If we think of the natural numbers as being denoted by x where x goes from 1 upwards as high as we please, then we may write the sum of the natural numbers as Σx where Σ is simply the Greek capital letter S (pronounced sigma), and means " the sum of all values of . . . ". Thus, Σ means " the sum of all values of x ". This, as it stands, is not very helpful because it depends on the number of values that we are prepared to let x assume. To remove this ambiguity we usually write the lower and upper values of x (or, sometimes, of some other variable, as we shall see later) immediately beneath and above the sign. For example,

$$\sum_{1}^{9} x \text{ means } 1 + 2 + 3 + \ldots + 9$$

$$\sum_{1}^{9} x^2 \text{ means } 1^2 + 2^2 + 3^2 + \ldots + 9^2$$

We will note that if $1 < a < b$ then, provided a and b are integers,

$$\sum_{1}^{b} x = \sum_{1}^{a} x + \sum_{a+1}^{b} x$$

and that

$$\sum_{a+1}^{b} x = \sum_{1}^{b} x - \sum_{1}^{a} x$$

In a similar notation we have that for the A.P., where the general term is $a + (p-1)d$, the sum of n terms may be written as

$$\sum_{p=1}^{n} [a + (p-1)d]$$

where the expression immediately following the Σ is the general term of the series, capable of yielding all the particular terms if we give p the values from 1 upwards. Since it is p that varies, and whose range determines the sum, it is the lower and upper values of p that we place beneath and above the Σ sign.

For the G.P. we have

$$\sum_{p=1}^{n} ar^{p-1} = \frac{a(r^n - 1)}{r - 1}$$

This notation is of great help in the solution of problems involving summation. Suppose, for example, that we wish to sum the series

$$1 + 3 + 7 + 13 + 21 + \ldots, \text{ to 10 terms}$$

This series is neither an A.P. nor a G.P. It may be noticed, however that this series may be written as

$$1 + (4 - 1) + (9 - 2) + (16 - 3) + (25 - 4) + \ldots$$

i.e., as $(1^2 - 0) + (2^2 - 1) + (3^2 - 2) + (4^2 - 3) + (5^2 - 4) + \ldots$

from which we see that the general term is

$$p^2 - (p - 1)$$

which may be written as

$$p^2 - p + 1$$

The sum of the series to 10 terms is therefore given by

$$\sum_{p=1}^{10} (p^2 - p + 1)$$

Now each term inside this bracket is made up of three parts, and the sum of ten such terms will be the sum of the ten first parts *plus* the sum of the ten second parts *plus* the sum of the ten third parts. The sum of the ten first parts is simply the sum of the first ten squares of the natural numbers; the sum of the second parts is the sum of the first ten natural numbers, with a minus sign before it; the ten third parts are all equal to unity and therefore have a sum of ten. More easily we may write this as

$$\sum_{p=1}^{10} (p^2 - p + 1) = \sum_{1}^{10} p^2 - \sum_{1}^{10} p + \sum_{1}^{10} 1$$

$$= \frac{n(n + 1)(2n + 1)}{6} - \frac{n(n + 1)}{2} + n$$

where, in this case, $n = 10$ giving

$$\frac{10 \cdot 11 \cdot 21}{6} - \frac{10 \cdot 11}{2} + 10$$

$$= 385 - 55 + 10$$

$$= 340$$

EXERCISE 2.3

By first expressing the general term in the form given, sum the following series to 8 terms:

1. 2, 6, 12, 20, 30, ...   general term $r(r+1) = r^2 + r$
2. 1, 6, 15, 28, 45, ... $2r^2 - r$
3. 2, 5, 12, 31, 86, ... $r + 3^{r-1}$

4. Convergency and divergency

So far we have always specified a definite number of terms for our series. Sometimes, however, we may think of a series as going on for ever, and so we divide series into *finite series* and *infinite series*. Provided we know its general term, a finite series can always be summed, even if we have to resort to writing down all the terms and adding them. On the other hand, problems arise when we attempt to sum an infinite series, including that of defining " the sum of an infinite series ". We mention these matters here, but no more than hint at their solution until we reach Chapter VII. Meanwhile we may obtain some introduction to the answer, as well as to the problems of convergency, if we examine a few geometric progressions.

Consider first of all the series

$$1 + 2 + 4 + 8 + 16 + 32 + ...$$

Clearly, by taking enough terms, we may make this sum as large as we like. This is an example of a divergent series. We shall define these concepts shortly, meanwhile let us note that if we wish to make the sum of this series greater than 250,000,000 all that we have to do is to take the first twenty-nine terms, when we shall obtain the sum 268,435,455.

Let us now consider the series

$$1 + \tfrac{1}{2} + \tfrac{1}{4} + \tfrac{1}{8} + ...$$

We may notice that in this series the $(r+1)$th term is always one half of the difference between 2 and the total of the first r terms. For example, the fourth term is $\tfrac{1}{8}$ which is one half of $(2 - 1\tfrac{3}{4})$, where $1\tfrac{3}{4}$ is the sum of the first three terms. If we consider this for a moment we shall see that the sum of this series will always fall short of 2, for if, after a large number of terms, the sum of the series is $2 - h$ where h is very small, then the next term to be added to this sum will be precisely $h/2$, which will leave the new sum still a little less than 2. This argument can be continued indefinitely. It is rather like the man who decides to eat an apple by always biting off one half of what remains : there will be no end to his eating, and no end to the apple. On the other hand, by taking enough terms we can make the total as close to 2 as we please. If we wish to obtain a total of more than 1·999 999 999 999 9 we have simply to continue the series to the forty-fifth term when we will obtain the sum 1·999 999 999 999 943 ... . This is an example of a convergent series.

Definition :

> An infinite series is *convergent* when the sum of the first n terms, which we will denote by S_n, can be made as close as we please to some definite number S, if we take large enough values of n.

An infinite series which is not convergent is usually said to be *divergent*. Some writers prefer to define a divergent series to be one such that S_n can be made (numerically) as large as we please by taking sufficient terms. This would mean that the series $1 + 1 - 1 + 1 - 1 + ...$ would be neither convergent nor divergent. It would be useful to call such a series *non-vergent*, reserving divergent for those series which can be made to have explosively large sums.

This distinction between convergent and divergent series is of immense practical importance. Handling infinite series is often necessary, but it is also extremely dangerous. It can be shown, for example, that the expression $1/(x-1)$ may be represented by the infinite series

$$\frac{1}{x} + \frac{1}{x^2} + \frac{1}{x^3} + \frac{1}{x^4} + \frac{1}{x^5} + ...$$

provided that certain conditions hold. If we put $x = 10$ we have that

$$\frac{1}{9} = \frac{1}{10} + \frac{1}{10^2} + \frac{1}{10^3} + ...$$

$$= 0·1 + 0·01 + 0·001 + ...$$

$$= 0·1111...$$

which is clearly correct. On the other hand, if we put $x = \frac{1}{2}$, then we obtain

$$\frac{1}{\frac{1}{2}-1} = 2 + 2^2 + 2^3 + ...$$

i.e.,

$$-2 = 2 + 4 + 8 + ...$$

which is nonsense.

It can be shown, in fact, that $1/(x-1)$ can be represented by this infinite series *only if the series converges for the particular value of x concerned.* Unless an infinite series is convergent it is very unsafe to use. A great deal of work in theoretical and practical mathematics

was invalid because this point was not adequately appreciated, until
it was stressed by Euler (1707–83) and by Gauss (1777–1855). A
mathematician as great as Laplace (1749–1827), who had spent his
life analysing the movements of the entire solar system, had to
spend an anxious time checking the convergence of all the series he
had used after hearing Cauchy (1789–1857) lecture on the subject.
Fortunately for his *Mécanique céleste*, the series converged. In
economics the study of convergency is closely related to many
matters, especially to the study of oscillations, and the solution of
difference equations. The reader may care to go back to our example
about the multiplier and to look at the series of monthly incomes to
see if they converge.

   Although they are not really needed at this stage of the book,
some of the more important tests for the convergence and divergence
of series are now summarised. We do not attempt to prove these
tests in this book. It must be emphasised that the mere fact that a
series cannot be shown to be divergent does not necessarily mean
that it is convergent.

   (1) *An infinite series in which the terms alternate in sign is convergent if each term is numerically smaller than the preceding term.* For
example, the series

$$1 - \tfrac{1}{2} + \tfrac{1}{3} - \tfrac{1}{4} + \tfrac{1}{5} \ldots$$

converges.

   (2) *An infinite series in which all the terms are of the same sign is
divergent if each term is numerically greater than some finite quantity
of the same sign,* however small this quantity may be.
   As an example we have that

$$64{\cdot}00001 + 32{\cdot}00001 + 16{\cdot}00001 + 8{\cdot}00001 + 4{\cdot}00001 + 2{\cdot}00001 + \\ 1{\cdot}00001 + 0{\cdot}50001 + 0{\cdot}25001 + 0{\cdot}12501 + 0{\cdot}06251 + \ldots$$

is divergent, since each term is greater than 0·00001, the pth term
being

$$64(\tfrac{1}{2})^{p-1} + 0{\cdot}00001$$

   (3) *An infinite series is convergent if, from and after some fixed term,
the ratio of each term to the preceding term is numerically less than some
quantity* which is itself numerically less than unity, *provided that all
the terms are positive.*
   This means that, quite independently of anything that may
happen in the earlier part of the series, if there exists a term such
that *subsequently* the ratio of each term to its preceding one is
numerically less than some quantity which is itself less than unity,

then the series converges. Before considering the meaning of this more closely let us note a simple example. The series

$$1 + 2(\tfrac{1}{2}) + 3(\tfrac{1}{2})^2 + 4(\tfrac{1}{2})^3 + \ldots + n(\tfrac{1}{2})^{n-1} + \ldots$$

is such that, if we denote the nth term by u_n and the $(n+1)$th term by u_{n+1} then the ratio of the one term to its preceding term is

$$\frac{u_{n+1}}{u_n} = \frac{(n+1)}{n} \frac{(\tfrac{1}{2})^n}{(\tfrac{1}{2})^{n-1}}$$

$$= \frac{n+1}{2n}$$

and so, provided we choose $n > 1$ the ratio of u_{n+1} to u_n is less than (say) 0·751, (it being equal to 0·75 when $n = 2$ but thereafter always less than this). 0·751 is less than unity, and so the condition for convergency holds.

The point about stipulating " less than some quantity which is itself less than unity " is that otherwise we might find the ratio being less than unity, but approaching indefinitely close towards it. For example, the series

$$\frac{1}{1} + \frac{1}{2} + \frac{1}{3} + \frac{1}{4} + \ldots + \frac{1}{p} + \ldots$$

is such that

$$\frac{u_{n+1}}{u_n} = \frac{n}{n+1}$$

As n increases, this ratio will always be less than unity: but by choosing a very large n we make this ratio as close to unity as we wish it to be, and it is impossible to find *a quantity less than unity* such that for all values of n the ratio is less than *this quantity*. Because of that, the series does not pass our test for convergency. On the other hand, we should not therefore assume that it is definitely divergent : it is simply that we have failed to prove its convergence, rather than that we have proved its divergence.

(4) *An infinite series in which all the terms are of the same sign is divergent if, from and after some fixed term, the ratio of each term to the preceding term is greater than unity or equal to unity.*

The truth of this, which is almost the converse of the previous rule, is self-evident. On the other hand, it must be noted carefully that neither rule says anything about what happens if the ratio of each term to the preceding term is numerically less than unity but approaches it indefinitely closely.

There remains one other very important rule, differing from the previous ones (in that it compares one series with another) but being of tremendous practical importance.

(5) *If two infinite series have all their terms positive, and if the ratio of the corresponding terms in the two series is always finite, then the two series are either both convergent or both divergent.*

Consider for example, the series

$$\frac{10}{2} + \frac{20}{2^2} + \frac{10}{2^3} + \frac{20}{2^4} + \frac{10}{2^5} + \cdots$$

Let us compare this with

$$\frac{1}{2} + \frac{1}{2^2} + \frac{1}{2^3} + \frac{1}{2^4} + \frac{1}{2^5} + \cdots$$

We see that if we divide any term of the second series into the corresponding term of the first then the resulting ratio will be finite. Now the second series can easily be shown to be convergent (by, for example, the use of Rule (3)). Therefore the original series is convergent.

Two consequences of this rule, although they are not obviously so, are so important that it may be worthwhile, here, to consider them as additional rules. They are:

(6) *The infinite series*

$$\frac{1}{1^p} + \frac{1}{2^p} + \frac{1}{3^p} + \frac{1}{4^p} + \cdots$$

is always divergent except when p is positive and greater than unity, when it is convergent.

(7) *The infinite series*

$$x - \frac{x^2}{2} + \frac{x^3}{3} - \frac{x^2}{4} + \cdots$$

is convergent if x is numerically less than unity. It is also convergent if $x = +1$ but not if $x = -1$.

It must be emphasised that there is a great deal more that could be said on this subject. At this stage, however, the only important point for the student to grasp is that before using an infinite series he should decide whether it converges: if it does not, he should not use it.

EXERCISE 2.4

Test the following series for convergence :

1. $4 - 3 + 2 - 1 + 0 + 1 - 2 + 3 \ldots$

2. $1 - x - (1 - x)^2 + (1 - x)^3 - (1 - x)^4 \ldots \quad 1 > x > 0$

3. $(1 + x) + (1 + x^2) + (1 + x^3) + (1 + x^4) + \ldots \quad 1 > x > 0$

4. $1 + \dfrac{x}{1} + \dfrac{x^2}{1 \cdot 2} + \dfrac{x^3}{1 \cdot 2 \cdot 3} + \ldots$

5. $1 + 2(\tfrac{1}{2}) + 4(\tfrac{1}{2})^2 + 6(\tfrac{1}{2})^3 + 8(\tfrac{1}{2})^4 + \ldots$

6. $8 + 8(\tfrac{1}{2}) + 8(\tfrac{1}{2})^2 + 6(\tfrac{1}{2})^3 + 4(\tfrac{1}{2})^4 + \tfrac{10}{4}(\tfrac{1}{2})^5 + \ldots$

CHAPTER III

PERMUTATIONS, COMBINATIONS AND THE BINOMIAL THEOREM

And when they would not let him arrange
The fish in the boxes
He stroked those which were already arranged.
POUND : *The Study in Aesthetics*

1. Combinations of events

There are a few very simple ideas which are quite indispensable to our later work, and form part of the theory of permutations and combinations. This theory is extensive, and can become very complicated, but only the basic ideas are necessary here. The first of these is a rule concerning the combination of events. It may be stated as follows :

If one event can happen in a different ways, and another happen in b different ways, then the two events can happen in ab different ways, provided that the events are independent of each other.

For example, I can go to work in any one of 3 different suits and any one of 2 different pairs of shoes. There are, in all, six possibilities.

Suit	Shoes
Brown	Brown
Grey	Brown
Blue	Brown
Brown	Black
Grey	Black
Blue	Black

It is sometimes useful to represent these possibilities in a diagram, dividing a square horizontally (for suits) and vertically (for shoes). The number of separate cells then represents the number of possibilities. We shall later see the use of this.

Brown Suit

Grey Suit

Blue Suit

Brown Black
Shoes Shoes

A consequence of this rule is that

If there are several events, of which one can happen in a different ways, a second in b ways, a third in c ways, etc., then there are

$$abc \dots$$

different possible combinations of events, provided that the events are independent of each other.

For example, if I have 3 suits, 2 pairs of shoes and 4 hats then there are $3 \times 2 \times 4$ possible ways of dressing, and of these 6 will involve wearing a cap, 6 a bowler, and so on.

There is an extension of this rule which applies when each independent event can happen in the same number of ways. For example, a football match can result in a win, a loss or a draw, which we may denote by 1, 2 and X respectively. A second match may also end in any of three different ways, again denoted by 1, 2 and X. A moment's consideration shows that the two matches may therefore end in any of nine different ways, these being 12, 1X, 11; 22, 2X, 21; X2, XX, X1, where the first symbol represents the result of one match and the second symbol that of the second match.

If we now have a third match, then this may end in any one of three ways *for each of the nine ways in which the first two matches may end*, and so the three matches may end in 27 ways. Similarly, we will obtain that four matches may result in 81 different combinations of results, and so on. If we have n matches, each being capable of ending in any one of three ways, then the n results may combine in 3^n different ways.

A similar argument will show that:

/ *If each of n independent events can occur in r different ways, then the n events can combine to occur in r^n different ways.* /

Exercise 3.1

1. I toss a penny and throw a six-sided die simultaneously. How many different results may be obtained?

2. I toss a penny, throw a six-sided die and throw an eight-sided die simultaneously. How many different results may be obtained?

3. I have to predict the results of ten football matches. How many different predictions will I have to make in order to be certain of being correct?

4. It is agreed to call any combination of three letters a " word " provided that the middle letter is vowel. How many three-letter " words " are possible?

5. In considering a credit squeeze, the Government feels that it can put the rate of interest at one of 3 levels, adopt one of 4 alternative policies towards hire purchase, and adopt one of 2 different policies towards bank lending. In how many different ways may the squeeze be effected?

6. To be acceptable to a certain economist a growth model must have acceptable assumptions about (i) the consumption function

(ii) the investment decision, (iii) the production function (iv) the availability of factors of production. If he considers that there are 3 acceptable assumptions in each of these four areas, and it takes him one week to work out each model, how long can he work on growth models before exhausting the subject?

2. Permutations of objects

Many very important problems are basically identical with the problem of arranging a number of objects in a row. To illustrate the solution of such problems we may consider the very simple question of arranging five different books on a shelf. Clearly, the first book, to go on the left end of the row, may be chosen in five ways. Let us make such a choice, denoting the particular book we have chosen by A. Four books will remain. The one of these to go next to A may be chosen in four ways. This means that *if A is the first book*, then the first two books may be any of AB, AC, AD and AE. We could, however, have chosen B as the first book, and in this case, the combinations BA, BC, BD and BE could occur. Choosing C as the first book will lead to another four possibilities, and choosing D and E to another four in each case. There will thus be 5×4 different ways of arranging the first two books.

We thus have that the first book may be chosen in 5 ways, and the second in 4 for each of the 5, giving 5×4 different ways of choosing the first two. The third may be chosen in 3 different ways for each of these 20 ways thereby indicating 60 different ways of choosing the first three. For each of these sixty different ways there will be 2 ways of choosing the fourth book, giving 120 different ways of choosing the first four. Clearly, when the first four have been chosen, the fifth is automatically chosen, and so we have that the five books may be arranged in 120 different ways.

It will be noted that his figure of 120 has been obtained by performing the multiplication $5 \times 4 \times 3 \times 2 \times 1$ and that this represents the number of ways in which 5 different objects may be *ordered in a row*. More briefly, we say that this is the number of ways of *permuting* 5 different objects. It is clear that:

n different objects may be permuted in

$$n(n-1)(n-2)(n-3)...3 \times 2 \times 1$$

different ways. In the interests of economy we denote this number by n! (which is sometimes written as $\lfloor n$*), and call it " factorial n " or " n factorial ".*

We have, for example, that eight different men may be lined against a wall in 8! ways, i.e., in $8 \times 7 \times 6 \times 5 \times 4 \times 3 \times 2 \times 1 = 40{,}320$ ways.

As we can see by considering the case of the five books, if we have to fill r different positions by choosing r out of n objects, then the problem can be solved in $n!/(n-r)!$ different ways. The first two books could be placed on the shelf in 5×4 ways, which is, of course, $(5 \cdot 4 \cdot 3 \cdot 2 \cdot 1)/(3 \cdot 2 \cdot 1) = 5!/3!$. In such a case we say that we are *permuting r objects out of n.* It follows that *we may permute r objects out of n different objects in $n!/(n-r)!$ ways, which we denote by $_nP_r$.*

For example, we may distribute first, second and third prizes to three of ten competitors in

$$_{10}P_3 = \frac{10!}{(10-3)!}$$

$$= \frac{10 \cdot 9 \cdot 8 \cdot 7 \cdot 6 \cdot 5 \cdot 4 \cdot 3 \cdot 2 \cdot 1}{7 \cdot 6 \cdot 5 \cdot 4 \cdot 3 \cdot 2 \cdot 1}$$

$$= 10 \cdot 9 \cdot 8$$

$$= 720 \text{ ways}$$

If the objects are not all different the above result has to be modified. It can be shown that :

If there are n objects of which p are exactly alike of one kind, q exactly alike of another kind, and r exactly alike of a third kind, then the n objects may be arranged amongst themselves in

$$\frac{n!}{p!\, q!\, r!} \quad ways$$

For example, the letters of the word BALE may be permuted in 4! ways : but the letters of the word BALL may be permuted in only $4!/2!$ ways. This will be seen if the 24 permutations of BALE are written out, and then the E is replaced by a second L.

Similarly we have that GERALD has 6! permutations, but GORDON has only $6!/2!$. HOLMWOOD has $8!/3!$.

The name WILLIAMS has two letters which are duplicated, and so the number of permutations is $8!/2!\, 2!$. MACMILLAN has three duplicated letters, resulting in the reduction of the number of permutations to $9!/2!\, 2!\, 2!$.

DEVERELL has one duplicated letter and one triplicated. The number of permutations is therefore $8!/2!\, 3!$.

EXERCISE 3.2

1. In how many different ways may the letters of the word BASKET be arranged?

2. Two vacancies occur in an organisation, the one being better paid than the other. Eight applications are received. In how many different ways can the vacancies be filled?

3. In how many different ways may the letters of the following words be arranged?

    (i) BISCUIT   (iv) STATISTICALLY
    (ii) EVEREST   (v) DIFFERENTIATED
    (iii) BANANA   (vi) ABRACADABRA

3. Combinations of objects

When we permute a number of objects we are really performing two separate operations on them. We are first of all *choosing* the objects, and then we are *arranging* them in order. Sometimes, however, we wish to perform only the former of these operations. If, for example, we wish to choose three out of ten people in order to knight them, or five firms out of fifty so that they may be investigated by a commission, the question of arrangement does not occur. Let us take the question of choosing three people out of ten. We have already seen that we may permute three people out of ten in $_{10}P_3 = 720$ ways. Some of these ways involve the same people arranged in different orders. For example, the figure of 720 will include ABC as well as ACB. In fact, since there are three people in each permutation, and since a *given* three people may be permuted amongst themselves in six different ways (ABC, ACB, BCA, BAC, CAB and CBA) it follows that the 720 different permutations will include the six possible permutations of every possible selection of three people. To put this another way, we see that the 720 permutations include every possible *combination* of three people arranged in their six different orders. It follows that the number of *different combinations*, independently of order, is obtained by dividing 720 by 6. More generally, we have that :

The number of *different* combinations *of r objects drawn from n objects is* $\dfrac{_nP_r}{r!}$ *which is* $\dfrac{n!}{(n-r)!\,r!}$. *This we denote by* $_nC_r$ (*or sometimes by the alternative notation* $\dbinom{n}{r}$)

Thus we may choose five firms out of fifty in

$$_{50}C_5 = \frac{50!}{45!\,5!} = \frac{50 \cdot 49 \cdot 48 \cdot 47 \cdot 46}{5 \cdot 4 \cdot 3 \cdot 2 \cdot 1} = 2{,}118{,}760 \text{ ways.}$$

If we wish to list five firms out of fifty in some order of merit, then we have to take each of the 2,118,760 combinations of five firms and arrange each such combination in its 5! possible orders, obtaining a result of 254,251,200 possible arrangements, which is, of course, $_{50}P_5$.

It should be noted that when we choose three people out of ten in order to give them a knighthood we also choose seven people out of ten who are not to receive this honour. Clearly, the number of ways of choosing the three people out of ten is identical with the number of ways of choosing the seven.

i.e., $$_{10}C_3 = {}_{10}C_7$$

or, more generally,

$$_nC_r = {}_nC_{n-r}$$

We may also note there is one, and only one, way of choosing n people out of n, which is to take them all. This is also the one, and only one, way of choosing no people out of n. In the above notation, the number of ways of choosing n people out of n would be

$$_nC_n = \frac{n!}{(n-n)!\,n!} = \frac{n!}{0!\,n!} = \frac{1}{0!}$$

This looks a little strange, but we must remember that its value is unity : there is only one way of choosing everybody. We thus have that

$$\frac{1}{0!} = 1$$

which can be true only if $0! = 1$. Hitherto, we have not defined 0!. If we decide to define it as unity, then we have the satisfaction of knowing that many puzzles such as how to identify $_nC_n$ with unity are immediately resolved, and also that it is not unreasonable to suggest that there is one way of arranging no objects in a row. As we have said, this may appear to be a little puzzling, but a careful consideration of the matter will show that there is nothing contradictory or slick in this decision to *define* 0! as unity.

EXERCISE 3.3

1. A shopkeeper decides that he has surplus stocks of 9 different kinds of goods. He decides to make up gift parcels, each to contain four different goods, which he will sell at a special price. How many different kinds of parcel may he produce?

2. 40 people apply for admission to a class which is restricted to 10 students. In how many different ways may the class be composed?

4. The Binomial Theorem

It is easily shown by multiplication of the left-hand side that

$$(x+a)^2 = x^2 + 2ax + a^2$$

and that
$$(x+a)^3 = x^3 + 3ax^2 + 3a^2x + a^3$$

The result for $(x+a)^2$ was known to Euclid (c. 300 B.C.) and about fourteen hundred years later Omar Khayýam stated the results for $(x+a)^4$, $(x+a)^5$ and $(x+a)^6$. A little later, around the year 1300, Chinese writers gave a diagrammatic method for deriving the terms in the expansion of $(x+a)^n$ where n could, in theory, be any positive integer. In 1654 Pascal stated and proved the result we print below for n a positive integer. Another proof was given by Newton in 1676. Newton also extended the theorem so that it could be used (as illustrated below) for negative and fractional values of n, although it was not until almost a hundred years later that adequate proofs were given by Euler; and Gauss (1777–1855) and Abel (1802–29) later extended these to the perfectly general case.

We shall not give a proof of the theorem here, since the proofs for n a positive integer are readily available in algebra books, and the more general proofs are very complicated. On the other hand, it must be emphasised that this problem, which exercised mathematicians for two thousand years, is a highly important one, and that time and again we shall be assuming familiarity with it.

The Binomial Theorem states that, for n a positive integer,

$$(x+a)^n \equiv x^n + {}_nC_1 x^{n-1}a + {}_nC_2 x^{n-2}a^2 + {}_nC_3 x^{n-3}a^3 + \ldots + {}_nC_{n-1}xa^{n-1} + a^n$$

For n a negative or fractional number, the above finite expansion is replaced by the infinite expansion shortly to be given. We have, for example, that

$$(x+a)^5 \equiv x^5 + 5x^4a + 10x^3a^2 + 10x^2a^3 + 5xa^4 + a^5$$

and that

$$(x+a)^6 \equiv x^6 + 6x^5a + 15x^4a^2 + 20x^3a^3 + 15x^2a^4 + 6xa^5 + a^6$$

In evaluating these cases, of course, we obtain the numerical co-efficients by remembering that

$$_nC_r = \frac{n!}{(n-r)!\, r!} = \frac{n(n-1)(n-2)\ldots(n-r+1)}{1.2.3.4\ldots r}$$

In order to consider the result for n a negative or fractional number it is convenient to consider only the expansion of $(1+y)^n$, noting that it can easily be transformed into the expansion of $(x+a)^n$ by multiplying the result by a^n.

We have that

$$(x+a)^n = a^n\left(1+\frac{x}{a}\right)^n$$

which may be written as $\qquad a^n(1+y)^n$

where $\qquad\qquad\qquad y = \dfrac{x}{a}$

The Binomial Theorem now states that

$$(1+y)^n = 1 + ny + \frac{n(n-1)}{1\cdot2}y^2 + \frac{n(n-1)(n-2)}{1\cdot2\cdot3}y^3 + \ldots$$

where the right-hand side is now an infinite series. (It cannot stop, since if n is negative or fractional no bracket in the numerator of any of its successive terms can be zero.) As with all infinite series, the result must be qualified. It is true only if the right-hand side converges, and the condition for this is that y *shall be numerically less than* unity.

We may consider a few examples. Noting that

$$\frac{1}{1+y} = (1+y)^{-1} = 1 + \frac{(-1)}{1}y + \frac{(-1)(-2)}{1\cdot2}y^2 + \frac{(-1)(-2)(-3)}{1\cdot2\cdot3}y^3 + \ldots$$

$$= 1 - y + y^2 - y^3 + \ldots$$

we may substitute $y = \tfrac{1}{2}$ and obtain

$$\tfrac{2}{3} = \frac{1}{1\frac{1}{2}} = (1+\tfrac{1}{2})^{-1} = 1 - \tfrac{1}{2} + \tfrac{1}{4} - \tfrac{1}{8} + \ldots$$

which a few more terms will quickly show to be reasonable. We may also try putting $y = -\tfrac{1}{2}$ when we obtain

$$2 = \frac{1}{1-\frac{1}{2}} = (1-\tfrac{1}{2})^{-1} = 1 + \tfrac{1}{2} + \tfrac{1}{4} + \tfrac{1}{8} + \ldots$$

which converges to 2. But if we substitute $y = -2$ we obtain

$$-1 = \frac{1}{1-2} = (1-2)^{-1} = 1 + 2 + 2^2 + 2^3 + \ldots$$

which is not so.

The same is true in the case of fractional values. For example, if we put $n = \tfrac{1}{2}$ we have

$$(1+y)^{\frac{1}{2}} = 1 + \tfrac{1}{2}y + \frac{\frac{1}{2}(-\frac{1}{2})}{1\cdot2}y^2 + \frac{\frac{1}{2}(-\frac{1}{2})(-1\frac{1}{2})}{1\cdot2\cdot3}y^3 + \ldots$$

$$-1 + \tfrac{1}{2}y - \tfrac{1}{8}y^2 + \tfrac{1}{16}y^3 - \ldots$$

which is true if y is numerically less than unity.

It follows, of course, that the expansion of $(x+a)^{1/2}$ is given by

$$a^{1/2}\left(1+\frac{x}{a}\right)^{1/2} = a^{1/2}(1+\tfrac{1}{2}y-\tfrac{1}{8}y^2+\ldots) \quad \text{where } y=\frac{x}{a}<1, \text{ if } x<a$$

but by

$$x^{1/2}\left(1+\frac{a}{x}\right)^{1/2} = x^{1/2}(1+\tfrac{1}{2}y-\tfrac{1}{8}y^2+\ldots) \quad \text{where } y=\frac{a}{x}<1, \text{ if } a<x.$$

EXERCISE 3.4

1. Write down the expansion of $(x+1)^7$.

2. By noting that $11=10+1$, write down a series which gives the value of 11^7.

3. Obtain an expansion for $\dfrac{1}{(1+y)^2}$, and (by applying one of the tests of Chapter II) consider whether the expansion is valid for *any* value of y, or only for certain restricted values.

5. Probability

Although we shall say nothing about it anywhere else in this book, we might as well use the above remarks to indicate some of the more elementary ideas involved in the theory of probability.

Consider the letters AHMOST. These may be permuted in 6! ways. If the letters are put into a hat and then drawn out one at a time (or, more satisfactorily, selected by a more foolproof method), we will have no idea of the order in which they will appear. There are 6! ($=720$) possible orders, and each of these is as likely to turn up as any other. One possible order is THOMAS ; and this has a certain chance of appearing. There are 719 other possible orders, each *with the same chance of appearing.* Clearly, one of the orders is bound to occur : and as they all have the same chance of occurring, the chance of our obtaining THOMAS in a single draw is 1 in 720. The chance of obtaining THOMSA is also 1 in 720 ; so is the chance of OMHTAS, and so on.

More precisely, we say that *if n events are equally likely and only one of them may happen, then the probability of that event is* $1/n$.

The probability of getting a head when we toss a coin is $\tfrac{1}{2}$ since the two possible results are equally likely.

The probability of getting a " five " when we throw a six-sided die is $\tfrac{1}{6}$, since the six possible results are equally likely.

Sometimes, however, the alternative possibilities are not equally likely. If we choose a letter at random from the alphabet, the chance

of it being a vowel is not $\frac{1}{2}$. It is true that there are only two possibilities—vowel or consonant : but these possibilities are not
equally likely. If we choose a letter at random the chance of its
being an A is $\frac{1}{26}$. The chance of E is $\frac{1}{26}$. The chance of I is $\frac{1}{26}$. The
chance of O is $\frac{1}{26}$, and the chance of U is $\frac{1}{26}$. Although intuition is a
most dangerous guide in these matters, we may be forgiven in a brief
note like this if we suggest that it is intuitively obvious that the
chance of getting A, E, I, O or U is therefore $\frac{5}{26}$, while the chance of
getting a consonant is (similarly) $\frac{21}{26}$. It will be noticed that the
probability of a vowel ($\frac{5}{26}$) plus the probability of a consonant ($\frac{21}{36}$)
total unity. This is because we are bound to obtain one thing or
the other.

There is a very great deal more than this to probability theory,
and the reader who is interested in it should consult one of the
works listed in the Bibliography. We may, however, cite a few
results of considerable importance.

(1) *If two events are* mutually independent, *and the probability of
the one is* p_1, *while that of the other is* p_2, *the probability of the two events
occurring is the product* $p_1 p_2$.

For example, if I toss a coin the probabilty of a head is $p_1 = \frac{1}{2}$.
If a throw a die, the probability of obtaining a " five " is $p_2 = \frac{1}{6}$.
Tossing the coin and throwing the die are mutually independent
actions. Neither affects the other. Consequently the probability of
obtaining both a head and a " five " if I toss the coin once and throw
the die once is

$$\frac{1}{2} \times \frac{1}{6} = \frac{1}{12}$$

We may also demonstrate this result by using a diagram. It
would in some ways be better to defer it until Chapter XXVIII,
where we look at Set Theory, but it may usefully be done here. Let
us suppose that the chance of my wearing a dark suit is $p_1 = \frac{1}{3}$ and
that the chance of the first man I meet having black hair is $p_2 = \frac{1}{7}$.
Suppose that there is no relationship between these events. Let the
inside of the square in Diagram 3.1 represent everything that can
happen. Divide it vertically by AB so that the area to the left of AB
is one third of the whole, and so represents the probability $p_1 = \frac{1}{3}$ of
my wearing a dark suit. Also divide it horizontally by CD so that
the area above CD is $\frac{1}{7}$th of the whole and represents the chance of the
man having black hair. The shaded area is that part of the area to the
left of AB which is also above CD. It represents dark suit plus black
hair. The area of the square is 21 times this shaded area. If we let
unity be the total probability of everything, then the separate

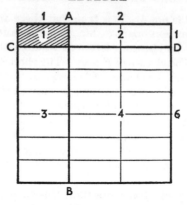

Fig. 3.1

probabilities of the four possible outcomes are indicated by the areas, 1, 2, 3, 4 expressed as fractions of unity.

1. Dark Suit + Black Hair　　　　　$\frac{1}{3} \times \frac{1}{7} = \frac{1}{21}$
2. Not Dark Suit + Black Hair　　　$\frac{2}{3} \times \frac{1}{7} = \frac{2}{21}$
3. Dark Suit + Not Black Hair　　　$\frac{1}{3} \times \frac{6}{7} = \frac{6}{21}$
4. Not Dark Suit + Not Black Hair　$\frac{2}{3} \times \frac{6}{7} = \frac{12}{21}$
　　Total　　　　　　　　　　　　$\frac{21}{21}$

(2) *If two events are mutually exclusive and the probability of the one is p_1, while of that of the other is p_2, the probability of either the one event or the other occurring is the sum $p_1 + p_2$.*

For example, if I throw a single six-sided die, I cannot obtain, in one throw, a " five " and a " six ". The occurrence of the event " five " makes the occurrence of the event " six " impossible, and vice versa. The probability of obtaining a " five " or a " six " is therefore

$$\tfrac{1}{6} + \tfrac{1}{6} = \tfrac{1}{3}$$

We have, in fact, a problem very similar to that of the vowels and consonants. The drawing of a single letter means that only one letter can possibly appear. We cannot have A and E. The events corresponding to the appearance of A, E, I, O and U are therefore mutually exclusive, and the probability of one of them occurring is

$$p_a + p_e + p_i + p_o + p_u$$
$$= \tfrac{1}{26} + \tfrac{1}{26} + \tfrac{1}{26} + \tfrac{1}{26} + \tfrac{1}{26}$$
$$= \tfrac{5}{26}$$

(3) *If the probability of an event is p, and $q = 1 - p$, then if there are n independent occasions on which the event can occur, the probability of it occurring r times is given by the term containing p^r in the binomial expansion of*

$$(p+q)^n.$$

This is a result which follows from an application of result (1) above. It may be demonstrated as follows.

Suppose we spin a coin and denote the result by H or T according to whether it is a head or a tail. We have possible results

$$H, T.$$

and if these same letters denote the chance of a head or of a tail then

$$H + T = 1.$$

Now take two identical coins and spin them. Writing the result for the first coin with subscript 1 and that for the second with subscript 2, we have possible results

$$H_1 H_2, \; H_1 T_2, \; T_1 H_2, \; T_1 T_2$$

and, since the coins are identical, the chances

$$H_1 = H_2; \; T_1 = T_2$$

and so as the total of all chances equals unity

$$H_1 H_2 + H_1 T_2 + T_1 H_2 + T_1 T_2 = 1$$

means

$$H^2 + 2HT + T^2 = 1$$

If we had had three coins the various possibilities would have been

$$H_1 H_2 H_3, \; H_1 H_2 T_3, \; H_1 T_2 H_3, \; T_1 H_2 H_3, \; H_1 T_2 T_3, \; T_1 H_2 T_3,$$
$$T_1 T_2 H_3, \; T_1 T_2 T_3$$

yielding

$$H^3 + 3H^2 T + 3HT^2 + T^3 = 1$$

and so on.

The expressions

$$H + T = 1$$

$$H^2 + 2HT + T^2 = 1$$

and

$$H^3 + 3H^2 T + 3HT^2 + T^3 = 1$$

are obviously capable of being written as

$$(H + T)^n = 1$$

with $n = 1$, 2 and 3 successively, and it is clear that in the case of n coins, with n having any value, the same result would hold.

Furthermore, if we go back to the case of two coins, we see that the chance of getting two heads is (by rule (1) above)

$$H_1H_2$$

which, as the coins are identical, is H^2.

The chance of one head and one tail is (by rules (1) and (2) above)

$$H_1T_2 + T_1H_2$$

which, because the coins are identical, is $2HT$.

Similarly the chance of two tails (or no heads) is T^2.

Arguing in this way we would find that in the case of three coins

the chance of 3 heads is	H^3
2	$3H^2T$
1	$3HT^2$
0	T^3

Example :

Suppose, to take a different example, that the chance of a building contract costing the builder more than expected is 0.3. If we consider ten independent contracts, what is the chance of (1) exactly 4 of them costing more than was expected, (2) 4 or more of them costing more than was expected?

The chance of exactly 4 out of 10 costing more than was expected is the coefficient of p^4 in the expansion of

$$(p+q)^{10}$$

where

$$p = 0.3 \text{ and } q = 0.7$$

The term containing p^4 is, from

$$_nC_rp^{n-r}q^r$$

given by

$$_{10}C_6p^4q^6 = {}_{10}C_4p^4q^6$$

$$= \frac{10 \cdot 9 \cdot 8 \cdot 7}{1 \cdot 2 \cdot 3 \cdot 4}(0.3)^4(0.7)^6$$

$$= 210 \times 0.0081 \times 0.117649$$

$$= 1.701 \times 0.117649$$

$$= 0.2001$$

The chance of 4 or more costing too much is best obtained by working out, separately, the chances of 0, 1, 2 and 3 costing too much. The total of these, subtracted from unity, will give the chances of 4 or more costing too much.

i.e., the chances of 4 or more costing too much are

$$1 - (q^{10} + {}_{10}C_1 q^9 p + {}_{10}C_2 q^8 p^2 + {}_{10}C_3 q^7 p^3)$$

where $p = 0 \cdot 3$ and $q = 0 \cdot 7$.

TRIGONOMETRY AND GEOMETRY

These chapters assume no previous knowledge of the subjects. Chapter IV defines a few basic ideas which are necessary later in the book, and concludes with a list of results which will be useful for reference.

Chapter V also starts from simple ideas, and the reader is advised to work on the exercises very carefully. Some simple economic illustrations are provided. The contents of this chapter are essential knowledge for a proper understanding of the rest of the book.

Chapter VI develops some slightly more difficult ideas. A cursory knowledge of its first half is adequate for a first reading.

CHAPTER IV

BASIC TRIGONOMETRY

So by the shadow cast, he had the wit
To judge that Phoebus, shining clear and bright
Had climbed some forty-five degrees in height.
CHAUCER : *The Man of Law's Tale*

1. Angles under 90 degrees

Although trigonometry is a branch of mathematics which, at first sight, appears to be exclusively concerned with the study of the sides and angles of triangles, it is, in fact, much more fundamental than that, and a knowledge of it is quite indispensable in many branches of economics and mathematics which few would consider to deal simply with triangles. In particular, it is important in studying cyclical fluctuations. In this chapter we shall introduce the simpler ideas, and then list some of the more important results, without bothering to encumber the reader with proofs.

We must begin with a definition of similar triangles.

Definition :

Two triangles are *similar* if the angles of one are equal to the angles of the other.

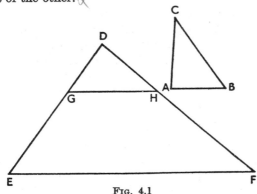

FIG. 4.1

For example, in Diagram 4.1, triangles *ABC* and *DEF* are similar. This is because their angles are equal. The smaller triangle can be picked up and fitted into the corner of the larger one, with the two bases remaining parallel. In the diagram, *ABC* has been fitted into the position *DGH*.

It can be shown (and is in any case fairly obvious) that because these triangles are similar then the following relationships between the lengths of their sides are true.

$$\frac{AB}{DE} = \frac{BC}{EF} = \frac{CA}{FD}$$

$$\frac{AB}{AC} = \frac{DE}{DF} \; ; \; \frac{BC}{BA} = \frac{EF}{ED} \; ; \; \frac{CA}{CB} = \frac{FD}{FE}$$

$$\frac{DG}{GE} = \frac{DH}{HF}$$

In trigonometry we begin by considering right-angled similar triangles, i.e., similar triangles having one right angle. Two such

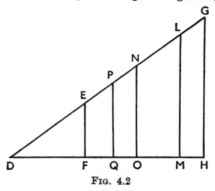

Fig. 4.2

triangles are DGH and DEF in Diagram 4.2. Let us consider the ratio GH/DH. From our knowledge of similar triangles we know

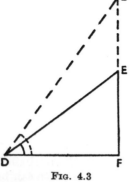

that this is equal to EF/DF. In fact, if all the other triangles shown in this diagram are right-angled, then

$$\frac{GH}{DH} = \frac{LM}{DM} = \frac{NO}{DO} = \frac{PQ}{DQ} = \frac{EF}{DF}$$

This means that, quite independently of the size of the triangle, the ratio of the side opposite the angle D (e.g., the side GH) to the side adjacent to the angle D (e.g., the side DH) is constant, provided that the angle D remains unchanged.

If, however, the angle D were to increase,

Fig. 4.3

then the ratio of the opposite side (EF in Diagram 4.3) to the adjacent side (DF) would also increase (to $E'F/DF$).

Another way of saying this is to say that the ratio of the opposite side to the adjacent side depends only on the angle concerned. We shall put this a little more formally in a moment, but it is as well to look at the matter from an alternative point of view.

Suppose that we draw a circle of radius one inch, as shown in Diagram 4.4. Let OP be a radius and TP be a tangent from a point T. Suppose that we know the angle TOP which we denote by θ (theta) and that we wish to know the length of the tangent TP.

Clearly the ratio TP/OP will depend on the angle θ, just as, previously, the ratio EF/DF depended on the angle D. The larger the angle θ the longer is the tangent TP.

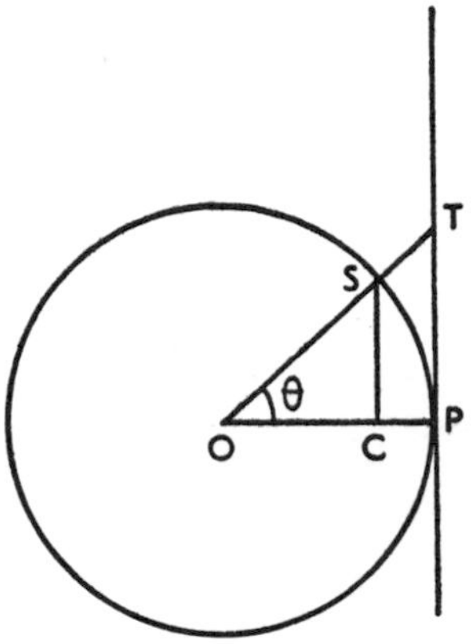

FIG. 4.4

Definition :

We shall call the ratio TP/OP the *tangent* of the angle TOP and denote it by tan θ.

Since, in this particular case, OP is one inch, then tan θ gives the length of the tangent TP.

It also follows that in any triangle similar to triangle TOP the ratio of the side corresponding to TP over the side corresponding to OP will be constant and equal to tan θ.

For example, it will be found by measurement (or by considering the geometry of an equilateral triangle) that if $\theta = 60°$ then, whatever the size of the triangle,

$$\tan 60° = \frac{\text{opposite side}}{\text{adjacent side}} = 1 \cdot 73 \text{ (approximately)}$$

while if $\theta = 45°$ we would find that

$$\tan 45° = 1 \cdot 00$$

Tables giving values of tan θ for θ varying from $0°$ to $90°$ are given at the back of the book, and will be referred to later.

An elementary example of the practical use to which this idea may be put is provided by the task of estimating the height of a tower. If the observer knows that he is standing (say) 20 feet from the base of the tower and on the same level as it, and that his eye is 5′ 6″ from the ground, then he may estimate the height of the tower by measuring (with a theodolite) the angle ADE in Diagram 4.5. We have that the height of the tower is

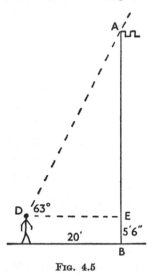

$$AE + EB = AE + 5\tfrac{1}{2} \text{ feet}$$

But $\tan 63° = \dfrac{AE}{DE} = \dfrac{AE}{20}$

The table at the back of the book shows that $\tan 63° = 1\cdot9626$ and so we have that

$$AE = 20 \times 1\cdot9626 \text{ feet}$$

$$= 39\cdot252 \text{ feet} = 39'\ 3'' \text{ (roughly)}$$

The tower is thus

$$39'\ 3'' + 5'\ 6'' = 44'\ 9'' \text{ high.}$$

Before leaving the subject of the tangent of an angle we should notice that in Diagram 4.4

$$\tan\theta = \frac{SC}{OC}$$

FIG. 4.5

Just as the angle θ defines the ratio of the opposite side to the adjacent side, so does it define two other ratios, which we know as the sine and cosine of the angle. Referring once again to Diagram 4.4 we have the following definitions.

Definition :

We shall call the ratio SC/OS the *sine of the angle SOC* and denote it by $\sin\theta$.

Definition :

We shall call the ratio OC/OS the *cosine of the angle SOC* and denote it by $\cos\theta$.

If the circle has unit radius, then $\sin\theta$ will give the length of SC and $\cos\theta$ the length of OC.

These definitions may conveniently be remembered as

$$\sin\theta = \frac{\text{opposite}}{\text{hypotenuse}}$$

and

$$\cos\theta = \frac{\text{adjacent}}{\text{hypotenuse}}$$

Tables at the back of the book give values of sin θ and cos θ. Meanwhile the following values of sin θ, cos θ and tan θ may be noted. They have been derived by considering the elementary geometry of the triangles shown in Diagram 4.6

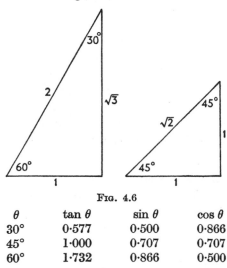

FIG. 4.6

θ	tan θ	sin θ	cos θ
30°	0·577	0·500	0·866
45°	1·000	0·707	0·707
60°	1·732	0·866	0·500

The relationships we have just discussed occur so frequently that it is most important not to confuse them. There is a useful, if frightful, mnemonic, whose continued existence is permissible only because it is so bad that it cannot be forgotten.

$$\text{Tan} = \text{Opposite/Adjacent} \quad ; \quad T = O/A \; ; \; TOA$$
$$\text{Sin} = \text{Opposite/Hypotenuse} \; ; \quad S = O/H \; ; \; SOH$$
$$\text{Cos} = \text{Adjacent/Hypotenuse} \; ; \quad C = A/H \; ; \; CAH$$

which may be summarised in

" I am going on a *TOA* and *SOH* I buy a *CAH*."

Sometimes it is convenient to deal, not with cos θ, sin θ and tan θ but with their reciprocals, which are known by the following names :

$$\text{cotangent} = \frac{1}{\text{tangent}} ; \quad \cot \theta = \frac{1}{\tan \theta}$$

$$\text{cosecant} = \frac{1}{\text{sine}} ; \quad \text{cosec } \theta = \frac{1}{\sin \theta}$$

$$\text{secant} = \frac{1}{\text{cosine}} ; \quad \sec \theta = \frac{1}{\cos \theta}$$

EXERCISE 4.1

1. Check from the tables that :

sin 51° = 0·7771 ; sin 51° 30′ = 0·7826 ; sin 51° 32′ = 0·7830
cos 51° = 0·6293 ; cos 51° 30′ = 0·6225 ; cos 51° 32′ = 0·6220
tan 51° = 1·2349 ; tan 51° 30′ = 1·2572 ; tan 51° 32′ = 1·2587

2. $\triangle ABC$ is such that the angle B is a right angle. Given the following data, estimate the angle or length required :

(i) $AB = 3''$, $\widehat{A} = 50°$.   Estimate BC and AC
(ii) $AC = 3''$, $\widehat{A} = 50°$.   Estimate BC and AB
(iii) $BC = 3''$, $\widehat{A} = 50°$.   Estimate AB and AC
(iv) $BC = 3''$, $AB = 2''$.   Estimate $\widehat{A}$ and AC
(v) $BC = 3''$, $AC = 4''$.   Estimate $\widehat{A}$ and AB

2. Angles exceeding 90 degrees

So far we have considered only angles of between 0 and 90 degrees. Even in this consideration, we have, in fact, omitted actual reference to the extreme values of 0° and 90°, and we shall continue to

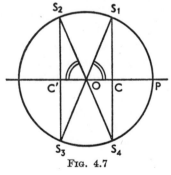

FIG. 4.7

ignore the problems associated with the definition of the trigonometric functions of these angles until a little later. We cannot, however, leave the subject of angles exceeding 90°.

Let us examine the Diagram 4.7 which is a modification of Diagram 4.4. We may think of a line originally occupying the position OP as moving around, in an anti-clockwise direction so that it successively occupies the positions OS_1, OS_2, OS_3 and OS_4. The angles between the line OP and the line OS will be denoted by θ which takes all possible values between 0° and 360° as the point S moves around the circle.

Consider, first, the angle θ_1 corresponding to a position S_1. We have already seen that tan θ_1 is the ratio of the lengths $S_1 C$ and OC. We call $S_1 C$ the *perpendicular* from S_1 onto OP and the point C is known as the *projection of the point S_1 on OP*. The distance OC is the *projection of the distance OS_1 on OP*. We see that, in this terminology, we may define the tangent of θ_1 to be

$$\tan \theta_1 = \frac{\text{perpendicular from } S_1 \text{ onto } OP}{\text{projection of } OS_1 \text{ onto } OP}$$

Now let the point S move around to occupy position S_2. The angle θ changes to θ_2 (where $\theta_2 = \widehat{S_2 O P}$) which lies between 90° and 180°. The perpendicular is now given by $S_2 C'$ and the projection by OC'. We may therefore consider the tangent of the angle θ_2 to be given by

$$\frac{\text{perpendicular}}{\text{projection}} = \frac{S_2 C'}{OC'}$$

Similarly, if we consider the angle θ_3 we may have that the tangent is given by $S_3 C'/OC'$ while that of the angle θ_4 is given by $S_4 C/OC$. This extends the idea of the tangent of an angle from the simple property of a triangle, as shown in Diagrams 4.2 and 4.3, and the length of a line TP, as shown in Diagram 4.4, to an idea associated with the rotation of a line through any angle up to 360°. This is an

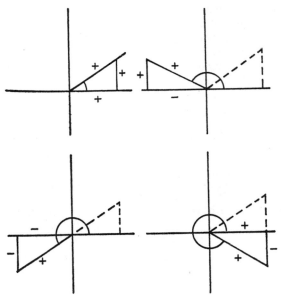

Fig. 4.8

example of how the mathematician often starts with a definition that is of restricted application, and then finds a more general definition, of which his earlier definition was a special case.

At the same time, we introduce a convention which, although not essential, has proved of great value. We shall consider all horizontal lines drawn to the right of O to have a positive sign, and all horizontal lines to the left of O to have a negative sign. Thus we may

D

think of OC as being a positive length, while OC' is a negative length. Similarly, we will attribute a positive sign to all vertical lines drawn above O and a negative sign to all vertical lines drawn below O. This makes S_1C and S_2C' positive, but S_3C' and S_4C negative. Finally we shall always consider all lines drawn outwards from O to be positive (unless they happen to coincide with angles of 90°, 180°, 270° or 360° when the whole position will have to receive special treatment, as we shall see later). These ideas are illustrated in Diagram 4.8.

We may now summarise our ideas about the tangent of an angle by saying that

$$\tan \widehat{SOP} = \frac{\text{perpendicular from } S \text{ onto } OP}{\text{projection of } OS \text{ onto } OP}$$

where the two lines concerned are given their appropriate sign. This means that the tangent of an angle lying between 0° and 90° is of positive sign, that of an angle between 90° and 180° negative (being a positive length divided by a negative one), that of an angle between 180° and 270° positive, (being the quotient of two negative lengths) and that of an angle between 270° and 360° negative (being a negative length divided by a positive one).

This leads us to a very important set of results. We have drawn Diagram 4.7 such that $S_1\widehat{OC} = S_2\widehat{OC'}$, which means that $\theta_2 = 180° - \theta_1$ and so $\theta_1 = 180° - \theta_2$. Now $\tan \theta_2 = \dfrac{S_2C'}{C'O}$ where S_2C' is positive but $C'O$ is negative and equal numerically to CO. We thus have that

$$\frac{S_2C'}{C'O} = -\frac{S_2C'}{CO} = -\frac{S_1C}{CO} = -\tan S_1\widehat{OC}$$

In other words, we have that

$$\tan \theta_2 = -\tan \theta_1$$

whence $\tan (180° - \theta_1) = -\tan \theta_1$

By considering the angles corresponding to the successive positions of S in a similar way, we may obtain the following results :

     for θ between   90° and 180°, $\tan \theta = -\tan (180° - \theta)$
     for θ between 180° and 270°, $\tan \theta = +\tan (\theta - 180°)$
     for θ between 270° and 360°, $\tan \theta = -\tan (360° - \theta)$

It must be emphasised that these results are simple consequences of our decision to define the tangent of an angle to be the ratio of the perpendicular to the projection with a certain convention about signs. We could have had a different convention (such as, for example, letting all lengths be positive) but it is found that the above convention has important advantages in more advanced work.

In a similar way, we may define the cosine and sine of an angle greater than 90°, letting it be the projection or the perpendicular (respectively) upon the radius (OS). A careful examination of Diagram 4.7 will show the validity of the following results. Diagram 4.8 and the subjoined examples make this matter clearer :

for θ between   90° and 180°, $\sin \theta =$ $\sin (180° - \theta)$
for θ between 180° and 270°, $\sin \theta = - \sin (\theta - 180°)$
for θ between 270° and 360°, $\sin \theta = - \sin (360° - \theta)$
for θ between   90° and 180°, $\cos \theta = - \cos (180° - \theta)$
for θ between 180° and 270°, $\cos \theta = - \cos (\theta - 180°)$
for θ between 270° and 360°, $\cos \theta =$ $\cos (360° - \theta)$

For example,
$\sin 120° =$ $\sin 60°$; $\sin 210° = - \sin 30°$; $\sin 300° = - \sin 60°$
$\cos 120° = - \cos 60°$; $\cos 210° = - \cos 30°$; $\cos 300° =$ $\cos 60°$
$\tan 120° = - \tan 60°$; $\tan 210° =$ $\tan 30°$; $\tan 300° = - \tan 60°$

EXERCISE 4.2

1. Check the following results :
$\sin 110° =$ $0 \cdot 9397$; $\cos 110° = - 0 \cdot 3420$; $\tan 110° = - 2 \cdot 7475$
$\sin 240° = - 0 \cdot 8660$; $\cos 240° = - 0 \cdot 5000$; $\tan 240° =$ $1 \cdot 7321$
$\sin 310° = - 0 \cdot 7660$; $\cos 310° =$ $0 \cdot 6428$; $\tan 310° = - 1 \cdot 1918$

3. Radians

So far we have measured all of our angles in degrees, letting a right angle be 90 degrees, so that a complete rotation is 360 degrees. There is, however, a more fundamental way of measuring an angle.

We saw, in Diagram 4.7 that as S moved around the circle the line OS swept out an angle θ. For a circle of any given radius, the size of this angle is proportional to the length of the arc PS which the moving point has described. An obvious question is whether it is possible to calculate the size of the angle if the length of the arc is known. Part of the answer to this is given by our knowledge of one of the fundamental properties of the circle, namely that a circle of radius r has a circumference of $2\pi r$, where π is a numerical constant whose value is approximately $3 \cdot 1416$. Now the circumference of the circle is the length of the arc traced out by a point that moves through 360°. Another way of saying this is that an *arc* of length $2\pi r$ subtends an angle of 360° at O (just as the arc PS_1 subtends an angle of θ_1 at O). It follows that an *arc* of length r (i.e., an arc of length equal to the radius) subtends an angle at O of $\dfrac{360}{2\pi} = \dfrac{180}{\pi}$

degrees. This is approximately $57 \cdot 3$ degrees.

Definition :

> We define a *radian* to be the angle subtended at the centre of a circle by an arc equal in length to the radius. 1 radian = 57·3 degrees approximately. π radians = 180 degrees exactly.

Many results in trigonometry, and mathematics more generally, are true only if the angles are measured in radians. Unless the contrary is clearly indicated, *we shall henceforth assume that our angles are so measured.*

We may notice that a right angle, i.e. 90°, is measured in radian measure as $\pi/2$.

4. Trigonometric series

We shall make no attempt to prove the results of this section, and it will be some time before we use them, but now is a convenient time to meet some highly interesting, and very important, results.

It can be shown that, if x is an angle measured in radians, then $\cos x$ and $\sin x$ are given by the infinite series

$$\cos x = 1 - \frac{x^2}{2!} + \frac{x^4}{4!} - \frac{x^6}{6!} + \dots$$

and

$$\sin x = x - \frac{x^3}{3!} + \frac{x^5}{5!} + \frac{x^7}{7!} + \dots$$

Proofs of these series are indicated in Chapter XVII.

The continuation of these series is quite obvious. There is a similar result for $\tan x$ but the coefficients of the powers of x are much more complicated, and it is difficult to state it beyond the first few terms, which are given by

$$\tan x = x + \frac{2x^3}{3!} + \frac{16x^5}{5!} + \frac{272x^7}{7!} + \dots$$

$$= x + \frac{x^3}{3} + \frac{2x^5}{15} + \frac{17x^7}{315} + \dots$$

There are also various series, and products, for giving the value of π. Some of these are given below.

$$\frac{4}{\pi} = \frac{3}{2} \cdot \frac{3}{4} \cdot \frac{5}{4} \cdot \frac{5}{6} \cdot \frac{7}{6} \cdot \frac{7}{8} \cdot \frac{9}{8} \cdot \frac{9}{10} \dots \quad \text{(Wallis 1655)}$$

$$\frac{\pi}{4} = 1 - \frac{1}{3} + \frac{1}{5} + \frac{1}{7} + \frac{1}{9} - \dots \quad \text{(Leibniz 1673)}$$

$$\pi = 6 \left(\frac{1}{2} + \frac{1 \left(\frac{1}{2}\right)^3}{2 \cdot 3} + \frac{1 \cdot 3 \left(\frac{1}{2}\right)^5}{2 \cdot 4 \cdot 5} + \dots \right)$$

Although subsequent reference will be made to these series, they are given here for interest rather than use. Meanwhile it is instructive to consider the rapidity with which it is possible to obtain π by using these results.   Noting that, correct to 10 places of decimals,

$$\pi = 3 \cdot 141\ 592\ 653\ 6$$

then we have that

$$\frac{\pi}{4} = 0 \cdot 785\ 398\ 163$$

and

$$\frac{4}{\pi} = 1 \cdot 273\ 239\ 544$$

Use of the formula due to Wallis gives us, as successive approximations to $4/\pi$

after one term	1·500 000 000
after two terms	1·125 000 000
after three terms	1·406 000 000
after four terms	1·171 875 000
after five terms	1·367 187 500

Leibniz's formula gives us, for $\pi/4$, the values

after one term	1·000 000 000
after two terms	0·666 666 667
after three terms	0·866 666 667
after four terms	0·723 809 524
after five terms	0·834 920 635

In both of these cases the values are alternating around the correct value and slowly getting nearer to it.

The last formula gives us the following values for π

after one term	3·000 000 000
after two terms	3·125 000 000
after three terms	3·139 062 500
after four terms	3·141 155 134
after five terms	3·141 511 172

We see that the last series converges to the correct value much more rapidly than either of the other expressions.

5. Trigonometric graphs

In this section we consider an important graphical representation of sine, cosine and tangent. The reader who has no knowledge of graphs would do well to read the first section of the next chapter before attempting this section. The results of this section are not used until Chapters XXV and XXVI.

If we consider Diagram 4.4 once again, we can see that if the circle is of unit radius, then the sine of any angle is given by the length of the perpendicular SC. It is clear from the diagram that when θ is zero, the length of this line is also zero; and that when $\theta = \pi/2$

(i.e., **90°**) the length of this line is unity. Similarly, for $\theta = \pi$ the perpendicular is of zero length, while for $\theta = 3\pi/2$ the perpendicular is of length *minus* unity (since SC is then below the horizontal through O).

Let us now consider Diagram 4.9 in which the horizontal axis denotes values of θ from 0 upwards, and the vertical axis denotes

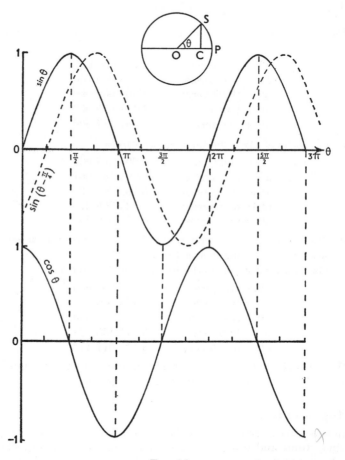

FIG. 4.9

the value of sine θ which is obtained by considering the length of the perpendicular corresponding to the angle concerned. Careful examination of Diagrams 4.4 and 4.9 will show that they are, in fact, merely different representations of the same idea. Instead of going around the circle of Diagram 4.4 in an anti-clockwise direction, and measuring the height SC, we go along the axis $O\theta$, and for each value of θ mark off a height corresponding to the appropriate length of SC. When $\theta = 0$, this length is zero. When $\theta = \dfrac{\pi}{2}$ it is unity. It then decreases to zero as θ approaches π, after which it becomes negative. If we proceed along this axis beyond the value of 2π we are simply performing the second (or subsequent) rotation of the circle.

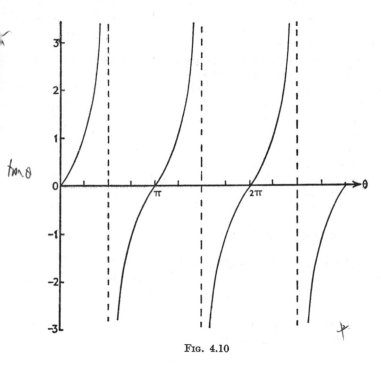

FIG. 4.10

The second part of Diagram 4.9 shows a graph of cosine θ. This has been obtained in a similar way by considering the length of CO which represents the cosine of the angle when the circle is of unit radius.

The graph of $\tan \theta$ is not so obvious. We have seen that the tangent of an angle is represented in magnitude by the ratio of SC to CO. When the angle is zero, SC is zero, but CO is unity, and so their quotient is also zero. When $\theta = \pi/6$ (which is 30°) then, as we have seen, this ratio has a value of 0·5774. For $\theta = \pi/4$, SC and CO are equal, giving a ratio of unity. These values are shown in Diagram 4.10

As θ increases towards $\pi/2$, so SC increases in length towards unity, and CO decreases towards zero. The quotient, consisting of SC, which is increasing, divided by CO, which is decreasing, therefore increases. For an angle of 70° it has a value (which may be checked by drawing) of 2·4775. For an angle of 80° it is 5·671. For 85° the ratio is 11·43, while for 89° it is 57·29. For 89° 30′ (i.e., for 89½°) the value is 114·6, and for 89° 54′ (i.e., for 89·9°) it is 573·0. For $\theta = 90°$ the value of tan θ is infinitely great. The reader who is bothered by the idea of something being infinitely great is asked to be patient for a while, until the concept can be discussed a little further later in the book.

Once we pass 90°, however, we are in a field where tan θ is negative. We have seen that, in fact, for θ between 90° and 180° the value of tan θ is given by $-\tan(180° - \theta)$. Accordingly the part of the graph corresponding to this range of values of θ must appear as a negative version of the previous part, beginning with an extremely high negative value, for θ just greater than 90°, and then falling to a less negative value as θ approaches 180°, when it reaches zero. For angles just exceeding 180° the value of tan θ is again positive, and increases until it reaches very high values for angles just short of 270°, after which it passes abruptly to very high negative values, for angles just exceeding 270° and then moves once again toward zero.

All of this is shown in Diagram 4.10 which will repay careful study

Sometimes we need to graph slightly more complicated functions such as

$$y = \sin(\omega x - \epsilon)$$

where ω and ϵ are constants.

For example, we might have

$$y = \sin\left(2x - \frac{\pi}{3}\right)$$

where x is measured in radians.

This problem presents no difficulty. We can begin by thinking of the graph of

$$y = \sin \omega x$$

This graph is very similar to that of $y = \sin x$, but its turning points*

* Turning points are defined in Chapter XI. For the moment we may think of them as being the highest and lowest points, i.e., the points where the curve begins to turn downwards or upwards.

occur when $\omega x = \dfrac{\pi}{2}, \dfrac{3\pi}{2}, \ldots$, rather than when x has these values.
The graph is shown in Diagram 4.11, which shows that the effect of the ω is to compress the curve horizontally if $\omega > 1$ and to extend it if $0 < \omega < 1$.

Let us now consider the graph of

$$y = \sin (x - \epsilon)$$

This is just like the graph of $y = \sin x$ but has turning points where $x = \epsilon + \dfrac{\pi}{2}, \ \epsilon + \dfrac{3\pi}{2}, \ldots$, rather than at $\dfrac{\pi}{2}, \dfrac{3\pi}{2}, \ldots$. The effect of ϵ is to

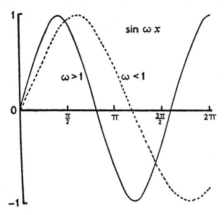

Fig. 4.11

shift the whole curve a distance ϵ along the x-axis, as shown in Diagram 4.9, where $\epsilon = \pi/4$.

It follows that the graph of

$$y = \sin (\omega x - \epsilon)$$

is like the graph of $y = \sin x$, but is compressed (or extended) by the presence of ω, and shifted bodily by the presence of ϵ. It has turning points when $\omega x - \epsilon = \dfrac{\pi}{2}, \dfrac{3\pi}{2}, \ldots$

i.e., when
$$x = \dfrac{\epsilon + \dfrac{\pi}{2}}{\omega}, \ \dfrac{\epsilon + \dfrac{3\pi}{2}}{\omega}, \ldots$$

The graph of $y = \sin x$ consists of a certain pattern which is repeated as we move along the x-axis. The " pattern " concerned is that part of the graph between $x = 0$ and $x = 2\pi$, and the rest of the graph consists of repetitions of this section, which is of length 2π. Because of this we describe the function $y = \sin x$ as a *periodic function*, the magnitude of the *period* being 2π.

The graph of $y = \sin \omega x$ has a period of $2\pi/\omega$, since it consists of a pattern of this length which is repeated regularly.

The pattern itself is called a *cycle*.

The graph of $y = \sin (x - \epsilon)$ also exhibits cycles of period 2π; but in this case they have been shifted bodily a distance ϵ. We indicate this by saying that the function $y = \sin (x - \epsilon)$ has a *phase* ϵ. Two periodic functions whose turning points coincide are said to be *in phase*. Clearly, if two functions which have the same period have a phase difference equal to that period, then the one graph has been shifted by an exact number of cycles over the other, and the two functions are once again in phase.

The function $y = \sin (\omega x - \epsilon)$ has a period of $2\pi/\omega$ and a phase ϵ.

6. The inverse form

If $\sin \theta = y$ we often refer to θ as " the angle whose sine is y ". This is written more briefly as " $\sin^{-1} y$ ". Here the -1 is not to be interpreted in any sense which we have previously met. The statement

$$\theta = \sin^{-1} 0.5$$

is simply another way of saying

" θ is the angle whose sine is 0.5 ".

This angle is, of course, $30°$.

We also use $\cos^{-1} y$ and $\tan^{-1} y$, to which we give similar meanings. We refer to these expressions as inverse trigonometrical forms.

7. Some useful results

In this chapter we have presented the basic ideas, and a few of the more important points which will be useful to us later on. From these ideas it is possible to develop an unlimited number of important identities. Often the derivation of these results is of great interest and beauty, but to attempt proofs in a book of this kind would lead us too far afield. Instead we list here some of the more important results so that a student who finds himself reading some trigonometrical argument may have some idea of the path by which it is proceeding. Any time spent with an elementary trigonometry book establishing these identities will be profitable to the reader who wishes to become something of a mathematician himself.

✕ (1) *Results that are true for all values of θ*

$$\tan \theta \equiv \frac{\sin \theta}{\cos \theta}$$

$$\sin^2 \theta + \cos^2 \theta \equiv 1$$

Note : by $\sin^2 \theta$ we mean " the square of the sine of θ ".
(e.g. $\sin^2 30° = (\sin 30°)^2 = (0{\cdot}5)^2 = 0{\cdot}25$)

$$\sin 2\theta \equiv 2 \sin \theta \cos \theta$$

$$\cos 2\theta \equiv \cos^2 \theta - \sin^2 \theta \equiv 1 - 2 \sin^2 \theta \equiv 2 \cos^2 \theta - 1$$

$$\tan 2\theta \equiv \frac{2 \tan \theta}{1 - \tan^2 \theta}$$

e.g., $\qquad \tan 60° = \dfrac{2 \tan 30°}{1 - \tan^2 30°} = \dfrac{1{\cdot}1548}{1 - 0{\cdot}3333} = 1{\cdot}7321$

(2) *Results that are true for any values of A and B*

$$\sin (A + B) = \sin A \cos B + \cos A \sin B$$

e.g. $\qquad \sin 240° = \sin (180 + 60)°$

$$= \sin 180° \cos 60° + \cos 180° \sin 60°$$

$$= (0 \times 0{\cdot}5) + (-1 \times 0{\cdot}866)$$

$$= -0{\cdot}866$$

$$\sin (A - B) = \sin A \cos B - \cos A \sin B$$

$$\cos (A + B) = \cos A \cos B - \sin A \sin B$$

$$\cos (A - B) = \cos A \cos B + \sin A \sin B$$

$$\tan (A + B) = \frac{\tan A + \tan B}{1 - \tan A \tan B}$$

$$\tan (A - B) = \frac{\tan A - \tan B}{1 + \tan A \tan B}$$

$$2 \sin A \cos B = \sin (A + B) + \sin (A - B)$$

$$2 \cos A \sin B = \sin (A + B) - \sin (A - B)$$

$$2 \cos A \cos B = \cos (A + B) + \cos (A - B)$$

$$2 \sin A \sin B = \cos (A - B) - \cos (A + B)$$

$$\sin A + \sin B = 2 \sin \frac{A + B}{2} \cos \frac{A - B}{2}$$

$$\sin A - \sin B = 2 \cos \frac{A + B}{2} \sin \frac{A - B}{2}$$

$$\cos A + \cos B = 2 \cos \frac{A + B}{2} \cos \frac{A - B}{2}$$

$$\cos A - \cos B = -2 \sin \frac{A + B}{2} \sin \frac{A - B}{2} \qquad$$ (*Note* the minus sign)

(3) Results that are true when A, B and C are the angles of a triangle (and therefore add up to $180°$) and a, b and c are the lengths of the sides opposite these angles.

$$\frac{\sin A}{a} = \frac{\sin B}{b} = \frac{\sin C}{c}$$

$$\cos A = \frac{b^2 + c^2 - a^2}{2bc}$$

with similar formulae for $\cos B$ and $\cos C$

Area of triangle $= \frac{1}{2}bc \sin A = \frac{1}{2}ac \sin B = \frac{1}{2}ab \sin C$

$$= \sqrt{s(s-a)(s-b)(s-c)}$$

where $s = \frac{1}{2}(a + b + c) =$ semi-perimeter of the triangle.

CHAPTER V

CO-ORDINATE GEOMETRY

With figured symbols weaving
Truth so easily,
SIDNEY KEYES : *William Yeats in Limbo*

1. The location of a point

To-day, an economics book that contains no diagram is something of a surprise. This chapter is concerned with the algebra of diagrams and graphs. On the one hand, we shall find that many diagrams have properties which it would be hard to find except by an algebraic approach. On the other hand we shall find that many algebraic equations, and expositions, become simpler to handle and to understand when put into a graphical form. It is also through this study of algebraic geometry that we shall introduce certain ideas of importance in the calculus.

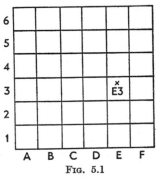

FIG. 5.1

Diagrams and graphs usually consist of lines ; and lines consist of points. It is with the point that our study must begin.

Most people are able to locate a town on a map if they are told that it is in square E3. The map would have the letters A, B, C, ... , along the lower edge and the numbers 1, 2, 3 ... , along the left-hand edge. The town would be located in the column E and the row 3, as shown in Diagram 5.1.

A similar principle is used when we wish to locate a point in co-ordinate geometry. We take a pair of lines at right angles, which we may compare with the edges of the map. We call these lines the *axes*, and usually denote the horizontal one by Ox and the vertical one by Oy (although this is not necessary). The point where they intersect is called the *origin* and is usually denoted by O.

Now suppose that we have a point P as shown in Diagram 5.2. Let us measure the distances OA and OB, denoting them by x and y respectively. Then the values of x and y enable us to locate the

93

point in the same way as we located the town. The point P lies on a vertical line that is a distance x to the right of Oy and on the horizontal line that is a distance y above Ox. We could, in fact, travel from the origin to the point P by going along Ox for a distance

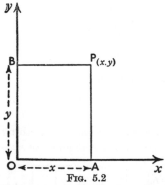

FIG. 5.2

x, and then along AP, parallel to Oy, for a distance y. Furthermore, if we performed such a journey, we could reach P and no other point. A given pair of values of x and y could lead us to one point and one only. In other words, the location of the point P is uniquely defined by the pair of values (x, y). We refer to x and y as *co-ordinates* of the point P. More precisely, since x and y are distances measured parallel to axes which are at right-angles to each other, we refer to

them as *rectangular co-ordinates*. As we shall see later on, there are other kinds of co-ordinates. This particular system of co-ordinates, the earliest of many, was introduced into mathematics by Descartes (1586–1650) who may be described as the inventor of algebraic geometry. Because of this, rectangular co-ordinates are often called Cartesian co-ordinates, and the algebraic geometry that uses them is called Cartesian Geometry.

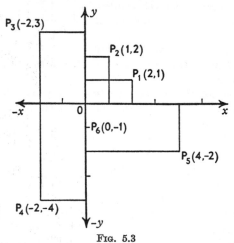

FIG. 5.3

If we have several points we often refer to them as $P_1, P_2, \dots$, etc., and then the co-ordinates are denoted by $(x_1, y_1), (x_2, y_2), \dots$, etc. If the point is to the left of the vertical axis, we indicate this by prefacing the x co-ordinate with a minus sign. If it is beneath the horizontal axis, we preface the y co-ordinate with a minus sign. This is illustrated in Diagram 5.3, which shows the points

$$P_1 = (2, 1); \qquad P_2 = (1, 2); \qquad P_3 = (-2, 3);$$
$$P_4 = (-2, -4); \; P_5 = (4, -2); \; P_6 = (0, -1)$$

The horizontal co-ordinate (x) is sometimes called the *abscissa* of the point, while the vertical co-ordinate is called the *ordinate*.

2. The equation of a straight line

When a point moves it traces out a line. In studying graphs and diagrams we are studying these lines, or, as we more commonly say, these " curves ". Many of these curves have important properties, and because of this we shall study a few of them more closely.

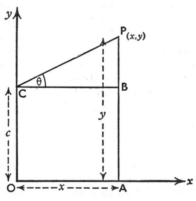

We begin with the straight line. Suppose that we have such a line passing through a point C on the y-axis, as shown in Diagram 5.4, such that the distance OC is equal to c. Another way of saying this is that it passes through the point $(0, c)$; yet another way is to say that the line makes an *intercept* of length c on the y-axis.

FIG. 5.4

Suppose, too, that the line makes an angle θ with the direction of Ox (i.e., it has a slope or *gradient* of tan θ). Let tan θ be denoted by m.

Now let us take any point P on this line. We do not know its co-ordinates, but we will call them (x, y). If we consider the diagram for a moment we can see that x must measure the distance OA and y the distance PA.

Since triangle CPB is right-angled, we have that

$$\tan \widehat{PCB} = PB/BC$$

i.e.
$$m = PB/BC$$
$$= (PA - BA)/BC$$
$$= (y - c)/x$$

Cross multiplication of this result gives us

$$mx = y - c$$

which we may write as $\qquad y = mx + c \qquad\qquad$ (5.1)

Let us consider this equation for a moment. There are four

quantities in it. Two of these, m and c, are given, and they define the position and direction of the line. We refer to them as the *parametric constants* of the line concerned. They are constant, in the sense that for any given line there will be only one (given) m and only one c. But a different line will have a different m and c, and so a knowledge of the m and the c of two lines will enable us to compare the lines with one another. Because of this we say that m and c are parametric, i.e., they facilitate comparison. Sometimes, when we are referring to the properties of a particular line, we refer to m and c as being simply the *constants* of the line. At other times, when we are thinking of one line in relation to another, we may refer to m and c as the *parameters* of the line. This is a distinction which will become clear as we proceed.

There are two other quantities in the equation : x and y. These are not given. They tell us nothing about the line. They do, however, tell us something about the points on the line. We have chosen x and y to be the co-ordinates of *any* point on the line through C with a gradient of m. The equation tells us that there is a certain relationship between x and y. For example, if (x, y) is any point on the line for which $c = 3$ and $m = \frac{1}{2}$, then x and y must be such that

$$y = \tfrac{1}{2}x + 3$$

By letting x vary through all possible values, we can calculate corresponding values of y which satisfy this equation : and these pairs of values of x and y will be the co-ordinates of all the points on the line.

In other words, the equation $y = mx + c$ is the equation which must be satisfied by the co-ordinates of *any*, and therefore *every*, point on the line. Because of this we may think of it as being the equation of all points on the line, or, more briefly, the equation of the line. More formally we have that

/   *The straight line whose intercept on the y-axis is c and whose gradient is m has the equation*

$$y = mx + c \quad / $$

We thus have, in the terms of our example, that

$$y = \tfrac{1}{2}x + 3$$

is the equation of the straight line through the point $(0, 3)$ with a gradient of $\frac{1}{2}$. Any point whose co-ordinates (x, y) satisfy this equation will lie on the line concerned. For example, the points $(2, 4)$ and $(-8, -1)$ lie on it. On the other hand, the point $(6, 4)$ clearly does not.

The cost curve

The above remarks may be illustrated by a simple economic example. Suppose that a firm is producing a certain good under the following conditions. There are fixed overhead costs, which have to be met however much, or however little, of the good is produced. We shall denote these costs by c. In order to produce x units of the good, a proportional amount of raw material and labour has to be used, costing an amount mx. The total costs of producing x units will therefore be $c + mx$, where c represents the fixed costs and mx the variable costs. If we denote the total cost by y we will have that

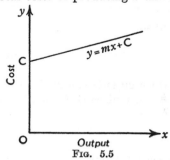

FIG. 5.5

$$y = mx + c$$

We recognise this as the equation of a straight line, having an intercept on the y-axis of c, and a slope of m as shown in Diagram 5.5. All points on it have co-ordinates which relate y, the cost of production, to x the quantity produced. The slope of the line tells us how much is added to costs for each unit of additional output. It is, in fact, a simple example of a *cost curve*. Later we shall meet cost curves that are not straight lines.

EXERCISE 5.1

Write down the equations of the following straight lines :

1. Through the point $(0, 1)$ with slope $\frac{1}{3}$
2. Through the point $(0, 5)$ with slope 2
3. Through the point $(0, -3)$ with slope 4
4. Through the point $(0, -\frac{1}{2})$ with slope -2
5. Through the point $(0, 2)$ with slope $-\frac{2}{3}$
6. Through the point $(0, -\frac{7}{8})$ with slope $-\frac{3}{4}$

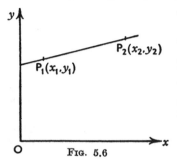

FIG. 5.6

Another equation for the straight line

It is often useful to be able to write down the equation of the line that joins two given points. There is a very simple formula enabling us to do this. The derivation of the formula is rather instructive, and we accordingly give it.

Consider the equation $y = mx + c$ which we have already met. Let there be two points $P_1(x_1, y_1)$ and $P_2(x_2, y_2)$

which are on this line, as shown in Diagram 5.6. Since they are on the line, the values of their co-ordinates must satisfy the equation of the line, and so

$$y_1 = mx_1 + c \qquad (5.2)$$

and
$$y_2 = mx_2 + c \qquad (5.3)$$

Subtracting the second equation from the first gives us

$$y_1 - y_2 = mx_1 - mx_2$$
$$= m(x_1 - x_2)$$

and so,

$$m = \frac{y_1 - y_2}{x_1 - x_2} \qquad (5.4)$$

(which gives us the gradient of the line joining the two points). If we now replace m in equation (5.2) by this value we have that

$$y_1 = \left(\frac{y_1 - y_2}{x_1 - x_2}\right) x_1 + c$$

whence we obtain, by multiplying throughout by $(x_1 - x_2)$

$$x_1 y_1 - x_2 y_1 = x_1 y_1 - x_1 y_2 + c x_1 - c x_2$$

which reduces to

$$c = \frac{x_1 y_2 - x_2 y_1}{x_1 - x_2} \qquad (5.5)$$

(which gives the value of the intercept on the y-axis).

Substitution of the values of m and c in equations (5.4) and (5.5) in the equation (5.1)—or, more simply, subtraction of (5.2) from (5.1) and then subtraction of (5.3) from (5.2)—will yield the equation

$$\frac{y - y_1}{y_2 - y_1} = \frac{x - x_1}{x_2 - x_1} \qquad (5.6)$$

which relates the co-ordinates of the point $P(x, y)$ which may be *any* point on the line, to the co-ordinates of the two given points $P_1(x_1, y_1)$ and $P_2(x_2, y_2)$, through which the line passes. It is, in fact, the equation of the straight line through these two points P_1 and P_2.

The cost curve again

To illustrate this form of the equation of the straight line we may consider the firm that is able to produce 100 units of a good for £300 and 500 units for £600. Suppose that we are told that the cost curve is a straight line (i.e., as we shall see later, that the marginal costs are constant) and that we are asked to estimate the cost of producing 400 units.

Here we have two fixed points, whose co-ordinates are (100, 300) and (500, 600). They are shown in Diagram 5.7. If we substitute these co-ordinates in the equation (5.6) we will have that the equation of the straight line is

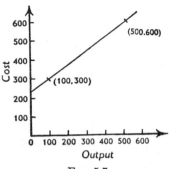

$$\frac{y-300}{600-300} = \frac{x-100}{500-100}$$

which yields

$$4y = 3x + 900$$

after multiplication and cancellation. This may be written as

$$y = \tfrac{3}{4}x + 225$$

Fig. 5.7

If we compare this with equation (5.1) we see that it is the equation of a straight line whose intercept on the y-axis is 225, which is the value, in pounds, of the overhead costs. In order to produce each single unit, there is an additional cost of £$\tfrac{3}{4}$. The cost of producing 400 units is

$$y = £\left(\tfrac{3}{4} \times 400\right) + £225$$
$$= £525$$

Notice that the average cost is falling as we expand output. The cost of producing 400 units is £525 which averages just over £1·31 per unit. But to produce 800 units would cost

$$y = £\left(\tfrac{3}{4} \times 800\right) + £225$$
$$= £825$$

which averages just over £1·03 per unit. Algebraically, if the cost curve is

$$y = mx + c$$

then the average cost associated with the production of x units is

$$\frac{y}{x} = m + \frac{c}{x}$$

As x increases, $\dfrac{c}{x}$ decreases, and so the average cost also decreases. The only cases where this is not true are when $c = 0$ or when c is negative. In the former case the graph goes through the origin. Costs rise proportionally with output. A negative c would imply that for low levels of output there are negative production costs, which would be a very special case.

The general equation of the straight line

We have already considered two equations for the straight line. We now consider a third, which is much more commonly met, and which has many advantages.

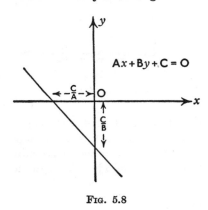

FIG. 5.8

Consider the equation

$$Ax + By + C = 0 \qquad (5.7)$$

where A, B and C are any values, positive, negative or zero. It can be written as

$$By = -Ax - C$$

whence

$$y = -\frac{A}{B}x - \frac{C}{B} \qquad (B \neq 0)$$

which may be compared with

$$y = mx + c$$

This shows that equation (5.7) is the equation of a straight line whose intercept on the y-axis is $-C/B$ and whose slope is $-A/B$. This is shown in Diagram 5.8. If A, B or C is negative, then this must be taken into account when evaluating the results. Thus the equation $3x + 2y - 7 = 0$ is of a straight line with a positive intercept on Oy of

7/2 and a negative (i.e., downward) slope of 3/2. This is illustrated in Diagram 5.9.

We call equation (5.7) the *General Equation of the Straight Line*. We should note that it is linear in x and in y. For this reason it is sometimes called the *General Equation of the First Degree in x and y*.

It may be noted that if $A = 0$ so that the equation is of the form $By + C = 0$ then the line is parallel to the x-axis, having a zero slope. If $B = 0$, the line $Ax + C = 0$ is parallel to the y-axis, having infinite slope. If C is zero, the line $Ax + By = 0$ goes through the origin.

FIG. 5.9

EXERCISE 5.2

1. Write down the equations of the lines joining the following pairs of points :

(i)   (1, 3) to (7, 8)   (ii)   (1, 3) to (−7, 8)

(iii)   (1, 3) to (7, −8)   (iv)   (1, 3) to (−7, −8)

(v)   (−1, 3) to (7, 8)   (vi)   (1, −3) to (7, 8)

(vii) (−1, −3) to (7, 8)   (viii) (−1, 3) to (−7, 8)

(ix)   (1, −3) to (−7, 8)   (x) (−1, −3) to (−7, 8)

(xi)   (−1, 3) to (7, −8)   (xii) (1, −3) to (7, −8)

(xiii) (−1, −3) to (7, −8) (xiv) (−1, 3) to (−7, −8)

(xv)   (1, −3) to (−7, −8) (xvi) (−1, −3) to (−7, −8)

2. Express your answers to question 1 in the forms $y = mx + c$ and $Ax + By + C = 0$.

3. Write down the gradients and the values of the intercepts on the x- and y-axes of the following lines :

(i) $3x + 2y - 7 = 0$ (ii) $3x + 2y + 7 = 0$

(iii) $3x - 2y - 7 = 0$ (iv) $3x - 2y + 7 = 0$

(v) $-3x + 2y - 7 = 0$ (vi) $-3x + 2y + 7 = 0$

(vii) $-3x - 2y - 7 = 0$ (viii) $-3x - 2y + 7 = 0$

4. A firm produces 200 units for a total cost of £730, and 500 units for a total cost of £970. It is known that the cost curve is a straight line. Derive the equation of this line, and use it to estimate the cost of producing 400 units.

3. The distance between two points

We often require the distance between two points whose co-ordinates are given. This is easily obtained in Cartesian Geometry by a simple application of Pythagoras' Theorem.

If we denote the two points by $P_1(x_1, y_1)$ and $P_2(x_2, y_2)$ as shown in Diagram 5.10, then we have that

$$P_1P_2{}^2 = P_1L^2 + P_2L^2$$
$$= (x_2 - x_1)^2 + (y_2 - y_1)^2$$

from which we see that the distance between P_2 and P_1 is given by

$$P_1P_2 = l = \sqrt{(x_2 - x_1)^2 + (y_2 - y_1)^2}$$

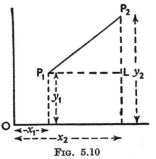

FIG. 5.10

EXERCISE 5.3

1. Find the distances between the following pairs of points:

(i)   (5, 4) and (4, 5)   (ii) $(-2, 3)$ and (2, 3)
(iii) $(-2, -2)$ and (0, 0)   (iv) $(-2, -2)$ and (3, 3)
(v) $(\frac{1}{2}, -\frac{3}{4})$ and $(1\frac{2}{3}, -2\frac{1}{3})$

2. A triangle has its vertices at the points (2, 3), (4, 5), and (8, 1). Find its perimeter.

3. A triangle has vertices at (2, 3), (6, 6) and $(-2, 0)$. Find its perimeter and then, by considering the results listed at the end of Chapter IV, find its area.

4. The intersection of curves and the solution of equations

We have already seen that the equation

$$Ax + By + C = 0$$

is the equation of a straight line. To be more precise, it is an equation such that all points whose x and y co-ordinates satisfy it lie on a straight line. This may be seen in practice by giving x a series of values, working out the corresponding values of y from the equation, and then plotting the results on some graph paper. For example, if we have the equation

$$3x - y + 4 = 0$$

then we may rewrite it as

$$y = 3x + 4$$

It is then convenient to compile a table, as below, inserting a series of values of x in the top row, and then building up the value of y by successive stages as shown.

x	-3	-2	-1	0	1	2	3	4
$3x$	-9	-6	-3	0	3	6	9	12
$y = 3x + 4$	-5	-2	$+1$	4	7	10	13	16

The points $(-3, -5)$, $(-2, -2)$, $(-1, 1)$, $(0, 4)$, ..., may then be plotted, when it will be found that they lie on a straight line This is shown in Diagram 5.11.

We may then proceed similarly if we have equations that are not linear. For example, if we have the equation

$$y = 3x^2 - 4$$

we may obtain a series of values of y by considering some arbitrary values of x. We may begin by considering the integral values of x from -3 to $+3$, as shown below.

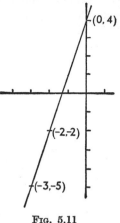

Fig. 5.11

x	-3	-2	-1	0	1	2	3
x^2	9	4	1	0	1	4	9
$3x^2$	27	12	3	0	3	12	27
$y = 3x^2 - 4$	23	8	-1	-4	-1	8	23

The points $(-3, 23)$, $(-2, 8)$, ..., may now be plotted. These clearly do not lie on a straight line, but they do appear to lie on some smooth curve, and if we take some intermediate values of x such as $\frac{1}{3}$, $\frac{2}{3}$, $1\frac{1}{2}$ or, indeed, any other values, we will find that the points so obtained lend support to the belief that the above equation is satisfied by a set of values of x and y which correspond to a smooth curve. The points considered above, including $(\frac{1}{3}, -3\frac{2}{3})$, $(\frac{2}{3}, -2\frac{2}{3})$, and $(1\frac{1}{2}, 2\frac{3}{4})$, are shown in Diagram 5.12.

Later on we shall consider this matter a little more formally, but for the moment it will be sufficient to say that if the points corresponding to a series of solutions to an equation lie on a smooth curve, which is such that *all* points on it satisfy the equation, then we say that that equation is the *equation of the curve*. Various diagrams in this book show a few curves, along with their equations. Some of these are quite simply obtained; others have more complicated equations. A more rigorous course would explain how some of these curves are drawn, and how it is possible to deduce the shape of a curve by examining the equation, without actually drawing anything. We shall shortly touch on these ideas ourselves, but for the time being the important point to notice is that it is possible to draw a graph of almost every equation that we are likely to come across in this book, and that theoretically it is possible to find an equation for every reasonably smooth curve.

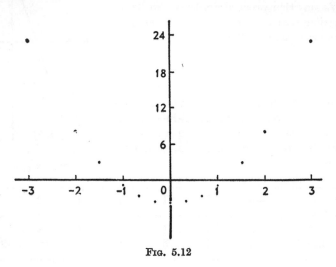

Fig. 5.12

Because so many equations can be graphed, it is often possible to solve an equation by graphing it. For example, suppose that we wish to solve the equation

$$3x^3 - 9.26x^2 - 36.3x + 2.84 = 0$$

We may take a series of values of z and evaluate the corresponding value of

$$y = 3x^3 - 9.26x^2 - 36.3x + 2.84$$

just as when we attempt a solution by the Remainder Theorem. The corresponding points may then be plotted on a piece of graph paper, and joined to form a graph. Where this graph cuts the x-axis, the value of y is zero, and so the corresponding value of x gives one root of the above equation. Since this is purely a graphical method, the root may be only approximate, but with sufficiently careful drawing, which may involve taking several values of x in the near vicinity of the root, considerable accuracy may be obtained. The solution to the above equation is obtained graphically in Diagram 5.13.

It should be noticed that this method of solution is simply a graphical application of the Remainder Theorem, the change in sign of the Remainder Theorem corresponding to the crossing of the x-axis.

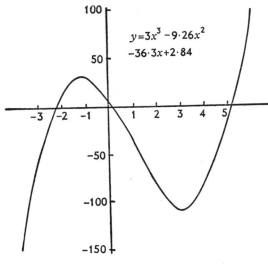

$$y=3x^3-9{\cdot}26x^2-36{\cdot}3x+2{\cdot}84$$

FIG. 5.13

A similar method may be used to solve simultaneous equations. We illustrate this by referring to a problem in economics.

Supply and demand curves

Suppose that we have an upward sloping line whose equation is
$$y=4x+30$$
and that this represents a *supply curve*, y denoting the price of a good, and x the amount that manufacturers are prepared to supply at that price. As the price rises, so does the amount that manufacturers are prepared to supply. If we know the ruling price, then an examination of the curve (or of its equation) will enable us to determine how much of the good will be supplied. The curve itself is shown in Diagram 5.14.

Let us also have a second equation
$$y=50-2x-0{\cdot}05x^2$$
This curve slopes downwards for $x>0$, as in the top right hand quadrant of Diagram 5.14. If we let y denote the price, and x the amount demanded by the public at that price, then the equation may be regarded as a downward sloping demand curve. As the price increases, so the amount that the public is prepared to buy decreases. If we know the ruling price, then we may read off the amount that will be demanded.

Now clearly at any point on the supply curve S the first of the above equations is satisfied; and at any point on the *demand curve* D the second equation is satisfied. It follows that at the point P, where the two curves intersect, both equations must be satisfied,

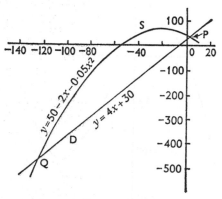

FIG. 5.14

since this point lies on both curves. This means that the co-ordinates of the point P must provide the solution to the two equations taken simultaneously, and indicate the price and quantity at which demand equals supply.

Solution of these two equations by algebraic methods, as explained in Chapter I, will yield the solution

$$x = 3\cdot21 \quad or \quad -123\cdot24$$

$$y = 42\cdot96 \quad or \quad -462\cdot96$$

Clearly, we are not interested in the negative solutions, which would indicate negative prices and quantities (at Q). Indeed the " demand curve " is a meaningful concept only for positive prices, and we have drawn the graph of its equation for non-positive values of x only to demonstrate the mathematics. There is no economic justification for the existence of the left hand portion of the diagram. Consequently we are interested in only the positive solution which shows that at a price of 42·96 units supply and demand are equal, being 3·24 units.

The reader may try to graph the above equations, and see how closely the co-ordinates of the point of intersection correspond to (3·24, 42·96).

The existence of real solutions

If we have two equations such as

$$y = 4x^2 + 4x - 3$$

and

$$y = 2x^3 + 4x^2 + 2x - 3$$

then we may graph them as shown in Diagram 5.15. To do this we compile tables as follows :

x	-3	-2	-1	$-\frac{1}{2}$	0	$\frac{1}{2}$	1	2	$-\frac{3}{4}$	$-\frac{1}{4}$	$\frac{1}{4}$
$4x$	-12	-8	-4	-2	0	2	4	8	-3	-1	1
$4x^2$	36	16	4	1	0	1	4	16	$2\frac{1}{4}$	$\frac{1}{4}$	$\frac{1}{4}$
$4x^2 + 4x - 3$	21	5	-3	-4	-3	0	5	21	$-3\frac{3}{4}$	$-3\frac{3}{4}$	$-1\frac{3}{4}$

and

x	-3	-2	-1	$-\frac{1}{2}$	0	$\frac{1}{2}$	1	2	$-\frac{3}{4}$	$-\frac{1}{4}$	$\frac{1}{4}$
$2x$	-6	-4	-2	-1	0	1	2	4	$-1\frac{1}{2}$	$-\frac{1}{2}$	$\frac{1}{2}$
$4x^2$	36	16	4	1	0	1	4	16	$2\frac{1}{4}$	$\frac{1}{4}$	$\frac{1}{4}$
$2x^3$	-54	-16	-2	$-\frac{1}{4}$	0	$\frac{1}{4}$	2	16	$-\frac{27}{32}$	$-\frac{1}{32}$	$\frac{1}{32}$
$2x^3 + 4x^2 + 2x - 3$	-26	-7	-3	$-3\frac{1}{4}$	-3	$-\frac{3}{4}$	5	33	$-3\frac{3}{32}$	$-3\frac{9}{32}$	$-2\frac{7}{32}$

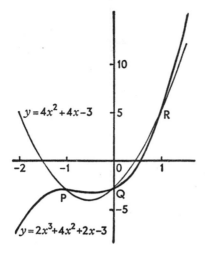

Fig. 5.15

It will be seen that the two curves intersect at three points, P, Q and R, corresponding to $x = -1$, 0 and $+1$. A moment's examination of the equations will show that if we extend the curves indefinitely they will never again intersect, since the curve corresponding to the equation of the third degree (i.e., the cubic curve) grows more steeply than the second degree curve to the right of R, while to the left of P the one curve will become more and more negative while the other becomes more and more positive.

By arguing as before, we may see that the co-ordinates of the points P, Q and R provide the simultaneous solutions of the two equations. We can check this by writing the two equations equal to each other, obtaining

$$2x^3 + 4x^2 + 2x - 3 = 4x^2 + 4x - 3$$

which yields

$$2x^3 - 2x = 0$$

i.e.,

$$2x(x-1)(x+1) = 0$$

giving solutions

$$x = 0, \quad x = +1 \quad \text{and} \quad x = -1$$

If, however, we consider the equations

$$y = 4x^2 + 4x + 1$$

and

$$y = 2x^3 + 4x^2 + 2x - 3$$

then we will find that, as shown in Diagram 5.16, the two curves cross at only one point. This indicates that there is only one real

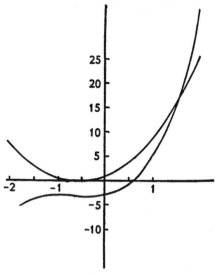

Fig. 5.16

value of x which satisfies both equations, and an attempt to solve the equations algebraically by the method of Appendix 4 will show that this is so.

The two equations

$$y = 4x^2 + 4x + 1$$

and $$y = 2x^2 - 5$$

have graphs which never intersect; and this shows that the two equations have no common real root; i.e., there is no real value of x which will satisfy both equations simultaneously. There are, however, other solutions which are not real. We take up this subject in Chapter XXIV.

EXERCISE 5.4

Solve the following equations by graphical methods:

1. $2x^2 - 5x - 3 = 0$
2. $2x^2 - 5x - 3{\cdot}1 = 0$
3. $3x^3 - 2x^2 - x - 1 = 0$
4. $3x^3 + 2x^2 - x - 1 = 0$
5. $16x + 3 = y$; $y = 13 - 3x$
6. $y = 16x + 3$; $y = x^2$
7. $y = 16x + 3$; $y = -x^2$
8. $y = 16x^2 + 3x + 1$; $2x^2 + x + 10$
9. $y = 16x^2 + 3x - 10$; $y = 10 + x - 2x^2$
10. $y = 2x$; $y = \sin x$ (measure x in radians)
11. $y = \dfrac{x}{2}$; $y = \sin x$ (measure x in radians)

CHAPTER VI

CONIC SECTIONS AND POLAR CO-ORDINATES

> They was all one to me.
> Well, all but two was all one to me.   And they,
> Strange enough, was two who kept recurring.
>
> CHRISTOPHER FRY : *A Phoenix Too Frequent*

1. The general equation of the second degree

We shall devote the first part of this chapter to a brief study of some of the properties of a single equation, which is of great importance. / The equation concerned is the *general equation of the second degree,* which is usually written as

$$ax^2 + 2hxy + by^2 + 2gx + 2fy + c = 0 \quad /$$

It will be noticed that it contains a constant term c, terms in the first powers of x and y, terms in x^2 and y^2, and a product term in xy. The letters a, h, b, g, f and c denote constants and some of them are prefaced by 2 for reasons that will become apparent later on.

Except in certain special cases to be noted later, the above equation can always be graphed. The shape of the graph depends on the values of the constants, but it can be shown that the graph will always be one of the following forms :

   i   A pair of straight lines (which may, in a special case, coincide)
   ii A circle
  iii A parabola
  iv An ellipse
   v A hyperbola

Before examining the conditions under which we may obtain one or the other of these curves, it is as well to consider certain properties of the curves themselves, and in particular to consider their relationship to the cone.

2. Sections of a cone

Let us consider Diagram 6.1, which is much less complicated than it looks. It shows a cone, vertex A, on a circular base BC. The cone has been extended upwards above A so that we have what is, in effect, a double cone. We could, in fact, think of a long thin stick,

---

* This Chapter is included because sometimes one needs to know some of its contents. But we make no further reference to them in this book, and the chapter can be omitted.

represented by *DAC*, fixed to a bearing at *A*, and allowed to move
so that the end *C* traces out a circle. The stick would then trace out

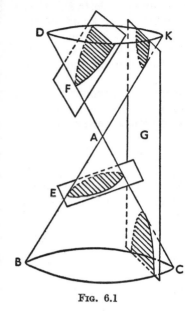

the double cone, whose vertex is
at *A*. It will be convenient to
suppose that we have a very long
stick, so that the cone extends indefinitely in each direction beyond
the vertex *A*.

We now consider various ways of
cutting this double cone. One way
would be to slice it from top to bottom through the vertex, and in the
plane of the paper. In this case we
would be presented with the two
straight lines shown (*BAK* and
DAC). Another way would be to
slice it through *A* with a plane inclined to the paper in the same way
as a book-rest is inclined to a table.
This would result in a different pair
of straight lines containing a somewhat narrower angle. We now go
on to consider other sections, stating the results, and hoping that

FIG. 6.1

they will become clear after a little thought, even though they may
seem difficult at first.

Let us move to some other part of the cone, away from the

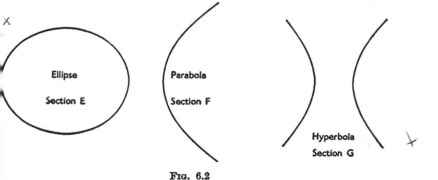

FIG. 6.2

vertex. If we cut across it in a horizontal plane then we shall obtain
a circular section, as at *BC*.

If we now tilt the plane of the section a little, so that it appears as at E, then the curve in which the plane is cut by the cone will be an ellipse.

If the plane is tilted so much that it becomes parallel to BK, then it will clearly never cut BK. This is shown in the case F and the curve that results will be a parabola (such as that shown in Diagram 6.2).

If the plane is tilted further, so that it cuts both parts of the cone, then, as shown in G, the curve will have two open branches, forming a hyperbola.

Diagram 6.2 shows these different sections of the cone. It will be worthwhile spending some time attempting to visualise their derivation. The single straight line is obtained when we take a section through the vertex, along KAB.

3. The conic sections as loci

Another way of considering these curves is to think of them as loci of a point moving in a particular way. We now indicate some of the uses of this approach, without offering any proof.

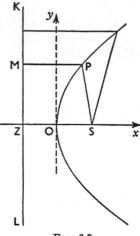

Let us take a fixed point S which we shall call a *focus*, and a fixed straight line KL which we shall call a *directrix*. Consider now a movable point P which is such that its distance from the focus S is always equal to its distance from the directrix KL. This is shown in Diagram 6.3, where $PS = PM$. It can be shown that the path which the point P must follow if PM is always to equal PS is the path shown, which is a parabola, identical in form with the section of the cone made by the plane F in Diagram 6.1. It can also be shown that if we let the shortest distance from S to KL be SZ, and denote this by $2a$, then the equation of the parabola is

Fig. 6.3

$$y^2 = 4ax$$

referred to the mid point of SZ as origin, with axes as shown.

The ellipse may be obtained similarly. Once again we take a focus S and a directrix KL. Instead, however, of arranging that P shall move so that $PS = PM$ we arrange for it to move so that the ratio PS/PM is constant and less than unity. We will denote this ratio by e, where e may have any positive value less than unity. It can be shown that the path now described by the point P will be an ellipse, as shown in Diagram 6.4.

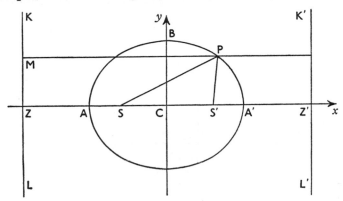

FIG. 6.4

Denoting the distance SZ by l, we mark off a point C such that

$$SC = \frac{e^2 l}{1 - e^2}$$

Taking this point as origin, and the line through S and C as the x-axis, we can obtain the equation of the ellipse in the form

$$\frac{x^2}{a^2} + \frac{y^2}{b^2} = 1$$

where

$$a = \frac{el}{1 - e^2} = CA = e\,CZ$$

and

$$b = a\sqrt{1 - e^2} = CB$$

The ellipse, being perfectly symmetrical, could also be considered as the path of a point moving such that its distance from the focus S' is e times its distance from $K'L'$.

An important property is that the sum of the two distances SP, $S'P$ is a constant for all positions of P, being equal to $2a$ which is the length AA'.

We call e the *eccentricity* of the ellipse.

The hyperbola is obtained similarly, but the eccentricity e is greater than unity. We allow the point P to move in such a way that the ratio of the distance PS to the distance PM is constant and greater than unity.

The path of such a point is shown in Diagram 6.5. All points, such as P_1, on the right-hand curve, are such that the ratio of the

E

distance from the focus S to their distance from the directrix through Z is a constant greater than unity. It is also true, however, that all points on the left-hand curve, such as P_2, have this same property

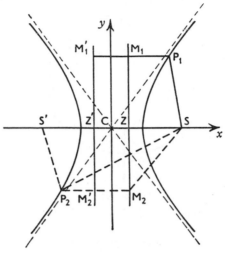

FIG. 6.5

in respect of the distances measured from the same focus S and the same directrix through Z. In addition to this, we may take a second focus S' and a second directrix, through Z', and obtain a similar result. If we take P_2 as a typical point it can be shown that

$$\frac{P_2 S'}{P_2 M_2'} = e = \frac{P_2 S}{P_2 M_2}$$

where e is a constant greater than unity. This result is true whether P_2 be on the left-hand or the right-hand branch of the curve.

If we take the point mid-way between the two foci as origin, and denote it by C, then it can be shown that

$$SC = \frac{e^2 l}{e^2 - 1}$$

and the hyperbola has the equation

$$\frac{x^2}{a^2} - \frac{y^2}{b^2} = 1$$

where
$$a = \frac{el}{e^2 - 1} = CA = e\,CZ$$

and
$$b = a\sqrt{e^2 - 1}$$

4. The classification of a conic section

The information in the last section was rather compressed, and will not be properly appreciated until later on. It must, however, be noted that the equations given in that section are the equations of the conic sections when written in their easiest forms : but they

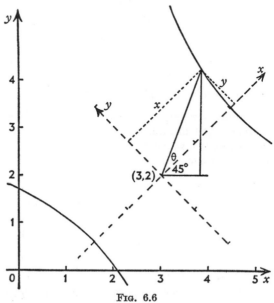

FIG. 6.6

are by no means the only ones. They are, in fact, the equations obtained when we measure our x and y from axes which are chosen with the sole purpose of simplifying the equations. The curve shown in Diagram 6.6 is a hyperbola. Referred to the axes of the diagram, its equation is

$$5x^2 + 5y^2 + 26xy - 82x - 98y + 149 = 0$$

which is rather complicated. If, however, we had taken the dotted lines as axes, then the equation would have been obtained in the standard form

$$\frac{x^2}{4} - \frac{y^2}{9} = 1$$

We refer to the equations written in these simplified forms as the *canonical equations*, and as soon as an equation has been put into its canonical form it is quite simple to identify the kind of conic section with which we are dealing. On the other hand, we often have to examine a second degree equation which is not in its canonical form, and this is rather less easy. Fortunately, however, there are some simple rules which enable us to determine the shape of the curve. These are now given, without any proof.

(1) If $a = b$ and $h = 0$ so that the equation reduces to

$$ax^2 + ay^2 + 2gx + 2fy + c = 0$$

then the curve is a circle with centre at $\left(-\dfrac{g}{a}, \ -\dfrac{f}{a} \right)$ and radius of

$\dfrac{\sqrt{g^2 + f^2 - ac}}{a}$. If, in addition, $g^2 + f^2 = ac$, then the circle is of zero radius, and becomes a point. If $g^2 + f^2 < ac$ then the circle has an imaginary radius. The meaning of this will be shown in a later chapter ; meanwhile it may be said that in such a case it is impossible to represent the equation by a graph.

(2) If $abc + 2fgh - af^2 - bg^2 - ch^2 = 0$ then the graph is of two real or imaginary straight lines, which are parallel if $h^2 = ab$.

(3) If the expression given in (2) is not zero, then (a) if $h^2 < ab$, the graph is an ellipse ; (b) if $h^2 = ab$, the graph is a parabola ; (c) if $h^2 > ab$, the graph is a hyperbola, which is rectangular (see next section) if $a + b = 0$.

The rectangular hyperbola

The hyperbola has one very important property which we must now note. As we have seen, it has two branches, corresponding to the two halves of the double cone. If we draw the two straight lines

$$y = \frac{b}{a} x \quad \text{and} \quad y = -\frac{b}{a} x$$

we shall find that the two branches of the hyperbola get closer and closer to these two lines. If we were to take very large values of x and y we would find it very difficult to distinguish between the lines and the hyperbola itself. Diagram 6.5 attempts to show this. It can be shown that, although the hyperbola and these lines never quite coincide, the lines represent with increasing accuracy the position and direction of the hyperbola. We call these lines the *asymptotes* to which the hyperbola tends, or, more simply, the asymptotes of the hyperbola.

If $a = -b$ then, as we shall shortly see, the two asymptotes are at right angles to each other. In this case we refer to the hyperbola as being *rectangular*. If we turn the curve through 45 degrees, so that these asymptotes occupy the vertical and horizontal directions commonly associated with the axes, then it can be shown that, referred to these new axes, the equation may be written as

$$xy = \text{constant}$$

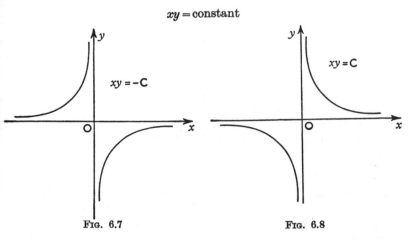

FIG. 6.7   FIG. 6.8

where the constant may be negative (for a rectangular hyperbola as shown in Diagram 6.7) or positive (for a rectangular hyperbola as shown in Diagram 6.8).

EXERCISE 6.1

Classify the following conic sections according to their shapes:

1. $3x^2 + 4xy + 5y^2 - 2x + 2y + 5 = 0$
2. $4x^2 + 10xy + y^2 + 2x - 2y + 5 = 0$
3. $4x^2 + 2xy - y^2 + 2x - 2y + 5 = 0$
4. $3x^2 + 4xy - 3y^2 - 2x + 2y - 7 = 0$
5. $3x^2 + 6xy + 3y^2 + 3x + y - 4 = 0$
6. $x^2 + 4xy + 4y^2 + 2x + 4y + 1 = 0$
7. $x^2 + 4xy + 6y^2 + 2x + 4y + 1 = 0$
8. $x^2 + y^2 + 2x + 4y + 1 = 0$
9. $5x^2 + 5y^2 + 2x + 4y + 1 = 0$
10. $10x^2 + 10y^2 + 2x + 4y + 2 = 0$

5. Economic examples

⋊ *The constant returns curve* ⋎

If x units of a good are sold at a price of £y per unit then the returns of the sale amount to £xy. This obvious fact enables us to represent the returns very simply on a demand curve. If, for

example, we have a demand curve as shown in Diagram 6.9, then the
returns obtained by selling a quantity OA at a price OB will be given
by $OA \times OB$ which is the area of
the rectangle OAP_1B. The returns obtained by selling at a
price P_2 will be given by the
area OCP_2D and comparisons of
these areas immediately shows
that the second price gives a
higher return.

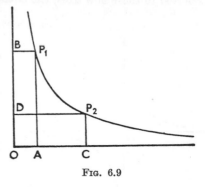

In this connection, the rectangular hyperbola is particularly interesting.   Because its
equation is

$$xy = C$$

FIG. 6.9

it follows that C represents the area of the rectangles such as $OAPB$
which are all equal.   This means that if the demand curve is a rectangular hyperbola the returns are constant and independent of price.
Later we shall see that this result is of great importance in the study
of elasticities.

The location of the firm

Now that we have done a fair amount of mathematics we are in a
position to tackle a more complicated problem than those we have
considered so far.   In choosing a problem connected with the location
of a firm, we are choosing one which can, nevertheless, be introduced
at a very simple level.   It also serves as a useful introduction to the
idea of indifference curves, which are not to be thought of as restricted
to demand theory.   On the other hand, it must not be ignored that
once the basic principles of the subject have been acquired, mathematics will still prove useless unless continually used and revised :
and unless the reader feels really happy about the earlier part of this
book he would do well to re-read it, and to work out some examples,
before proceeding.   It should also be remembered that there is still a
great deal of mathematics to be done ; and the problem we now
consider must still be introduced in a very simplified form.   Later,
we shall make it a little more realistic.

Let us begin by supposing that we are going to build a factory.
Certain transport costs will be incurred in operating it.   We have to
transport labour and materials to the factory.   We have to transport
our finished goods from the factory.   To simplify the mathematics,
and at the same time to keep quite close to a very familiar practice,
we shall suppose that the transport charges on the finished goods are

paid by the buyer, and that we may therefore exclude them from an analysis of the firm's costs.   Even so, other transport costs arise, and we have to decide where to locate the factory so that these costs may be as low as possible.

The simplest case arises when the labour and materials are both concentrated at one and the same point O.   The transport costs will be zero if the factory can be built at the same point.   But if this is impossible then the factory should be built as close to O as is possible ; and unless there are some other considerations, it is immaterial whether the factory is four miles to the east of O or four miles to the west.   In fact we can imagine a whole series (or family) of circles

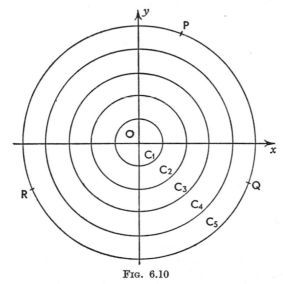

Fig. 6.10

drawn around O as in Diagram 6.10, such that the transport costs arising when the factory is at any point on circle C_4 will be identical to the costs arising when the factory is built at any other point on the same circle ; but more than the costs arising at any point on C_3 or C_2, etc. ; and less than the costs arising from any point on C_5 or C_6, etc.   Whether the factory is built at P, Q or R is clearly a matter of indifference, as far as transport costs are concerned.

Now all of this is very obvious, but in order to familiarise ourselves with the basic mathematical ideas involved, we might as well consider it a little more formally—although we will simply prove the obvious.

Let us take O, the point at which the raw material and labour are concentrated, and draw two axes through it at right angles as in

Diagram 6.10. Now suppose that the factory is built at a point $P(x, y)$. Then the distance of the factory from the point O is given by

$$d = \sqrt{(x - 0)^2 + (y - 0)^2}$$
$$= \sqrt{x^2 + y^2}$$

If the cost of transporting one unit of labour one mile is £1, and the cost of transporting one unit of raw material is the same, then the total cost of transporting m units of each will be

$$c = £2m\sqrt{x^2 + y^2}$$

If the transport costs are, in fact, to be exactly equal to £T then P must be such that

$$2m\sqrt{x^2 + y^2} = T$$

whence

$$4m^2(x^2 + y^2) = T^2$$

which may be written as

$$4m^2x^2 + 4m^2y^2 - T^2 = 0$$

Comparing this with the general equation of the second degree, and the notes on page 93 we see that this is the equation of a circle whose centre is at $(0, 0)$ and whose radius is $T/2m$. If we allow T to increase (or to decrease) then we may move from one circle to another.

We have, in fact, obtained mathematically the result that was obvious from the very beginning. Let us now turn to a slightly more complicated case.

Instead of letting the labour and materials be concentrated in the same point, let us suppose that labour is concentrated at a point L and materials at a point M. Let us, again, suppose that to transport one unit of labour a distance of one mile costs £1 ; that the same is true of materials ; and that labour and materials are used unit for unit.

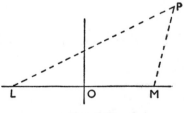

Fig. 6.11

We choose axes as shown in Diagram 6.11 and consider the costs of transport if the factory is built at P. Now before we become involved in more complicated mathematics, let us note the similarity between this diagram and Diagram 6.4. This may encourage us to remember that the cost of transport of m units will be

$$£(mLP + mMP) = £m(LP + MP)$$

and if this is to be equal to some given amount T then it means that $LP + MP$ will be equal to the constant T/m. Now we noted on page 113 that in an ellipse whose foci are S and S' the sum of the distances SP and $S'P$ is constant. It is obvious that if we draw an ellipse with L and M as foci, such that the distance AA' is equal to T/m, then at all points on this ellipse the transport costs will be T.

Alternatively we may be more analytical in our approach. Let us denote the distance OM by p. Then the distances over which the labour and materials have to be carried are given by

$$LP = \sqrt{(x - (-p))^2 + y^2} = \sqrt{(x + p)^2 + y^2}$$

and
$$MP = \sqrt{(x - p)^2 + y^2}$$

The transport costs will be the sum of these two distances multiplied by m, and if this is to equal T, we have

$$m(\sqrt{(x - p)^2 + y^2} + \sqrt{(x + p)^2 + y^2}) = T$$

which is the condition that x and y must satisfy if the transport costs arising from a factory at the point P are to be exactly equal to T.

To solve this equation we first re-arrange it by using a procedure similar to that mentioned in Chapter I (page 34). We square each side, obtaining

$$[(x - p)^2 + y^2] + [(x + p)^2 + y^2] + 2\sqrt{(x - p)^2 + y^2}\,\sqrt{(x + p)^2 + y^2} = \frac{T^2}{m^2}$$

which may be rewritten as

$$2\sqrt{[(x - p)^2 + y^2][(x + p)^2 + y^2]} = \frac{T^2}{m^2} - 2(x^2 + y^2 + p^2)$$

Squaring again, expanding and then collecting terms we find that a great deal cancels out, and we are left with

$$4\frac{T^2}{m^2}(x^2 + y^2 + p^2) - 16p^2 x^2 - \frac{T^4}{m^4} = 0$$

which gives

$$x^2\left(\frac{4T^2}{m^2} - 16p^2\right) + \frac{4T^2}{m^2}y^2 + \frac{4T^2 p^2}{m^2} - \frac{T^4}{m^4} = 0$$

This is easily written as

$$\frac{x^2}{\dfrac{T^2}{4m^2}} + \frac{y^2}{\dfrac{T^2 - 4p^2 m^2}{4m^2}} = 1$$

which is clearly the equation of an ellipse. Of course, if $T^2 - 4p^2 m^2$ were negative then the equation would represent a hyperbola, but this cannot be so. Except at L and M and at points on the straight line joining L to M, and lying between L and M, the total transport costs T must be greater than the cost of transporting labour from L to M or materials from M to L ... and this comes to $2mp$ which means that the value T^2 must be more than $4p^2m^2$.

Once again, we have proved a conclusion we had already reached, but we have done so in an instructive way. It is a conclusion that is usefully illustrated by Diagram 6.12. At any point on the ellipse E_2 the transport costs will be the same ; and they will be more than at any point on the ellipse E_1 (which may be drawn from the above equation by taking a smaller value of T) and less than the costs corresponding to points on the ellipse E_3. Whether to build the factory

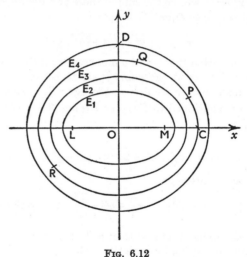

FIG. 6.12

at P, Q or R is a matter of indifference, as far as transport costs are concerned, but C is a preferable point to D.

We shall now go two stages nearer to reality, and immediately into some more complicated (but not difficult) mathematics.

Let us suppose, first of all, that instead of one unit of labour combining with one unit of raw material, each costing £1 to transport a mile, we have a situation in which we have to spend £l transporting the labour that combines with material that can be transported for £m. In order to produce a given number of units of finished product, we need to spend an amount

$$£(l\ LP + m\ MP)$$

which, in given circumstances, must equal £T.

The solution to this problem is not so obvious. The basic equation to be satisfied is

$$l\sqrt{(x+p)^2+y^2}+m\sqrt{(x-p)^2+y^2}=T$$

This may be solved in the same way as the previous equation, but the existence of l as well as m makes it more complicated, and eventually we obtain a quartic equation (i.e., one involving x^4 and y^4). This equation has an egg-shaped graph as shown in Diagram 6.13, the bulge occurring around M if materials cost more than labour to transport, but around L if labour transport is the more expensive. There is a purely graphical method for obtaining the curve, which involves no algebra or calculation. At the moment, however, the important thing to notice is that, although the procedure we have outlined above leads to a complicated equation, it does provide the answer ; and it shows that, once again, there is a whole family of curves such that any point on one of them is (a) to be preferred to any point outside that curve, (b) not as good as any point inside the curve, (c) just as good as any other point on the same curve. There is no preference between P, Q and R, but C is to be preferred to D.

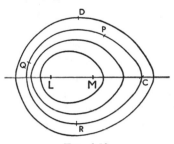

FIG. 6.13

Finally we come to the case where labour is concentrated at a point L and materials come from two different points M and N. To illustrate the method, we will suppose that in order to produce our goods we need the same number of units of labour, material M and material N and that to transport from L, M or N costs £1 per unit per mile.

The solution of this problem is probably best obtained by graphical method. We can, however, proceed in the same way as before. We will choose an origin mid-way between L and M and suppose that N has co-ordinates (a, b). Then if we locate the firm at (x, y) we have that the distances LP and MP are the same as before, while there is also the distance NP which is

$$NP=\sqrt{(x-a)^2+(y-b)^2}$$

and so the total transport costs, which are to equal T, come to

$$\sqrt{(x+p)^2+y^2}+\sqrt{(x-p)^2+y^2}+\sqrt{(x-a)^2+(y-b)^2}=T$$

Solution of this equation proceeds as previously, squaring first of all, then re-arranging so that all the square root terms are on one side and all the other terms on the other side. We then square again; and find that in order to reach the solution we have to square yet again. The resulting equation of the family of curves involves x^6. There is little point in working it out especially since there are easier graphical methods but, cumbersome though it may be, a solution will be obtained if we proceed in this manner.

The curves obtained look something like those shown in Diagram 6.14. Once again, each curve is such

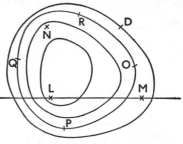

Fig. 6.14

that at all points on it transport costs are the same, being smaller for the inner curves than for the outer ones. It is a matter of indifference whether to locate the factory at P, Q or R, but O is preferred to D.

6. Polar co-ordinates

So far we have used only Cartesian co-ordinates. There are, however, other systems of co-ordinates and one of these is very widely used in quite simple work. Many quite common problems are dealt with more easily by the use of polar co-ordinates than by the use of Cartesian co-ordinates. As before, we take a fixed point, which we may denote by O. We call this point

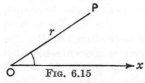

Fig. 6.15

the pole. Instead of taking two axes through it at right angles, we take only one axis (*which we call the initial line*) as shown in Diagram 6.15. Let us now draw a straight line from any point P to the pole. Let this line be of length r, and let it make an angle of θ with the initial line, where θ is measured from Ox in a counterclockwise direction as indicated in the diagram. We call r and θ the *polar co-ordinates* of the point P with respect to the pole O and the initial line Ox. It is clear that if we are given that O is the pole and Ox is the initial line, then by drawing a line of

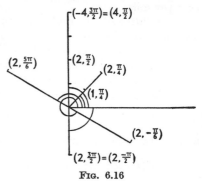

Fig. 6.16

length r at an angle of θ to Ox we are bound to reach P and no other point. In fact, the values of r and θ define the position of P in polar co-ordinates just as precisely as the values of x and y define the position in Cartesian co-ordinates. Diagram 6.16 shows a number of points with their polar co-ordinates. We may note that a negative value of r is easily interpreted. For example, the point $(-4, 3\pi/2)$ coincides with $(4, \pi/2)$.

The simplest example of a case in which the equation of a curve is less complicated when expressed in polar co-ordinates is the circle. The general co-ordinates of any point are usually written, in polar form, as (r, θ). The equation relates r and θ just as in the Cartesian case the equation links x and y. In a very simple case, however, there is a Cartesian equation involving no y at all—the equation

$$x = \text{constant}$$

which denotes a straight line parallel to the y-axis. There is also a simple polar equation involving no θ. This is

$$r = a \quad (\text{where } a = \text{constant})$$

and, as a moment's consideration will show, it is the equation of circle radius a and centre O. This can be seen quite easily when we note that the successive points on the circle have co-ordinates (r, θ) where r is *always* equal to radius of the circle and θ varies from 0 to 2π as the point goes around the circle.

We shall not often have cause to use polar co-ordinates in this book, although the particular problems dealt with in the last section would have been solved more easily by this method. As an indication of the general approach to their use we may consider the problem of finding the equation of a straight line in polar co-ordinates.

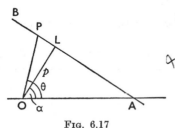

FIG. 6.17

We consider the line AB in Diagram 6.17. Letting OL be the perpendicular from the pole O to the line AB, we denote its length by p. Let the angle LOA be α.

Denoting any point on the line by P, we let its co-ordinates be (r, θ). The equation of the line will relate r and θ to p and α

From the triangle POL we have that

$$p = r \cos \widehat{LOP}$$
$$= r \cos (\alpha - \theta)$$
$$= r \cos (\theta - \alpha)$$

yielding the final equation

$$r \cos (\theta - \alpha) = p$$

which is the polar equation of the line concerned. Successive points on the line may be obtained by giving θ all values from 0 to 2π and working out the corresponding values of r.

The equation of the conic section is easily obtained in polar form if the pole is taken at the focus of the section concerned. It can be shown that the equation is

$$r = \frac{l}{1 - e \cos \theta} \qquad l \text{ a positive numerical constant}$$

where e is the eccentricity of the conic section. If $e = 1$ then the equation represents a parabola. If $e < 1$ then the equation represents an ellipse. If $e > 1$ then the curve is a hyperbola.

Exercise 6.2

1. By giving θ the values 0, 10°, 20°, 30°, ... , compile a table of values of r for the equation

$$r = \frac{2}{1 - \frac{1}{2} \cos \theta}$$

and use this to draw a graph of the function. Compile a similar table for the equation

$$r = \frac{1}{\cos \left(\theta - \dfrac{\pi}{4} \right)}$$

and graph this on the same axes. Read off the polar co-ordinates at the points of intersection.

2. Show, by analysis, that the above curves intersect where θ is such that $(3 + \sqrt{2}) \cos^2 \theta - 4(1 + \sqrt{2}) \cos \theta + 2 = 0$ and check that the solutions of question 1 satisfy this equation. (Use the formula for the expansion of $\cos (A - B)$ and the fact that $\cos^2 \phi + \sin^2 \phi \equiv 1$.)

SECTION III

CALCULUS: FUNCTIONS OF ONE VARIABLE

Chapter VII is the most important in the book, and the student is advised to be certain that he understands it thoroughly before he attempts to proceed. The relevance to Economics of the ideas put forward in this chapter are discussed in Chapter VIII. The next two chapters can be thought of as being introductory to Chapter XI which deals with monopoly problems.

Chapter XII considers certain summation problems, and will be seen to follow quite naturally from the preceding chapters. It poses certain problems which can be answered only if the reader knows the basic properties of logarithms, which are dealt with in Chapter XIII. The answers to the problems appear in Chapters XIV–XVI. Some of the proofs which should appear in a mathematics book are omitted from these chapters, which means that it is all the more important to read the argument, and the qualifications, very carefully.

Chapters XVII quotes a few useful results.

CONTINUITY AND LIMITS

Great patience must be ours ere we may know
The secrets held by labyrinthine time :
The ways are rough, the journeying is slow,
The perils deep,—

DRINKWATER : *The Loom of the Poets*

1. Introduction

Although this chapter contains very little actual mathematics, it is probably the most important in the book. It aims at introducing two fundamental ideas. Each is quite simple, and easy to grasp ; but most of the difficulties that beset the newcomer to mathematics arise out of a failure to appreciate one or the other of these points. Consequently this chapter should be read very carefully, and no attempt should be made to hurry to later chapters.

2. Continuity

It is impossible for any car to change its speed suddenly from 30 m.p.h. to 60 m.p.h. However fast it may be able to accelerate, at some time or the other before reaching 60 m.p.h., it must be travelling (if only for a moment) at 50 m.p.h. At some time it must do 40 m.p.h. In fact, there must be some moment when the speedometer needle registers any speed we wish to select between 30 m.p.h. and 60 m.p.h. At one moment it shows 38·5 m.p.h. At another 39·0 m.p.h. And at some time between these moments it will show 38·8 m.p.h. ; and at another time 38·84 m.p.h.

On the other hand, it is impossible for the retail price of a pound of sugar to change anything but suddenly. It is true that it need not change from 7p. to 10p. in one jump. It may change in stages of a halfpenny. But, although the price per pound may be 7p. on one day and 7½p. another, and conceivably 7¼p. at some time in between (by a sales arrangement of 7p. for this pound and 7½p. for the next), it is quite inconceivable that there is any one moment when you can go into a shop and buy one pound of sugar for 7·38p.

Now both the speed of a car and the retail price of sugar change ; they are variable quantities. But, as we have seen, they vary in different ways. The speed of a car varies *continuously*, changing from 30 m.p.h. to 60 m.p.h. by passing through all the intermediate speeds. The retail price of sugar varies *discontinuously*, changing from 7p. to $7\frac{1}{2}$p. abruptly.

It is possible to illustrate these ideas graphically as in Diagram 7.1. With the car, we may choose any speed at all between 30 m.p.h. and 60 m.p.h. and determine from the diagram the time at which the car was travelling at that speed. For example, it was travelling at 42·3 m.p.h. after $6\frac{1}{4}$ minutes.

But if in Diagram 7.2 we choose a price of $8\frac{1}{2}$p. or of 10p., we see that *at no time* did that price operate. The jumps from 8p. to 9p. and from 9p. to 11p. were abrupt ones; they were discontinuous.

Since both of these variables have values which depend on time, we say that they are *functions of time*. The speed is a *continuous function* of time; the price is a *discontinuous function* of time.

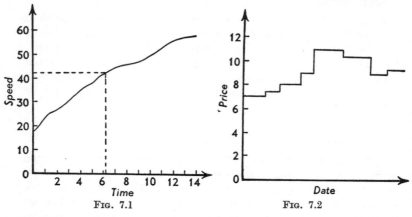

FIG. 7.1   FIG. 7.2

3. Limits

Eating an apple

We have already had an informal introduction to the idea of the limit in Chapter II, where we considered the sum of an infinite series. It is now time to consider the matter a little more formally, although we shall once again begin by referring to the problem of eating an apple. The problem, we may remember, is the practical one of how to consume the whole of an apple if each bite bites off one half of what is left after the previous bite. We may tabulate the process thus :

after 1 bite, we have eaten $\frac{1}{2}$ of the apple ; $\frac{1}{2}$ remains
after 2 bites, we have eaten $\frac{1}{2}+\frac{1}{4}=\frac{3}{4}$ of the apple ; $\frac{1}{4}$ remains
after 3 bites, we have eaten $\frac{3}{4}+\frac{1}{8}=\frac{7}{8}$ of the apple ; $\frac{1}{8}$ remains
after 4 bites, we have eaten $\frac{7}{8}+\frac{1}{16}=\frac{15}{16}$ of the apple ; $\frac{1}{16}$ remains
and so on.

No matter how many bites we take, there will always remain a portion equal in size to the last bite. By taking enough bites we may make this remaining portion as small as we please : but there will never be *nothing* left. After one bite, 0·5 of the apple remains. After two bites 0·25 remains. After five bites 0·03125 remains. After ten bites as little as 0·009775 remains, while after twenty bites there is only 0·000 000 95. But always something remains. *We can get as close to zero as we wish : but never quite reach it.* We say, in fact, that *zero is the Limit to which the amount of apple remaining tends as the number of bites become indefinitely great.*

The sum of an infinite series

Let us now consider the G.P.

$$1 + \tfrac{1}{2} + \tfrac{1}{4} + \tfrac{1}{8} + \dots +$$

We know from Chapter II that the sum of terms of this series is given by

$$\frac{a(1 - r^n)}{1 - r} = \frac{1(1 - (\tfrac{1}{2})^n)}{1 - \tfrac{1}{2}}$$

$$= 2(1 - (\tfrac{1}{2})^n)$$

Now as the number of terms increases, this last expression becomes closer and closer to 2, because the part of it arising from $(\tfrac{1}{2})^n$ becomes closer and closer to zero. On the other hand, it is never exactly zero, and so the sum of the series never quite reaches 2 : but if we take enough terms we can get as close to 2 as we wish. Because of this, we say that 2 *is the limit to which the sum of the G.P. tends as the number of terms increases indefinitely.* Furthermore, we say that 2 *is the sum of the infinite series.* This is a distinction which is most important, and which will become clearer as we proceed. The point to notice is that the sum of a *finite* series can be obtained quite simply by adding up the terms. As the number of terms increases, so the sums alter. If the series is convergent, the successive sums will tend to a limit. We define the sum of an infinite (convergent) series to be this limit.

Achilles and the tortoise

We shall now attempt to make these ideas a little clearer by considering a few more examples drawn from different fields. The student should be certain that he understands each example thoroughly. We begin with the time-honoured paradox of Achilles and the tortoise.

Achilles, on one of his lamer days, was still able to travel at ten times the speed of a tortoise. Feeling magnanimous, he offered to race the plodding quadruped, giving it a thousand yards start. The challenge was accepted, and the two started their race in the positions A_0 and T_0, as shown in Diagram 7.3. Now Achilles, going quite slowly, was still able to do 100 yards in a minute, and after ten minutes he had reached the point T_0 from which the tortoise started.

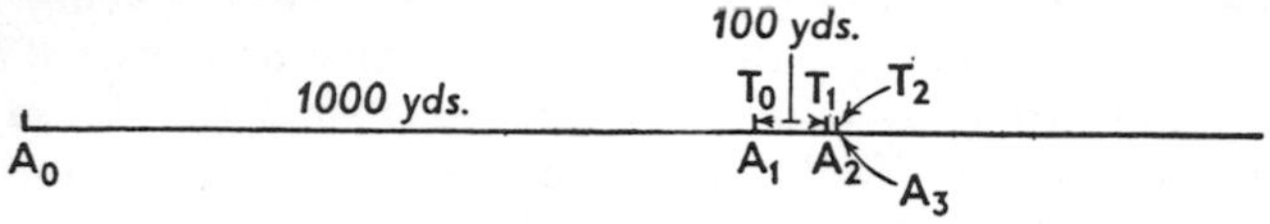

FIG. 7.3

By now, of course, the tortoise had proceeded to the point T_1, and Achilles was still behind. By the time that he had reached T_1 the tortoise had lumbered on to T_2. By the time that Achilles had reached this point, Testudo had crept another few inches to T_3, and so on. Whenever Achilles reached the point where the tortoise was a moment before, the tortoise had proceeded a little further. It follows that Achilles never caught the tortoise.

The solution to this hoary paradox is apparent if we set out the successive stages of the race in a table as follows.

Position of		Time (Minutes)	Additional Distance (in yards) Gone by		Total Distance (in yards) Gone by	
Achilles (1)	Tortoise (2)	(3)	Achilles (4)	Tortoise (5)	Achilles (6)	Tortoise (7)
A_0	T_0	0	0	0	0	1000
$A_1 = T_0$	T_1	10	1000	100	1000	1100
$A_2 = T_1$	T_2	11	100	10	1100	1110
$A_3 = T_2$	T_3	11·1	10	1	1110	1111
$A_4 = T_3$	T_4	11·11	1	0·1	1111	1111·1
$A_5 = T_4$	T_5	11·111	0·1	0·01	1111·1	1111·11

The first row of this table shows that Achilles is at A_0 and the Tortoise at T_0, these positions being separated by 1000 yards. After 10 minutes, Achilles is at A_1 (which coincides with T_0) and the Tortoise is at T_1. As is shown in column (4), Achilles has travelled 1000 yards and the Tortoise 100 yards during this time. We now come to the third row. Achilles has reached A_2 (coinciding with T_1) and the Tortoise is at T_2. These positions are reached after a total time of 11 minutes. During the single minute that has elapsed since the previous positions, Achilles has travelled 100 yards and the Tortoise 10 yards. The total distances travelled are 1100 yards and 1110 yards, showing a separation of 10 yards. Now it takes Achilles

0·1 minutes to cover this distance ; and during this time the tortoise runs another yard. To go *this* extra distance takes Achilles 0·01 minutes, during which the tortoise has gone another 0·1 yards. The successive stages of the race are shown by successive rows of the table. It is clear that, just as the apple was never quite eaten, the part remaining always being equal to the last bite taken, so *by this argument* the tortoise is never overtaken, the separating distance always being one-tenth of what it was when the last calculation was made.

But this is not the whole story. Just as this process will never allow column (6) to equal column (7), so will it never allow column (3) to show a time of anything like 12 minutes. Actually, in twelve minutes the tortoise would have travelled a distance of 120 yards, to be added to his start of 1000 yards ; but Achilles would have travelled 1200 yards, and therefore would be 80 yards in front. The form of argument we have adopted never allows us to consider what is going to happen after $11\frac{1}{9}$ minutes. As an elementary calculation will show, at $11\frac{1}{9}$ minutes, Achilles and the tortoise are abreast. When expressed as a decimal $11\frac{1}{9}$ is equal to

$$11 \cdot 111111111111 \dots \text{ ad infinitum.}$$

Clearly, only if we go through an infinite number of stages will the time shown in column (3) reach this value. Our argument is fallacious because it does not (and cannot) go far enough.

The distance travelled by the tortoise is given by

$$S_T = 1000 + 100 + 10 + 1 + 0 \cdot 1 + 0 \cdot 01 + 0 \cdot 001 + \dots \text{ yards}$$

while the distance gone by Achilles is

$$S_A = 0 + 1000 + 100 + 10 + 1 + 0 \cdot 1 + 0 \cdot 01 + \dots \text{ yards}$$

in a time given by

$$t = 10 + 1 + 0 \cdot 1 + 0 \cdot 01 + 0 \cdot 001 + 0 \cdot 0001 + 0 \cdot 00001 + \dots \text{ minutes.}$$

Now it is clear that if an equal number of terms is taken in each case. then the sum S_T *for that number of terms* will exceed the sum S_A, but as the number of terms increases indefinitely, so do both S_T and S_A tend to a common limit of $1111\frac{1}{9}$ yards. The sum denoting the time, also tends to the limit of $11\frac{1}{9}$ minutes. We say that, in the limit, at a time of exactly $11\frac{1}{9}$ minutes (which is a time that can never quite be reached by the above series, but to which it tends indefinitely close), Achilles and the tortoise will be abreast, having travelled $1111\frac{1}{9}$ yards (which is a distance that can never quite be reached by summing the above series, but is the limit to which they tend indefinitely close).

The area of a circle

Let us now consider a geometrical example.

Suppose that we wish to find the area of a circle of radius one inch. We decide to do so by approximating to it as follows. Consider a square drawn around the circle, and also the square drawn inside it,

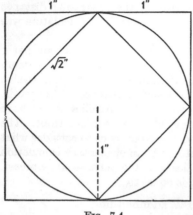

as shown in Diagram 7.4. The area of the outer square is clearly four square inches; that of the inner is two square inches. The area of the circle lies between these values.

Now take a pair of octagons. The area of the outer one will be 3·3136 square inches. (We shall indicate how this may be calculated later on; for the moment the method does not matter.) The area of the inner one will be 2·8284 square inches. The area of the circle lies between these values. Next take

FIG. 7.4

a pair of regular sixteen-sided polygons. The area of the inner one will be 3·0616 square inches, while that of the outer one will be 3·1824 square inches. The area of the circle will lie between these values.

We could proceed in this manner and draw up a table of areas thus :

Number of Sides	Area of	
	Outer Polygon	Inner Polygon
4	4·0000	2·0000
8	3·3136	2·8284
16	3·1824	3·0616
32	3·1488	3·1216
⋮	⋮	⋮

Now the area of the circle clearly lies between the final values of these columns. But we can say more than that. As the number of sides increases, so the outer polygon gets closer and closer to the circle. This is illustrated by Diagram 7.5. If we give it a couple of hundred sides it will be very difficult to distinguish it from the circle itself. In fact, *by giving it as many sides as we please*, we may make *the area it contains approach the area of the circle as closely as we please :* but the area will always be slightly more than the area of the circle. In other words, the area of the circle is the *limit* to which the area of the outer polygon tends as the number of sides increases indefinitely. Since the areas tend downwards towards this limit, we say that the area of the circle is the *lower limit* to which the area of the outer polygon tends.

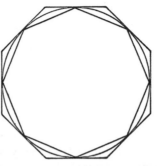

FIG. 7.5

Similarly we may make the area of the inner polygon approach that of the circle as closely as we please : but it will always be a little less than the area of the circle. We say that the area of the circle is the *upper limit* to which the area of the inner polygon tends.

There is an important consequence to this. We know that, in fact, the area of a circle of radius one inch is given by π square inches where π is approximately equal to 3·1416. Furthermore, it can be shown that if we have a polygon of n equal sides inscribed in a circle of radius one inch, then its area will be

$$\frac{n}{2} \sin \left(\frac{360}{n} \right)^{\circ}$$

while the area of the larger, circumscribing, polygon, will be

$$n \tan \left(\frac{180}{n} \right)^{\circ}$$

(For example, if we have a square, then $n = 4$. The above formulae show that the area of the inscribed square is

$$2 \sin 90^{\circ} = 2 \text{ square inches ;}$$

while the area of the circumscribing square will be

$$4 \tan 45^{\circ} = 4 \text{ square inches.)}$$

This means that the values in the above table are all values of these functions, with the appropriate values of n.

We may, therefore, make two equivalent statements :

(i) The area of the unit circle is the lower limit to which the area of the circumscribing regular polygon tends as the number of sides

increases indefinitely ; and the upper limit to which the area of the
inscribed polygon tends under the same conditions.

(ii) π is the limit to which $\dfrac{n}{2} \sin \left(\dfrac{360}{n}\right)$ and $n \tan \left(\dfrac{180}{n}\right)^{\circ}$ both tend

as n increases indefinitely. This may be put more concisely as

$$\pi = \underset{n\to\infty}{\text{Limit}}\ \frac{n}{2}\sin\left(\frac{360}{n}\right)^{\circ}$$

$$= \underset{n\to\infty}{\text{Limit}}\ n \tan\left(\frac{180}{n}\right)^{\circ}$$

These limits are of very great importance in some work. Even
more important, however, is a much simpler trigonometric limit.
We now consider this.

The limit of $\dfrac{\sin \theta}{\theta}$ as $\theta \to 0$

Diagram 7.6 shows the circle which we used to define the sine of
an angle.

Let us now draw the perpendicular PC as shown. The length of
PC will clearly be $r \tan \theta$ since (by definition) $\tan \theta = PC/OP$.

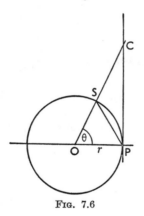

FIG. 7.6

We now consider some areas. It can be
shown that the area of the triangle POS is
given by $\frac{1}{2}r^2 \sin \theta$. We shall not prove this
result here, although it follows from one of
the results listed at the end of Chapter IV.
It can also be shown that the area of a
sector of a circle of radius r and angle θ is
given by $\frac{1}{2}r^2\theta$ which is, therefore, the area
of the sector OPS (bounded by the two
straight lines OP and OS and the arc PS).
Finally, we have that the area of the triangle CPO is

$$\frac{1}{2}OP \times CP$$
$$= \tfrac{1}{2}r \times r \tan \theta$$
$$= \tfrac{1}{2}r^2 \tan \theta$$

It is obvious from the diagram that the area of the sector OPS
lies between the area of the triangle POS and the area of the triangle
CPO. We may write this as

Area $\triangle OPS$ < area sector POS < area $\triangle CPO$

from which it follows that

$$\tfrac{1}{2}r^2 \sin \theta < \tfrac{1}{2}r^2\theta < \tfrac{1}{2}r^2 \tan \theta$$

and so

$$\sin \theta < \theta < \tan \theta$$

Dividing each term by $\sin \theta$ we have

$$1 < \frac{\theta}{\sin \theta} < \frac{1}{\cos \theta}$$

whence

$$1 > \frac{\sin \theta}{\theta} > \cos \theta$$

Now as θ becomes indefinitely small, the value of $\cos \theta$ tends to unity. We thus have that the ratio

$$\frac{\sin \theta}{\theta}$$

lies between unity and a quantity which tends to unity. As we make θ progressively smaller and smaller, the value of the ratio must there-fore tend to unity.

We thus have the very important result that the limit to which $\dfrac{\sin \theta}{\theta}$ tends as θ tends to zero is unity, provided that θ is measured in radians.

The tangent to a curve

Finally we come to another geometric example, concerning a chord drawn between two points on a curve. We will consider the

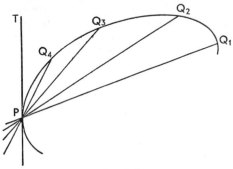

FIG. 7.7

point P to be fixed, while the point Q moves along the curve towards P as shown in Diagram 7.7. As this happens the chord PQ gets closer and closer to the tangent PT. As long as Q is a separate point from P, then PQ remains a chord : but, by moving Q as close as we please to P, we may make this chord approach the tangent as closely as we please. The tangent occupies a position which is the *limit* to which the chord tends as the point Q approaches indefinitely close to P.

Speed at a point

Apples, Achilles, tortoises, circles and tangents do not often appear together. They have been introduced here in order to emphasise that the concept of limit has wide and varied applications, but is basically a very simple idea. Upon this idea is built the whole of the calculus, and a good deal of the rest of this book. While some of these examples have been illustrative, and in a sense interruptive, we now proceed to another example, which also develops the idea a little further, and paves the way for the next chapter.

We often say that a car is travelling at a certain speed. It is useful to consider this statement a little more carefully than we do in ordinary speech. First of all, let us ask the meaning of the word " speed ".

If a car travels sixty miles in two hours, we say that it is travelling at thirty miles per hour. We obtain its speed by dividing the distance by the time. But we do not consider for one moment that the car has been travelling at exactly thirty miles per hour for the whole of the time ; possibly it has even stopped ; at times it has gone much faster. All we can say is that the *average speed* has been thirty miles per hour.

We now suppose that we want to obtain some information about the speed of the car when it passed a certain point . . . say a particular lamp post, or a line on the road. Clearly we would have little justification for saying that the car was doing thirty miles per hour, simply because it was doing an average of thirty miles per hour over a distance of sixty miles. It would be more reasonable to take a stretch of road for about a hundred yards on either side of the point concerned, and to measure the time taken to travel that distance. But even this will provide only the average speed over two hundred yards, which at best will be but an approximation to the speed of the car at the point concerned. Given a quick eye, or a delicate instrument, we could measure off a distance of one yard on either side of the line, and see how long it takes for the car to travel that distance. We might find that the average speed over the two miles was 30 m.p.h., the average over two hundred yards 33·2 m.p.h., while the average over the two yards was 32·6 m.p.h. But even this is not enough. The car could be accelerating, and travelling faster when leaving the two yard stretch than when entering it. It is highly unlikely that the speed would differ very much from 32·6 m.p.h. : but it is possible. To be more precise we could measure a distance of one foot on either side of the line, and obtain an average speed of, say, 32·5 m.p.h. over that distance. But this is still not a speed *at a point*. It is essentially an *average speed over a distance*. Now the speed of a car is a continuous variable, and if we know that the average speed of the car over a distance of two yards was 32·6 m.p.h. while that over a distance of two feet was 32·5 m.p.h., we are not going to expect great (or sudden) variations in the speed within this distance. Let us take a narrower range. Mark off six inches on either side of the line, and time this journey. We may find an average speed of 32·47 m.p.h. ; but, although it is measured over a distance as small as a foot, it is still an average speed over a distance.

We can go on indefinitely. As a *first approximation* to the speed of the car at the point, we can take the average speed over two miles, of 30 m.p.h. As a closer approximation, in which we have greater confidence, we can take the average speed of 33·2 m.p.h. over the two hundred yards. Even more faith can be put in the average over two yards of 32·6 m.p.h., and still more in the average of 32·5 m.p.h. over a distance of two feet. The average of 32·47 m.p.h. over an even smaller distance is even more reliable. If we could time the car in its passage through a distance of one inch, we should clearly be getting pretty close to the speed of the car at the point concerned : but we would still be dealing with an average speed over a distance. There is no end to the subdivision that could take place, each subdivision leading closer and closer to the speed *at the point*, but never quite getting there.

There is, of course, a very good reason why it will never quite get there. Speed is defined as distance divided by time. A point is of no size at all ; it has no distance. However small it may be, once we measure a distance to see how long it takes for a car to traverse it, we have departed from the idea of measuring speed at a point, and have made yet another measure of average speed over a distance.

Despite this, the concept of speed at a point has its uses ; and it is a concept that is less paradoxical than the above account suggests. It is entirely a matter of definition.

We define the *average speed over a distance* to be the quotient of the distance and the time taken to travel it, i.e.,

$$\text{average speed} = \text{distance} \div \text{time}$$

We define the *speed at a point* to be the limit to which the average speed over a distance surrounding the point tends as the distance tends to zero, e.g., the speed at the point P is the limit to which the average speed over LM tends as both L and M move up to P.

There is another way of looking at this matter. Suppose that we draw a graph showing the distance travelled by the car after successive moments of time. Thus, in Diagram 7.8, after a time of one hour the car has travelled a distance OA, etc.

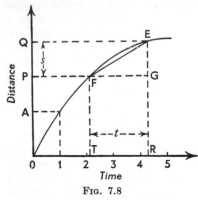

FIG. 7.8

Now consider the positions P and Q. To travel the distance OP takes the time OT; and to travel the distance OQ takes the time OR. Therefore, to travel from P to Q takes the time represented by TR. Let us call the distance PQ, s feet, and the time TR, t seconds. Then we have, that, in its journey from P to Q the car travels a distance of s feet in t seconds, and so its average speed is s/t feet per second.

Consider the triangle EFG. The slope of the line EF is given by EG/FG. But $FG = TR = t$; and $EG = QP = s$. The slope EG/FG is therefore equal to s/t which is the average speed between P and Q. We thus have the very important result that the *average speed* between two points is given by the *gradient of the chord* joining the corresponding points on the space-time curve.

If we want to measure the speed of the car *at the point* P we may begin by taking an approximation to it as the average speed over the distance PQ. Now let us take successive approximations by letting Q move gradually up towards P. Then the speed at the point P is approximated to by the average speed over the distance PQ, which is represented by the slope of the chord EF. We now come to the heart of the matter. As Q approaches P so does the average speed approach the speed at the point, which is (by definition) its limiting value. But as Q approaches P so does the slope of the chord approach the slope of the tangent at F which is (by definition) its limiting value. And we have seen that the slope of the chord EF gives the average speed over the distance PQ. It follows that the speed at the point P is given by the slope of the tangent at F. The speed at P is the limit to which the average speed over PQ tends as Q approaches P; and this is the limit to which the slope of the chord EF (which measures the average speed over PQ) tends; and this is the slope of the tangent at P.

The speed of the car at any point may therefore be obtained by measuring the slope of the tangent to the space-time graph at that point.

We have spent some time considering these aspects of the concept of limit in order to lay a sure foundation for the study of the calculus. Possibly it has been noticed that none of these examples has been drawn from economics. Before proceeding any further we must consider the reason for this.

CHAPTER VIII

RATES OF CHANGE OF ECONOMIC VARIABLES

But the age of chivalry is gone. That of
sophisters, economists, and calculators,
has succeeded; and the glory of Europe
is extinguished for ever.

BURKE : *Reflections on the Revolution in France*

1. Continuity of economic variables

We have already seen that while the average speed of a car is a
continuous variable, the price of a pound of sugar is a discontinuous
one. As we shall shortly see, the calculus is concerned with the
behaviour of continuous variables; and that is why our main example in the last chapter concerned the speed of a car, rather than
the price of sugar.

The first object of this short chapter is to consider to what extent
other economic variables are continuous, or discontinuous, and
whether it is permissible, under certain conditions, to treat some of
the discontinuous variables as though they were continuous. It must
be said at the outset that a thorough analysis of this problem would
lead us into realms of mathematical logic far outside the scheme of
this book ; but we can, at least, make ourselves aware of the problems
involved.

Let us begin by considering the level of employment. At first
sight this may seem to be a clear case of a discontinuous variable.
The labour force cannot very well increase by less than one man. If
the level of employment is 20,000,000 then the smallest possible
change will place it at 20,000,001 and this is a discontinuous change.
On the other hand, it is a change which is exceedingly small when
compared with the basic level of 20,000,000. Because of this there
are grounds for treating the change as a continuous one. After all,
in our measurement of a continuous variable, such as speed, we could
hardly hope to be correct to one place in twenty million, being able
to express a speed such as 20·000 001 m.p.h. with some confidence
in the accuracy of the last figure. Furthermore, the level of employment is very often used simply as a convenient substitute for man-
hours worked. And if we use this latter idea, taking into account

142

short time and overtime, and considering a single man-hour to be the smallest unit, then we may well have a position (in a forty-hour week) of the number of man-hours rising from 800,000,000 to 800,000,001. It is difficult indeed to consider such a measurement to be one of a discontinuous variable.

But when we consider the price of a pound of sugar we are faced with a very different problem. It is true that the price need not increase from 9p. to $9\frac{1}{2}$p. in one jump. One could sell sugar at $18\frac{1}{2}$p. for two pounds, giving a price of $9\frac{1}{4}$p. per pound ; or at $27\frac{1}{2}$p. for three pounds, giving a slightly lower price. But where we do draw the line? Some householders may buy sugar by the stone, and to sell a stone at a price of £1·$26\frac{1}{2}$ instead of £1·26 would mean raising the price from 9p. to 9·036p. per pound. On the other hand, if we attempt to go much beyond here we will find ourselves talking not of the retail but of the wholesale price. Of course, for expensive goods, such as a car costing £2000, the retail price is virtually continuous. The difference between £2000 and £2000·01 being a difference of about five in a million, should hardly condemn a variable to the ranks of the discontinuous. But the retail price of a cheap good cannot be looked at so easily in this way.

We may think of any other economic variable. In almost every case we are likely to find that *strictly speaking*, we are faced with a discontinuous variable ; but in some cases the discontinuities are so small (as in the case of a single man-hour in a large labour force) that there is a great temptation to ignore them. Is it safe to yield to this temptation?

Ricardo felt that it was. It is true that he did not embark on the journey into calculus, but he accustomed men to thinking of increasing (or decreasing) the labour force by very small amounts ; and to treating the acreage of land under cultivation similarly. If an attempt is made to translate the Law of Diminishing Returns into calculus it is quite successful. Problems arising out of the choice between different methods of production, or the utilisation of differing quantities of certain factors of production, are very easily treated by the calculus, and the right answers are obtained : but only under certain circumstances. Although the discontinuities in the labour supply, or the level of employment, are real, they are also very small ; and no *logical* reason exists why they should not, *theoretically* be smaller. There is no logical reason why we should not think of man-minutes, or man-seconds, thereby reducing the size of the discontinuity. Nor is there any logical reason why ten thousand people should not combine to buy their sugar between them, offering £375·01 instead of £375, reducing the discontinuity in the retail, price to one-ten thousandth of a penny. In such cases as these, correct conclusions will be reached by the application of the calculus. It may be necessary, when interpreting them to bear in mind that the calculus is concerned with continuous variables, and then to ask whether it is necessary to convert an answer telling us something about a point, into an answer telling us something about a range— such as a speed at a point into an average speed over a distance. On the other hand, if we are concerned with the problem facing modern firms, of the ratio in which to combine two or three methods of production that involve costly units then we cannot proceed by these methods. If the factor of production is not land but a complete steel rolling mill, the problem is not whether to add one square yard to the several thousand already in use, but whether to add a third mill to the two already existing : and this can hardly be treated as anything but a discontinuity—to be analysed by other methods.

With these considerations in mind, we may pass to the examination of an economic problem of great importance.

2. Elasticity of demand

Suppose that when the price of a good is p (pence per pound) the amount of the good that is demanded is x (pounds). The amount of money spent on the good is therefore px (pence). Now suppose that the price increases. Since the good is dearer, there will probably be a fall in demand. Suppose that the price increase is one of h (pence), so that the new price is $p+h$, and that the amount demanded falls by an amount k so that the new amount demanded is $x-k$. Then the new expenditure on the good will be

$$(p+h)(x-k) = px + hx - pk - hk$$

The question that arises is " Is this new expenditure greater or less than the previous expenditure? " Clearly the answer depends on the sign of the quantity

$$hx - pk - hk$$

by which the new expenditure differs from the old. If the sign of this quantity is positive, then the new expenditure is greater than the old ; if it is negative, then the new expenditure is less.

Already we are getting into algebraic confusion. With x and p we may be able to cope, but when h and k are in the picture we begin to wonder what they are. Mathematicians, being lazy, like one symbol to remind them of a related symbol. Now the symbol h has a very special meaning. It refers to price, which we have already denoted by p. More specifically, it refers to a change in price.

Because of this, it is convenient to employ a special notation. Denoting price by p, let us denote the value of a change in price by Δp where Δ (delta) means, quite simply, " the change in " or " an increment of ". Δp does *not* mean Δ multiplied by p, or even a particular fraction of p. Δ is not a symbol in its own right ; by itself, it has no meaning. It is part of a symbol pair Δp in which Δ performs an operation on the variable p, the operation being the taking of an increment. Similarly Δx may be used to denote an increment of x which in this case will be negative.

We thus have the following :

The price changes from $\qquad p$ to $p + \Delta p$
The amount changes from $\qquad x$ to $x + \Delta x \qquad$ where Δx is negative
The expenditure changes from px to $(p + \Delta p)(x + \Delta x)$
$$= px + x\,\Delta p + p\,\Delta x + \Delta x \,.\, \Delta p$$

If the latter expenditure is greater than the former, so that an increase in price results in an increased expenditure, we consider the demand to be *relatively inelastic*. If the opposite is true, so that an increase in price results in a decrease in expenditure, we consider the demand to be *relatively elastic*.

Although this may do to introduce the basic idea of elasticity of demand, it leaves several questions unanswered. The first point we must notice is that the above note refers to elasticity of demand over a certain *range*. When the price changes from p to $p + \Delta p$ one is reminded of the distance travelled by a car changing from s to $s + \Delta s$, where Δs denotes the increment of distance. We can hardly think of the elasticity, as introduced above, to be elasticity *at a price*. It is elasticity *over a price range*. Suppose, for example, that when the price is 9p. the amount demanded is 1000 pounds. Now if the price increases to 10p. the amount purchased may fall to 910 pounds. In the original case the expenditure was 9000 pence ; after the price change it becomes 9100 pence. We can say that there is a relatively inelastic demand *over the range* 9p. to 10p. But we cannot say that the elasticity of demand is inelastic *at the price* 9p. Suppose that the price had risen from 9p. to 11p., and that the amount purchased fell from 1000 pounds to 800 pounds, (which is quite compatible with the information we have previously considered). In this case the amount spent would be 8800 pence. In other words, the increase in price from 9p. to 11p. results in a fall in the value of sales. There is a comparatively elastic demand *over the price range* 9p. to 11p. This does not contradict the previous statement that the demand over the price range 9p. to 10p. is inelastic, any more than the statement that the average speed of an aeroplane between London and Paris was slow, contradicts the statement that its average speed between

London and Berlin, via Paris, was fast. It would, however, result in a contradiction if we could say, in the first case, that elasticity of demand at the price 9p. was inelastic; for in this case we would also be justified in saying, from consideration of the second data, that elasticity at the price 9p. was elastic.

It is useful to have a more precise definition of elasticity of demand.

Definition :

> If the demand changes from x to $x + \Delta x$ (where Δx may be positive or negative) when the price changes from p to $p + \Delta p$ then we define the *average elasticity of demand over the price range p to $p + \Delta p$* to be the ratio of the proportionate change in demand $\left(\dfrac{\Delta x}{x}\right)$ to the proportionate change in price $\left(\dfrac{\Delta p}{p}\right)$,

i.e.,
$$\frac{\Delta x}{x} \div \frac{\Delta p}{p}$$

$$= \frac{p}{x} \frac{\Delta x}{\Delta p}$$

It must be emphasised that this definition is essentially concerned with an average elasticity over a price range p to $p + \Delta p$. In saying this we have to emphasise not only the *range* but also the *direction* of change. Consider, for example, a rise of price from 9p. to 10p. leading to a decline in sales from 1000 pounds to 910 pounds. We have

$$\frac{p}{x}\frac{\Delta x}{\Delta p} = \frac{9}{1000} \times \frac{910 - 10000}{10 - 9}$$

$$= \frac{9}{1000} \times \frac{-90}{1}$$

$$= -0 \cdot 81$$

But now let us start at the other end, with an initial price of 10p. and sales of 910 units. Reduce the price to 9p. and let sales expand to 1000 units. In this case, $p = 10$ and $x = 910$, so

$$\frac{p}{x}\frac{\Delta x}{\Delta p} = \frac{10}{910} \times \frac{1000 - 910}{9 - 10}$$

$$= \frac{1}{91} \times \frac{90}{-1}$$

$$= \frac{-90}{91} = -0 \cdot 99$$

which is not the same as before.

This difficulty may be overcome if we think of a price elasticity *at a particular price*. Although this may seem to be an odd idea, it is in fact no odder than the idea of speed at a point. Just as elasticity implies a price range over which it can be measured, so speed implies a distance over which it can be measured. And just as we define the speed at a point to be the limit to which the average speed over a distance tends as that distance becomes smaller and smaller, so do we have the following definition.

Definition :

> If the average elasticity over a range p to $p + \Delta p$ tends to a limit as the price increment Δp tends to zero, then we call this limit the *elasticity at the price p*.

This definition, the concept of elasticity, and the mathematical ideas contained in this chapter will receive more detailed attention in the next few chapters. But we should note now that the definition presupposes that the increment of price *can* tend to zero ; and this is tantamount to supposing that price is a continuous variable, or, at least, can be treated as one. When talking of average elasticity over a range it was not necessary to specify continuity. But we can allow Δp to tend to zero only if price can be treated as a continuous variable. Elasticity of demand *at* a price means nothing unless the price concerned is capable of changing by amounts so small that it is virtually continuous. Elasticity over a price range, however, is not so restricted.

CHAPTER IX

DIFFERENTIATION

—or Little by Little.

DEAN FARRAR

1. A new notation

We have really been talking about differentiation in the last two chapters, and the present one contains no really new ideas. It merely introduces a simple device which enables the ideas of the preceding chapters to be stated more concisely and easily.

Let us begin by considering two points P and Q on a curve. Let the co-ordinates of P be (x, y), and let Q have co-ordinates $(x + \Delta x,$

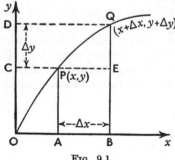

FIG. 9.1

$y + \Delta y)$, so that the distance AB has magnitude Δx and the distance QE has magnitude Δy, as shown in Diagram 9.1.

Then the slope of the chord PQ is given by QE/PE which is equal to $\Delta y/\Delta x$.

Now the slope of the tangent at P is defined as the limit to which the slope of the chord PQ tends as the point Q approaches P. This means that it is the limit to which the ratio $\Delta y/\Delta x$ tends as the value of Δx tends to zero. There may seem to be a difficulty here. " When Δx reaches zero, will not this ratio have an infinite value? " it may be asked. The answer is that Δy is also zero and we are faced with the ratio $0/0$. If we insist on considering the problem in this way we will find it difficult to reach a meaningful conclusion. But the argument we have considered in past chapters shows that the ratio $\Delta y/\Delta x$ gives the slope of the chord, and that as the two points get closer and closer (which is the same as saying " as Δx tends to zero ") the slope of this chord tends to the slope of the tangent. It is clear from the diagram that the tangent has a non-zero slope, and also one that is not infinite. Clearly the limit to which $\Delta y/\Delta x$ tends as Δx tends to zero is measurable. Clearly, too, it is likely to have different values in different cases. If the curve is very steep, then this limit will be high. This is because, if the curve is steep, y increases more quickly than does x, and so the increment of y is greater than the corresponding increment of x, even when both are tending to zero.

Now we are going to come across this idea very often. Speeds, elasticities and a host of other concepts are going to be expressed in terms involving the gradients of curves. If we have constantly to write out " the limit to which the ratio $\Delta y/\Delta x$ tends as Δx tends to zero " whenever we refer to the slope of a curve we are going to become very tired. We therefore write it more simply as

$$\underset{\Delta x \to 0}{\text{Limit}} \frac{\Delta y}{\Delta x}$$

Let us consider two examples. A car travels from P to Q. It takes a time t to travel the distance s up to the point P ; and a further time Δt to travel the additional distance Δs from P to Q. The *average speed of the car over the distance PQ* is $\Delta s/\Delta t$. The *speed at the point P* is given by $\underset{\Delta t \to 0}{\text{Limit}} \dfrac{\Delta s}{\Delta t}$. This is simply re-writing an argument that has already been considered in detail.

If a change in price from p to $p + \Delta p$ results in a change in demand from x to $x + \Delta x$ then the *average elasticity over the price range p* to $p + \Delta p$ is defined to be $\dfrac{p}{x} \dfrac{\Delta x}{\Delta p}$. We define *the elasticity at the price p* to be $\underset{\Delta p \to 0}{\text{Limit}} \dfrac{p}{x} \dfrac{\Delta x}{\Delta p}$ which, since p and x are definite and fixed quantities, is the same as

$$\frac{p}{x} \underset{\Delta p \to 0}{\text{Limit}} \frac{\Delta x}{\Delta p}$$

Already we are probably feeling that it would be a good idea to have a short way of writing $\underset{\Delta x \to 0}{\text{Limit}} \dfrac{\Delta y}{\Delta x}$. It is a laborious and clumsy thing to keep writing. At times, when it is necessary to be more than usually explicit, we may have to write it out fully : but more usually we denote

$$\underset{\Delta x \to 0}{\text{Limit}} \frac{\Delta y}{\Delta x} \quad \text{by} \quad \frac{dy}{dx}$$

Great care is necessary in interpreting this. The symbols d, x and y are not used in their ordinary sense. $\Delta y/\Delta x$ is a pure ratio of two small but measurable quantities Δy and Δx that have existences of their own. But dy/dx does not mean dy divided by dx. It means quite simply the value to which the ratio $\Delta y/\Delta x$ tends as Δx tends to zero. It is not a ratio itself (because we have not given any meaning to dy and dx). It is the limit to which a ratio tends. We could equally well denote this limit by L. But that would be just one more symbol telling us nothing about itself. The notation we have used reminds us that it is something to do with $\Delta y/\Delta x$. But it must be emphasised that dy/dx appears as a ratio only to remind us that it is the limit of a ratio : it is not a ratio itself. dy and dx have no separate meaning or existence. Sometimes these symbols appear on their own in mathematics books, and this may lead the student to consider that they have a separate existence. If he examines their use carefully, he will find that either they are being used in a different way (as in Chapter XXIII) having been differently defined ; or they are being used in a lazy way, that may be justifiable but may also lead to trouble. These ideas follow fairly closely the ideas first put forward by Newton (1642–1727) and Leibniz (1646–1716). The symbolism used here is essentially that of Leibniz.

2. The process of differentiation

We must now consider how to find the value of dy/dx in any case with which we are presented, i.e., how to find the slope of the tangent to a given curve. Clearly this slope depends on two things. It depends on the shape of the curve itself (and therefore on its equation) and it depends on the point on the curve at which the tangent is drawn. Before considering the procedure to be followed in a more general case, we will consider a couple of concrete examples.

If we drop a stone into a well, and assume that it has fallen a distance of y feet in a time of x seconds, then x and y are fairly accurately related by the equation

$$y = 16x^2$$

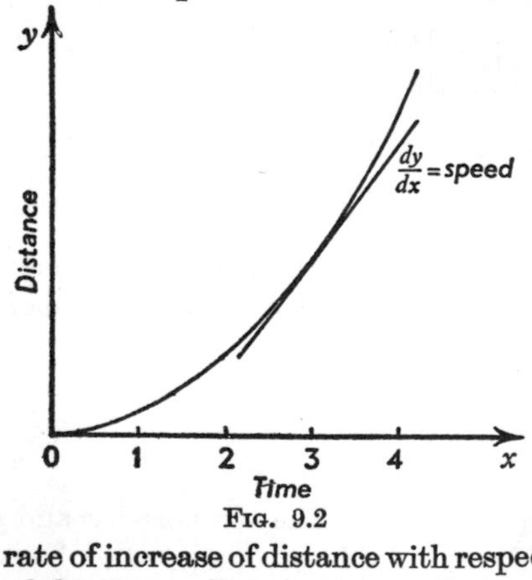

We may recognise this as being the equation of a parabola, and the relevant part of it is shown in Diagram 9.2. This graph, showing distance travelled (y) in time (x) is, of course, a space-time graph, and the tangent to it, measuring the

Fig. 9.2

rate of increase of distance with respect to time, will measure the speed of the stone. Let us suppose that we want to find the speed after exactly three seconds. We could find this by drawing the tangent at the point shown : but an easier and more accurate method is at hand. To begin with, we shall go into this fully ; but we shall soon see that there is a very simple rule which allows one to answer a question of this kind very quickly indeed, often without having to write anything down.

Let us consider the equation

$$y = 16x^2$$

Take any point on this curve. Let its co-ordinates be (x_1, y_1). Then these co-ordinates must satisfy the equation of the curve, so that

$$y_1 = 16x_1^2$$

Now let us recall that when we considered the slope of the tangent we began by taking another point Q. We do the same now, letting its co-ordinates be $(x_1 + \Delta x, y_1 + \Delta y)$. The co-ordinates of Q must also satisfy the equation of the curve, and so

$$y_1 + \Delta y = 16 (x_1 + \Delta x)^2$$
$$= 16x_1^2 + 32x_1 \, \Delta x + 16 (\Delta x)^2$$

But $\qquad y_1 \qquad = 16x_1^2$

and so $\qquad \Delta y = \qquad 32x_1 \, \Delta x + 16 (\Delta x)^2 \text{ (by subtraction)}$

Dividing both sides by Δx gives us

$$\frac{\Delta y}{\Delta x} = \qquad 32x_1 + 16\Delta x$$

and this, of course, measures the slope of the chord PQ. Now the slope of the tangent at P is given by the value of dy/dx, which is the limit to which the value of $\Delta y/\Delta x$ tends as Δx tends to zero. Clearly in this case we have that

$$\frac{dy}{dx} = \underset{\Delta x \to 0}{\text{Limit}} \frac{\Delta y}{\Delta x}$$
$$= \underset{\Delta x \to 0}{\text{Limit}} (32x_1 + 16\Delta x)$$
$$= 32x_1$$

since, as Δx tends to zero, $32x_1$ remains unchanged, but $16\Delta x$ also tends to zero.

This means that the slope of the tangent to the curve $y = 16x^2$ at the point P, whose abscissa is x_1, is given by $32x_1$; and so the speed of the stone after three seconds is obtained by substituting 3 for x_1 and obtaining

$$32 \times 3 = 96 \text{ feet per second.}$$

Before going any further we might as well consider briefly what we have just done. We considered a curve whose equation was given. By considering two close points P and Q we found the value of $\Delta y/\Delta x$, and then proceeded to find the limit to which this value tended as Δx tended to zero. This value, denoted by dy/dx, gave us the slope of the tangent at the point (x_1, y_1) as $32x_1$. Usually we do not bother about the subscripts, but simply say that the gradient of the curve $y = 16x^2$ at a point (x, y) is $32x$, it being understood that x can take any value.

The process of finding dy/dx when we are given y in terms of x is known as *differentiating y with respect to x*. The value of dy/dx is called the *derivative* of y with respect to x, and it measures the rate of increase of y with respect to x.

In the above example we have differentiated from first principles : but there is a simple rule which enables one to differentiate very quickly—often at sight—and to which we now come. It is based on a process very similar to that which we have just considered.

Suppose that we have an equation

$$y = ax^n$$

where a is any constant, negative or positive, and n is a positive integer.

Let P be a point on the curve $y = ax^n$ and let its co-ordinates be (x, y). Consider a close point Q whose co-ordinates are $(x + \Delta x, y + \Delta y)$. Then since this is also on the curve we have,

$$y + \Delta y = a(x + \Delta x)^n$$
$$= a(x^n + {}_nC_1x^{n-1}\,\Delta x + {}_nC_2x^{n-2}(\Delta x)^2 + {}_nC_3x^{n-3}(\Delta x)^3 + \ldots)$$

where this last expression, obtained by the use of the Binomial Theorem, contains terms which depend on successively higher powers of Δx. But

$$y = ax^n$$

and so subtraction yields

$$\Delta y = a\,({}_nC_1x^{n-1}\Delta x + {}_nC_2x^{n-2}(\Delta x)^2 + {}_nC_3x^{n-3}(\Delta x)^3 + \ldots)$$

whence

$$\frac{\Delta y}{\Delta x} = a\,({}_nC_1x^{n-1} + {}_nC_2x^{n-2}\Delta x + {}_nC_3x^{n-3}(\Delta x)^2 + \ldots)$$

where the remaining terms on the right contain cubes and higher powers of Δx.

Now as $\Delta x \to 0$, so $(\Delta x)^2 \to 0$ even more rapidly. If, for example,

$\varDelta x = 0.01$ then $(\varDelta x)^2 = 0.0001$ while $(\varDelta x)^3 = 0.000001$, etc.   We thus have that as $\varDelta x \to 0$

$$\frac{dy}{dx} = \operatorname*{Limit}_{\varDelta x \to 0} \frac{\varDelta y}{\varDelta x}$$
$$= \operatorname*{Limit}_{\varDelta x \to 0} a \left({}_nC_1 x^{n-1} + {}_nC_2 x^{n-2} \varDelta x + {}_nC_3 x^{n-3} (\varDelta x)^2 + \ldots \right)$$
$$= a \, {}_nC_1 x^{n-1} \qquad \text{(Since all remaining terms become zero)}$$
$$= anx^{n-1}$$

i.e., if
$$y = ax^n$$
then
$$\frac{dy}{dx} = anx^{n-1}$$

which gives the slope of the curve at any point.

Example :

If
$$y = 16x^2$$
then
$$\frac{dy}{dx} = 2 \times 16x^{2-1}$$
$$= 2 \times 16x$$
$$= 32x$$

and so the slope of the curve at the point on it where $x = 3$, (and $y = 144$) is 96, i.e., the tangent at this point goes up 96 inches in going one inch across.

If
$$y = 8x^5$$
then
$$\frac{dy}{dx} = 5 \times 8x^{5-1}$$
$$= 5 \times 8x^4$$
$$= 40x^4$$

etc.

In particular we should note that if $y = a$ where a is some constant, then $dy/dx = 0$. There are two ways of looking at this. Graphically we have that if $y = a$ then the graph is a horizontal straight line, whose slope is clearly zero. Algebraically we have that

$$a = a \times 1$$
$$= a \times x^0 \qquad \text{(since } x^0 = 1)$$
$$= ax^0$$

and so, if
$$y = a$$
then
$$y = ax^0$$
and
$$\frac{dy}{dx} = 0 \times ax^{-1} = 0 \times \frac{a}{x} = 0$$

The proof we have given is valid only for positive integers. We show in Appendix 9 that the following rule is true for all values of n, be they positive or negative, integral or fractional:

$$\text{If } y = ax^n \text{ then } \frac{dy}{dx} = anx^{n-1}$$

EXERCISE 9.1

1. The following basic results should be checked by the careful application of the rule just enunciated.

y	$\dfrac{dy}{dx}$
ax^n	nax^{n-1}
$4x^3$	$12x^2$
x^2	$2x$
x	1
$x^{3/2}$	$\frac{3}{2}x^{\frac{1}{2}} = \frac{3}{2}\sqrt{x}$
$x^{1/2}$	$\frac{1}{2}x^{-\frac{1}{2}} = \dfrac{1}{2\sqrt{x}}$
$\dfrac{1}{x} = x^{-1}$	$-1x^{-2} = \dfrac{-1}{x^2}$
$\dfrac{4}{x^2} = 4x^{-2}$	$-8x^{-3} = \dfrac{-8}{x^3}$

2. Differentiate from first principles, and then check by application of the above result, the following:

(i) $y = 3x^2$ (ii) $y = 7x^5$

(iii) $y = 5x^3$ (iv) $y = 4x^4$

(v) $y = -\dfrac{4}{x}$ (vi) $y = \dfrac{5}{x^2}$

(vii) $y = \dfrac{7}{x^7}$ (viii) $y = \dfrac{x^3}{5}$

(ix) $y = \sqrt{x^3}$ (x) $y = 3\sqrt{x^3}$

(xi) $y = \sqrt[3]{x^2}$ (xii) $y = \sqrt[3]{2x^2}$

3. Elasticity of supply

Just as we define the elasticity of demand to be the proportionate change in demand associated with a proportionate change in price, so we define the elasticity of supply to be the proportionate change in supply associated with a proportionate change in price. If, for example, when the price increases from 9p. to 10p. the amount that producers are prepared to supply increases from 1000 to 1050 units, then the proportionate increase in supply is 50/1000 while the proportionate increase in price is $\frac{1}{9}$p. The elasticity of supply over the range 9p. to 10p. is therefore

$$\frac{50/1000}{1/9} = 0 \cdot 45$$

We define the elasticity of supply at a price to be the limit to which the average elasticity of supply over a price range tends as the price range Δp tends to zero.

i.e., the elasticity of supply over the range Δp is $\dfrac{p}{x}\dfrac{\Delta x}{\Delta p}$

but the elasticity of supply at the price p is

$$\frac{p}{x} \operatorname*{Limit}_{\Delta p \to 0} \frac{\Delta x}{\Delta p} = \frac{p}{x}\frac{dx}{dp}$$

Suppose now that the amount that producers are prepared to supply (x) is related to the price (p) by the equation

$$x = 3p^2$$

which is the equation of the supply curve. (It may seem to be a most unlikely equation for a supply curve to have, but more realistic ones must be deferred until our mathematics has progressed a little.) Then the elasticity of supply when the price is 5p. is given by the following argument.

When $p = 5$, $x = 75$. Also, since $x = 3p^2$,

$$\frac{dx}{dp} = 6p = 6 \times 5 = 30$$

therefore, elasticity of supply $= \dfrac{p}{x}\dfrac{dx}{dp}$

$$= \frac{5}{75} \times 30$$

$$= 2$$

This means that a small proportionate increase in price will lead to an increase in supply twice as great.

EXERCISE 9.2

1. The relationship between the supply (x) and the price (p) of a particular commodity is

$$x = 15p^3$$

Find the price elasticity of supply over the price range 8p. to 9p., and at the price 8p.

2. If the relationship between supply (x) and price (p) is

$$x^2 = 8p^5$$

find the price elasticity of supply over the price range 3p. to 4p., and at the prices 3p. and 4p.

4. Differentiation of sums and differences

If we have an expression such as

$$y = 4x^2 + 7x^5 - 6x^8$$

we may differentiate it term by term, obtaining

$$\frac{dy}{dx} = 8x + 35x^4 - 48x^7$$

This procedure may be adopted whenever we have sums or differences to differentiate. It is not necessary to know the proof of this, but for those who are interested we give it. In any case, the proof is not difficult, and can be read with profit, demonstrating as it does the fundamental principles involved.

Let us denote $4x^2$ by u, $7x^5$ by v, and $6x^8$ by w so that

$$y = u + v - w$$

Let x change by a small increment Δx, so that u changes by an amount Δu, v by an amount Δv, and w by an amount Δw. Then y will also change ; let it do so by an amount Δy. It follows that

$$y + \Delta y = u + \Delta u + v + \Delta v - w - \Delta w$$

But
$$y \quad\quad = u \quad\quad + v \quad\quad - w$$

and so
$$\Delta y = \quad \Delta u \quad + \Delta v \quad - \Delta w$$

whence
$$\frac{\Delta y}{\Delta x} = \frac{\Delta u}{\Delta x} \quad + \frac{\Delta v}{\Delta x} \quad - \frac{\Delta w}{\Delta x}$$

and
$$\underset{\Delta x \to 0}{\text{Limit}} \frac{\Delta y}{\Delta x} = \underset{\Delta x \to 0}{\text{Limit}} \left(\frac{\Delta u}{\Delta x} + \frac{\Delta v}{\Delta x} - \frac{\Delta w}{\Delta x} \right)$$

We shall now assume that the sum of these three quantities tends to a limit which is the sum of the limits to which the three quantities tend individually. Strictly we should prove this. But here we shall just state that it can be proved to be a valid procedure in this particular case. We then have that

$$\underset{\Delta x \to 0}{\text{Limit}} \frac{\Delta y}{\Delta x} = \underset{\Delta x \to 0}{\text{Limit}} \frac{\Delta u}{\Delta x} + \underset{\Delta x \to 0}{\text{Limit}} \frac{\Delta v}{\Delta x} - \underset{\Delta x \to 0}{\text{Limit}} \frac{\Delta w}{\Delta x}$$

which gives

$$\frac{dy}{dx} = \frac{du}{dx} + \frac{dv}{dx} - \frac{dw}{dx}$$

which justifies the procedure we have adopted in our example.

EXERCISE 9.3

1. Differentiate :

(i) $y = 4x^3 + 3x^2$ (ii) $y = 4x^3 - 3x^2$

(iii) $y = 4x^3 + 3x^2 + 2x$ (iv) $y = 5x^8 - 7x + 2$

(v) $y = 14 + 9x^3 + x$ (vi) $y = 14x - 9x^3$

(vii) $y = \dfrac{14}{x} - \dfrac{9}{x^2}$ (viii) $y = \sqrt{x^3} + 3x^2$

2. If the supply (x) is related to the price (p) by the equation

$$x = 200 + 4p + 0 \cdot 1p^2$$

estimate the elasticity of supply at a price of (a) £0·10 and (b) £0·20.

3. If the demand (x) is related to the price (p) by the equation

$$x = 250 - p - 0 \cdot 05p^2$$

estimate the elasticity of demand at a price of (a) £0·10 and (b)£0·20.

4. If the supply is given by the equation of question 2 above and the demand by the equation of question 3, estimate the price at which supply equals demand, and then estimate the elasticities of supply and demand under equilibrium conditions.

5. Differentiation of products

Consider

$$y = (4x^2 + 3x)(8x^2 + 2x + 1)$$

One way of differentiating this is to expand it term by term, and then to differentiate it as above ; but there is an easier method, similar to the one just described. We shall just indicate the proof of this, leaving the detail to the reader as an exercise.

Replacing the first bracket by u and the second by v, and letting an increment of Δx in x result in increments of Δu, Δv and Δy we have that

$$y + \Delta y = (u + \Delta u)(v + \Delta v)$$

and
$$y = uv$$

whence, by expansion and subtraction, and division by Δx,

$$\frac{\Delta y}{\Delta x} = u\frac{\Delta v}{\Delta x} + v\frac{\Delta u}{\Delta x} + \frac{\Delta u}{\Delta x}\Delta v$$

The last term disappears in the limit, and we are left with

$$\frac{dy}{dx} = u\frac{dv}{dx} + v\frac{du}{dx}$$

In terms of our example,

$$u = 4x^2 + 3x \; ; \qquad v = 8x^2 + 2x + 1$$

$$\frac{du}{dx} = 8x + 3 \quad ; \quad \frac{dv}{dx} = 16x + 2$$

and so, $\quad \dfrac{dy}{dx} = (4x^2 + 3x)(16x + 2) + (8x^2 + 2x + 1)(8x + 3)$

More generally, if y is expressed as a product of several terms, r, s, t, u, v, w, etc., then

$$\frac{dy}{dx} = rstuv\frac{dw}{dx} + rstuw\frac{dv}{dx} + rstvw\frac{du}{dx} + rsuvw\frac{dt}{dx}$$

$$+ rtuvw\frac{ds}{dx} + stuvw\frac{dr}{dx}$$

EXERCISE 9.4

1. Differentiate :

   (i) $y = (4x^2 + 3x)(2x + 1)$ (ii) $y = (x + 3)(x^2 + 7x + 1)$

   (iii) $y = (4x^3 - 2x + 1)(3 - 5x^2)$ (iv) $y = x^7(8x^5 + 4)$

   (v) $y = (4x^2 + 3x)(2x^3 - x)(17 + x + x^5)$

2. The supply equation of a certain product is

$$x = (200 + 4p)(1 + 0 \cdot 1p)$$

while the demand equation is

$$x = (300 - 3p)(1 + 0 \cdot 1p)$$

Estimate the equilibrium price and the elasticities of supply and demand at this price.

6. Differentiation of a quotient

It may be similarly proved that in order to differentiate a quotient such as

$$y = \frac{4x^3 - 7x^2 - 2}{3x + 5}$$

it should first be written in the form $y = \dfrac{u}{v}$, and that then the derivative is given by

$$\frac{dy}{dx} = \frac{v\dfrac{du}{dx} - u\dfrac{dv}{dx}}{v^2}.$$

It is important to notice the order of the terms in the numerator of this expression. The proof is in Appendix 9.

In the example just cited

$$u = 4x^3 - 7x^2 - 2 \; ; \quad v = 3x + 5$$

$$\frac{du}{dx} = 12x^2 - 14x \; ; \quad \frac{dv}{dx} = 3$$

and so

$$\frac{dy}{dx} = \frac{(3x+5)(12x^2 - 14x) - (4x^3 - 7x^2 - 2)(3)}{(3x+5)^2}$$

EXERCISE 9.5

1. Differentiate :

(i) $y = \dfrac{4x^2 + 3x}{2x + 1}$ (ii) $y = \dfrac{x+3}{x^2 + 7x + 1}$

(iii) $y = \dfrac{4x^3 - 2x + 1}{3 - 5x^2}$ (iv) $y = \dfrac{x^7}{8x^5 + 4}$

2. The supply equation of a certain product is

$$x = \frac{200 + 4p^2}{1 + 10p}$$

while the demand equation is

$$x = \frac{250 - p - 4p^2}{1 + 10p}$$

Estimate the equilibrium price and the elasticities of supply and demand at this price.

7. The elasticity of certain demand curves

We are now in a position to see the peculiar property of the rectangular hyperbola, whose equation is $xy = C$. This may be written in the form

$$x = \frac{C}{y}$$

Suppose that the price (p) and the demand (x) of a good are related by

$$x = \frac{C}{p} = Cp^{-1}$$

Then

$$\frac{dx}{dp} = -Cp^{-2} = -\frac{C}{p^2}$$

The elasticity of demand is, therefore, given by

$$\frac{p}{x}\frac{dx}{dp} = -\frac{p}{x}\frac{C}{p^2}$$

$$= -\frac{C}{xp}$$

$$= -\frac{C}{C}$$

$$= -1$$

(where the minus sign is a feature of all downward-sloping demand curves)

In other words, at any point on the demand curve with this equation, the elasticity is minus unity.

Let us examine the implication of this more closely. We may recall from Chapter VIII that if a change in price from p to $p + \Delta p$ results in a change in demand from x to $x + \Delta x$ then the volume of sales changes from xp to $xp + x\Delta p + p\Delta x + \Delta p\Delta x$. (See page 145.)

If we denote sales by S, we have that the change in sales is given by

$$\Delta S = x\,\Delta p + p\,\Delta x + \Delta p\,\Delta x$$

and so the average rate of change of sales over the price range Δp is

given by
$$\frac{\Delta S}{\Delta p} = x + p\frac{\Delta x}{\Delta p} + \Delta x$$

Proceeding to the limit, we have that the rate of change of sales *at the price p* is

$$\frac{dS}{dp} = x + p\frac{dx}{dp}$$

since the last term must disappear, Δx tending to zero with Δp.

Now this expression for the rate of change of sales may be written as

$$\frac{dS}{dp} = x\left(1 + \frac{p}{x}\frac{dx}{dp}\right)$$

and the last term in the bracket is the elasticity. If, therefore, the elasticity is minus unity, the rate of change of sales is zero ; i.e., a small price change will have no effect on the volume of sales. Now if the demand curve is a rectangular hyperbola, the elasticity is minus unity at all points, and so, whatever the price may be, and however much it may change, the volume of sales will be the same. We have already seen this by considering the equation $xp = C$. But this is a very special case. There are many demand curves, with all sorts of equations, which have some point, or series of points, at which the elasticity is minus unity.

Consideration of the same expression tells us that if the curve is such that, at some particular point, the numerical value of the elasticity is less than unity, then dS/dp will be positive, which means that a price increase will result in a greater volume of sales ; but if the elasticity is more than unity, then dS/dp will be negative, indicating that a rise in price will lead to a fall in sales.

We may therefore consider the state of demand *at a particular price* to be inelastic if the elasticity at that price is numerically less than unity ; but elastic if the elasticity at that price is numerically more than unity.

CHAPTER X

DIFFERENTIATION (*Continued*)

So some strange thoughts transcend our wonted theams,
And into glory peep.

HENRY VAUGHAN

1. Introduction

This chapter is in two parts. The first deals with " second derivatives ", which were omitted from the last chapter in order to prevent confusion, but are of fundamental importance, and which must be understood before we proceed to the next chapter. The second part deals with the technique of differentiation, which is important, but not really essential to an understanding of the next chapter. The reader who is impatient to get on may therefore leave the second part of this chapter ; but the reader who is ever likely to attempt some differentiation for himself should not proceed until he has thoroughly grasped it.

2. Second derivatives

Let us return to the car. We saw that we could draw a graph showing the distance (s) travelled in a time (t). The average speed between two points was given by the slope of the chord joining the corresponding points on the curve. The speed at a point was given by the slope of the tangent at that point, i.e., by the value of ds/dt.

Now when we think of a moving car, we think not only of its speed, but also of its acceleration. In ordinary language we might think of acceleration as the rate at which it increases its speed. If a car increases its speed from 30 m.p.h. to 40 m.p.h. in one minute then it is accelerating twice as fast as if it takes two minutes to do this. We can in fact define acceleration in a way very similar to the way in which we defined speed ; and it is this definition that will lead us to the second derivative.

If the speed of the car at a point P is 44 feet per second, and it takes 10 seconds to travel to a point Q where its speed is 66 feet per second, we say that the average acceleration over the distance PQ is given by :

change in speed ÷ time taken to change speed
$$= (66 - 44 \text{ feet per second}) \div 10 \text{ seconds}$$
$$= \quad (22 \text{ feet per second}) \div 10 \text{ seconds}$$
$$= 2 \cdot 2 \text{ feet per second per second.}$$

Similarly we may define the acceleration at the point P to be the limit to which the average acceleration over the distance PQ tends as Q approaches P ; and another way of putting this is to express it as the limit to which the average acceleration over the distance PQ tends as the time tends to zero (since, as Q approaches P, the time taken to travel from P to Q tends to zero).

Now let us compare these definitions with those of speed ; and let us, in particular, compare the definition of speed at a point, (which is the limit to which the average speed over PQ tends as the time tends to zero) with the definition of acceleration at a point, which has just been given. Letting s denote distance and t denote time, we denoted the speed by

$$v = \frac{ds}{dt} = \underset{\Delta t \to 0}{\text{Limit}} \frac{\Delta s}{\Delta t}$$

In the same way, if we denote speed by v we may denote the average acceleration by

$$\frac{\text{increment of speed}}{\text{increment of time}} = \frac{\Delta v}{\Delta t}$$

and then the acceleration at the point P may be put as

$$a = \frac{dv}{dt} = \underset{\Delta t \to 0}{\text{Limit}} \frac{\Delta v}{\Delta t}$$

For example, if the speed of a car is such that the distance s travelled in a time t is given by

$$s = 40t + 3t^2$$

then the speed is given by

$$v = \frac{ds}{dt} = 40 + 6t$$

which means that after three seconds, the speed is 58 feet per second, etc., and the acceleration is given by

$$a = \frac{dv}{dt} = 0 + 6 = 6 \text{ feet per second per second}$$

at constant acceleration.

We now come to the important point of all this. We have expressed acceleration in terms of speed and time. We have expressed speed in terms of distance and time. It follows that we should be able to eliminate a stage, and express acceleration in terms of distance and time, without having to drag in speed. More mathematically we put it as,

a is a function of v and t (i.e., $a - f_1(v, t)$)

and v is a function of s and t (i.e., $v = f_2(s, t)$)

and so a is a function of s and t (i.e., $a = f_3(s, t)$)

where $a = f(v, t)$ is the conventional way of indicating that changes in the values of v and t may be expected to lead to changes in the value of a. The point is easily made; but how do we express it more precisely? We do it by a simple extension of the notation we have already used. In the example of the car, we differentiated s once, in order to get speed; and then we differentiated a second time in order to get acceleration. The acceleration is, in fact, the second derivative of distance with respect to time, just as the speed is its first derivative. The notation we use will show us that we have differentiated twice.

$$s = 40t + 3t^2$$

$$v = \quad \frac{ds}{dt} = 40 + 6t$$

$$a = \frac{dv}{dt} = \frac{d\left(\frac{ds}{dt}\right)}{dt} = \quad 6$$

and usually, to save space, we write this last rather cumbersome expression as $\dfrac{d}{dt}\dfrac{ds}{dt}$ or, even more compactly $\dfrac{d^2s}{dt^2}$. This is merely a convenient way of saying " s differentiated twice successively with respect to time t ".

We may now consider an important economic illustration of a second derivative.

Suppose that a country's existing stock of fixed capital (consisting of buildings, machines and so forth) is denoted by K. Let K be growing over time, in such a way that at time t

$$K = 10000 + 1800t + 100t^2 - t^3$$

where K is in £m. and t is in (say) weeks.

This growth has two components. There is new investment and there is also depreciation. Let us call the net result of these " net investment ".

The *rate of net investment* at a moment t is the rate at which K increases. It is given by

$$I = \frac{dK}{dt} = 1800 + 200t - 3t^2$$

But this rate of investment is altering. At time $t=1$, the rate of investment is

$$1800 + 200 - 3 = £1997\text{m. per week.}$$

At time $t=2$, the rate of investment is

$$1800 + 400 - 12 = £2188\text{m. per week.}$$

At what rate does the rate of investment change? The answer is given by

$$\frac{dI}{dt} = 200 - 6t$$

and this is measured in £m. per week per week. The *rate of net investment* is growing each week at a declining rate. Eventually, when $6t > 200$, it falls. There is net disinvestment.

It is very important to notice two things. One is that investment should always be expressed as a *rate*. It is investment per period of time. The other is that the rate of change of investment is $\frac{dI}{dt}$, which is $\frac{d^2K}{dt^2}$, since $I = \frac{dK}{dt}$. The rate of change of investment is the second derivative of the capital stock.

Although we have introduced the idea of the second derivative with the help of practical examples, we could have done so purely as a mathematical device. If we are asked to differentiate

$$p = x^3 + 3x^2 + 2$$

then we can easily write down

$$\frac{dp}{dx} = 3x^2 + 6x$$

But if we are now told that in fact the expression for p was obtained by differentiating

$$y = \frac{x^4}{4} + x^3 + 2x + 6$$

then we may ask about the relationship between y and $\frac{dp}{dx}$. Clearly we have obtained p by differentiating y once ; and we have obtained $\dfrac{dp}{dx}$ by differentiating the derivative of y. For this reason we may call it the second derivative of y, and denote it by $\dfrac{dp}{dx} = \dfrac{d\left(\dfrac{dy}{dx}\right)}{dx} = \dfrac{d^2y}{dx^2}$.

Similarly, we could obtain the third derivative, by differentiating the second derivative, to obtain

$$\frac{d^3y}{dx^3} = \frac{d}{dx}\left(\frac{d^2y}{dx^2}\right) = \frac{d}{dx}\left(\frac{dp}{dx}\right) = 6x + 6$$

and the fourth derivative would be

$$\frac{d^4y}{dx^4} = 6$$

Sometimes dy/dx is denoted by y' and d^2y/dx^2 by y''. In the same way, if we have $f(x)$ as some function of x then its first derivative is sometimes denoted by $f'(x)$ and its second by $f''(x)$. We can proceed further if necessary.

e.g.,
$$f(x) = 3x^3 - x^2 + x + 2$$
$$f'(x) = 9x^2 - 2x + 1$$
$$f''(x) = 18x - 2$$
$$f'''(x) = 18$$
$$f_{\mathrm{iv}}(x) = 0$$

The nth derivative is sometimes denoted by $f^{(n)}(x)$.

Just as $f'(x)$ measures the rate of increase of $f(x)$, so $f''(x)$ measures the rate of increase of $f'(x)$. (In terms of our example, the acceleration d^2s/dt^2 is the rate of increase of the velocity ds/dt.) But $f'(x)$ represents the slope of a curve, and so, if $f''(x)$ is positive, it means that the slope of the curve is increasing at a positive rate, i.e., the curve is becoming steeper. An example is the curve $y = x^3$, x taking only positive values. The slope of the curve is given by $y' = 3x^2$, while $y'' = 6x$ which is positive. The curve starts at the origin, where it is horizontal (i.e., parallel to the x-axis) and quickly shoots up at an ever-increasing rate. We say that such a curve is *convex from below*. In order for a curve to be convex from below, the second derivative must be positive. Furthermore, if the second derivative is positive, the curve must be convex from below.

This is an example of a *necessary and sufficient condition*. For convexity, the second derivative must be positive; and if the second derivative is positive nothing else matters.*

A similar argument shows that a negative second derivative indicates a curve which is *concave from below*, such as $y = -x^3$, for positive x. For negative x, $y'' > 0$ and so this part of the curve is convex from below.

The significance of a zero second derivative appears in the next chapter.

EXERCISE 10.1

1. Find the first and second derivatives of the following functions, and consider whether the curves are concave or convex from below in the region of $x = 1$:

(i) $y = 4x^5 - 3x^2 + 2x + 7$ (ii) $y = 4x^5 - 3x^2 + 8x - 9$

(iii) $y = \dfrac{x^2 + 3}{x - 2}$ (iv) $y = \dfrac{x^3 + 3x}{2x^2 - 7}$

2. A car moves in such a way that after t seconds from starting time it has travelled a distance of s feet, where s and t are related by

$$s = 40t + 4t^2 + 0 \cdot 2t^3$$

Find expressions giving the car's velocity and acceleration at any given moment, and evaluate these for $t = 10$.

3. The technique of differentiation

In the last chapter we saw how to differentiate powers of x, and expressions composed of the sums, differences, products or quotients of such powers. We now consider how to extend our rule that if $y = ax^n$ then $dy/dx = nax^{n-1}$ to cover other cases.

---

* The distinction between a necessary condition and a sufficient one is clear from the following example.

Consider the inequality $x + 4 \geqslant 14$.

For this to be true, it is *necessary* that $x \geqslant 0$, for there is no negative value of x which could possibly satisfy this inequality. But it is not sufficient for x to be $\geqslant 0$, since (for example) $x = 2$ does not satisfy it.

On the other hand, it is *sufficient* for $x \geqslant 100$, because for all such values of x the inequality is satisfied. This, however, is not a necessary condition, since there are values of x which are less than 100 but which satisfy the inequality.

Let us now consider the condition $x \geqslant 10$. This is clearly *necessary*, since no smaller value of x can satisfy the inequality. It is also *sufficient* since all such values of x are bound to satisfy the inequality.

We can therefore say that the *necessary and sufficient* condition for x to satisfy the inequality $x + 4 \geqslant 14$ is that $x \geqslant 10$.

Functions of a function

We begin by considering functions of a function. Consider the expression

$$y = 4x^2 + 3x - 2$$

This expresses y in terms of x; y is a function of x. Now consider the expression

$$y = (3x^2 + 7x + 8)^3 + (3x^2 + 7x + 8)^2$$

Once again, y is a function of x: but it is a somewhat complicated function. It could clearly be written as

$$y = u^3 + u^2$$

where $\qquad u = 3x^2 + 7x + 8$

In this case we have that y is a function of u, and u is a function of x, which means that y is a function of (a function of x). Usually we drop the brackets, and say that y is a function of a function of x.

When we have to differentiate a function of a function it is often simpler to use the following procedure.

It can be shown that, in all cases that we are likely to meet in reasonably elementary work,

$$\underset{\Delta u \to 0}{\text{Limit}} \frac{\Delta y}{\Delta u} \times \underset{\Delta x \to 0}{\text{Limit}} \frac{\Delta u}{\Delta x} = \underset{\Delta x \to 0}{\text{Limit}} \frac{\Delta y}{\Delta x}$$

and that, therefore,

$$\frac{dy}{du} \times \frac{du}{dx} = \frac{dy}{dx}$$

This is not as obvious as it may look, because we cannot automatically cancel the du's on the left-hand side (since they have no separate existence). Nevertheless, as more advanced considerations would show, it is permissible to write

$$\frac{dy}{dx} = \frac{dy}{du}\frac{du}{dx}$$

if y is a function of u, and u is a function of x.

We now proceed to differentiate the example we have quoted,

$$y = (3x^2 + 7x + 8)^3 + (3x^2 + 7x + 8)^2$$

Putting $\qquad u = (3x^2 + 7x + 8); \qquad y = u^3 + u^2$

we have $\qquad \dfrac{du}{dx} = 6x + 7 \qquad ; \qquad \dfrac{dy}{du} = 3u^2 + 2u$

and so $$\frac{dy}{dx} = \frac{dy}{du}\frac{du}{dx} = (3u^2 + 2u)(6x + 7)$$

which may be written as

$$[3(3x^2 + 7x + 8)^2 + 2(3x^2 + 7x + 8)](6x + 7)$$

Powers of brackets

The above method is of great importance when we have to differentiate powers of brackets, such as

$$y = (4x^3 + 2x + 3)^8$$

We put $\qquad y = u^8$ where $\qquad u = 4x^3 + 2x + 3$

$$\frac{dy}{du} = 8u^7 \qquad\qquad \frac{du}{dx} = 12x^2 + 2$$

and so

$$\frac{dy}{dx} = \frac{dy}{du}\frac{du}{dx} = 8(12x^2 + 2)(4x^3 + 2x + 3)^7$$

Examination of this result will show that we may express a more general rule thus :

If $y = u^n$ where $u = f(x)$ (i.e., where u is a function of x)

then $$\frac{dy}{dx} = nu^{n-1}\frac{du}{dx}$$

Irrational expressions

The method is also of great importance when we have to differentiate irrational expressions, i.e., expressions involving roots, such as

$$y = \frac{1}{\sqrt{3x^2 + 2x}}$$

Here we put $3x^2 + 2x = u$ so that then $y = \frac{1}{\sqrt{u}} = u^{-\frac{1}{2}}$

$$\frac{du}{dx} = 6x + 2 \qquad\qquad \frac{dy}{du} = -\tfrac{1}{2}u^{-3/2} = \frac{-1}{2u^{3/2}}$$

whence $$\frac{dy}{dx} = \frac{dy}{du}\frac{du}{dx} = \frac{-(6x + 2)}{2(3x^2 + 2x)^{3/2}} = \frac{-(3x + 1)}{(3x^2 + 2x)^{3/2}}$$

As a final example of this kind of problem, we may consider a more complicated case such as

$$y = \frac{(x^2 + 3x)^6 \sqrt{x^3 + 2x^2}}{(x^4 - 3x + 2)^3}$$

The easiest way of doing this is to write it as

$$y = (x^2 + 3x)^6 (x^3 + 2x^2)^{\frac{1}{2}} (x^4 - 3x + 2)^{-3}$$

We now differentiate this as a product of three terms, as explained in the last chapter; and then we note that each differentiation involves dealing with a function of a function, in the form of a power of a bracket.

We have $\dfrac{dy}{dx} = (x^2 + 3x)^6 (x^3 + 2x^2)^{\frac{1}{2}} \dfrac{d}{dx} (x^4 - 3x + 2)^{-3}$

$$+ (x^2 + 3x)^6 (x^4 - 3x + 2)^{-3} \frac{d}{dx} (x^3 + 2x^2)^{\frac{1}{2}}$$

$$+ (x^3 + 2x^2)^{\frac{1}{2}} (x^4 - 3x + 2)^{-3} \frac{d}{dx} (x^2 + 3x)^6$$

We now take each part of this separately, obtaining

$$(x^2 + 3x)^6 (x^3 + 2x^2)^{\frac{1}{2}} (-3) (x^4 - 3x + 2)^{-4} (4x^3 - 3)$$
$$+ (x^2 + 3x)^6 (x^4 - 3x + 2)^{-3} (\tfrac{1}{2}) (x^3 + 2x^2)^{-\frac{1}{2}} (3x^2 + 4x)$$
$$+ (x^3 + 2x^2)^{\frac{1}{2}} (x^4 - 3x + 2)^{-3} (6) (x^2 + 3x)^5 (2x + 3)$$

where the last part of each row has been obtained by applying the rule for the differentation of a power of a bracket. Some algebra will show that this may be expressed as

$$\frac{(x^2 + 3x)^5 x}{2\sqrt{x+2}(x^4 - 3x + 2)^4} [3x^6 - 23x^5 - 60x^4 - 63x^3 - 147x^2 + 50x + 168]$$

but it is far neater to leave the result as it stands. We have deliberately done an awkward example in order to illustrate the method employed. The reader should obtain plenty of practice, starting with the simpler examples, and progressing systematically to the more complicated cases.

EXERCISE **10.2**

Differentiate :

(i) $y = (4x^3 + 2x^2 - 3x + 8)^4$

(ii) $y = 7 (4x^3 - 3x + 8)^3 \times 3 (4x^3 - 3x + 8)^2$

(iii) $y = \sqrt{3x^2 + 2x}$

(iv) $y = \sqrt{4x^3 + 2x^2 - x + 1}$

(v) $y = \dfrac{1}{\sqrt{4x^3 + 2x^2 - x + 1}}$

(vi) $y = (x^2 + 2x)^4 (x^3 + 3x^2 + 7)^3$

(vii) $y = \dfrac{(x^2 + 2x)^4}{(x^3 + 3x^2 + 7)^3}$

(viii) $y = \dfrac{(x^2 + 2x)^4 (x^3 + 3x^2 + 7)^3}{(x^3 - 8)^2}$

The reciprocal of the derivative

We have seen that dy/dx measures the slope of the tangent to the curve $y = f(x)$, which we may measure by BC/AC, as in Diagram 10.1(a) which shows the graph of $y = x^2 + 3$.

The same graph can be redrawn in a form which gives x as a function of y. This simply means measuring y along the horizontal axis and the graph will appear as in Diagram 10.1(b) which shows $x = \sqrt{y - 3}$. The slope of this curve will be dx/dy which will be

$$B'D'/A'D' = A'C'/B'C'$$

It is obvious from the diagrams that

$$\frac{A'C'}{B'C'} = \frac{1}{\dfrac{BC}{AC}}$$

and so we have that

$$\frac{dx}{dy} = \frac{1}{\dfrac{dy}{dx}}$$

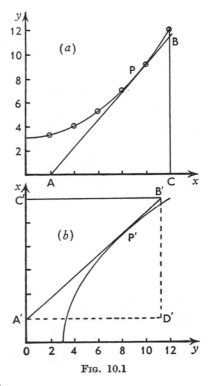

Fig. 10.1

Alternatively we have that if

$$y = x^2 + 3$$

then

$$\frac{dy}{dx} = 2x$$

But also if

$$y = x^2 + 3$$

then

$$x = \sqrt{y - 3}$$

and

$$\frac{dx}{dy} = \frac{1}{2\sqrt{y-3}} = \frac{1}{2x} = \frac{1}{\dfrac{dy}{dx}}$$

We shall have occasion to use this result later in the book.

We must, however, be careful to notice that

$$\frac{d^2x}{dy^2} \quad \text{is not} \quad \frac{1}{\dfrac{d^2y}{dx^2}}$$

as will be seen if we calculate these derivatives for some simple functions such as that already given.

We have

$$y = x^2 + 3 \qquad\qquad x = \sqrt{y - 3}$$

$$\frac{dy}{dx} = 2x \qquad\qquad \frac{dx}{dy} = \frac{1}{2\sqrt{y-3}}$$

$$\frac{d^2y}{dx^2} = 2 \qquad\qquad \frac{d^2x}{dy^2} = \frac{-1}{4(y-3)^{3/2}}$$

$$= -\frac{1}{4x^3}$$

and so

$$\frac{d^2x}{dy^2} = -\frac{1}{2x^3}\frac{1}{\dfrac{d^2y}{dx^2}}$$

in this particular case.

This can also be written as

$$\frac{d^2x}{dy^2} = -\frac{1}{4x^3} = -\frac{2}{8x^3}$$

$$= -\frac{2}{(2x)^3} = -\frac{d^2y}{dx^2}\Big/\left(\frac{dy}{dx}\right)^3$$

It is this result, that

$$\frac{d^2x}{dy^2} = -\frac{d^2y}{dx^2}\Big/\left(\frac{dy}{dx}\right)^3$$

which is of general application.

EXERCISE 10.3

1. Prove the validity of the above formula for dx/dy and d^2x/dy^2 in the following cases:

$$y = x^2 - 5 \qquad y = (x+3)^3 \qquad y = \sqrt{(x+1)}$$

Trigonometric functions

Suppose a car goes around a circular track at a constant speed. If we denote the initial position of the car by A in Diagram 10.2 and its position after a time t by T then the angle TOA is going to increase at a constant rate, because the speed of the car is constant. For instance, if it goes right around the track in an hour, it will move so that the angle TOA increases 6 degrees every minute making 360 degrees in an hour. / If we denote this angle by θ then the rate of increase of θ wil be $d\theta/dt$; / and under the circumstances we have stipulated this will be constant at 6 degrees per minute.

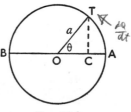

FIG. 10.2

Now suppose that a cyclist is riding back and fore along the line AB in such a way that the line joining him to the car is always perpendicular to AB. At what speed will he have to travel?

To answer this question, which may seem to be very divorced from economics but is, in fact, very important to us, we must reframe it. / The cyclist's distance from O is OC. At what rate does this change? (For clearly the rate at which it changes is the speed of the cyclist). Now $OC = a \cos \theta$ (since $\cos \theta = OC/OT$ and we let $OT = a$). In order to find the rate at which OC changes we have to find the rate at which $\cos \theta$ changes; i.e., we have to be able to differentiate $\cos \theta$.

By using the results

$$\sin (A + B) \equiv \sin A \cos B + \cos A \sin B$$

with

$$A = \theta \quad \text{and} \quad B = \delta\theta$$

and

$$\operatorname*{Limit}_{\theta \to 0} \frac{\sin \theta}{\theta} \to 1$$

the student should be able to prove that

if $\qquad y = \sin \theta \qquad\qquad$ then $\dfrac{dy}{d\theta} = \cos \theta$

Use of

$$\cos^2 \theta + \sin^2 \theta \equiv 1$$

and

$$\tan \theta \equiv \frac{\sin \theta}{\cos \theta}$$

should now enable him to prove that

if $\qquad y = \cos\theta \qquad$ then $\dfrac{dy}{d\theta} = -\sin\theta$

if $\qquad y = \tan\theta \qquad$ then $\dfrac{dy}{d\theta} = \sec^2\theta = \dfrac{1}{\cos^2\theta}$

where (of course) $\cos^2\theta$ means $(\cos\theta)^2$ and not $\cos(\theta^2)$.

Returning to our example, we have that if we denote the distance OC by x then the speed of the cyclist is given by

$$\frac{dx}{dt} = \frac{dx}{d\theta}\frac{d\theta}{dt} \quad\text{where}\quad x = a\cos\theta \quad\text{and}\quad \frac{d\theta}{dt} = 6 \text{ (given)}$$

If $x = a\cos\theta$ then $\dfrac{dx}{d\theta} = -a\sin\theta$ and so the speed of the cyclist is given by

$$-6a\sin\theta$$

Let us look at this for a moment. The minus sign indicates that the distance OC decreases as the car moves round the circle; but when the car has travelled past B, so that θ exceeds 180 degrees, then the $\sin\theta$ term brings in another minus sign. Now the cyclist will be travelling in the direction B to A, instead of A to B and this is indicated by the positive sign. The cyclist's speed depends on the angle TOA. It is greatest when $\theta = 90°$, for then the car is travelling parallel to him, and he has to equal the speed of the car; it is least when the car is at A or B for then the car is travelling momentarily at right angles to him, and he need not travel at all except to turn around.

We introduce this example simply to bring in a case where the derivative of a trigonometric function is important. Much later in the book we shall deal with some more important examples. Meanwhile it should be noted that the rules we have used above may be extended to cover trigonometric functions. For example, if $y = \sin^2\theta$ we may rewrite it as $y = u^2$ where $u = \sin\theta$. This gives that

$$\frac{dy}{d\theta} = 2u\frac{du}{d\theta} = 2u\cos\theta = 2\sin\theta\cos\theta = \sin 2\theta$$

But we must be careful *not* to think that if $y = \sin 2\theta$ then $dy/d\theta = \cos 2\theta$; for this is just not true. It is safer here to write $y = \sin x$ where $x = 2\theta$. Then

$$\frac{dy}{d\theta} = \frac{dy}{dx}\frac{dx}{d\theta} = \cos x \times 2 = 2\cos x = 2\cos 2\theta$$

More generally, if $y = \sin p\theta$ then $dy/d\theta = p \cos p\theta.$ / Similar results hold for the other trigonometric functions. We shall have little cause to use them for some time.

EXERCISE 10.4

Differentiate :

(i) $y = x^2 + \sin x$

(ii) $y = x^2 - \sin x$

(iii) $y = (x^2 + 2) \sin x$ (by writing $x^2 + 2 = u$ and $\sin x = v$)

(iv) $y = (x^3 + x) \cos x$

(v) $y = x^2 (\sin x - \cos x)$

(vi) $(x^2 + 2)(\cos x - \sin x)$

(vii) $y = x \tan x$

(viii) $y = (x \cos x)(\cos x + \tan x)$

(ix) $y = \dfrac{\sin 3x}{\cos x}$

(x) $y = \dfrac{\sin 3x}{\cos 2x}$

(xi) $y = \dfrac{x \sin x}{x + 1}$

(xii) $y = \cos^2 x \ (= (\cos x)^2)$

(xiii) $y = \sin^4 x$

(xiv) $y = \sqrt{\sin x}$

(xv) $y = \sqrt{\tan x}$

(xvi) $y = x \sqrt{\sin x}$

CHAPTER XI

MAXIMA, MINIMA AND POINTS OF INFLEXION

There is much ingenuity in his system which is
supported both by mathematical demonstration
and experience,

ADMIRAL LORD COLLINGWOOD : *Letter*, June 3rd, 1797

1. The monopolist : an example

One of the uses of the work we have just done is illustrated by an examination of the problems of a monopolist. These problems also introduce us to some highly important results of a more general application.

The monopolist's net revenue, which we may define as the difference between his receipts and his costs, depends on his output and the price at which it is sold. These two factors are, however, very closely related, and any attempt to fix the one will affect the value of the other. The monopolist may decide that he will produce (and sell) a hundred units of his product. But if he decides to do this, he has no control at all over the price at which this quantity will be sold. Alternatively, he may decide to sell his product at £5 per unit ; but in this case the quantity that will be demanded at that price will depend entirely on the demand schedule of his customers.

Wishing to make his net revenue as high as possible, the monopolist therefore has to hit upon some level of output which will sell at such a price that his net revenue is higher than it would be for any other level of output. Alternatively, he may decide upon a price at which to supply whatever quantity the market will buy, doing his best to choose the price which will result in a higher net revenue than other prices. In either case, the problem is essentially one of maximising the net revenue, which is a function of the level of output and the price.

We will begin to study this problem by considering the monopolist's costs. We shall suppose that he has some fixed overhead costs, which will not depend on the scale of production (unless he closes down altogether or expands so much that he needs a new factory), and that these overhead costs come to £5,000 a month.

There are also variable costs, depending on the scale of production which must be added to this sum of £5,000. We will suppose that for every x units of his product he has to pay for labour and materials a sum of £$(200x + x^2)$.* If production is low then the term in x^2 is unimportant, but as production increases the effect of squaring x increases. Thus, the average variable cost involved in producing 10

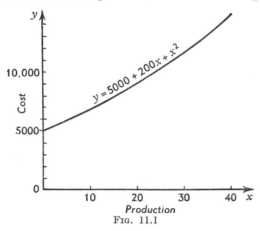

Fig. 11.1

units is £$(2000 + 100) \div 10 = £210$, but the average variable cost of producing 100 units is £$(20,000 + 10,000) \div 100 = £300$.

We have thus that the total cost of producing x units is given by

$$C = £(5,000 + 200x + x^2)$$

The approximate shape of the cost-curve is shown in Diagram 11.1. It is, of course, a parabola.

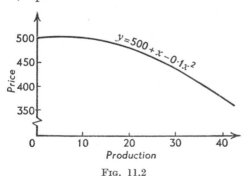

Fig. 11.2

Usually the greater the quantity sold the lower the price at which

* This sort of variable cost could, for example, arise if the basic cost of labour and material going into the production of x units is £$200x$, but an increasing demand for labour and raw material puts up the price by an amount £x as output increases.

G

it sells. We will suppose that in the case we are considering the quantity x is related to the price p by the equation

$$p = £(500 + x - 0\cdot1x^2)$$

In this case, if only one unit is produced it will sell for £500·9. If two units are produced each will sell for £501·6. Ten units would sell for £500 each, twenty for only £480 each, and fifty for only £300 each. The price curve is shown in Diagram 11.2.

The gross revenue obtained by selling the whole output x at a price p is given by

$$
\begin{aligned}
R &= \text{volume of sales} \times \quad \text{price} \\
&= \qquad x \qquad \times \qquad p \\
&= \qquad x \qquad \times (500 + x - 0\cdot1x^2) \\
&= 500x + x^2 - 0\cdot1x^3
\end{aligned}
$$

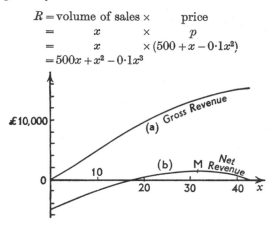

FIG. 11.3

The net revenue obtained by producing and selling a monthly amount x is therefore given by

$$
\begin{aligned}
P = R - C &= (500x + x^2 - 0\cdot1x^3) - (5,000 + 200x + x^2) \\
&= 300x - 0\cdot1x^3 - 5,000
\end{aligned}
$$

The curves corresponding to the gross and net revenue are shown in Diagram 11.3.

It will be noticed that we have expressed the net revenue entirely in terms of the volume of output. The equation

$$P = 300x - 0\cdot1x^3 - 5,000$$

has no p in it. This, of course, is because we have deliberately eliminated p by expressing it in terms of x. Later on we shall express P entirely in terms of p by eliminating x. For the moment, however, we shall content ourselves with noting that this can be done. Now that we have the net output in terms of the volume of production we shall consider the important question, " What output should the monopolist produce (and sell) in order to obtain a maximum net revenue? "

One way of answering this is to draw the Diagram 11.3 accurately and to read off the value of x corresponding to the point M at which the net revenue is highest.   But this is a tedious and inaccurate method.   There is a simpler method, involving differentiation, which often enables us to determine the answers to such questions as these very quickly indeed, and to which we now turn.   Before answering the problem, however, we should consider one or two general points whose relevance to this problem and to a wide range of other problems is immediately apparent.

2. Maxima, minima and points of inflexion

Consider Diagram 11.4, looking for the moment at only the upper part of it, which plots values of y against values of x.   Consider the point A.   We call this a *maximum point*, because, in a sense, y has a maximum value at this point.   It is true that the value of y is not higher than the value of y at the point F ; but it is a maximum in the sense that the value of y at A is higher than the value at a point on either side of A.

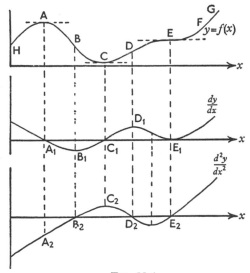

Fɪɢ. 11.4

Similarly we call C a *minimum point* because the value of y at C is lower than the value at a point on either side of C.

We should note that we do not call the points G and H maxima or minima, because although they are, respectively, the points associated with the highest and lowest values of y, it is possible (if not probable) that if we were to extend the graph a trifle in each direction then we would *immediately* come across higher and lower points. G and H are points that have been brought to our notice simply by our arbitrary decision to commence the graph at H and to terminate it at G. A and C are points which are fundamental to the shape of the graph. We call them *turning points*.

We shall consider the points B, D and E in a moment.

Now let us draw underneath this curve a second graph, on the same x-scale, but with the vertical scale measuring values of dy/dx instead of y. For instance the upper curve is such that at all points to the left of A it slopes upwards, so that the gradient is positive. This means that the value of dy/dx is positive for all values of x lying between O and A. But at A the curve is horizontal, and so the gradient (and the value of dy/dx) is zero. The part of the dy/dx curve corresponding to the section of the original curve lying between O and A will therefore be completely above the zero line, until it reaches the point corresponding to A when it will be zero. This is shown in the diagram.

Between A and C the curve slopes downwards, which means that the gradient (and therefore dy/dx) will be negative ; but it starts at a value of zero corresponding to the point A, and returns to a value of zero corresponding to the point C. The graph of dy/dx will therefore look something like the section $A_1 B_1 C_1$.

The curve is horizontal again at the point E, and so the gradient at this point is zero : but on both sides of E the curve slopes upwards. In fact, at all points to the right of C the value of dy/dx must be positive, except at the point E where it momentarily comes down to zero.

It is worthwhile to go over the preceding argument carefully, noting how the second graph gives, for every value of x the corresponding value of dy/dx.

In particular, we will notice that when y is a maximum, dy/dx is zero (see A and A_1) ; and that the same is true when y is a minimum (see C and C_1). It is also true at the point E, which we will consider shortly.

Let us now repeat the procedure. Let us draw a third graph which shows the slope of the second. A moment's thought will tell us that the graph is, in fact, a graph of values of the second derivative. For if we denote the values of dy/dx which we have drawn in the second graph by z, then the slope of this graph is going to be given by dz/dx which will be, as we saw in the last chapter,

$$\frac{d}{dx}\left(\frac{dy}{dx}\right) = \frac{d^2y}{dx^2}$$

By following the same sort of argument as before, we will find that at the points where dy/dx is a minimum (i.e., below B_1 and E_1), and at the points where it is a maximum (i.e., below D_1) then the values of d^2y/dx^2 are zero, and the third curve crosses the x-axis at these points.

Let us now study the whole diagram. Take the maximum A. We see by looking at the points A_1 and A_2 beneath it that dy/dx is zero and d^2y/dx^2 is negative (since A_2 is below the x-axis).

Take, now, the minimum C. We have that dy/dx is zero, and that d^2y/dx^2 is positive (since C_2 is above the x-axis).

Finally we may notice that at certain points, namely B, D and E, the values of d^2y/dx^2 are zero. We call such points, *points of inflexion*, for the reason which will be apparent in a moment. For the time being let us just note that at E, which we have just defined to be a point of inflexion, the value of dy/dx is also zero.

We may sum up our results thus :

(1) If $dy/dx = 0$, then y is at a maximum, minimum or point of inflexion.

(2) If $dy/dx = 0$, and $d^2y/dx^2 > 0$, then y is at a minimum.

(3) If $dy/dx = 0$, and $d^2y/dx^2 < 0$, then y is at a maximum.

(4) If $dy/dx = 0$, and $d^2y/dx^2 = 0$, then the above illustration indicates a point of inflexion. A more rigorous treatment would show that this need not be so. For example, if, in addition $d^3y/dx^3 = 0$ but $d^4y/dx^4 \neq 0$, there is a maximum if $d^4y/dx^4 < 0$, and a minimum if $d^4y/dx^4 > 0$. If, however, $d^3y/dx^3 \neq 0$ then there is a point of inflexion when $d^2y/dx^2 = 0$, even when $dy/dx \neq 0$. It must be admitted that these results have been demonstrated rather than proved ; but it is an adequate demonstration. A more rigorous treatment is given in R. G. D. Allen's *Mathematical Analysis for Economists*, pp. 179–95 and 459–61.

Example :

Consider
$$y = 20 + 24x - 2x^3$$
$$y' = 24 - 6x^2$$
$$y'' = -12x$$
$$y''' = -12$$

At the values of x which make $y' = 0$ there is a maximum, minimum, or point of inflexion.

These values are those which satisfy

$$24 - 6x^2 = 0$$
i.e., $x = +2 \text{ and } x = -2$

When $x = +2$, $y'' = -24$ and so, at $x = 2$ there is a maximum value of y.

When $x = -2$, $y'' = +24$ and so, at $x = -2$ there is á minimum value of y.

There is a point of inflexion where $y'' = 0$. This occurs where

$$-12x = 0$$

i.e.,
$$x = 0$$

At this value of x, $y''' = -12$. We therefore have that at $x = 0$

$$y'' = 0, \quad y''' \neq 0$$

and so there is a point of inflexion.

EXERCISE 11.1

1. Find the values of x for which the following functions have maxima, minima, and/or points of inflexion, distinguish between the cases, and then find the corresponding values of y. Examine the first few cases graphically to illustrate your results :

(i) $y = 40 + 3x - 2x^3$ (ii) $y = 3x^2 - 2x + 50$

(iii) $y = 40 + 3x - 2x^2 + \dfrac{x^3}{3}$ (iv) $y = 3x^3 - 4x^2 - x + 10$

(v) $y = \sin x$ (vi) $y = \sin 2x$

(vii) $y = \cos x$ (viii) $y = \tan x$

Before returning to the problem of the monopolist with which we began, let us deal with the point of inflexion. An easy way of

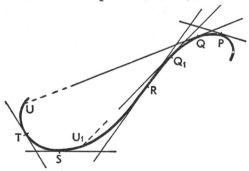

FIG. 11.5

defining this is to say quite simply that a point of inflexion is a point on the curve where the second derivative is zero. It is, however, useful to see more precisely what this means. Consider the curve shown in Diagram 11.5. The tangent at P is above the whole of the curve. The tangent at T is below the whole of it. The tangent at Q is not entirely above the curve, since it cuts it again at U, but it is above it in the sense that two points on the curve, very close to Q, one on either side of it, will lie below the tangent at Q. Similarly the tangent at S is below the curve in the sense that two points close to, and on either side of, S will lie above the tangent. If, however, we consider the point R, we see that the *tangent actually crosses the curve at that very same point*. A point on the one side of R will lie below the tangent ; a point on the other side will lie above it. This point, which is such that the tangent to the curve actually crosses the curve at the same point, is called a *point of inflexion*. It is the point on one side of which the tangents are above the curve, and on the other side below it. It is the point where the curvature of the curve changes from being convex to being concave. At this point d^2y/dx^2 is zero, although we have not proved this.

It is sometimes useful to look at it in yet another way. We have already seen that a tangent is the limit to which a chord tends as the two points move indefinitely close together on the curve. Now consider Q to be, in fact, two points which *have* moved together, turning a chord into a tangent. Now let us look at Q and U in the above diagram. Imagine Q sliding along the curve towards R. Then U would also slide towards R, as is seen by considering its position U_1 when Q reaches Q_1. All the time, the tangent at Q (in its successive positions) would be cutting the curve at U (in its successive positions). Eventually Q would reach R. At the same time U would reach R. The tangent at Q (now at R) would cut the curve at U (now at R). But Q itself is really made up of two coincident points ; and so R would be made up of three points . . . U and two from Q. Because of this we sometimes speak of the tangent at a point of inflexion as being a tangent of three point contact, whereas at other points a tangent has two point contact with the curve.

3. The monopolist again

We now return to our problem of the monopolist.

We found that if he produced and sold an amount x units per month then his net revenue would be given by

$$P = £(300x - 0\cdot 1x^3 - 5,000)$$

We are faced with the problem of choosing a value of x that will make this a maximum.

Now we know from the argument we have just pursued that for P to be a maximum the following conditions must hold :

(1) dP/dx must be zero

and   (2) d^2P/dx^2 must be negative

Let us begin by considering the first of these. To do so we must differentiate the above expression with respect to x getting

$$\frac{dP}{dx} = 300 - 0 \cdot 3x^2$$

If our first condition is to hold, this must be equal to zero. The first condition therefore reduces to finding the values of x that will allow the equation

$$300 - 0 \cdot 3x^2 = 0$$

to be true. This is a quadratic equation to which the solutions are given (see Chapter I) by

$$0 \cdot 3x^2 = 300$$
$$x^2 = 1000$$
$$x = \pm 10\sqrt{10}$$

Now clearly we are not concerned with negative outputs, and so we need only consider the solution $x = +10\sqrt{10}$. This tells us that if the manufacturer makes $10\sqrt{10}$ units a month then his net revenue will be at a maximum, a minimum or a point of inflexion. To decide which it is, we have to consider the second condition, concerning the sign of d^2P/dx^2. This means differentiating the value obtained above for dP/dx.

Since
$$\frac{dP}{dx} = 300 - 0 \cdot 3x^2$$

it follows that
$$\frac{d^2P}{dx^2} = -0 \cdot 6x$$

Now at the point where the first condition holds, $x = 10\sqrt{10}$.

Consequently
$$\frac{d^2P}{dx^2} < 0$$

This means that when $x = 10\sqrt{10}$ both conditions are satisfied, and therefore this output is such that the net revenue obtained is a maximum. Substitution in the equation will show us that, in fact, the net revenue will be about £1320.

It should be noted that this procedure yields a maximum in the sense of the argument we have pursued, namely a point like A. It does not preclude the possibility that there is a point such as F at which the net revenue would be greater . . . but if this is so there must be a point such as G at which it will be greater still, and so on ad infinitum. It is clear from a rough drawing of the profit diagram, or a cursory examination of the profit equation, that a very large output is bound to result in a loss : there is no point corresponding to F in this case.

Now let us return to the other approach—of selecting a selling price, and letting market conditions determine the quantity. We express P in terms of p rather than of x.

We know that

$$P = R - C$$
$$= px - (5000 + 200x + x^2)$$

We also have that

$$p = 500 + x - 0 \cdot 1x^2$$

This second equation can be written as

$$0 \cdot 1x^2 - x + (p - 500) = 0$$

which has solutions

$$x = \frac{1 \pm \sqrt{1 - 4(0 \cdot 1)(p - 500)}}{0 \cdot 2}$$
$$= \frac{1 \pm \sqrt{1 - 0 \cdot 4p + 200}}{0 \cdot 2}$$
$$= \frac{1 \pm \sqrt{201 - 0 \cdot 4p}}{0 \cdot 2}$$

It also enables us to write

$$0 \cdot 1x^2 = x + 500 - p$$

whence

$$x^2 = 10x + 5000 - 10p$$

Using these values of x and x^2 in our expression for P we have

$$P = px - (5000 + 200x + x^2)$$
$$= (p - 200)x - x^2 - 5000$$
$$= (p - 200)x - (10x + 5000 - 10p) - 5000$$
$$= (p - 210)x + 10p - 10{,}000$$
$$= (p - 210)\left(\frac{1 \pm \sqrt{201 - 0 \cdot 4p}}{0 \cdot 2}\right) + 10p - 10{,}000$$

For P to be a maximum, minimum or point of inflexion

$$\frac{dP}{dp} = 0$$

i.e.

$$-210\left(\frac{1 \pm \sqrt{201 - 0 \cdot 4p}}{0 \cdot 2}\right) + (p - 210)\left(\frac{\pm(-0 \cdot 4)}{2\sqrt{(201 - 0 \cdot 4p)} \times 0 \cdot 2}\right)$$
$$+ 10 = 0$$

This is a very messy expression, but it can be solved for p by using the method outlined on page 34. We have proceeded thus far partly to show the importance of eliminating the variable which will lead to the easiest solution but also for another reason.

We could have written

$$P = px - (5000 + 200x + x^2)$$

and, knowing that x is a function of p, we write

$$x = x(p)$$

$$\frac{dP}{dp} = x + p\frac{dx}{dp} - \left(200\frac{dx}{dp} + 2x\frac{dx}{dp}\right)$$

This is a procedure which some students may accept, but which others may prefer to see justified in Chapter XXIII. Basically, it hinges on our ability to write

$$\frac{d}{dp}(x^2) = \frac{d}{dx}(x^2) \cdot \frac{dx}{dp}$$

$$= 2x\frac{dx}{dp}\,.$$

If we accept this, then we have almost solved our problem, for we also know that

$$\frac{dx}{dp} = \frac{1}{\dfrac{dp}{dx}}$$

and since

$$p = 500 + x - 0{\cdot}1x^2$$

$$\frac{dp}{dx} = 1 - 0{\cdot}2x$$

$$\therefore \qquad \frac{dx}{dp} = \frac{1}{1 - 0{\cdot}2x}$$

Now $\dfrac{dP}{dp} = 0$ when

$$x + (p - 200 - 2x)\frac{dx}{dp} = 0$$

i.e.

$$x + \frac{(p - 200 - 2x)}{1 - 0{\cdot}2x} = 0$$

whence

$$x(1 - 0{\cdot}2x) + p - 200 - 2x = 0$$

We can now, if we wish, either insert

$$x = \frac{1 \pm \sqrt{201 - 0{\cdot}4p}}{0{\cdot}2}$$

in this equation and solve it for p, or we can write

$$p = 500 + x - 0 \cdot 1x^2$$

This will enable us to find the value of x which corresponds to that value of p which will make $\dfrac{dP}{dp} = 0$.

It would come from

$$x(1 - 0 \cdot 2x) + 500 + x - 0 \cdot 1x^2 - 200 - 2x = 0$$

i.e. $\qquad x - 0 \cdot 2x^2 + 500 + x - 0 \cdot 1x^2 - 200 - 2x = 0$

$$300 - 0 \cdot 3x^2 = 0$$

$$1000 - x^2 = 0$$

$$x^2 = 1000$$

$$x^2 = 10\sqrt{10}$$

This is the same value as before—when we sought the x which would make $\dfrac{dP}{dx} = 0$. The value of p corresponding to this is

$$500 + 10\sqrt{10} - 100$$

$$= 400 + 10\sqrt{10}$$

which has been obtained much more easily than if we had substituted

$$x = \frac{1 \pm \sqrt{201 - 0 \cdot 4p}}{0 \cdot 2}$$

There is often an advantage in working with a " wrong " variable until the last moment.

EXERCISE 11.2

1. By considering the derivatives, decide whether the demand curve $x = 4000 - 30p^2 + p^3$ is anywhere upward sloping (p = price).

2. Consider similarly the curve $x = 5000 - 25p^2 + p^3$.

3. A manufacturer with a monopoly in his product can produce a weekly output of x tons at a total cost of £$(0 \cdot 05x^2 + 2x + 300)$. If he sells at a price £p per ton the weekly demand will be $100 - 2p$ tons. Determine the price at which his net revenue will be a maximum.

4. The marginal theory of monopoly

It may be useful at this stage to apply these ideas to the problem of the monopolist in a more general sense. First of all let us note that if a producer finds that the cost of producing x units is C while the cost of producing $x + \Delta x$ units is $C + \Delta C$, then the ratio of the increment in cost to the increment in output is $\Delta C/\Delta x$ and this measures the *average* increase in cost per unit increase in output over the range of output x to $x + \Delta x$. The *average* rate of increase over this range tends to a *precise* rate of increase at the output x (just as an average speed over a distance tends to the speed at the point) as this range becomes smaller. In fact, it tends to a limit which is

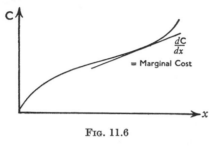

FIG. 11.6

denoted by $\dfrac{dC}{dx} = \underset{\Delta x \to 0}{\text{Limit}} \dfrac{\Delta C}{\Delta x}$ and it is this limit that we define to be the *Marginal Cost*, which is represented by the slope of the tangent to the cost curve, and measures the rate of increase of total cost with respect to a small increase in production.

We have similarly that if the revenue obtained from the sale of an output x is R, and this changes to $R + \Delta R$ when the output changes from x to $x + \Delta x$, then the *Marginal Revenue* is defined to be

$$\frac{dR}{dx}$$

Now suppose that we have a monopolist who decides that he will produce a certain quantity of the good concerned, leaving the price to be settled by the demand conditions. This means that the price will depend on, or be a function of, the output. Denoting the price by p we have that p is a function of x, which we write as

$$p = f(x)$$

The revenue, which we will denote by R, is also a function of x; and so is the production cost C. The former is, of course, the product of the output and the price, i.e.,

$$R = xp = xf(x)$$

The net revenue, defined to be the gross revenue R minus the production cost C, is to be maximised by choosing the proper output. If we denote the net revenue by N we have that

$$N = R - C$$

where R and C are functions of x, and we have to choose a value of x that will make

$$\frac{dN}{dx} = 0 \quad \text{and} \quad \frac{d^2N}{dx^2} < 0$$

The first of these conditions gives

$$\frac{dN}{dx} = \frac{dR}{dx} - \frac{dC}{dx} = 0$$

and this means that $dR/dx = dC/dx$, i.e., the *marginal revenue equals the marginal cost*. This alone does not guarantee a maximum net revenue; the same condition would hold for a minimum net revenue and at some points of inflexion. In order to be certain that we have a maximum we have also to have that

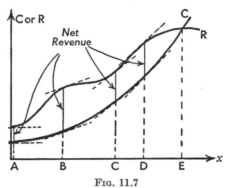

$$\frac{d^2N}{dx^2} = \frac{d^2R}{dx^2} - \frac{d^2C}{dx^2} < 0$$

which means that d^2R/dx^2 $< d^2C/dx^2$ showing that *the rate of increase of marginal revenue* (measured by $\dfrac{d^2R}{dx^2}$ which is $\dfrac{d}{dx}\left(\dfrac{dR}{dx}\right)$) *must be less than the rate of increase of marginal cost*. If, however, the sign of this in-

FIG. 11.7

equality is reversed, then the net revenue is a minimum; while if the two second derivatives are equal, the net revenue is at a point of inflexion. These three cases are shown in the Diagram 11.7. There is a minimum net revenue at A, a maximum at B and at D and another minimum at C. At E the net revenue is zero because total cost equals total gross revenue: but this is not formally a minimum. The points A, B, C and D occur where the tangents to the two curves are parallel (because the gradient $dR/dx =$ gradient dC/dx).

Analytically, we have that if the monopolist is going to fix his output letting the demand conditions determine the price, then since

$$R = xp$$

where p is a function of x the condition that

$$\frac{dR}{dx} = \frac{dC}{dx}$$

means that

$$\frac{dR}{dx} = p + x\frac{dp}{dx} = \frac{dC}{dx}$$

whence

$$x\frac{dp}{dx} = \frac{dC}{dx} - p$$

and so

$$x = \left(\frac{dC}{dx} - p \right) \frac{dx}{dp}$$

Now dC/dx measures the marginal production cost; and we know that $\frac{p}{x} \frac{dx}{dp}$ is the elasticity of the demand curve, while $\frac{dx}{dp}$ is the rate of change of demand with respect to price. It follows that the monopolist should fix his output in such a way that it is equal to the product of the difference between the marginal production cost and the price, and the rate of change of demand with respect to price. Under these conditions, the above equation shows that

$$p = \frac{p}{x} x = \frac{p}{x} \left(\frac{dC}{dx} - p \right) \frac{dx}{dp}$$

$$= \frac{p}{x} \frac{dx}{dp} \left(\frac{dC}{dx} - p \right)$$

i.e., the price, determined by the demand conditions, will be the product of the elasticity of demand and the difference between the marginal cost and the price itself.

In the same way we may consider the monopolist who fixes the price and lets the demand conditions determine the sales (and, therefore, the output). In this case we take output as a function of price.

$$x = f(p)$$

and

$$R = px = pf(p)$$

The first condition for a maximum net revenue gives $dR/dp = dC/dp$ and this means that

$$\frac{dR}{dp} = x + p \frac{dx}{dp} = \frac{dC}{dp}$$

This condition is just the same as before, for it may be written

$$x \frac{dp}{dx} + p \frac{dx}{dp} \frac{dp}{dx} = \frac{dC}{dp} \frac{dp}{dx}$$

whence

$$x \frac{dp}{dx} + p = \frac{dC}{dx}$$

which is our previous condition.

It follows that the monopolist who fixes his price, letting the demand conditions determine the output, achieves the same result as he who fixes the output, letting the demand conditions determine the price.

CHAPTER XII

INTEGRATION

Nature's true-born child, who sums his years
(Like me) with no arithmetic but tears.

BISHOP HENRY KING : *The Anniverse*

1. The area under a curve (1)

When we commenced our study of differentiation we used certain information about the distance travelled by a car in order to calculate the speed. We now consider this problem in an inverted form ; we are given the speed of the car at various moments and wish to determine the distance travelled.

Let us suppose that we know the speed of the car at every moment, and are able to draw a graph, as shown in Diagram 12.1 such that at a time Ot the speed is Ov. We see that the car was travelling at 30 m.p.h. at noon and at 60 m.p.h. an hour later.

Now consider the rectangle $OADC$. If the car had travelled at exactly 30 m.p.h. for one hour then the total distance travelled would have been

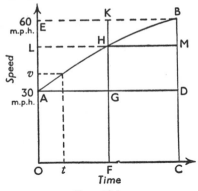

FIG. 12.1

$$30 \text{ m.p.h.} \times 1 \text{ hour}$$
$$= \quad OA \quad \times \quad OC$$

which is the area of the rectangle $OADC$. It may seem a little strange for an area to represent a distance, but we have to remember that we have let distances along the x-axis represent time, and distances along the y-axis represent speed, and it is because of this that an area, obtained by multiplying a speed (OA) by a time (OC), represents a distance.

Clearly in the case we are considering the car never travelled at less than 30 m.p.h., and most of the time was going more quickly than this, so that the actual distance travelled must be more than 30 miles—and consequently more than the area of $OADC$. Let us denote this distance actually travelled by S.

191

By a similar reasoning we have that the car travels less far than it would have travelled if it had been going at 60 m.p.h. the whole time. This means that the distance S is less than the distance represented by the area $OEBC$.

We can sum up our conclusions so far by writing

$$OEBC > S > OADC$$

i.e., $OEBC$ is greater than S which is greater than $OADC$.

Now let us take a point F, midway between O and C. The distance travelled in time OF is clearly more than that represented by the area $OAGF$ and less than that represented by the area $OLHF$.

The distance travelled in the time FC is more than that represented by the area $FHMC$ and less than that represented by the area $FKBC$.

If we add these two results we have that the distance S travelled in the time OC is more than that given by the sum of the areas $OAGF$ and $FHMC$, while it is less than that given by the sum of the areas $OLHF$ and $FKBC$.

i.e., $$OLHF + FKBC > S > OAGF + FHMC$$

We could similarly divide the time into four equal periods as in

FIG. 12.2

Diagram 12.2 and show that the distance travelled is less than the sum of the four larger rectangles

$$OQRN + NTHF \\ + FVWP + PYBC$$

and more than the sum of the four smaller rectangles

$$OASN + NRUF \\ + FHXP + PWZC$$

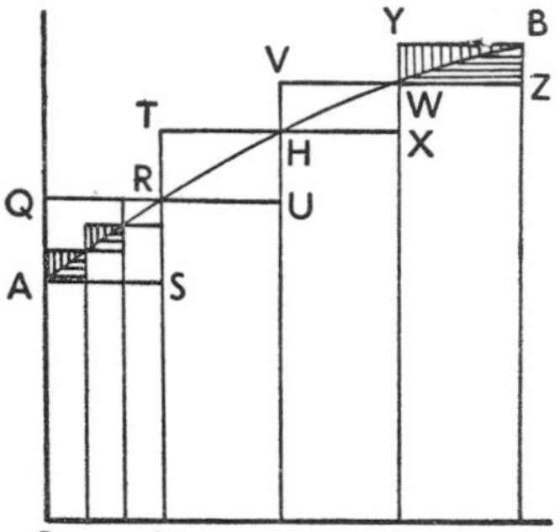

This process of subdivision could be carried on indefinitely, and each time we would find that the distance S would lie between the sum of the distances represented by the areas of the larger rectangles and the sum of the distances represented by the areas of the smaller rectangles.

It is clear that as the subdivision proceeds, the sum of the areas of the larger rectangles will get smaller and smaller until it becomes indistinguishable from the area under the curve. (This is shown quite clearly by the first two diagrams of this chapter.) Also the sum of the smaller areas will become larger and larger, tending once again to the area under the curve.

We therefore have a result which may be summarised thus :

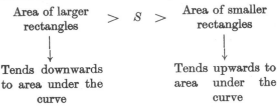

Area of larger rectangles $> S >$ Area of smaller rectangles

Tends downwards to area under the curve

Tends upwards to area under the curve

and it follows that S, the distance travelled, lying between two sums, one of which tends downwards to the area under the curve, and one

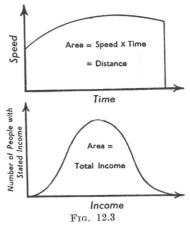

of which tends upwards to the same area under the curve, must, in fact, be identifiable with that area under the curve.

It follows that if the motion of the car is as shown in the diagram then the total distance travelled is equal to the area under the curve measured in the appropriate units.

Before proceeding to consider how we may find this area, we might note that, as is immediately obvious, the idea of the area under the curve has wide applications. For example, if we

FIG. 12.3

measure incomes along the x-axis, and along the y-axis the number of people with different incomes, then the area under the curve will be the total income of all the people, as demonstrated by Diagram 12.3.

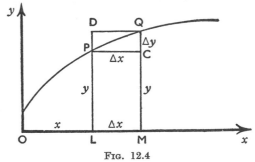

FIG. 12.4

Let us now consider the curve shown in Diagram 12.4 whose equation we will suppose to be

$$y = f(x)$$

We take a point P upon it, and let the co-ordinates of this point be (x, y). We take a close point Q, whose co-ordinates are $(x + \Delta x, y + \Delta y)$.

Now we have seen that the area under the curve (which we will still denote by S) is larger than the sum of the areas of the lower rectangles like $LPCM$.

Let us denote this element of area under the arc PQ by ΔS.

Then we have that ΔS is greater than the area of the rectangles $LPCM$. The area of this rectangle is $LP \times LM$ which is $y \, \Delta x$. We therefore have that

$$\Delta S > y \, \Delta x$$

Now we also have that the area under the arc PQ is less than the area of the rectangle $LDQM$, which is $LD \times LM$, which is $(y + \Delta y)\Delta x$, and so we have that

$$\Delta S < (y + \Delta y)\Delta x$$

If we combine these results we have that

$$(y + \Delta y)\Delta x > \Delta S > y \, \Delta x$$

This is a most important inequality. Let us divide it throughout by Δx. We obtain

$$y + \Delta y > \frac{\Delta S}{\Delta x} > y$$

Now if we continue the process of subdivision indefinitely, so that we have more and more rectangles, then $\Delta x \to 0$. Also, as examination of the diagram will clearly show, Δy tends to zero. Furthermore $\Delta S/\Delta x$ will tend to dS/dx. We may thus represent our position in the following terms :

$$
\begin{array}{ccc}
y + \Delta y & > \dfrac{\Delta S}{\Delta x} > & y \\
\downarrow & \downarrow & \downarrow \\
y & \dfrac{dS}{dx} & y
\end{array}
$$

which shows that dS/dx, being the limit of a quantity that lies between two quantities $(y + \Delta y)$ and y, each of which tends to the limit y, must equal y.

Suppose now that we wish to find the area under the curve between the two axes and the ordinate at the point x. This is the area within the heavy line in Diagram 12.5. Let us call this area S_x the subscript indicating that we are not concerned with the area to the right of the ordinate we have chosen.

Then the above argument shows that

$$\frac{dS_x}{dx} = y$$

and so, if we know y, the problem of finding the area S_x reduces to the problem of finding a function S_x which has y as its first deriva-

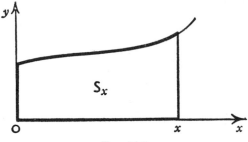

FIG. 12.5

tive. This was discovered independently by Newton (1642–1727) and Leibniz (1646–1716).

There is a complication. It arises out of the fact that the derivative of a constant is zero. Thus the derivative of $y = 8x^2$ is $16x$; but so is the derivative of $y = 8x^2 + 3$.

The way in which this complicates our task is best seen from a simple example. Consider the function $y = 3x$ which is, of course,

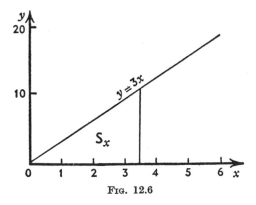

FIG. 12.6

represented by a straight line of gradient 3 as shown in Diagram 12.6. If we let the area to the left of the ordinate x be S_x then we have that S_x must be such that

$$\frac{dS_x}{dx} = 3x$$

Now we know that if S_x were equal to x^2 then dS_x/dx would be $2x$ which is two-thirds of what we want. It follows that if we put $S_x = 3x^2/2$ then $dS_x/dx = 3x$.

But the same would be true if we put $S_x = 3x^2/2 + C$, where C is any constant. There is thus a degree of indeterminacy in our answer.

To overcome this, we may note that the area between the y-axis and the particular value of x is given by $S_x = \dfrac{3x^2}{2} + C$, for all values of x. If we want the value of the area under the curve between $x = 7$ and $x = 4$ we may write:

Area between $x = 4$ and $x = 7$

equals (Area between $x = 0$ and $x = 7$)

less   (Area between $x = 0$ and $x = 4$)

which gives

$$S_7 - S_4$$

$$= \left[\frac{3(7)^2}{2} + C\right] - \left[\frac{3(4)^2}{2} + C\right]$$

$$= \frac{3(7)^2}{2} - \frac{3(4)^2}{2} = 49 \cdot 5$$

More generally we denote the area under the curve between $x = x_1$ and $x = x_2$ by $S_{x_2} - S_{x_1}$ which is sometimes written as

$$\left[S_x\right]_{x_1}^{x_2}$$

If we want the area under the curve $y = 3x$ between $x = 0$ and $x = 4$, then it is

$$\left[S_x\right]_0^4 = \left[\frac{3x^2}{2} + C\right]_0^4$$

$$= \left[\frac{3(4)^2}{2} + C\right] - \left[\frac{3(0)^2}{2} + C\right]$$

$$= \frac{3(4)^2}{2} - 0 = 24$$

The constant will always cancel out in a problem like this.

Exercise 12.1

1. Find the area under the curve $y = x^3/3$ between origin and $x = 2$
2. Find the area under the curve $y = x^4$ between origin and $x = 2$
3. Find the area under the curve $y = 3x^5$ between origin and $x = \frac{1}{2}$

2. Definite and indefinite integration

This process of finding the area under a curve is called *definite integration*. It rests upon our ability to find the function of x (here denoted by S_x) whose first derivative is y—whose value as a function of x defines the curve. Thus, to find the area under the curve $y = x^2$, between $x = 2$ and $x = 5$, we must first find the value of S_x such that $\frac{dS_x}{dx} = x^2$. This is a problem of *indefinite integration*. The answer will be a function of x involving an arbitrary constant C. It is important to retain C until it cancels out or has been evaluated.

It is fairly easy to see that in the simple case where $y = x^n$ then $S_x = \frac{x^{n+1}}{n+1} + C$, since in such a case $\frac{dS_x}{dx} = x^n$. But this is true only if n is not equal to -1. This is a special case to be considered later.

We now introduce a new notation, involving a symbol based on the old fashioned elongated S.

If we differentiate x^3 we obtain $3x^2$, and describe $3x^2$ as the derivative of x^3.

If we begin with $3x^2$ and try to find the function which has $3x^2$ as its derivative, we know that it is $x^3 + C$. We call $x^3 + C$ the *integral* of $3x^2$. We write it as

$$x^3 + C = \int 3x^2 \, dx$$

The elongated old-fashioned S reminds us of our S_{x_1} and our summation. The dx appears to remind us that $3x^2$ is derivative of $x^3 + C$ when we *differentiate it with respect to x*. The importance of this will become clear in later chapters.

We should note that we are not contradicting our earlier emphatic denials that dx has any meaning on its own. Here we have linked it quite firmly to the sign $\int$, which similarly has no meaning on its own. It is worthwhile to say a little more about this.

A few pages back we considered the area of a small rectangle of height y and breadth Δx, so that its area was $y \Delta x$. We summed these areas. Let us denote the sum by

$$\Sigma y \Delta x$$

where Σ, of course, means simply " the sum of all things like . . . ".

Then we have that, since the area S under the curve is given by the limit to which the sum of the areas of these rectangles tends as Δx tends to zero,

$$S = \underset{\Delta x \to 0}{\text{Limit}} \; \Sigma y \, \Delta x$$

To save labour we write this as

$$S = \int y \, dx$$

The symbol pair $\int \; dx$ enveloping y or some other function of x is to be taken as a convenient way of denoting the above limit.

EXERCISE 12.2

Show by differentiating the answers that the following results are true:

(i) $\displaystyle\int 3x^2 \, dx = x^3 + C$

(ii) $\displaystyle\int 7x^4 \, dx = \frac{7x^5}{5} + C$

(iii) $\displaystyle\int \cos x \, dx = \sin x + C$

If we are concerned with definite integration, then the limits between which we integrate (i.e., the values of x at the ordinates bounding the area concerned are written at either end of the integral sign, as below, the limit at the right-hand ordinate being at the top of the integral sign:

Example :

Area beneath curve $y = x^4$ and between ordinates $x = 2$ and $x = 6$ is

$$S = \underset{\Delta x \to 0}{\text{Limit}} \; \Sigma y \, \Delta x \text{ for } x \text{ between 2 and 6}$$

$$= \int_2^6 y \, dx$$

$$= \int_2^6 x^4 \, dx$$

$$= \left[\frac{x^5}{5} \right]_2^6$$

The area to the left of the ordinate $x = 6$ is $\dfrac{6^5}{5} + C$

The area to the left of the ordinate $x = 2$ is $\dfrac{2^5}{5} + C$

Therefore the areas between the ordinates is

$$\int_2^6 x^4 \, dx = \left[\frac{x^5}{5}\right]_2^6 = \left[\frac{6^5}{5} - \frac{2^5}{5}\right] = 1548 \cdot 8, \text{ as the } C\text{'s cancel.}$$

EXERCISE 12.3

1. Find the area under the curve $y = x^3/3$ between the ordinates $x = 1$ and $x = 2$

2. Find the area under the curve $y = x^4$ between the ordinates $x = 1$ and $x = 2$

3. Find the area under the curve $y = 3x^5$ between the ordinates $x = \frac{1}{2}$ and $x = 1$

EXERCISE 12.4

1. Find the area beneath the curve $y = x^5$ between $x = 2$ and $x = 3$

2. Find the area beneath the curve $y = \sqrt{x}$ between $x = 4$ and $x = 9$

3. Find the area beneath the curve $y = \sqrt[3]{x^2}$ between $x = 1$ and $x = 8$

3. The area under a curve (2)

Certain difficulties arise when we have to find areas bounded by a curve which lies wholly or partly beneath the x-axis.

Consider, for example

$$y = \int_0^6 (x - 3) \, dx$$

This should give the area shown in Diagram 12.7. Evaluation of the integral yields

$$y = \left[\frac{x^2}{2} - 3x\right]_0^6$$

$$= \left[\frac{36}{2} - 18\right] - \left[0\right]$$

$$= 0$$

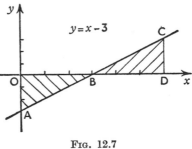

FIG. 12.7

The reason for this is that the area of $\triangle OAB$ by itself is obtained as

$$\int_0^3 (x - 3) \, dx$$

which is

$$\left[\frac{x^2}{2} - 3x\right]_0^3 = \frac{9}{2} - 9 = -\frac{9}{2}$$

while the area of $\triangle BCD$ appears as

$$\int_3^6 (x - 3) \, dx = \frac{9}{2}$$

When, therefore, the curve crosses the x-axis we must consider the problem in two separate parts unless we want the " negative area " below the axis to cancel (wholly or partly) the " positive area " above it.

There is no additional complication when the curve crosses the y-axis.

Example :

Consider the graph $y = x^2 - 4$ which is shown in Diagram 12.8. This cuts the x-axis when $x = 2$ and -2.

Suppose we wish to find the shaded area between $x = -3$ and $x = 4$. We proceed thus :

$$\text{Area} = \int_{-3}^{4} (x^2 - 4)\, dx$$

$$= \int_{-3}^{-2} (x^2 - 4)\, dx + \int_{-2}^{+2} (x^2 - 4)\, dx + \int_{2}^{4} (x^2 - 4)\, dx$$

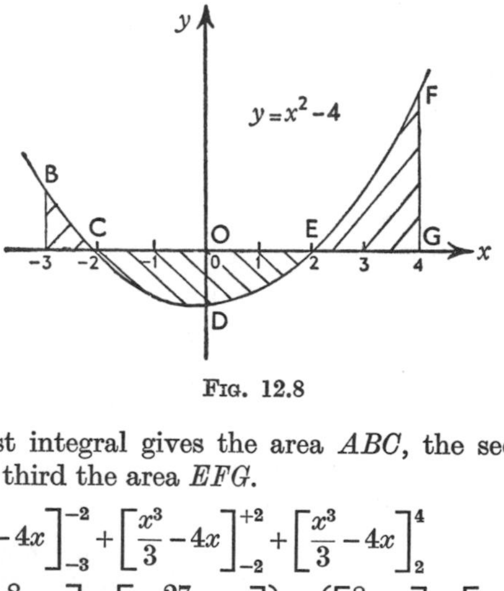

Fig. 12.8

where the first integral gives the area ABC, the second the area CDE and the third the area EFG.

$$\therefore \text{Area} = \left[\frac{x^3}{3} - 4x\right]_{-3}^{-2} + \left[\frac{x^3}{3} - 4x\right]_{-2}^{+2} + \left[\frac{x^3}{3} - 4x\right]_{2}^{4}$$

$$= \left\{\left[\frac{-8}{3} + 8\right] - \left[\frac{-27}{3} + 12\right]\right\} + \left\{\left[\frac{8}{3} - 8\right] - \left[\frac{-8}{3} + 8\right]\right\}$$
$$\quad\quad\quad\quad (ABC) \quad\quad\quad\quad\quad\quad\quad\quad\quad (CDE)$$

$$+ \left\{\left[\frac{64}{3} - 16\right] - \left[\frac{8}{3} - 8\right]\right\}$$
$$(EFG)$$

$$= \quad \left(\frac{7}{3}\right) \quad + \quad \left(-\frac{32}{3}\right) \quad + \quad \left(\frac{32}{3}\right)$$
$$\quad (ABC) \quad\quad\quad (CDE) \quad\quad\quad (EFG)$$

We now have a choice. If we wish to estimate the actual area, independently of sign, then it is

$$\tfrac{7}{3} + \tfrac{32}{3} + \tfrac{32}{3} = \tfrac{71}{3}$$

If, however, the graph indicates the rate of accumulation of profit over time, and is such that when the curve is beneath the x-axis there is a loss, then the area ABC represents a profit of $\tfrac{7}{3}$, the area CDE a loss of $\tfrac{32}{3}$, and the area EFG a profit of $\tfrac{32}{3}$. The net result is

$$\tfrac{7}{3} - \tfrac{32}{3} + \tfrac{32}{3}$$

which is a profit of $\tfrac{7}{3}$.

This result would have been obtained directly as

$$\int_{-3}^{4} (x^2 - 4)\, dx = \left[\frac{x^3}{3} - 4x \right]_{-3}^{4}$$

$$= \left[\frac{64}{3} - 16 \right] - \left[-\frac{27}{3} + 12 \right]$$

$$= \frac{16}{3} - \frac{9}{3}$$

$$= \frac{7}{3}$$

It is necessary to split up the process of integration only when we wish to ignore the sign of an area.

EXERCISE 12.5

1. Find the areas between the x-axis and the following curves between the ordinates stated (a) taking sign into account, and (b) ignoring sign :

(i) $y = x - 5$ between $x = -6$ and $x = 10$
(ii) $y = x^2 - 4x$ between $x = -6$ and $x = 6$
(iii) $y = \sin x$ between $x = 0$ and $x = \pi$
(iv) $y = \sin x$ between $x = 0$ and $x = 2\pi$
(v) $y = x^3$ between $x = -2$ and $x = 2$
(vi) $y = x^3 - x^2$ between $x = -2$ and $x = 2$

4. Some standard results

Although we have spent some time on the ideas behind integration the actual methods that are used for the evaluation of integrals have been hardly touched. We know that if

$$y = x^n$$

then

$$\frac{dy}{dx} = nx^{n-1}$$

and that if
$$y = x^n$$

then
$$\int y \, dx = \frac{x^{n+1}}{n+1} + C \qquad \text{except for } n = -1$$

and generally the process of integration may be regarded as the reverse of differentiation, in the sense that if p is the derivative of q then $q + C$ is the integral of p. But there are occasions when it is rather difficult to apply this rule. Some of these occasions are considered in the next few chapters. Meanwhile, however, the following results may be noticed. They are all obtained by reversing the findings tabulated in the chapters on differentiation, and can easily be checked by differentiating the integral in the right-hand column to see if we get the value in the left-hand column.

y	$\int y \, dx$
x	$\dfrac{x^2}{2} + C$
ax^n	$\dfrac{ax^{n+1}}{n+1} + C$
$4x^7$	$\dfrac{4x^8}{8} + C = \dfrac{x^8}{2} + C$
$(ax+b)^n$	$\dfrac{(ax+b)^{n+1}}{(n+1)a} + C$
	(to check this result consider the result on page 169)
$\cos x$	$\sin x + C$
$\sin x$	$-\cos x + C$

EXERCISE 12.6

1. The rate of investment in a certain economy is
$$I = 1000 + 80t + t^2$$

where I is measured in £m. and t is measured in months from $t=0$. Accepting that there may be an approximation if t is treated as a continuous variable, use the fact that capital stock K is such that

$$I = \frac{dK}{dt}$$

to evaluate the total addition to capital from $t=6$ to $t=12$. Then, refusing to accept this approximation, evaluate the total addition during this period by using formulae for the Σt and Σt^2 presented in Chapter II. Express the difference between these two answers as

a percentage of the second (" accurate ") answer. Repeat the whole
procedure for the time interval from $t = 6$ to $t = 18$, and compare
your results. Finally, consider whether the " accurate " answer is
likely to be really any more accurate than that obtained by considering t to be a continuous variable.

CHAPTER XIII

LOGARITHMS

By twos and threes they wandered off to rest.

CHAUCER : *The Squire's Tale*

We now have to spend a few chapters introducing some very important results. To do so will take us into some rather abstract fields of thought, which may appear to be very far removed from the problems of economics : but a familiarity with them is essential if we are to get very far, and certainly if we are to understand economics dealing with rates of growth or cycle theory.

We begin by saying a word or two about logarithms. The purpose of this chapter is not to instruct the reader in the mechanical processes of logarithmic multiplication and division, but to pave the way for more advanced ideas by giving an account of the underlying algebra.

Let us consider the following table of powers of 2.

n	2^n
0	1
1	2
2	4
3	8
4	16
5	32
6	64
7	128
8	256
9	512
10	1024
11	2048
12	4096
13	8192
14	16384
15	32768
16	65536
17	131072
18	262144
19	524288
20	1048576
21	2097152

We now know (from Chapter I) that $a^m \times a^n = a^{m+n}$ and so we have that

$$2^m \times 2^n = 2^{m+n}$$

which may be easily verified from the above table, since $2^3 = 8$ and $2^4 = 16$, while $2^7 = 128 = 8 \times 16$.

Suppose that we wish to multiply 512 by 4096, and that we have the above table in front of us. We could proceed as follows. From the table we see that $512 = 2^9$ while $4096 = 2^{12}$. The multiplication can therefore be written as

$$
\begin{aligned}
512 \times 4096 &= 2^9 \times 2^{12} \\
&= 2^{(9+12)} \\
&= 2^{21} \\
&= 2{,}097{,}152
\end{aligned}
$$

from the last line in the table. A tedious multiplication has been replaced by a simple addition.

This short-circuiting of work is possible, of course, simply because the numbers we wished to multiply are powers of 2. The table would be of little use if we wished to multiply 73 by 91. On the other hand, we have the opportunity to make the table more detailed. The square root of 2, which we can write as $2^{\frac{1}{2}}$ or even $2^{0 \cdot 5}$ is approximately 1·414. By virtue of this, the following results may be obtained :

$$2^{0 \cdot 5} = 1 \cdot 414 \; ; \quad 2^{1 \cdot 5} = 2^{1 \cdot 0} \times 2^{0 \cdot 5} = 2 \cdot 828 \; ; \quad 2^{2 \cdot 5} = 2^2 \times 2^{0 \cdot 5} = 5 \cdot 656 \; ;$$
$$2^{3 \cdot 5} = 11 \cdot 312 \; ; \quad 2^{4 \cdot 5} = 22 \cdot 624 \; ; \quad 2^{5 \cdot 5} = 45 \cdot 248 \; ; \quad 2^{6 \cdot 5} = 90 \cdot 576.$$

We have also that the cube root of 2, which may be written as $2^{1/3}$ or $2^{0 \cdot 333}$ is approximately equal to 1·2599, and so we have

$$2^{0 \cdot 333} = 1 \cdot 2599 \; ; \quad 2^{1 \cdot 333} = 2 \cdot 5198 \; ; \quad 2^{2 \cdot 333} = 5 \cdot 0396 \; ; \quad 2^{3 \cdot 333} = 10 \cdot 0792$$
$$2^{4 \cdot 333} = 20 \cdot 1584 \; ; \quad 2^{5 \cdot 333} = 40 \cdot 3168 \; ; \quad 2^{6 \cdot 333} = 80 \cdot 6336, \text{ etc.}$$

The fourth root of two, which may be put as $2^{0 \cdot 25}$ is approximately 1·189, and so

$$2^{0 \cdot 25} = 1 \cdot 189 \; ; \quad 2^{1 \cdot 25} = 2 \cdot 378 \; ; \quad 2^{2 \cdot 25} = 4 \cdot 756 \; ; \quad 2^{3 \cdot 25} = 9 \cdot 512$$
$$2^{4 \cdot 25} = 19 \cdot 024 \; ; \quad 2^{5 \cdot 25} = 38 \cdot 048 \; ; \quad 2^{6 \cdot 25} = 76 \cdot 096, \text{ etc.}$$

By proceeding in this way we could build up a table as follows ›

n	2^n
0·000	1·000
0·250	1·189
0·333	1·260
0·500	1·414
0·583 = 0·250 + 0·333	1·498 = 1·189 × 1·260
0·666 = 0·333 + 0·333	1·588 = 1·260 × 1·260
0·750 = 0·250 + 0·500	1·681 = 1·189 × 1·414
0·833 = 0·500 + 0·333	1·782 = 1·414 × 1·260
0·916 = 0·666 + 0·250	1·888 = 1·588 × 1·189
1·000	2·000
1·250	2·378
1·333	2·520
1·500	2·828
2·000	4·000
2·250	4·756
2·333	5·040
2·500	5·656
3·000	8·000
3·250	9·512
3·333	10·079
3·500	11·312
4·000	16·000

where, of course, the whole table could be quite as detailed as it is between 0·000 and 1·000 or even more so.

If we wish to multiply 1·189 by 9·512, we can express the terms as

$$1·189 \times 9·512 = 2^{0·250} \times 2^{3·250}$$
$$= 2^{3·500} = 11·312$$

Furthermore, we could multiply 118·9 by 0·9512 by using exactly the same items of the table, but noting that the multiplicand (118·9) is $100 \times 2^{0·250}$ while the multiplier (9·512) is $0·1 \times 2^{3·250}$ and so the product is $100 \times 0·1 \times 2^{0·250} \times 2^{3·250} = 10 \times 11·312 = 113·12$.

Now we could, if we wanted to, compile a table of powers of two in such a way that the *right*-hand side consisted of all the entries 1·000, 1·001, 1·002, 1·003, etc. To do this would be very tedious. It would mean taking hundreds of very small roots of two and building up the table very slowly indeed. But it could be done.

We could then multiply (say) 1·478 by 6·932, or any other pair of numbers, by proceeding as above.

As we have a numerical scheme based on powers of ten, it is more convenient to use a table of these powers, some of which are given below :

n	10^n
0·000	1·000
0·100	1·259
0·200	1·585
0·250	1·778
0·300	1·995
0·301	2·000
0·333	2·153
0·400	2·512
0·477	3·000
0·500	3·162
0·583	3·831
0·600	3·981
0·602	4·000
0·650	4·467
0·666	4·641
0·683	4·820
0·699	5·000
0·700	5·012
0·750	5·624
0·778	6·000
0·783	6·067
0·800	6·310
0·845	7·000
0·900	7·943
0·903	8·000
0·933	8·570
0·954	9·000
1·000	10·000
2·000	100·000
3·000	1000·000

This table can be derived in exactly the same way as the last table, although it is best obtained by more advanced and less tedious methods.

Now suppose that we wish to multiply 1·778 by 4·820. We see from this table that $1·778 = 10^{0·25}$ and that $4·820 = 10^{0·683}$ which shows us that the product is $10^{0·25} \times 10^{0·683} = 10^{0·933} = 8·570$.

When we use logarithms this is exactly what we do. Instead of writing it out as above, we say

$$\log 1{\cdot}778 = 0{\cdot}2500 \quad \text{(instead of } 1{\cdot}778 = 10^{0{\cdot}2500})$$
$$\log 4{\cdot}820 = 0{\cdot}6833 \quad \text{(instead of } 4{\cdot}820 = 10^{0{\cdot}6833})$$

$$0{\cdot}9333 \quad \text{(instead of adding the powers of 10, getting } 10^{0{\cdot}9333})$$

and then we find the number whose " antilogarithm " is $0{\cdot}9333$ (instead of finding the number which is $10^{0{\cdot}9333}$).

In fact, our ordinary logarithm tables are simply tables of powers of ten. When we say that " $\log_{10} 3 = 0{\cdot}4771$ " we mean quite simply that $10^{0{\cdot}4771} = 3$.

The student will probably recall the rules for using logarithms. If he does not then he should obtain a book on elementary algebra and get plenty of practice. It may be useful to note the following.

$$\log_{10} 3 = 0{\cdot}4771 \text{ means } 10^{0{\cdot}4771} = 3$$

Since $30 = 3 \times 10$, this means that

$$30 = 10^{0{\cdot}4771} \times 10 = 10^{1{\cdot}4771}$$

and so, $\log_{10} 30 = 1{\cdot}4771$

Similarly, $300 = 3 \times 10^2 = 10^{2{\cdot}4771}$

and so $\log_{10} 300 = 2{\cdot}4771$

These and similar results are summarised below in a way which will repay study

$$\log_{10} 3000 = 3{\cdot}4771$$
$$\log_{10} 300 = 2{\cdot}4771 = 3{\cdot}4771 - 1$$
$$\log_{10} 30 = 1{\cdot}4771 = 2{\cdot}4771 - 1$$
$$\log_{10} 3 = 0{\cdot}4771 = 1{\cdot}4771 - 1$$
$$\log_{10} 0{\cdot}3 = \qquad = 0{\cdot}4771 - 1$$

which we write as

$$\bar{1}{\cdot}4771$$

$$\log_{10} 0{\cdot}03 = \bar{1}{\cdot}4771 - 1$$
$$= \bar{2}{\cdot}4771$$

The meaning of $\bar{3}{\cdot}4771$ is $-3 + 0{\cdot}4771$. We *could* write

$$\log_{10} 0{\cdot}3 = 0{\cdot}4771 - 1$$
$$= -0{\cdot}5229$$

but then the relationship between $\log_{10} 0 \cdot 3$ and $\log_{10} 3$ or $\log_{10} 30$ would not be evident. To preserve the decimal part, $0 \cdot 4771$, which will facilitate later calculations, we write

$$\log_{10} 0 \cdot 3 = 0 \cdot 4771 - 1$$

which may conveniently be re-written as

$$\bar{1} \cdot 4771$$

where the bar indicates that the figure before the decimal point is to be subtracted

Note that

(i) $2 \cdot 6418 + \bar{1} \cdot 3121 = 1 \cdot 9539$

(ii) $2 \cdot 6418 + \bar{1} \cdot 4121 = 2 \cdot 0539$

(iii) $\frac{1}{2}(\bar{1} \cdot 6424) = \frac{1}{2}(\bar{2} + 1 \cdot 6424)$

$\qquad\qquad = \bar{1} + 0 \cdot 8212$

$\qquad\qquad = \bar{1} \cdot 8212$

Possibly the reader is slightly puzzled by the fact that we write " $\log_{10} 3$ " whereas he is used to writing simply " $\log 3$ ". The only reason for our doing this is that we wish to remind ourselves that the logarithms we are using are powers of 10, and not of 2 or 3 or anything else. Another way of saying this is to say that our logarithms are " to the base 10 ". Later on we shall see that it is often convenient to use logarithms which are powers of a numerical constant whose approximate value is $2 \cdot 178$, and that these logarithms, with this rather peculiar looking base, have certain extremely important properties. The logarithms to the base 10 are sometimes called " Common Logarithms " and the first table of them was computed by Henry Briggs (1561–1631). Logarithms were invented by John Napier (1550–1617).

It is convenient to sum up the above discussion by noting that

if $\qquad\qquad\qquad y = \log_a x \qquad$ (e.g. $0 \cdot 4771 = \log_{10} 3$

then $\qquad\qquad\qquad a^y = x \qquad\qquad\qquad 10^{0 \cdot 4771} = 3$)

We shall also have cause to use the important relationship that

$$\log_b x = \frac{\log_a x}{\log_a b}$$

e.g.,

$$\log_{16} 256 = \frac{\log_2 256}{\log_2 16}$$

H

as may be checked by noting that, since $256 = 16^2$, $\log_{16} 256 = 2$, while, since $256 = 2^8$, $\log_2 256 = 8$, and $\log_2 16 = 4$.

This is easily proved:

Write $\qquad\qquad \log_a x = A, \quad \log_b x = B, \quad \log_a b = C$

Then $\qquad\qquad a^A = x, \qquad b^B = x, \qquad a^C = b$

Thus $\qquad\qquad a^{CB} = (a^C)^B = b^B = x = a^A$

$\therefore \qquad\qquad CB = A$

whence $\qquad (\log_a b)(\log_b x) = \log_a x$

which proves our result.

CHAPTER XIV

EXPONENTIALS

" The appearance of that letter had nearly
broke my heart, but I still trusted that facts
wou'd stand their ground— "
ADMIRAL LORD COLLINGWOOD : *Letter*, June 30th, 1794

1. Introduction

In this chapter we introduce another very important result. It is related to the subject of the last chapter, and no basically new ideas are involved. We begin by noting a difficulty that arises out of our work in integration, and then we notice that a certain function (which we introduce by considering compound interest) has certain properties which provide the answer to this problem, and introduce us to a whole field of possibilities. To do all this rigorously would involve us in some very complicated mathematics. To avoid this, we shall simply indicate the lines of the proof in many places, and at times even make statements without offering any proof at all. This is essentially a case where it is the result that matters as far as we are concerned.

2. A problem in integration

Consider first of all the following table showing some of the results of simple differentiation :

y	$\dfrac{dy}{dx}$
$\dfrac{x^3}{3}$	x^2
$\dfrac{x^2}{2}$	$x = x^1$
x	$1 = x^0$
?	$\dfrac{1}{x} = x^{-1}$
$-x^{-1} = -\dfrac{1}{x}$	$\dfrac{1}{x^2} = x^{-2}$
$-\dfrac{x^{-2}}{2} = -\dfrac{1}{2x^2}$	$\dfrac{1}{x^3} = x^{-3}$

We notice that the table is incomplete : and yet it seems reasonable that there should be a function such that its derivative is $1/x$. There is a natural temptation to write this as x^{-1} and then to attempt an integration in the ordinary way. After all, in order to integrate $1/x^2$ we write it as x^{-2} and then proceed

$$\int \frac{1}{x^2} \, dx = \int x^{-2} \, dx = \frac{x^{-2+1}}{-1} = -x^{-1} = -\frac{1}{x}$$

using the rule that

$$\int x^n \, dx = \frac{x^{n+1}}{n+1}$$

But if we attempt this procedure in the case we are considering we get

$$\int \frac{1}{x} \, dx = \int x^{-1} \, dx = \frac{x^{-1+1}}{-1+1} = \frac{x^0}{0} = \frac{1}{0} \text{ which is } \textit{indeterminate} \text{ infinitely great.}$$

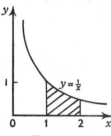

Now this is clearly wrong. Let us look at Diagram 14.1 which shows the graph of this function. If we want to find the area between the ordinates $x=1$ and $x=2$ we have that it is, according to the above analysis,

$$\int_1^2 \frac{1}{x} \, dx = \left[\frac{x^0}{0} \right]_1^2 = \frac{1}{0} - \frac{1}{0}$$

FIG. 14.1

and whatever value we give to this result it is clearly going to be difficult to identify it with the area shown in the diagram.

The truth is that here is a case where our rule for integration breaks down, and a more rigorous derivation of the rule would have shown it. Because of that we are still faced with the problem, what is the function that has $1/x$ as its first derivative? We shall now show that a certain function has this property. We turn to a problem in compound interest, which received considerable attention in the seventeenth and eighteenth centuries.

3. A problem in compound interest

A man has £1 which he is prepared to lend for a period of one year. He receives several different offers of interest, which we consider below.

(1) Simple interest at the rate of 100% payable at the end of the year. Money will amount to £2 by the end of year.

(2) Compound interest at the rate of 100% per annum, payable at the end of each six months. At end of first six months money amounts to £1·5. Interest is then paid on *this* sum during the second half year, bringing the accumulated sum to £2·25.

(3) Compound interest at the rate of 100% per annum, payable at the end of every four months. At end of first four months the money amounts to £1·333. This receives interest during the second period, making a total of £1·777. And during the third period the whole of this receives interest at the rate of 100% per annum, making a total of £2·369.

Let us look at this a little more generally, starting with the second case.

Initial amount (principal)	$= £1$
Amount at end of first half-year	$= £(1 + 0·5)$
Amount at end of second half-year	$= £(1 + 0·5) + 0·5(1 + 0·5)$
	$= £(1 + 0·5)^2 = £2·25$

Now take the third case.

Principal	$= £1$
Amount at end of first third-year	$= £(1 + 0·333)$
Amount at end of second third-year	$= £(1 + 0·333)$
	$\qquad + 0·333(1 + 0·333)$
	$= £(1·333)^2$
Amount at end of third third-year	$= £(1·333)^2 + 0·333(1·333)^2$
	$= £(1·333)^3 = £2·369$

Similarly if the interest were payable on four dates the amount would reach $£(1·25)^4$ while if it were due on five dates it would reach $£(1·20)^5$. More generally, if the interest were payable on n evenly spaced dates throughout the year, the amount at the end of the year would be

$$£ \left(1 + \frac{1}{n} \right)^n$$

The table below shows some of these values.

Number of times that interest is payable during the year.	Amount to which £1 accumulates during one year.
1	2·00
2	2·25
3	2·37
4	2·44
5	2·49
6	2·52
7	2·55
8	2·56
9	2·58
10	2·59
100	2·70
1,000	2·717
10,000	2·718

We can see quite clearly from the way in which this table is behaving that the man would never succeed in reaching £3 if he invested his money in this manner, no matter how many times in the year the compound interest might be added to the principal. In fact he would never reach as much as £2·719. It can be shown, by a method that we shall not consider, that this sequence, in the right-hand column, tends to a definite limit, which is approximately 2·71828. This limit is denoted by e, the symbol being introduced by Euler (c. 1727).

Definition :

$$e = \operatorname*{Limit}_{n \to \infty} \left(1 + \frac{1}{n}\right)^n = \text{(approximately)} \ 2\cdot71828$$

4. Some properties of e

Having introduced e, we must consider a few of its properties. It can be shown, by using the Binomial Theorem and a few theorems about limits which we have not discussed that

$$e = \operatorname*{Limit}_{n \to \infty} \left(1 + \frac{1}{n}\right)^n = 1 + \frac{1}{1!} + \frac{1}{2!} + \frac{1}{3!} + \dots, \text{ad infinitum}$$

(where, of course, $3! = 3 \times 2 \times 1$, etc.).

The terms of this series decrease rapidly and because of this we are able to use the series to calculate the value of the limit e correct to as many decimal places as we please. Thus, if we want it correct to seven decimal places we can proceed as follows :

$$1 = 1\cdot000\ 000\ 000\ 0$$

$$\frac{1}{1!} = 1\cdot000\ 000\ 000\ 0$$

$$\frac{1}{2!} = \frac{1}{2}\frac{1}{1!} = 0\cdot500\ 000\ 000\ 0$$

$$\frac{1}{3!} = \frac{1}{3}\frac{1}{2!} = 0\cdot166\ 666\ 666\ 6$$

$$\frac{1}{4!} = \frac{1}{4}\frac{1}{3!} = 0\cdot041\ 666\ 666\ 6$$

$$\frac{1}{5!} = 0\cdot008\ 333\ 333\ 3$$

$$\frac{1}{6!} = 0\cdot001\ 388\ 888\ 8$$

$$\frac{1}{7!} = 0\cdot000\ 198\ 412\ 6$$

$$\frac{1}{8!} = 0 \cdot 000\ 024\ 801\ 5$$

$$\frac{1}{9!} = 0 \cdot 000\ 002\ 755\ 7$$

$$\frac{1}{10!} = 0 \cdot 000\ 000\ 275\ 5$$

$$\frac{1}{11!} = 0 \cdot 000\ 000\ 025\ 0$$

$$\frac{1}{12!} = 0 \cdot 000\ 000\ 002\ 0$$

$$2 \cdot 718\ 281\ 827\ 6$$

The last couple of digits are not very accurate because of the incomplete division : but the first eight decimal places are quite reliable, and it is clear that however many more terms we may take, their sum will not be so great as to affect the seventh decimal figure. We can therefore say that, to seven decimal places, the value of e is given by $2 \cdot 718\ 281\ 8$.

Before relating this constant to the question with which we began this chapter, we must note a few more of its properties.

Let us return for a moment to the compound interest question. Suppose that the lender is paid compound interest at the rate of $100x$ per cent, instead of at 100 per cent. If, for example, $x = 2$ and the man waits for the end of the year, receiving interest only once, then his money will total £3 (instead of £2 which it was in the previous case when x was 1). More generally, the formula giving the amount to which one pound will accumulate by the end of the year if the compound interest is paid n times during the year will change from

$$£\left(1 + \frac{1}{n}\right)^n \quad \text{to} \quad £\left(1 + \frac{x}{n}\right)^n$$

Now we have seen that the first of these expressions tends to the limit e as n becomes indefinitely great. It is natural to ask whether the second expression tends to a limit, and, if so, to try to find this.

Let us write $\dfrac{x}{n} = \dfrac{1}{m}$ so that $n = mx$. Then we have that

$$\left(1 + \frac{x}{n}\right)^n = \left(1 + \frac{1}{m}\right)^{mx} = \left[\left(1 + \frac{1}{m}\right)^m\right]^x$$

Now we know that the limit to which $\left(1 + \dfrac{1}{n}\right)^n$ tends as n tends to infinity is e. Consequently the expression inside the square bracket tends to e. The fact that we have written m in one case and n in another cannot make any difference; all that matters is that m and/or n should tend to infinity.

This suggests (but it does not prove) that

$$\underset{n\to\infty}{\text{Limit}}\left(1+\frac{x}{n}\right)^n = \underset{m\to\infty}{\text{Limit}}\left[\left(1+\frac{1}{m}\right)^m\right]^x$$
$$= \left[\underset{m\to\infty}{\text{Limit}}\left(1+\frac{1}{m}\right)^m\right]^x$$
$$= e^x$$

It can be shown by more rigorous methods that this is true.

Furthermore, expansion of the left-hand side by the Binomial Theorem will yield the series

$$\underset{n\to\infty}{\text{Limit}}\left(1+\frac{x}{n}\right)^n = 1+\frac{x}{1!}+\frac{x^2}{2!}+\frac{x^3}{3!}+\dots$$

and so we have the very important result that while

$$e = 1+\frac{1}{1!}+\frac{1}{2!}+\frac{1}{3!}+\dots$$
$$e^x = 1+\frac{x}{1!}+\frac{x^2}{2!}+\frac{x^3}{3!}+\dots$$

5. The derivatives of e^x and $\log_e x$

We are at last in a position to deal with the question with which this chapter began. e is the limit to which a certain expression tends as n tends to infinity. It can be expressed as the sum of an infinite series, in which the terms get smaller and smaller, tending very quickly to zero (i.e., it is a highly convergent series), enabling us to obtain an approximate value of e which will be as accurate as we care to make it. Even more important to us at present, however, is that e^x can be expressed as the sum of an infinite series and that this series also converges to a limit.

Now e^x is, of course, a function of x. If we draw a graph (as in Diagram 14.2) of e^x we see that it increases pretty quickly as x increases. It is natural to ask at what rate it increases with respect to x. This can be answered by differentiating e^x with respect to x. Here there is a slight problem. We have differentiated x raised to some power; but we have never differentiated a constant raised to the power x, which is very different. We must tackle it from first principles.

Consider a point (x, y) on the curve whose equation is $y = e^x$. Let us take a close point whose co-ordinates are $(x + \varDelta x, y + \varDelta y)$. Then, since this is a point on the curve, we have that

$$y + \varDelta y = e^{x + \varDelta x}$$

and so

$$\varDelta y = e^{x + \varDelta x} - y$$
$$= e^{x + \varDelta x} - e^x$$
$$= e^x (e^{\varDelta x} - 1)$$

whence

$$\frac{\varDelta y}{\varDelta x} = \frac{e^x (e^{\varDelta x} - 1)}{\varDelta x}$$

$$= \frac{e^x \left(1 + \dfrac{\varDelta x}{1!} + \dfrac{(\varDelta x)^2}{2!} + \dfrac{(\varDelta x)^3}{3!} + \ldots - 1 \right)}{\varDelta x}$$

$$= \frac{e^x \left(\dfrac{\varDelta x}{1!} + \dfrac{(\varDelta x)^2}{2!} + \dfrac{(\varDelta x)^3}{3!} + \ldots \right)}{\varDelta x}$$

$$= e^x \left(\dfrac{1}{1!} + \dfrac{\varDelta x}{2!} + \dfrac{(\varDelta x)^2}{3!} + \ldots \right)$$

As $\varDelta x$ tends to zero, the left-hand side tends to dy/dx. At the same time all but the first of the terms inside the bracket on the right-hand side tend to zero. It happens that there is an infinite number of these terms, and strictly speaking we should consider very carefully whether, in this case, we are justified in assuming that since each term tends to zero, the sum of an infinite number of them will also tend to zero. But to do this would involve us in considerations outside the plan of this book ; we shall simply state that it can be shown that, *in this particular case*, we are justified in making this assumption. And so we have that the bracket on the right-hand side tends to unity, and the whole of the right-hand side becomes e^x.

We thus obtain the somewhat startling result that

if $\qquad\qquad y = e^x \qquad$ then $\quad \dfrac{dy}{dx} = e^x$

Also if $\qquad\quad y = A e^x \qquad$ then $\quad \dfrac{dy}{dx} = A e^x$

This is the only case of a function being equal to its own first derivative.

It is obvious that d^2y/dx^2, and all subsequent derivatives, also equal e^x (and therefore equal y).

Now we may remember from the last chapter that another way of writing $b = a^x$ is to say $\log_a b = x$. It follows that if (to reverse the customary roles of x and y for a moment)

$$x = e^y$$

then $\qquad\qquad \log_e x = y$

It follows that
$$\frac{d}{dx} \log_e x = \frac{dy}{dx} = 1 \Big/ \frac{dx}{dy}$$

But as $x = e^y$,
$$\frac{dx}{dy} = e^y = x$$

Therefore
$$\frac{d}{dx} \log_e x = 1 \Big/ \frac{dx}{dy} = \frac{1}{x}$$

It follows that $\displaystyle\int \frac{1}{x} dx = \log_e x + C = \log_e x + \log_e A$

$$= \log_e Ax \text{ where } C = \log_e A = \text{arbitrary constant,}$$

and here we have the answer to the all-important question. The integral of $1/x$ is $\log_e Ax$. The function that has $1/x$ as its first derivative is $\log_e Ax$.

6. Summary of results

Having at last answered the question with which this chapter began, we shall now list some of the more important points that we have discovered en route, before passing, in the next few chapters, to consider some of the consequences of this result.

(1) $\displaystyle e^x = \operatorname*{Limit}_{n \to \infty} \left(1 + \frac{x}{n}\right)^n$

$\displaystyle \quad = \operatorname*{Limit}_{n \to \infty} \left(1 + \frac{1}{n}\right)^{nx}$

$\displaystyle \quad = 1 + \frac{x}{1!} + \frac{x^2}{2!} + \frac{x^3}{3!} + \dots$

(2) e is as above with $x = 1$, and is approximately $2 \cdot 71828 \dots$

(3) If $y = e^x$ then $\log_e y = x$

(4) If $y = e^x$ then $dy/dx = e^x = y$

$\qquad d^2y/dx^2 = e^x = y$, etc.

(5) $\displaystyle\int \frac{1}{x} dx = \log_e x + C = \log_e x \log_e A = \log_e Ax$

$\qquad$ where A and C are arbitrary constants.

(6) $\displaystyle\frac{d}{dx}\left(\log_e Ax\right) = \frac{1}{x}$

7. Hyperbolic functions

There are, in addition, a few further results to be noted.

Diagram 14.2 shows the curve $y = e^x$. It also shows the curve $y = 1/e^x$ which may be written $y = e^{-x}$. These two functions are of

immense importance in practical mathematics and statistics, as well
as in theory.

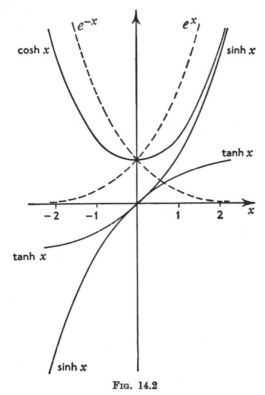

FIG. 14.2

There is also a set of functions, known as hyperbolic functions,
derived from these curves. They are defined as follows :

Definition :

 Hyperbolic Sine, written *sinh*, is defined to be

$$\sinh x = \frac{e^x - e^{-x}}{2}$$

 Hyperbolic Cosine, written *cosh*, is defined to be

$$\cosh x = \frac{e^x + e^{-x}}{2}$$

 Hyperbolic Tangent, written *tanh*, is defined to be

$$\tanh x = \frac{\sinh x}{\cosh x} = \frac{e^x - e^{-x}}{e^x + e^{-x}}$$

These functions are graphed in Diagram 14.2. We shall be using them later, when we will see their similarity to the trigonometric functions.

The names of these functions are similar to those of the trigonometric functions, and they have, in some respects, similar properties. But we must be careful to note some of the differences.

For example, our definitions show that

$$\cosh^2 x \equiv \left(\frac{e^x + e^{-x}}{2}\right)^2$$

$$\equiv \frac{e^{2x} + 2e^x \cdot e^{-x} + e^{-2x}}{4}$$

$$\equiv \frac{e^{2x} + 2 + e^{-2x}}{4}$$

Similarly

$$\sinh^2 x \equiv \frac{e^{2x} - 2 + e^{-2x}}{4}$$

It follows that

$$\cosh^2 x - \sinh^2 x \equiv \frac{e^{2x} + 2 + e^{-2x}}{4} - \frac{e^{2x} - 2 + e^{-2x}}{4}$$

$$\equiv \frac{4}{4} \equiv 1$$

which may be contrasted with

$$\cos^2 x + \sin^2 x \equiv 1$$

We may easily differentiate these functions :

$$\frac{d}{dx} \sinh x = \frac{d}{dx}\left(\frac{e^x - e^{-x}}{2}\right)$$

$$= \frac{e^x + e^{-x}}{2}$$

$$= \cosh x$$

Similarly

$$\frac{d}{dx} \cosh x = \sinh x$$

while

$$\frac{d}{dx} \tanh x = \frac{d}{dx} \frac{\sinh x}{\cosh x}$$

$$= \frac{\cosh^2 x - \sinh^2 x}{\cosh^2 x}$$

$$= \frac{1}{\cosh^2 x}$$

which we may find written as $\operatorname{sech}^2 x$.

These results may be compared with

$$\frac{d}{dx}\sin x = \cos x$$

$$\frac{d}{dx}\cos x = -\sin x$$

$$\frac{d}{dx}\tan x = \frac{1}{\cos^2 x} = \sec^2 x$$

We define

$$\operatorname{sech} x = (\cosh x)^{-1}$$

$$\operatorname{cosech} x = (\sinh x)^{-1}$$

and

$$\coth x = (\tanh x)^{-1}$$

There is an important set of logarithmic expressions for the hyperbolic functions.

If

$$x = \sinh y = \tfrac{1}{2}(e^y - e^{-y})$$

then

$$2x = e^y - e^{-y}$$

whence

$$2xe^y = e^{2y} - 1$$

and so

$$e^{2y} - 2xe^y - 1 = 0$$

This is a quadratic equation in e^y, and its solution is

$$e^y = x \pm \sqrt{x^2 + 1}$$

Thus

$$y = \log\{x + \sqrt{x^2 + 1}\}$$

(since we cannot use the negative root without becoming involved in the logarithm of a negative quantity).

But if

$$x = \sinh y$$

then

$$y = \sinh^{-1} x$$

and so

$$\sinh^{-1} x = \log\{x + \sqrt{x^2 + 1}\}$$

It can similarly be shown that for $x > 1$

$$\cosh^{-1} x = \log\{x \pm \sqrt{x^2 - 1}\}$$

and that for $x^2 < 1$

$$\tanh^{-1} x = \tfrac{1}{2}\log\frac{1+x}{1-x}$$

while for $x^2 > 1$

$$\coth^{-1} x = \tfrac{1}{2}\log\frac{x+1}{x-1}$$

CHAPTER XV

SOME TECHNIQUES OF DIFFERENTIATION

But he did for them both—

1. Introduction

Now that we have become acquainted with e we are able to consider many practical problems which would otherwise be extremely difficult if not quite impossible. This is because we are able to make use of a number of techniques and results of differentiation and integration which depend on the basic properties of e. In this and the next chapter we shall consider some of these methods, along with some others. Later in the book we shall see how they may be used to answer certain economic problems ; but for the moment we shall devote ourselves strictly to the mathematics.

2. $y = \log f(x)$

We know that if
$$y = \log_e x$$

then
$$\frac{dy}{dx} = \frac{1}{x}$$

Now suppose that we have
$$y = \log_e u$$
where $u = f(x)$, e.g., $y = \log_e (3x^2 + 2x - 1)$ or $y = \log_e (4 \sin x)$
We know from our rule for the differentiation of a function of a function that

$$\frac{dy}{dx} = \frac{dy}{du}\frac{du}{dx}$$

In this case, where $y = \log_e u$ we have that $\dfrac{dy}{du} = \dfrac{1}{u} = \dfrac{1}{f(x)}$ and so

$$\frac{dy}{dx} = \frac{1}{u}\frac{du}{dx} = \frac{1}{f(x)}\frac{d}{dx}f(x)$$
$$= \frac{f'(x)}{f(x)}$$

where we use $f'(x)$ to denote $\dfrac{d}{dx}f(x)$

This is a highly important result, which we shall illustrate with a few examples.

Examples :

(i) $y = \log_e (3x^2 + 2x - 1)$

Here we have that $\qquad f(x) = (3x^2 + 2x - 1)$

and so $\qquad f'(x) = \dfrac{d}{dx} f(x) = 6x + 2$

Therefore, $\qquad \dfrac{dy}{dx} = \dfrac{f'(x)}{f(x)} = \dfrac{6x + 2}{3x^2 + 2x - 1}$

(ii) $y = \log_e 4 \sin x$

Here $\qquad\qquad f(x) = 4 \sin x$

$\qquad\qquad\qquad f'(x) = 4 \cos x$

Therefore, $\qquad \dfrac{f'(x)}{f(x)} = \dfrac{4 \cos x}{4 \sin x} = \cot x$

i.e., if $\qquad y = \log_e 4 \sin x, \quad \dfrac{dy}{dx} = \cot x$

(iii) $y = \log_e (4x^2 - 3x)^5$

There are two ways of doing this. We can proceed as above, writing

$$f(x) = (4x^2 - 3x)^5$$

whence $\qquad f'(x) = 5(8x - 3)(4x^2 - 3x)^4 \qquad$ (see page 169)

and so $\qquad \dfrac{dy}{dx} = \dfrac{5(8x - 3)}{(4x^2 - 3x)}$

Alternatively we may note that

$$\log_e (4x^2 - 3x)^5 = 5 \log_e (4x^2 - 3x)$$

and so if we put $\qquad y = 5 \log_e (4x^2 - 3x)$

we immediately have, as in Example (i) that

$$\dfrac{dy}{dx} = \dfrac{5(8x - 3)}{4x^2 - 3x}$$

3. $y = f_1(x) f_2(x) f_3(x) \ldots$

One of the most useful properties of logarithms is that they enable us to differentiate a complicated product or quotient very simply. For instance, suppose that

$$y = \dfrac{(3x^3 - 2x + 1) \sqrt{2x - 3}}{(5x^2 + 4)^3 \sin x}$$

To differentiate this by the methods described in Chapter X would be possible, but rather tedious and cumbersome. There is a simpler method, which begins by taking the logarithm of each side, bearing in mind that the logarithm of a product is the sum of the logarithms of the components, and that the logarithm of a quotient is the logarithm of the dividend minus the logarithm of the divisor.

We have

$$\log y = \log (3x^2 - 2x + 1) + \tfrac{1}{2} \log (2x - 3) \\ - \{3 \log (5x^2 + 4) + \log \sin x\}$$

We can now differentiate this, term by term, with respect to x. On the left-hand side we have

$$\frac{d}{dx} (\log y) = \frac{1}{y} \frac{dy}{dx} \qquad \text{(see Section 2 above)}.$$

On the right-hand side we have four separate terms of the kind $f'(x)/f(x)$ which yield, when added together, and equated to the left-hand side,

$$\frac{1}{y} \frac{dy}{dx} = \frac{6x - 2}{3x^2 - 2x + 1} + \frac{1}{2x - 3} - \frac{30x}{5x^2 + 4} - \frac{\cos x}{\sin x}$$

from which we have that the value of dy/dx is the right-hand side of this expression multiplied by the value of y.

4. Logarithms to any base

Although we have seen how to differentiate a logarithm to the base e, we have not yet considered how to differentiate a logarithm to some other base. This is done quite simply by remembering that

$$\log_a x = \frac{\log_e x}{\log_e a}$$

It follows immediately from this that

$$\frac{d}{dx} \log_a x = \frac{1}{x \log_e a}$$

And it is not difficult to deduce that

$$\frac{d}{dx} \log_a f(x) = \frac{f'(x)}{f(x) \log_e a}$$

For example, if we have that $y = \log_{10} (3x^2 + 5x)$, then

$$\frac{dy}{dx} = \frac{6x + 5}{(3x^2 + 5x) \log_e 10}$$

5. e^x, $e^{f(x)}$, a^x and $a^{f(x)}$

We know already that if $y = e^x$ then $\dfrac{dy}{dx} = e^x$.

It follows from our rule for the differentiation of a function of a function that if

$$y = e^u \quad \text{where } u = f(x)$$

then
$$\frac{dy}{dx} = \frac{dy}{du}\frac{du}{dx}$$
$$= e^u f'(x)$$
$$= e^{f(x)} f'(x)$$

For example,

if
$$y = e^{3x^2 + 2x + 4}$$

then
$$\frac{dy}{dx} = (6x + 2)e^{3x^2 + 2x + 4}$$

It can also be shown, quite easily, although we shall not do so, that

if
$$y = a^x$$

then
$$\frac{dy}{dx} = a^x \log_e a$$

and that

if
$$y = a^{f(x)}$$

then
$$\frac{dy}{dx} = a^{f(x)} f'(x) \log_e a$$

For example,

if
$$y = a^{3x^2 + 2x + 4}$$

then
$$\frac{dx}{dy} = a^{3x^2 + 2x + 4}(6x + 2) \log_e a$$

EXERCISE **15.1**

1. Differentiate :
   (i) $y = \log_e (2x^3 - x^2 + 3)$
   (ii) $y = 3 \log_e (2 \cos x)$
   (iii) $y = (7x^5 - 3x^3)^8$
2. Differentiate : $y = \dfrac{(3x^2 - 2x + 1) \sin x}{(x^5 + 3x^2)^3}$
3. Differentiate : $y = \log_{10} (2x^5 + 3x)$
4. Differentiate :
   (i) $y = e^{2x^3 - 3x} + e^x$
   (ii) $y = 4^{2x^2 + x} + 2x^2 + x$

5. Differentiate :

    (i) $y = (x^2 + 3x)(x^3 - 4x^2 + x) \sin x$

    (ii) $y = (x^2 + 3x)(x^3 - 4x^2 + x)^2 \sqrt{x^2 + 2} \sin x$

6. Differentiate :

    (i) $y = \log_{10}(x^2 + 4x + 2)$

    (ii) $y = \log_{10} \sin x$

    (iii) $y = \log_4 \sin x$

7. Differentiate :

    (i) $y = e^{x^2 - \sin x} \sqrt{x^2 - \sin x}$

    (ii) $y = e^{e^x}$

8. Examine for maxima, minima and points of inflexion the functions :

    (i) $y = \log(3x^3 - x)$

    (ii) $y = e^{4x - x^2}$

    (iii) $y = x^2 e^x$

    (iv) $y = \frac{1}{2}(e^x + e^{-x})$

CHAPTER XVI

SOME TECHNIQUES OF INTEGRATION

—by his plan of attack
SIEGFRIED SASSOON : *The General*

1. Some standard integrals

In our chapter on integration it was stated that the process of integration may be considered as the reverse of the process of differentiation, and that if p is the derivative of q, then the integral of p will be $q + C$, where C is an arbitrary constant. This statement was followed by a short table of results. Below we print a slightly longer table of Standard Integrals. In each case the integral of a function in the first column has been obtained by asking the question, " What is the function that has this entry in the first column as its derivative? " Differentiation of the entries in the second column, by one of the methods set forth in one of the chapters on differentiation, will yield the corresponding entry in the first column.

	y	$\int y \, dx$	
(1)	a	$ax + C$	
(2)	ax^n	$\dfrac{ax^{n+1}}{n+1} + C$	$n \neq -1$
(3)	$(ax+b)^n$	$\dfrac{(ax+b)^{n+1}}{(n+1)a} + C$	$n \neq -1$
(4)	$f'(x)[f(x)]^n$	$\dfrac{[f(x)]^{n+1}}{n+1} + C$	$n \neq -1$
(5)	e^{ax}	$\dfrac{e^{ax}}{a} + C$	
(6)	$\dfrac{1}{x}$	$\log_e x + C = \log_e Ax$	
(7)	$\dfrac{f'(x)}{f(x)}$	$\log_e f(x) + C = \log_e Af(x)$	
(8)	$f'(x)e^{f(x)}$	$e^{f(x)} + C$	
(9)	$a^{f(x)}f'(x)$	$\dfrac{a^{f(x)}}{\log_e a} + C$	
(10)	$\sin x$	$-\cos x + C$	
(11)	$\cos x$	$\sin x + C$	
(12)	$\sec^2 x$	$\tan x + C$	

The following examples illustrate how a knowledge of these Standard Integrals enables other cases to be integrated.

(i) $(3x^2 + 4x)(x^3 + 2x^2 + 7)^3$ is of the form $f'(x)[f(x)]^n$ since the first bracket is the derivative of the second. We, therefore, have from the Standard Integral (4) that

$$y = \frac{(x^3 + 2x^2 + 7)^4}{4} + C$$

(ii) $\dfrac{2x + 7}{x^2 + 7x}$ is of the form $\dfrac{f'(x)}{f(x)}$ since the top is the derivative of the bottom. The integral is, therefore, given by (7), i.e.,

$$y = \log_e (x^2 + 7x) + C = \log_e A(x^2 + 7x)$$

(iii) $\dfrac{3x^2 - 4}{x^3 - 4x + 1}$ is of the same form, and the result is

$$y = \log_e (x^3 - 4x + 1) + C = \log_e A(x^3 - 4x + 1)$$

(iv) $\dfrac{x + 4}{3x^2 + 24x}$ is equal to $\dfrac{1}{6}\dfrac{6x + 24}{3x^2 + 24x}$ which is of the same form once again, and so we have

$$y = \tfrac{1}{6} \log_e (3x^2 + 24x) + C = \log_e (3x^2 + 24x)^{1/6} + C$$
$$= \log_e A (3x^2 + 24x)^{1/6}$$

(v) $2xe^{x^2}$ is of the form $f'(x)e^{f(x)}$ and so the integral is (from (8))

$$y = e^{x^2} + C$$

(vi) $2xa^{x^2+7}$ is of the form $f'(x)a^{f(x)}$ and has an integral (from (9))

$$y = \frac{a^{x^2+7}}{\log_e a} + C$$

(vii) $\cos x \cdot e^{\sin x}$ is of the form $f'(x)e^{f(x)}$ and so (from (8))

$$y = e^{\sin x} + C$$

The subject of integration is vast, and we can barely touch on it here. The first method that must be tried whenever an integration is to be performed is that of writing the integrand in such a way that it becomes a standard form. There are many more standard forms than those we have just listed, and we shall look at some of them later on ; but it will be useful at this stage to illustrate the methods employed by restricting ourselves to integrals which can be reduced to one of the above forms.

EXERCISE 16.1

1. Integrate the following with respect to x :

(i) $(2x+3)(x^2+3x+5)^4$

(ii) $\dfrac{3x^2+14x}{x^3+7x^2+2}$

(iii) $3x^2e^{x^3}$

(iv) $(2x+5)4^{x^2+5x+1}$

(v) $\sin x \cdot e^{\cos x}$

(vi) $\dfrac{x^2+2x+2}{x^3+3x^2+6x}$

(vii) $(x+4)(x^2+8x+3)^3$

(viii) $(x^3+2x)e^{x^4+4x^2+3}$

(ix) $\dfrac{4}{x}$

(x) $\dfrac{\sec^2 x}{\tan x}$

(xi) $\dfrac{1}{\cos x \sin x}$

(xii) $a^{\tan x}\sec^2 x$

(xiii) $\cos x \sin^2 x$

(xiv) $(2x+5)(x^2+5x+1)^4$

(xv) $(3x^2+\sin x)3^{x^3-\cos x}$

(xvi) $\cos x \cdot e^{\sin x}$

2. Change of variable

Often the integrand may be rewritten in a more tractable form by means of some algebraic or trigonometric substitution. The following examples will illustrate this method, although they do not justify it.

(1) $I = \int (ax+b)^n\,dx$. Put $ax+b=u$ so that $dx/du=1/a$. Then the integral becomes

$$I = \int (ax+b)^n\frac{dx}{du}\,du \qquad \text{(The validity of this step must be taken on trust)}$$

$$= \int u^n\frac{1}{a}\,du$$

$$= \frac{1}{a}\int u^n\,du$$

$$= \frac{u^{n+1}}{a(n+1)}+C$$

$$= \frac{(ax+b)^{n+1}}{a(n+1)}+C$$

(2) $I = \int \dfrac{1}{\sqrt{a^2-x^2}}\,dx$. Put $x=a\sin\theta$ so that $\dfrac{dx}{d\theta}=a\cos\theta$

The integral becomes
$$I = \int \frac{1}{\sqrt{a^2 - a^2 \sin^2 \theta}} \frac{dx}{d\theta} d\theta$$
$$= \int \frac{1}{\sqrt{a^2 \cos^2 \theta}} a \cos \theta \, d\theta$$
$$= \int d\theta$$
$$= \theta + C$$
$$= \sin^{-1} \frac{x}{a} + C$$

This is a result that we shall later add to our list of standard forms. By a similar substitution, which we shall leave the reader to find, we can show that

$$\int \frac{dx}{x^2 + a^2} = \frac{1}{a} \tan^{-1} \frac{x}{a} + C.$$

(3) Integration of trigonometric functions can often be very complicated. Generally a rational function of sin x and cos x (and therefore of tan x, since tan $x = \sin x / \cos x$) may be integrated by using the following very important change of variable.

Let $\tan \frac{x}{2} = t$, so that $\frac{1}{2} \sec^2 \frac{x}{2} = \frac{dt}{dx}$.

Then if we use the formulae for sin 2θ, cos 2θ and tan 2θ on p. 91 we will find that

$$\sin x = \frac{2t}{1 + t^2}, \quad \cos x = \frac{1 - t^2}{1 + t^2}.$$

and

$$\frac{dx}{dt} = \frac{2}{1 + t^2}$$

The integral may now be transformed into the integral of a rational function of t. For example,

$$I = \int \operatorname{cosec} x \, dx$$
$$= \int \frac{1}{\sin x} \frac{dx}{dt} dt$$
$$= \int \frac{1 + t^2}{2t} \frac{2}{1 + t^2} dt$$
$$= \int \frac{1}{t} dt$$
$$= \log t + C$$
$$= \log \left(\tan \frac{x}{2} \right) + C = \log \left(A \tan \frac{x}{2} \right)$$

Some of the standard forms listed at the end of this chapter have been obtained by similar methods.

EXERCISE 16.2

1. Integrate the following with respect to x :

(i) $(3x+5)^4$

(ii) $\dfrac{1}{\sqrt{9-x^2}}$

(iii) $\sec x$

(iv) $\dfrac{1}{\sqrt{9-4x^2}}$

(v) $\sin 2x$

(vi) $\dfrac{3}{9+x^2}$

(vii) $\dfrac{3}{x^2+2x+10}$

(viii) $\dfrac{4}{x^2+4x+20}$

3. Integration by parts

It can be shown that if we have two functions of x denoted by u and v, then

$$\int u \frac{dv}{dx}\, dx = uv - \int v \frac{du}{dx}\, dx$$

This result is of great use in the integration of certain functions, as is demonstrated in the example below.

Examples :

(i) Consider $\qquad I = \displaystyle\int x \cos x\, dx$

Since $\dfrac{d}{dx} \sin x = \cos x$ we may write this as

$$I = \int x \frac{d}{dx} (\sin x)\, dx$$

and compare it with $\qquad I = \displaystyle\int u \frac{dv}{dx}\, dx$

We have that $u = x$ and $v = \sin x$.   The result is

$$I = uv - \int v \frac{du}{dx}\, dx$$

$$= x \sin x - \int \sin x\, dx$$

$$= x \sin x + \cos x + C$$

(ii) Consider

$$I = \int \log x \, dx$$

$$= \int u \frac{dv}{dx} \, dx \qquad \text{where } u = \log x, \ v = x$$

$$= x \log x - \int x \frac{d(\log x)}{dx} \, dx$$

$$= x \log x - \int x \left(\frac{1}{x} \right) dx$$

$$= x \log x - \int dx$$

$$= x \log x - x + C$$

(iii) Consider $I = \int \sqrt{a^2 - x^2} \, dx \qquad \text{where } a^2 > x^2$

$$= \int u \frac{dv}{dx} \, dx \qquad \text{where } u = \sqrt{a^2 - x^2}, \ v = x$$

$$= x\sqrt{a^2 - x^2} - \int x \frac{d\sqrt{a^2 - x^2}}{dx} \, dx$$

$$= x\sqrt{a^2 - x^2} - \int x \frac{-x}{\sqrt{a^2 - x^2}} \, dx$$

$$= x\sqrt{a^2 - x^2} + \int \frac{x^2}{\sqrt{a^2 - x^2}} \, dx$$

Now $\int \frac{x^2 \, dx}{\sqrt{a^2 - x^2}} = \int \left[-\frac{(a^2 - x^2)}{\sqrt{a^2 - x^2}} + \frac{a^2}{\sqrt{a^2 - x^2}} \right] dx$

$$= \int -\sqrt{a^2 - x^2} \, dx + \int \frac{a^2}{\sqrt{a^2 - x^2}} \, dx$$

$$= -\int \sqrt{a^2 - x^2} \, dx + a^2 \sin^{-1} \frac{x}{a} + C$$

from Example 2 of the last section.
    We thus have that

$$I = x\sqrt{a^2 - x^2} - \int \sqrt{a^2 - x^2} \, dx + a^2 \sin^{-1} \frac{x}{a} + C$$

$$= x\sqrt{a^2 - x^2} - I + a^2 \sin^{-1} \frac{x}{a} + C$$

whence $I = \frac{x}{2} \sqrt{a^2 - x^2} + \frac{a^2}{2} \sin^{-1} \frac{x}{a} + C$

(iv) Consider
$$I = \int \frac{1}{\sqrt{x^2 + a^2}}\, dx$$

Put
$$x = a \sinh \theta$$

$$\frac{dx}{d\theta} = a \cosh \theta$$

$$\sqrt{x^2 + a^2} = \sqrt{a^2(1 + \sinh^2 \theta)}$$
$$= \sqrt{a^2 \cosh^2 \theta}$$
$$= a \cosh \theta$$

Thus
$$I = \int \frac{1}{a \cosh \theta} \cdot a \cosh \theta\, d\theta$$
$$= \int d\theta$$
$$= \theta + C$$
$$= \sinh^{-1} \frac{x}{a} + C$$

We may note from the definition of $\sinh x$ that
$$\sinh x = \sinh (-x)$$
and so our integral, having the value θ where $\sinh \theta = \dfrac{x}{a}$, is really

$$\pm \sinh^{-1} \frac{x}{a} + C$$
$$= \log \frac{x \pm \sqrt{x^2 + a^2}}{a} + C$$

because of the result derived on page 221.

EXERCISE 16.3

1. Integrate:

(i) xe^{-x}	(ii) $x^2 e^{-x}$	(iii) $xe^{x/a}$
(iv) $x \log x$	(v) $x \sin x$	(vi) $x \sin x \cos x$
(vii) $x \sec^2 x$	(viii) $\log 2x$	(ix) $\sec^4 x$

2. Prove the formula for Integration by Parts by considering the formula for the differentiation of a product, viz.,

$$\frac{d}{dx}(uv) = u\frac{dv}{dx} + v\frac{du}{dx}$$

4. Some standard forms

In addition to those listed earlier, we have the following results for reference. Some have already been proved as examples. The others may be proved by similar methods and are left to the reader as an exercise.

y	$\int y \, dx$
	(the arbitrary constant C is omitted)
$\tan x$	$\log (\sec x)$
$\operatorname{cosec}^2 x$	$-\cot x$
$\cosh x$	$\sinh x$
$\sinh x$	$\cosh x$
$\operatorname{sech}^2 x$	$\tanh x$
$\dfrac{1}{\sqrt{a^2 - x^2}}$	$\sin^{-1} \dfrac{x}{a} \text{ or } \cos^{-1} \dfrac{x}{a}$
$\dfrac{1}{x^2 + a^2}$	$\dfrac{1}{a} \tan^{-1} \dfrac{x}{a}$
$\dfrac{1}{\sqrt{x^2 + a^2}}$	$\log \dfrac{x \pm \sqrt{x^2 + a^2}}{a} = \pm \sinh^{-1} \dfrac{x}{a}$
$\dfrac{1}{\sqrt{x^2 - a^2}}$	$\log \dfrac{x \pm \sqrt{x^2 - a^2}}{a} = \pm \cosh^{-1} \dfrac{x}{a}$
$\dfrac{1}{a^2 - x^2}$	$\dfrac{1}{2a} \log \dfrac{a + x}{a - x} = \dfrac{1}{a} \tanh^{-1} \dfrac{x}{a} \quad \text{for } \lvert x \rvert < a$
	and $\dfrac{1}{2a} \log \dfrac{a + x}{x - a} = \dfrac{1}{a} \coth^{-1} \dfrac{x}{a} \quad \text{for } \lvert x \rvert > a > 0$
$\sqrt{x^2 \pm a^2}$	$\dfrac{x}{2} \sqrt{x^2 \pm a^2} \pm \dfrac{a^2}{2} \log (x + \sqrt{x^2 \pm a^2})$
$\sqrt{a^2 - x^2}$	$\dfrac{x}{2} \sqrt{a^2 - x^2} + \dfrac{a^2}{2} \sin^{-1} \dfrac{x}{a}$

5. Some economic examples

(i) A machine uses up expendable parts at a rate which depends on the time t for which it has been running. If the rate at time t is

$$y = t^2 e^t$$

then the value of total depreciation after a time T is

$$\int_0^T t^2 e^t \, dt$$

We evaluate this by integrating by parts. We have

$$\int t^2 e^t \, dt = \int t^2 \, d(e^t)$$
$$= t^2 e^t - \int e^t \, d(t^2)$$
$$= t^2 e^t - 2\int e^t t \, dt$$
$$= t^2 e^t - 2(te^t - \int e^t \, dt)$$
$$= t^2 e^t - 2(te^t - e^t)$$
$$= e^t (t^2 - 2t + 2) + C$$

Therefore, after time T, the value of depreciation is

$$e^T (T^2 - 2T + 2) - e^0 (0 - 0 + 2)$$
$$= e^T (T^2 - 2T + 2) - 2$$

(ii) A number of equally desirable plots of land become available for sale for building at the rate of one a month. The price p of each plot depends on the number n already sold, in such a way that

$$p = Ae^{(n-N) \times 0.01}$$

where N is the total number available. Estimate the total income from the sale of 100 plots, if (a) $N = 100$, (b) $N = 200$.

(a) The income from the sale of 100 plots is given by the following integral in which we treat n as a continuous variable capable of taking all values between 0 and 200 :

$$I = \int_0^{100} Ae^{(n-N) \times 0.01} \, dn$$

$$= \left[Ae^{(n-N) \times 0.01} \right]_0^{100} \times \frac{1}{0.01}$$

$$= A \left[e^{(100-N)0.01} - e^{-0.01N} \right] \times 100$$

$$= 100A \left[\frac{e-1}{e^{N/100}} \right]$$

When $N = 100$, total income is $100A \left(\dfrac{e-1}{e} \right)$

$$= 100A \left(1 - \frac{1}{e} \right)$$

$$= 63.22A$$

(b) When $N = 200$, the income from the sale of the first 100 plots is

$$100A \left(\frac{e-1}{e^2} \right) = 100A \left(\frac{1}{e} - \frac{1}{e^2} \right)$$

$$= 23.25A$$

But if the 200 plots had been sold the total income would have been

$$\left[A e^{0.01\ (n-N)} \right]_0^{200} \times \frac{1}{0\cdot01}$$

$$= 100A \left[e^{\frac{200-N}{100}} - e^{-\frac{N}{100}} \right]$$

$$= 100A \left[\frac{e^2 - 1}{e^{N/100}} \right]$$

which, when $N = 200$, gives

$$100A \left[\frac{e^2-1}{e^2} \right] = 100A \left[1 - \frac{1}{e^2} \right]$$

$$= 86\cdot47A$$

It follows that from the sale of the second hundred plots the income is $86\cdot47A - 23\cdot25A = 63\cdot22A$. It will be noticed that this sum is (as might have been expected) the same as that obtained by selling 100 when only 100 are available.

(iii) (a) A firm has a marginal revenue function given by

$$M.R. = \frac{a}{x+b} - c$$

where x is the output and a, b, c are constants.

We want to find the demand equation.

Denoting revenue by R, and price by p we write

$$R = px$$

$$M.R. = \frac{dR}{dx} = \frac{a}{x+b} - c$$

Therefore
$$R = \int \left(\frac{a}{x+b} - c \right) dx$$

and to obtain the total revenue obtained by selling a quantity x_1 we evaluate this integral between $x = 0$ and $x = x_1$. This gives

$$R = \int_0^{x_1} \left(\frac{a}{x+b} - c \right) dx$$

$$= \left[a \log_e (x+b) - cx \right]_0^{x_1}$$

$$= a \log_e (x_1+b) - a \log_e b - cx_1$$

$$= a \log_e \left(\frac{x_1+b}{b} \right) - cx_1$$

$$= \log_e \left(\frac{x_1+b}{b} \right)^a - cx_1$$

Thus

$$px = \log_e \left(\frac{x+b}{b}\right)^a - cx$$

where we have used $R = px$, and dropped the arbitrary subscript.

This may be written

$$p = \frac{1}{x} \log_e \left(\frac{x+b}{b}\right)^a - c$$

which, relating price and quantity saleable at that price, is the demand equation.

(b) If the marginal revenue function is

$$M.R. = \frac{a}{(x+b)^2} - c$$

then

$$R = \int_0^{x_1} \left(\frac{a}{(x+b)^2} - c\right) dx$$

$$= \left[\frac{-a}{(x+b)} - cx\right]_0^{x_1}$$

$$= \left(\frac{-a}{x_1 + b} + \frac{a}{b} - cx_1\right)$$

whence

$$px = \left(\frac{a}{b} - \frac{a}{x+b} - cx\right)$$

and so

$$p = \frac{1}{x}\left(\frac{a}{b} - \frac{a}{x+b}\right) - c$$

$$= \frac{1}{x}\left(\frac{ax + ab - ab}{b(x+b)}\right) - c$$

$$= \frac{a}{b(x+b)} - c$$

which can be rewritten

$$pb(x+b) = a - cb(x+b)$$

$$pbx + cbx = a - pb^2 - cb^2$$

$$x(p+c)b = a - (p+c)b^2$$

$$x = \frac{a}{b(p+c)} - b$$

which expresses the demand equation with x as the dependent variable.

6. Integration and statistics

The student of economics will probably be familiar with the arithmetic mean and the standard deviation. If there is a variable x, taking values x_1, x_2, ... , with frequencies f_1, f_2, ... , then the arithmetic mean of x is defined to be

$$\frac{\Sigma f_i\, x_i}{\Sigma f_i}$$

For example, if we have the following data

$x=$ no. of children in family	$f=$ no. of families with this no. of children	$f \cdot x$
0	10	0
1	15	15
2	25	50
3	20	60
4	10	40
5	5	25
6	3	18
7	2	14
	$\Sigma f = 90$	$\Sigma fx = 222$

then the arithmetic mean of x is

$$\frac{\Sigma fx}{\Sigma f} = \frac{222}{90}$$

$$= 2 \cdot 467$$

which is the "average" number of children per family. Usually this is denoted by $\bar{x}$ where the bar above the x has no relationship at all to the bar used in logarithms. It is simply an easily written device for indicating the arithmetic mean.

Let us represent this diagrammatically. The horizontal axis is divided in to a number of spaces, each representing a whole number of children. On the vertical scale we measure, first, the number of families with the various numbers of children. Each column represents a value of f. The total number of families is the total of the lengths of these columns; and as they are all of unit width this means that the *total area* of the columns gives the number of families. This, of course, is the total area beneath the stepped line in Diagram 16.1.

Now let us construct a similar graph, but this time show the values of fx (from column (3) of the table) on the vertical axis. This will be another column graph, and the area underneath it will represent Σfx.

It follows that the arithmetic mean is the ratio of the areas beneath the two column graphs.

Now suppose that x becomes a continuous variable (such as time). To be more specific about it, suppose that we make a large number of observations about the time that it takes one man to walk to work.

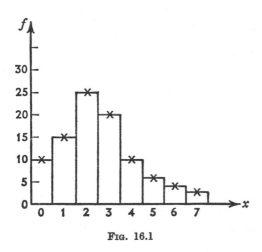

FIG. 16.1

We may find that, provided we measure time very accurately, the results can be summarised in

$$f = bx + cx^2$$

where x measures time (in seconds) and f is the number of days on which the man took x seconds. For example, then the number of days on which x takes various values is given by

$$\begin{aligned}
x &= 0 & f_0 &= b \times 0 + c \times 0 \\
x &= 1 & f_1 &= b + c \\
x &= 2 & f_2 &= 2b + 4c \\
& & &\text{etc.}
\end{aligned}$$

This is unlikely to be a realistic case, but it is, for the moment, mathematically convenient for our exposition.

Let us now consider graphing f against x. Instead of a column graph we obtain a continuous graph, and by pursuing an argument. similar to that with which we introduced integration, we would find that the area under this graph would represent the total number of days (i.e. Σf). This area would be given by

$$\int f \, . \, dx$$

$$= \int (bx + cx^2) \, dx$$

$$= \frac{bx^2}{2} + \frac{cx^3}{3} + C$$

If x extends from 0 to T then the area under the curve is

$$\left[\frac{bx^2}{2} + \frac{cx^3}{3} + C \right]_0^T$$

$$= \frac{bT^2}{2} + \frac{cT^3}{3}$$

which represents the total number of days on which the man took between 0 and T seconds.

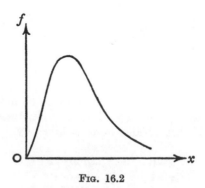

Fig. 16.2

Now consider a graph of ft plotted against x (which compares with the plotting of column (3)). On this occasion the area under the graph will correspond to Σfx in our earlier example. It will be

$$\int fx \, dx$$

$$= \int_0^T (bx + cx^2)x \, dx$$

$$= \int_0^T (bx^2 + cx^3) \, dx$$

$$= \left[\frac{bx^3}{3} + \frac{cx^4}{4} \right]_0^T$$

$$= \frac{bT^3}{3} + \frac{cT^4}{4}$$

which represents the total time taken (being obtained by summing the product of time per day and days taking that time).

The arithmetic mean time will be obtained by dividing these results, getting

$$\bar{x} = \frac{\displaystyle\int_0^T fx \, dx}{\displaystyle\int_0^T f \, dx}$$

$$= \frac{\frac{1}{3}bT^3 + \frac{1}{4}cT^4}{\frac{1}{2}bT^2 + \frac{1}{3}cT^3}$$

$$= \frac{(4bT^3 + 3cT^4) \div 12}{(3bT^2 + 2cT^3) \div 6}$$

$$= \frac{1}{2}\frac{(4b + 3cT)T^3}{(3b + 2cT)T^2}$$

$$= \frac{(4b + 3cT)T}{2(3b + 2cT)}$$

provided that $T \neq 0$. This is the average time of a journey to work on those days that it takes not more than T seconds.

We can now consider a different problem, more shortly. If the number of people (f) earning £x per annum is given by

$$f = a + bx + cx^2 - dx^3$$

for $0 < x < X$, where X is the highest salary in which we are interested, what is the arithmetic mean income?

I

It is

$$\bar{x} = \frac{\int_0^X fx \, dx}{\int_0^X f \, dx}$$

where

$$\int_0^X fx \, dx = \int_0^X (ax + bx^2 + cx^3 - dx^4) \, dx$$

$$= \frac{aX^2}{2} + \frac{bX^3}{3} + \frac{cX^4}{4} - \frac{dX^5}{5}$$

and

$$\int_0^X f \, dx = \int_0^X (a + bx + cx^2 - dx^3) \, dx$$

$$= aX + \frac{bX^2}{2} + \frac{cX^3}{3} - \frac{dX^4}{4}$$

and so, multiplying each expression by 60 to get rid of fractions, and then dividing, we have

$$\bar{x} = \frac{(30a + 20bX + 15cX^2 - 12dX^3)X}{60a + 30bX + 20cX^2 - 15dX^3}$$

unless $X = 0$.

We also define the *standard deviation* of x to be

$$\sqrt{\frac{\Sigma f(x - \bar{x})^2}{\Sigma f}}$$

in the case of a discontinuous variable. It is a measure of the spread of individual values of x around the mean.

For a continuous variable we use the equivalent definition

$$\sqrt{\int f(x - \bar{x})^2 \, dx \Big/ \int f \, dx}$$

Suppose, for example, that we have a distribution in which

$$f = a + bx$$

We need to evaluate three integrals, each between the same limits, which we may conveniently choose to be 0 and X.

(i)

$$\int_0^X f \, dx = \int_0^X (a + bx) \, dx$$

$$= \left[ax + \frac{bx^2}{2} + C \right]_0^X$$

$$= aX + \frac{bX^2}{2}$$

This is the total frequency.

(ii)
$$\int_0^X fx\,dx = \int_0^X (ax + bx^2)\,dx$$

$$= \frac{aX^2}{2} + \frac{bX^3}{3}$$

which is the total of the values of x.

(iii) These two integrals show that

$$\bar{x} = \left(\frac{aX^2}{2} + \frac{bX^3}{3}\right) \div \left(aX + \frac{bX^2}{2}\right)$$

$$= \frac{(3a + 2bX)X}{(2a + bX)3} \qquad \text{(unless } X = 0\text{)}$$

We use this in the evaluation of

$$I = \int_0^X f(x - \bar{x})^2\,dx$$

$$= \int_0^X fx^2\,dx - 2\bar{x}\int_0^X fx\,dx + \bar{x}^2\int_0^X f\,dx$$

$$= \int_0^X fx^2\,dx - 2\bar{x}\left(\bar{x} \times \int_0^X f\,dx\right) + \bar{x}^2\int_0^X f\,dx$$

where the bracketed expression follows from the definition of $\bar{x}$. We thus have that

$$I = \int_0^X fx^2\,dx - 2\bar{x}^2\int_0^X f\,dx + \bar{x}^2\int_0^X f\,dx$$

$$= \int_0^X fx^2\,dx - \bar{x}^2\int_0^X f\,dx$$

Denoting the standard deviation by σ we have

$$\sigma = \sqrt{I \Big/ \int_0^X f\,dx}$$

or
$$\sigma^2 = I \Big/ \int_0^X f\,dx$$

$$= \frac{\displaystyle\int_0^X fx^2\,dx}{\displaystyle\int_0^X f\,dx} - \bar{x}^2$$

from the expression for I which we have just found. We call σ^2 the *variance* of x. It is now easy to show that in our example

$$\sigma^2 = \frac{\left(\dfrac{4aX^3 + 3bX^4}{12}\right)}{\left(\dfrac{2aX + bX^2}{2}\right)} - \left(\frac{3a + 2bX}{2a + bX}\,\frac{X}{3}\right)^2$$

$$= \frac{4aX^2 + 3bX^3}{6(2a + bX)} - \left(\frac{X}{3}\,\frac{3a + 2bX}{2a + bX}\right)^2$$

For a fuller development of these ideas the reader must consult a book on statistics. Our purpose here is simply to show how integration enters into statistical theory.

7. Reduction formulae

Sometimes it is useful to express an integral in terms of a similar integral containing a lower exponent of a variable.

Suppose, for example, that we wish to integrate

$$I_n = \int x^n\, e^x\, dx$$

where we will soon see the significance of the subscript in I_n.

Integrating by parts we have

$$I_n = \int x^n\, e^x\, dx = e^x\, x^n - \int e^x \cdot nx^{n-1}\, dx$$

$$= x^n\, e^x - n \int x^{n-1}\, e^x\, dx$$

$$= x^n\, e^x - nI_{n-1}$$

where

$$I_{n-1} = \int x^{n-1}\, e^x\, dx$$

Thus, if we wish to integrate

$$I_4 = \int x^4\, e^x\, dx$$

we can write it as

$$I_4 = x^4\, e^x - 4\, I_3$$
$$= x^4\, e^x - 4[x^3\, e^x - 3\, I_2]$$
$$= x^4\, e^x - 4[x^3\, e^x - 3(x^2\, e^x - 2\, I_1)]$$
$$= x^4\, e^x - 4[x^3\, e^x - 3(x^2\, e^x - 2[xe^x - I_0])]$$
$$= x^4\, e^x - 4[x^3\, e^x - 3(x^2\, e^x - 2[xe^x - e^x])] + C$$
$$= (x^4 - 4x^3 + 12x^2 - 24x + 24)e^x + C$$

Now consider a second example.

$$I_{m,n} = \int \sin^m \theta \, \cos^n \theta \, d\theta$$

$$= \int \sin^m \theta \, \cos^{n-1} \theta \, d(\sin \theta)$$

which, on integration by parts, yields

$$I_{m,\,n} = \sin^{m+1} \theta \, \cos^{n-1} \theta - \int \sin \theta \, d(\sin^m \theta \, \cos^{n-1} \theta) \, d\theta$$

$$= \sin^{m+1} \theta \, \cos^{n-1} \theta$$

$$- \int \sin \theta \, [(m \, \sin^{m-1} \theta \, \cos^n \theta - (n-1) \sin^{m+1} \theta \, \cos^{n-2} \theta] \, d\theta$$

$$= \sin^{m+1} \theta \, \cos^{n-1} \theta - \int m \sin^{m} \theta \, \cos^n \theta + (n-1) \int \sin^{m+2} \theta \, \cos^{n-2} \theta \, d\theta$$

$$= \sin^{m+1} \theta \, \cos^{n-1} \theta - m \, I_{m,\,n} + (n-1) \int \sin^2 \theta \, (\sin^m \theta \, \cos^{n-2} \theta) \, d\theta$$

$$= \sin^{m+1} \theta \, \cos^{n-1} \theta - m \, I_{m,\,n} + (n-1) \int (1 - \cos^2 \theta) (\sin^m \theta \, \cos^{n-2} \theta) \, d\theta$$

$$= \sin^{m+1} \theta \, \cos^{n-1} \theta - m \, I_{m,\,n} + (n-1) \int \sin^m \theta \, \cos^{n-2} \theta \, d\theta$$

$$- (n-1) \int \sin^m \theta \, \cos^n \theta \, d\theta$$

$$= \sin^{m+1} \theta \, \cos^{n-1} \theta - m I_{m,\,n} + (n-1) I_{m,\,n-2} - (n-1) I_{m,\,n}$$

whence, by transposing the I terms to the L.H.S.

$$(m+n) I_{m,\,n} = \sin^{m+1} \theta \, \cos^{n-1} \theta + (n-1) I_{m,\,n-2}$$

This means that if, for example, $m = 6$ and $n = 3$ we could use the above result as follows

$$(m+n) I_{m,\,n} = \sin^{m+1} \theta \, \cos^{n-1} \theta + (n-1) I_{m,\,n-2}$$

whence

$$I_{m,\,n} = \frac{\sin^{m+1} \theta \, \cos^{n-1} \theta}{m+n} + \frac{n-1}{m+n} I_{m,\,n-2}$$

$$\therefore \qquad I_{6,\,3} = \frac{\sin^7 \theta \, \cos^2 \theta}{9} + \frac{2}{9} I_{6,\,1}$$

where

$$I_{6,\,1} = \int \sin^6 \theta \, \cos \theta \, d\theta$$

$$= \frac{\sin^7 \theta}{7} + C$$

Similarly, if $m=6$ and $n=5$,

$$I_{6,\,5}=\frac{\sin^7\theta\,\cos^4\theta}{11}+\frac{4}{11}I_{6,\,3}$$

where $I_{6,\,5}$ has the value just found, and so

$$\int\sin^6\theta\,\cos^5\theta\,d\theta=\frac{\sin^7\theta\,\cos^4\theta}{11}+\frac{4}{11}\left[\frac{\sin^7\theta\,\cos^2\theta}{9}+\frac{2}{63}\sin^7\theta\right]+C$$

If n is even then the answer is rather different, as is shown by

$$I_{5,\,2}=\int\sin^5\theta\,\cos^2\theta\,d\theta$$
$$=\frac{\sin^6\theta\,\cos\theta}{7}+\frac{1}{7}I_{5,\,0}$$

and now we have to evaluate

$$\int\sin^5\theta\,d\theta$$

There are a few ways of proceeding. Two of them involve reduction formulae. We can write

$$V_n=\int\sin^n\theta\,d\theta$$

and, proceeding as before, integrate by parts and show that

$$V_n=-\frac{1}{n}\cos\theta\,\sin^{n-1}\theta+\frac{n-1}{n}V_{n-2}$$

which will eventually, by successive applications, allow us to evaluate $\int\sin^5\theta\,d\theta$.

Alternatively we could note that, by writing

$$I_{m,\,n}=\int\sin^m\theta\,\cos^n\theta\,d\theta$$
$$=-\int\sin^{m-1}\theta\,\cos^n\theta\,d(\cos\theta)$$

we can obtain the alternative reduction formula

$$I_{m,\,n}=-\frac{1}{m+n}\sin^{m-1}\theta\,\cos^{n+1}\theta+\frac{m-1}{m+n}I_{m-2,\,n}$$

instead of

$$I_{m,\,n}=\frac{1}{m+n}\sin^{m+1}\theta\,\cos^{n-1}\theta+\frac{n-1}{m+n}I_{m,\,n-2}$$

Use of one or the other of these equations, or both, will always enable us to reduce

$$I_{m, n}$$

to one of the following forms

$$I_{1, 1} = \int \sin \theta \, \cos \theta \, d\theta = \tfrac{1}{2} \sin^2 \theta$$

$$I_{0, 0} = \int d\theta = \theta$$

$$I_{0, 1} = \int \cos \theta \, d\theta = \sin \theta$$

$$I_{1, 0} = \int \sin \theta \, d\theta = - \cos \theta$$

provided that m and n are positive integers.

Exercise 16.4

1. If $I_n = \int \tan^n \theta \, d\theta$ prove that $I_n = \dfrac{1}{n-1} \tan^{n-1} \theta - I_{n-2}$

2. If $I_n = \int \cot^n \theta \, d\theta$ prove that $I_n = -\dfrac{1}{n-1} \cot^{n-1} \theta - I_{n-2}$

3. If $I_n = \int \cos^n \theta \, d\theta$ prove that $I_n = \dfrac{1}{n} \sin \theta \cos^{n-1} \theta + \dfrac{n-1}{n} I_{n-2}$

4. If $I_n = \int x^n \cosh x \, dx$ and $J_n = \int x^n \sinh x \, dx$

prove that $I_n = x^n \sinh x - nJ_{n-1}$

and that $J_n = x^n \cosh x - nI_{n-1}$

CHAPTER XVII

MACLAURIN'S AND TAYLOR'S THEOREMS

Briefly, there never was upon the face
Of earth so much within so small a space.
CHAUCER : *The Knight's Tale*

1. Maclaurin's theorem

Although we did not give a rigorous proof, we have already noted that e^x may be represented by the sum of an infinite series containing powers of x. We found that this series converged very rapidly. In this chapter we draw attention to a method that enables us to express very many functions of x as infinite series of a similar kind. This is particularly useful because, if the series concerned converges rapidly, we are enabled to represent the function with reasonable accuracy by taking just the first few terms of the series, which are often easier to manipulate (or to estimate numerically) than the function itself. This will be made clear in the examples.

Maclaurin's theorem, published in 1742, states that if $f(x)$ is differentiable any number of times then

$$f(x) = f(0) + \frac{x}{1!}f'(0) + \frac{x^2}{2!}f''(0) + \dots + \frac{x^n}{n!}f^{(n)}(0) + \dots$$

where $f(0)$ indicates the value of $f(x)$ when $x = 0$, $f'(0)$ the value of $f'(x)$ when x is put equal to zero after differentiating, and $f^{(n)}(0)$ is the value of the nth derivative of x, with x put equal to zero at the last stage. This is true only if the resulting series converges.

Examples :

(i) Suppose that we wish to express e^x as a series. We have that in this case

$$f(x) = e^x \quad \text{and so} \quad f(0) = 1$$
$$f'(x) = e^x \qquad f'(0) = 1$$
$$f''(x) = e^x \qquad f''(0) = 1$$
$$f'''(x) = e^x \qquad f'''(0) = 1$$
$$f^{(n)}(x) = e^x \qquad f^{(n)}(0) = 1$$

248

Substitution of these results in the above statement of the theorem shows that

$$e^x = 1 + \frac{x}{1!} + \frac{x^2}{2!} + \frac{x^3}{3!} + \ldots + \frac{x^n}{n!} + \ldots$$

Application of one of the convergency tests mentioned in Chapter II shows that this is a convergent series.

(ii) Consider $f(x) = \sin x$

In this case we have

$$
\begin{array}{lll}
f(x) = \sin x & \text{and so} & f(0) = 0 \\
f'(x) = \cos x & & f'(0) = 1 \\
f''(x) = -\sin x & & f''(0) = 0 \\
f'''(x) = -\cos x & & f'''(0) = -1 \\
f^{\mathrm{iv}}(x) = \sin x & & f^{\mathrm{iv}}(0) = 0
\end{array}
$$

$$\text{etc.}$$

Substitution gives the result

$$\sin x = 0 + \frac{x}{1!} + \frac{0x^2}{2!} + (-1)\frac{x^3}{3!} + \frac{0x^4}{4!} + \frac{x^5}{5!} + \ldots$$

$$= x - \frac{x^3}{3!} + \frac{x^5}{5!} - \frac{x^7}{7!} + \ldots$$

Suppose now that we wish to solve an equation such as

$$x = 2 \sin x$$

where we are told that x is measured in radians. This, in fact, is typical of a kind of problem that quite often arises. An accurate solution is rather difficult, but as we have already pointed out (in Chapter I) there are occasions when a knowledge of an approximate solution enables us to obtain a more accurate one by a method such as that of Appendix 5. Now the terms in this series are becoming progressively smaller after the first few, whatever the value of x. It follows that as a first approximation to the solution of the equation

$$x = 2 \sin x$$

we may consider the solution to the equation

$$x = 2x - \frac{2x^3}{3!}$$

which has solutions $x = 0$ and $x = \pm\sqrt{3}$. These results may now be used as the basis of successive approximations to a more accurate solution as shown in Appendix 5.

(iii) It may be shown similarly that

$$\cos x = 1 - \frac{x^2}{2!} + \frac{x^4}{4!} - \frac{x^6}{6!} + \ldots$$

(iv) For the sake of completeness we record that

$$\tan x = x + \frac{x^3}{3} + \frac{2x^5}{15} + \frac{17x^7}{315} + \ldots$$

which may be proved by a similar (but rather more tedious) method. It is difficult to state the general term. A proof of this series appears in a number of the better books, including Horace Lamb's *Infinitesimal Calculus* (C.U.P.) which is very much worth possessing.

(v) The reader should be able to prove for himself that, provided $|x| < 1$ then

$$\log_e (1 + x) = x - \frac{x^2}{2} + \frac{x^3}{3} - \frac{x^4}{4} + \frac{x^5}{5} - \ldots$$

2. Taylor's theorem

This theorem, published in 1716, is really a more general form of Maclaurin's theorem. It states that

$$f(a + x) = f(a) + xf'(a) + \frac{x^2}{2!} f''(a) + \ldots + \frac{x^n}{n!} f^{(n)}(a) + \ldots$$

where $f(a)$ is the value of $f(a + x)$ when $x = 0$, and the dashes have their usual meanings. An example will make it clear.

Example :

Suppose that we wish to express $\sin\left(\tfrac{1}{4}\pi + x\right)$ as a series containing powers of x. In this case we have that

$$f(a + x) = \sin\left(\frac{\pi}{4} + x\right) \quad \text{and so} \quad f(a) = \sin\frac{\pi}{4} = \frac{1}{\sqrt{2}}$$

$$f'(a + x) = \cos\left(\frac{\pi}{4} + x\right) \qquad f'(a) = \cos\frac{\pi}{4} = \frac{1}{\sqrt{2}}$$

$$f''(a + x) = -\sin\left(\frac{\pi}{4} + x\right) \qquad f''(a) = -\sin\frac{\pi}{4} = -\frac{1}{\sqrt{2}}$$

$$f'''(a + x) = -\cos\left(\frac{\pi}{4} + x\right) \qquad f'''(a) = -\cos\frac{\pi}{4} = -\frac{1}{\sqrt{2}}$$

$$f^{iv}(a + x) = \sin\left(\frac{\pi}{4} + x\right) \qquad f^{iv}(a) = \sin\frac{\pi}{4} = \frac{1}{\sqrt{2}}$$

etc.

Substitution of these results in the above equation shows that

$$\sin\left(\frac{\pi}{4}+x\right)=\frac{1}{\sqrt{2}}+\frac{x}{\sqrt{2}}-\frac{x^2}{2!\sqrt{2}}-\frac{x^3}{3!\sqrt{2}}+\frac{x^4}{4!\sqrt{2}}+\frac{x^5}{5!\sqrt{2}}-\cdots$$

Although we ourselves shall not be using these theorems very much, many of the mathematical proofs deliberately omitted from this book are in some way or the other indebted to one or the other of these theorems. We shall use some of the series just derived in later work.

EXERCISE 17.1

1. Prove the validity of the above results for cos x and $\log_e (1+x)$.

2. Consider (by application of the tests in Chapter II) whether the series for e^x, sin x, cos x and $\log_e (1+x)$ converge for all values of x

3. Express as a series, and examine the conditions for the convergence of the series,

$$y=\frac{1}{1-x}$$

4. Repeat question 3 for the functions

$$\frac{1}{1+x}, \quad \frac{1}{(1+x)^{\frac{1}{2}}}, \quad \frac{1}{(1-x)^{\frac{1}{2}}}$$

5. Express as a series in powers of x

$$\sin\left(x-\frac{\pi}{4}\right), \quad \cos\left(x+\frac{\pi}{4}\right), \quad \cos\left(x-\frac{\pi}{4}\right)$$

SECTION IV

CALCULUS: FUNCTIONS OF MANY VARIABLES

This section is the most difficult in the book. A knowledge of it is not essential to an understanding of Section V. Students should read Appendix 6 before commencing Chapter XX.

The first two chapters extend the ideas of Chapters IX and X to cover cases in which there is more than one independent variable, thereby enabling us to deal with less artificial problems than hitherto. The most common kind of problem involves some attempt at maximisation, and the next two chapters consider this subject. These are difficult chapters, but the mathematical theory has been reduced to a minimum, and anyone who has thoroughly grasped Section III of this book should, with a little extra effort, be able to understand Section IV.

Chapter XXII quotes certain results which are frequently used in mathematical economics. Chapter XXIII describes some other forms of integration which are used extensively in advanced statistical theory and to some extent in economic theory. Chapter XXIV discusses the validity of treating dx and dy as separate entities.

CHAPTER XVIII

PARTIAL DIFFERENTIATION

He never supposed
That he might be truth, himself, or part of it,
That the things he rejected might be part
And the irregular turquoise, part, the perceptible blue
Grown denser, part, the eye so touched, so played
Upon by clouds, the ear so magnified
By thunder, parts, and all these things together,
Parts, and more things, parts.

WALLACE STEVENS : *Landscape with Boat*

1. Functions of two or more variables

So far, the whole of this book has been concerned with functions of a single variable. We have considered the effect on the demand for a good of a change in the price of that good : but we have not considered the effect on that demand of simultaneous changes in the prices of that good and of substitute goods. What happens to the demand for beef if a change in its price coincides with a change in the price of mutton? What happens if, at the same time, money wages increase? Possibly the rate of interest will also change, exerting an overall effect on spending. We could go on indefinitely adding to the list of variables which in some way or the other affect the demand for beef. In this chapter we shall see how the ideas that we have already considered may be modified to deal with this more realistic problem.

Let us continue with our example involving beef and mutton. We shall suppose, for the moment, that the supply of each commodity has a definite price schedule, and that the demand schedules are also fixed. To begin with, we shall also suppose that a change in the price of either commodity is likely to affect the supply of and demand for each : but that nothing else will influence these supplies and demands. For example, if we denote the supply by S, the demand by D, and the price by p, with subscripts b and m to denote beef and mutton, we may have equations such as

$$D_b = 100 - 0.3p_b + 0.1p_m$$
$$D_m = 50 - 0.2p_m + 0.2p_b$$
$$S_b = 90 + 0.2p_b - 0.1p_m$$
$$S_m = 40 + 0.3p_m - 0.1p_b$$

255

The first of these equations indicates that if the price of beef (p_b) increases, then the demand for beef (q_b) will fall ; while if the price of mutton (p_m) increases, the demand for beef will increase. A similar interpretation applies to the second equation. The third indicates that if the price of beef increases, the amount that suppliers will be prepared to supply (S_b) will increase ; but that if the price of mutton rises, then the suppliers of beef will reduce the amount of beef that they are prepared to supply at the prevailing price, (possibly because they become more attracted to the idea of supplying mutton instead).

These equations are, of course, purely hypothetical ones. The important point to notice is that, in all of them, the left-hand side is expressed as a function of two variables that appear on the right-hand side. A more general form of the first equation, in which D_b is expressed as a function of p_b and p_m, would be

$$D_b = f(p_b, p_m)$$

and this is a typical member of the class of equations expressing a dependent variable (z) as a function of two independent variables (x, y) :

$$z = f(x, y)$$

If we imagine ham to be yet another substitute for beef, then conceivably an increase in the price of ham (p_h) would lead to an increase in the demand for beef ; and in this case the demand for beef would be a function of three variables :

$$D_b = f(p_b, p_m, p_h)$$

More generally, if a variable is a function of n different variables, we may denote these n independent variables by $x_1, x_2, \ldots , x_n$ and write

$$y = f(x_1, x_2, \ldots , x_n)$$

which shows that the value of y depends on the values of the n different x's (of which, for example, x_1 may be the price of sugar, x_2 the price of fruit, x_3 the wage-rate in the jamming industry, x_4 the rate of interest, etc.).

In what follows we shall talk, usually, of functions of two variables ; but this is more general than it may seem, for it is a simple matter to extend the results that we shall obtain to cases of n variables.

It will be useful to consider a geometrical interpretation of a function of two variables. Suppose that we let x represent the price of beef, y the price of mutton, and z the demand for beef; and that we express z as a function of x and y thus:

$$z = f(x, y)$$

Now when we used to express y as a function of x, we found it useful to take a graph, with x measured along one axis and y along the other. Here we will replace the graph by a three-dimensional diagram (Diagram 18.1). The three axes OX, OY and OZ are at right angles (like the adjacent edges of a cube). Now suppose that the price of beef is given by Ox as shown; and that the price of mutton is given by Oy. This particular price combination may be identified with the point P' in the plane of OX and OY. Suppose that at this price combination the demand for beef is given by Oz. Then if we draw a line through P' parallel to Oz, and of the same length, we will reach a point P. This point will be identified with the prices Ox and Oy and the demand Oz. We can, in fact, speak of the values of x, y and z as being the co-ordinates of the point P, just as x and y are the co-ordinates of the point P'.

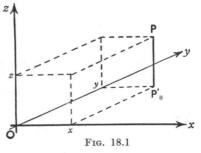

Fig. 18.1

If we let P' have somewhat different co-ordinates, then the demand will change, and so z will be different. Consequently the point P corresponding to the new P' will be at a different level than the point P corresponding to the old P'. There will, of course, be a point P corresponding to every single point P' and these points P may exist at a variety of different levels. Usually, however, just as a graph in two dimensions is fairly smooth, the points will lie on some sort of smooth surface; and the equation $z = f(x, y)$ will be the equation of this surface (just as the equation $y = f(x)$ was the equation of a curve).

2. Partial derivatives

Diagram 18.2 illustrates the kind of surface that could well represent the demand (z) for beef as a function of the price of beef (x) and the price of mutton (y).

Once again values of x are measured horizontally across the page. Values of z are measured vertically, while values of y are measured downwards through the page. The axes Ox, Oy, Oz are at right angles.

We may note that at all points in the plane of the page, such as

O, A, B and C, the value of y is zero. The curve CB relates values of z to values of x, provided that y is zero.

If we go down through the page we may reach the plane containing G, F, E and D. At all points on this plane the value of y is the same, for this plane is parallel to the axial plane containing $OABC$. Let us call this value y_1. The curve DE shows the relationship between values of z and values of x when the value of y is y_1.

Suppose, for example, that

$$z = 1000 - \sqrt{x} + y^2 + xy$$

When $y = 0$, the relationship between z and x is given by

$$z = 1000 - \sqrt{x}$$

which would have a graph of which CB could be part.

But if we put y at (say) 10, then the relationship becomes

$$z = 1000 - \sqrt{x} + 100 + 10x$$
$$= 1100 + 10x - \sqrt{x}$$

which would have a graph of which the different curve DE could be part.

In a similar way, the curves CD and BE show the relationships between z and y for two different values of x (of which one is zero).

Now let us consider the curve DE. This shows the way in which the demand for beef (measured by z) will respond to changes in price x *when the price of mutton* (y) *is kept at the value OG*. The slope of the tangent to this curve will give the rate of increase (or, since it slopes downwards, decrease) of demand with respect to a change in price of beef, *when the price of mutton is kept constant at the value OG*.

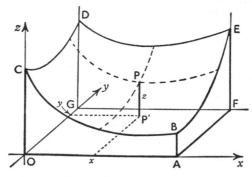

Fig. 18.2

It is very important to keep this final qualification in mind, because it may well be that a change in the price of beef will affect the price of mutton, or that some other variable (such as size of population) acts in such a way that it increases the price of mutton at the same time as it increases the price of beef. The slope of the curve DE tells us nothing about the way in which the demand for beef will change when the price of beef changes if we also allow the price of mutton to change at the same time. That is something to which we will come later.

Let us take a concrete example. We will suppose that the demand for beef is related to the price of beef and the price of mutton by the equation

$$z = 1000 - \sqrt{x} + y^2$$

where x and y are the prices, respectively, of beef and mutton measured in pence per pound. This equation represents a surface which will be similar to that drawn in the diagram.

Now suppose that the price of mutton is 36p. a pound, and that beef costs the same amount. Then the demand for beef will be given by

$$z = 1000 - 6 + 1296 = 2290 \text{ units}$$

Suppose now that the price of beef changes slightly, and *that the price of mutton is kept constant at* 36p. *per pound.* Then since the price of mutton does not vary as far as this particular problem is concerned, we can rewrite the equation as

$$z = 1000 - \sqrt{x} + 1296$$
$$= 2296 - \sqrt{x}$$

and the elasticity of demand for beef with respect to a change in the price of beef, *when the price of mutton is kept constant* at 36p. per pound, will be given by

$$\frac{x}{z} \frac{\partial z}{\partial x} = \frac{x}{z}\left(-\frac{1}{2\sqrt{x}}\right)$$

where we write $\dfrac{\partial z}{\partial x}$ instead of $\dfrac{dz}{dx}$ to indicate that we are differentiating z with respect to x *keeping everything else* (which in this case means y) *constant.*

If we now substitute the values of $x = 36$ and $z = 2296 - \sqrt{x}$ in the right-hand side of the above equation we will have that the price elasticity of demand for beef *with respect to a change in its own price only* is

$$\frac{x}{z}\frac{dz}{dx} = \frac{36}{2290}\left(-\frac{1}{12}\right) = -\frac{3}{2290}$$

To be more precise, we have that when the price of beef is 36p. a pound and *the price of mutton is maintained at 36p. a pound*, the elasticity of demand for beef with respect to a change in its own price only is -0.0013.

Now suppose that mutton costs 64p. a pound, instead of 36p., but that beef still costs 36p. a pound. Then if we keep the price of mutton steady at this higher figure, the demand equation becomes

$$z = 1000 - \sqrt{x} + 4096$$
$$= 5096 - \sqrt{x}$$

When $x = 36$, $z = 5090$, and the elasticity of demand for beef with respect to a change in its own price only will be, from the equation above,

$$-\frac{36}{5090 \times 12} = -\frac{3}{5090}$$

This demonstrates that the elasticity of demand for beef with respect to a change in the price of beef only will depend on the level at which the price of mutton is maintained. If we wish to be a little more general about it we may write

$$z = 1000 - \sqrt{x} + y^2$$

and so

$$\frac{\partial z}{\partial x} = -\frac{1}{2\sqrt{x}} \qquad \text{(since we treat } y \text{ as constant)}$$

The elasticity of demand with respect to a change in the price of beef only is therefore given by

$$\frac{x}{z}\frac{\partial z}{\partial x} = \frac{x}{1000 - \sqrt{x} + y^2} \times \frac{-1}{2\sqrt{x}}$$
$$= \frac{-\sqrt{x}}{2(1000 - \sqrt{x} + y^2)}$$

showing that this elasticity clearly depends on the level of x and of y, even though we are allowing only x to vary.

We call $\partial z/\partial x$ the *partial derivative of z with respect to x*. It is obtained by differentiating z with respect to x, treating everything else as constant. We call the elasticity we have just found the *partial price elasticity of demand*. The full significance of the adjective " partial " will appear later.

Having examined the way in which the demand for beef responds to a change in the price of beef when we keep the price of mutton constant, we may now consider how the demand for beef responds to a change in the price of mutton when we keep the price of beef constant. This will be given by

$$\frac{y}{z}\frac{\partial z}{\partial y} = \frac{y}{z}\,2y \qquad \left(\text{since } \frac{\partial z}{\partial y} = 2y \text{ when } x \text{ is constant}\right)$$

$$= \frac{2y^2}{1000 - \sqrt{x} + y^2}$$

We thus have two partial elasticities of demand for beef. The one, denoted by $\dfrac{x}{z}\dfrac{\partial z}{\partial x}$ is the partial elasticity of demand for beef with respect to the price of beef. The other, denoted by $\dfrac{y}{z}\dfrac{\partial z}{\partial y}$ is the partial elasticity of demand for beef with respect to the price of mutton. The former shows that the demand for beef falls when the price of beef rises and the price of mutton is kept constant; and that the higher the price of mutton the less does the demand for beef fall as its price increases. The latter shows that the demand for beef rises as the price of mutton rises, and that the higher the price of mutton the greater is the rise in demand for beef associated with any given proportional increase in the price of mutton. This may be seen by putting $x = 1$ and $y = 1$ in the second elasticity; and then by putting $x = 1$ and $y = 2$.

Although we have restricted our discussion to a single economic example, the idea of the partial derivative is of wide application. If we have z, a function of several variables, and wish to measure the rate of increase of z with respect to an increase in *one* of those variables, *with all the others held constant*, then we need a partial derivative. We now consider a few examples of the technique involved.

Examples

(i) $$z = 4x^2 + 2y^3 - 7.$$

To obtain $\dfrac{\partial z}{\partial x}$, we differentiate, treating y as a constant, and obtain

$$\frac{\partial z}{\partial x} = 8x$$

Similarly, treating x as constant, we obtain

$$\frac{\partial z}{\partial y} = 6y^2$$

(ii) If
$$z = 4xy^2$$

$$\frac{\partial z}{\partial x} = 4y^2 \qquad \text{(since } y \text{ is held constant)}$$

$$\frac{\partial z}{\partial y} = 8xy \qquad \text{(since } x \text{ is held constant)}$$

EXERCISE 18.1

Evaluate $\partial z/\partial x$ and $\partial z/\partial y$ when z has the following values:

(i) $3x + 4y - 3$ (ii) $x^2 + xy + 2y^2$

(iii) $(x^2 + y^2)(x^3 - y^3)$ (iv) $\dfrac{x^2 + y^2}{x^3 - y^3}$

(v) $e^{x^2 + 3y^2}$ (vi) $\log(x^2 + 3y^2)$

3. Second order partial derivatives

This last paragraph will be more easily understood if we look at it in a slightly different way. $\partial z/\partial x$ gives the slope of the curves like DE, just as in the case where there is no y to complicate things, the slope would be given by dz/dx. Corresponding to d^2z/dx^2, which measures the rate at which the gradient changes as x changes, there is $\partial^2z/\partial x^2$. Similarly we may calculate a $\partial^2z/\partial y^2$ which will indicate the rate at which $\partial z/\partial y$ increases when we increase y, still keeping x constant; $\partial z/\partial y$ gives the slope of curves such as CD.

But as we see from the diagram, the curve CB differs from the curve DE. Consequently the value of $\partial z/\partial x$ (which gives the slope of such a curve) for any particular value of x will differ from curve to curve. The value of $\partial z/\partial x$ does, in fact, depend on the value of y, just as the elasticity depends on the values of x and y. And since the value of $\partial z/\partial x$ depends on the value of y, we can usefully ask at what rate $\partial z/\partial x$ will increase as we increase y. The answer to this question is best obtained by considering a definite case.

Suppose that

$$z = 50 + 3x^2y + 2xy^2 - y^3$$

The rate at which z increases with respect to x, when y is held constant, is given by

$$\frac{\partial z}{\partial x} = 6xy + 2y^2$$

which depends on both x and y. The rate at which z increases with respect to y when x is held constant is given by

$$\frac{\partial x}{\partial y} = 3x^2 + 4xy - 3y^2$$

which also depends on both x and y.

Let us now consider $\partial z/\partial x$ which gives the slope of the curve such as DE.

This slope changes as x changes (i.e., as we move along DE), and the rate at which it does is given by

$$\frac{\partial}{\partial x}\left(\frac{\partial z}{\partial x}\right) = \frac{\partial}{\partial x}(6xy + 2y^2) = 6y$$

i.e.,
$$\frac{\partial^2 z}{\partial x^2} = 6y$$

We have similarly that the function $\partial z/\partial y$ changes with respect to changes in y at a rate $\partial^2 z/\partial y^2$, which is

$$\frac{\partial}{\partial y}\left(\frac{\partial z}{\partial y}\right) = \frac{\partial}{\partial y}(3x^2 + 4xy - 3y^2)$$

$$= 4x - 6y$$

But also the value of $\partial z/\partial x$ changes as y changes, and the rate at which it does so is given by

$$\frac{\partial^2 z}{\partial y\,\partial x} = \frac{\partial}{\partial y}\left(\frac{\partial z}{\partial x}\right) = \frac{\partial}{\partial y}(6xy + 2y^2)$$

$$= 6x + 4y$$

This may be interpreted as giving the rate at which the slope (at a given value of x) of a curve such as CB changes as the curve is shifted towards the position DE.

Similarly we may define $\dfrac{\partial^2 z}{\partial x\,\partial y}$ to be $\dfrac{\partial}{\partial x}\dfrac{\partial z}{\partial y}$ which is the rate at which $\dfrac{\partial z}{\partial y}$ changes as we change x (or the rate at which the gradient of the curve CD changes as it moves towards the position BE).

We have that

$$\frac{\partial^2 z}{\partial x\,\partial y} = \frac{\partial}{\partial x}\left(\frac{\partial z}{\partial y}\right) = \frac{\partial}{\partial x}(3x^2 + 4xy - 3y^2)$$

$$= 6x + 4y$$

which, we may notice, is the same as the value of $\dfrac{\partial^2 z}{\partial y\,\partial x}$.

We call $\dfrac{\partial^2 z}{\partial y\,\partial x}$, $\dfrac{\partial^2 z}{\partial x\,\partial y}$, $\dfrac{\partial^2 z}{\partial x^2}$ and $\dfrac{\partial^2 z}{\partial y^2}$, *second order partial derivatives*, the first two being called *mixed* or *cross* derivatives.

It can be shown that provided certain conditions about continuity hold (and they almost always do) then

$$\frac{\partial^2 z}{\partial y\,\partial x} = \frac{\partial^2 z}{\partial x\,\partial y}$$

It is a good exercise to write out a few equations such as that which we have just looked at in our example and to derive the two first order and four second order derivatives, checking the results by seeing that the two mixed derivatives are identical.

4. Examples

We now do a few examples of this kind, introducing at the same time a less cumbersome notation, in which the ∂'s are omitted, and the variables with respect to which the differentiation has been performed are denoted by subscripts, thus:

(i) $$z = 3x^2 + xy + 2y$$

$$z_x = \frac{\partial z}{\partial x} = 6x + y \qquad z_y = \frac{\partial z}{\partial y} = x + 2$$

$$z_{xx} = \frac{\partial^2 z}{\partial x^2} = 6 \qquad z_{yy} = \frac{\partial^2 z}{\partial y^2} = 0$$

$$z_{yx} = \frac{\partial^2 z}{\partial y\,\partial x} = 1 \qquad z_{xy} = \frac{\partial^2 z}{\partial x\,\partial y} = 1$$

(ii) $$z = y \sin x + x \cos y$$

$$z_x = y \cos x + \cos y \qquad z_y = \sin x - x \sin y$$

$$z_{xx} = -y \sin x \qquad z_{yy} = -x \cos y$$

$$z_{yx} = \cos x - \sin y \qquad z_{xy} = \cos x - \sin y$$

(iii) $$z = e^{x+y}(\cos x + \cos y)$$

$$z_x = e^{x+y}(\cos x + \cos y) - e^{x+y}\sin x$$

$$z_y = e^{x+y}(\cos x + \cos y) - e^{x+y}\sin y$$

$$z_{xx} = e^{x+y}(\cos x + \cos y - \sin x) - e^{x+y}(\sin x - \cos x)$$

$$z_{yy} = e^{x+y}(\cos x + \cos y - \sin y) - e^{x+y}(\sin y - \cos y)$$

$$z_{yx} = e^{x+y}(\cos x + \cos y - \sin x) - e^{x+y}\sin y$$

$$z_{xy} = e^{x+y}(\cos x + \cos y - \sin y) - e^{x+y}\sin x$$

EXERCISE 18.2

1. Obtain z_x, z_y, z_{xx}, z_{yy}, z_{xy} and z_{yx} for the following functions, checking that $z_{xy} = z_{yx}$:

(i) $z = x^2 + xy + y^2$

(ii) $z = (x^2 + 3xy + y^2)(x^2 + y^2)$

(iii) $z = 4x^3 y^2 + 3xy^3$

(iv) $z = (x^2 + y^2)(x^3 + 5y^3)$

(v) $z = x^2 \sin y$

(vi) $z = (x^2 + y^3)(\cos x + \cos y)$

(vii) $z = e^{x^2 + y^2}$

(viii) $z = e^{x^2 + 2xy + 3y^2}$

(ix) $z = \log(x^2 + y^2)$

(x) $z = \log(x^3 + y^4)$

(xi) $z = \log(\sin x + \cos y)$

(xii) $z = \log \sin x - \log \cos y$

(xiii) $z = \dfrac{x^3 + 4y^2}{y \sin x}$

(xiv) $z = \dfrac{(x^2 + y^2) \sin x}{(x^3 + y^3) \sin y}$

CHAPTER XIX

PARTIAL DIFFERENTIATION (*Continued*)

I am sorry
if you have
a green pain
gnawing your brain away.
I suppose
quite a lot of it is
gnawed away
by this time.

G. K. CHESTERTON: *To a Modern Poet*

1. Total derivatives

In the last chapter we saw that if z is a function of x and y, i.e., if $z = f(x, y)$, then $z_x = \dfrac{\partial z}{\partial x}$ measures the rate of increase of z with respect to x when y is held constant; and that $z_y = \dfrac{\partial z}{\partial y}$ measures the rate of increase of z with respect to y if x is held constant.

In this chapter we consider the problem of finding how the value of z behaves if x and y vary *simultaneously* at given rates.

It may help us to present the main ideas of the chapter if we think of a very common kind of three-dimensional surface—a mountain.

Any point on it has three co-ordinates—its latitude, measured in terms of its (angular) distance north or south of the Equator, its longitude (East-West), and its altitude (above sea level).

If we stand on the mounta   and try to measure the slope of the ground the answer will normally depend on where we stand and the direction in which we measure the slope. At a given point there could be a steep slope in the North-South direction, but only a gentle gradient in the East-West direction.

As part of a systematic study of the slope of the mountain, we could (but need not) ask the following questions:

If I walk along the mountain at a constant distance north of the Equator (say, along the 49th Parallel) how does its East-West slope change?

How does the North-South slope change from point to point along the 49th Parallel?

How do the answers to these questions alter if I repeat the measurements 100 yards north of the 49th Parallel?

The idea that emerges is that we may be interested not only in the extent to which the gradient in the East-West direction may depend on our position, but also in the rate at which this gradient may change as we alter our position by moving either East-West or North-South, or, presumably, in some other direction.

An analogous example in price theory is easily stated. We know that a change in the price of beef is likely to affect the demand for beef. We are interested in knowing to what extent this effect may change as beef becomes dearer, or cheaper. We would also like to know to what extent this effect on the demand for beef of a change in the price of beef may alter as we alter the price of mutton.

For example, if both the price of beef and the price of mutton are functions of time, then clearly the demand for beef will be a function of time. Suppose that, as before, the demand z is related to the price of beef (x) and the price of mutton (y) by the equation

$$z = 1000 - \sqrt{x} + y^2$$

Now suppose that the price of beef varies over time in such a way that

$$x = A^t(\alpha + \cos 2\pi t)$$

where t is the number of days measured from a fixed date. (Examination of this equation will show that there is a steady rise in the price of beef accompanied by a periodic fluctuation.) We also suppose that the price of mutton fluctuates in a similar way, being

$$y = B^t(\beta + \sin 2\pi t)$$

Now there is nothing to stop us from substituting these values of x and y in the equation for z which will give us the daily demand for beef, namely,

$$z = 1000 - \sqrt{A^t(\alpha + \cos 2\pi t)} + [B^t(\beta + \sin 2\pi t)]^2$$

Here we have z as a function of a single variable t, and we could obtain the rate of increase of z with respect to t by differentiating this function in the ordinary way. To do so, however, is rather messy. There is a neater and easier solution to the problem. Instead of performing this substitution and then differentiating we work out four derivatives and then combine them, using the fact that

$$\frac{dz}{dt} = \frac{\partial z}{\partial x}\frac{dx}{dt} + \frac{\partial z}{\partial y}\frac{dy}{dt}$$

which is a formula that we shall discuss in a moment. Let us, meanwhile, take it on trust and use it in this case before us.

We have

$$z = 1000 - \sqrt{x} + y^2$$

and so

$$\frac{\partial z}{\partial x} = -\frac{1}{2\sqrt{x}} \quad \text{and} \quad \frac{\partial z}{\partial y} = 2y$$

Remembering that

$$\frac{d}{dt}(uv) = u\frac{dv}{dt} + v\frac{du}{dt}$$

we have

$$\frac{dx}{dt} = A^t(\alpha + \cos 2\pi t) \log_e A - A^t \cdot 2\pi \sin 2\pi t$$

and

$$\frac{dy}{dt} = B^t(\beta + \sin 2\pi t) \log_e B + B^t \cdot 2\pi \cos 2\pi t$$

Substitution of these results in the formula for $\dfrac{dz}{dt}$ yields

$$\frac{dz}{dt} = -\frac{1}{2\sqrt{x}}[A^t(\alpha + \cos 2\pi t) \log_e A - A^t \cdot 2\pi \sin 2\pi t]$$

$$+ 2y[B^t(\beta + \sin 2\pi t) \log_e B + B^t \cdot 2\pi \cos 2\pi t]$$

which yields, when the values of x and y are substituted into this equation, the rate of increase of demand (z) with respect to time (t)

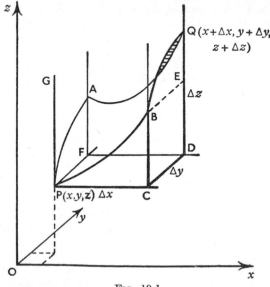

FIG. 19.1

as the result of changes in the two prices. Admittedly it is a pretty complicated expression ; but it is a pretty complicated question. Certainly to obtain this solution by substituting first and then differentiating would have been less straightforward than the method we have adopted of performing four reasonably simple differentiations and then substituting in the very last stage. There would also have been more opportunities for making errors.

The meaning and validity of the formula we have just used may be demonstrated by means of a three-dimensional diagram.

Suppose that we have a point whose co-ordinates are (x, y, z) and which is denoted by P. Let Q be another point on the same surface as P such that its co-ordinates are $(x + \Delta x, y + \Delta y, z + \Delta z)$. These points are shown in Diagram 19.1 along with part of the surface concerned. The lines PC, PF and PG have been drawn parallel to the axes. It is clear from the diagram that PC denotes Δx, CD denotes Δy, and DQ denotes Δz.

Now through B draw a line BE parallel to CD cutting DQ in E. Then we have that

$$\Delta z = DQ = DE + EQ = BC + EQ$$

Let the slope of the straight line PB be denoted by m. Then the length BC is equal to $m \, \Delta x$.

Also let the slope of the chord BQ be denoted by n. Then we have, by considering the triangle BQE, that QE is of length $n \, \Delta y$.

We thus have that

$$\Delta z = BC + EQ$$
$$= m \, \Delta x + n \, \Delta y$$

Suppose now that it takes a time Δt for the value of z to increase to $z + \Delta z$. Then the average rate of increase of z during time Δt will be given by

$$\frac{\Delta z}{\Delta t} = m \, \frac{\Delta x}{\Delta t} + n \, \frac{\Delta y}{\Delta t}$$

Now let Δt tend to zero. Then Δx, Δy and Δz will also tend to zero (since the position Q will tend to the position P). This means that $\Delta z / \Delta t$ will tend to dz/dt (by definition), $\Delta x / \Delta t$ to dx/dt, and $\Delta y / \Delta t$ to dy/dt.

Furthermore the slope m of the chord PB will tend to the slope of the tangent to the curve PB at the point P, which is given by $\partial z / \partial x$; and the slope of the chord BQ will tend to the slope of the tangent to the curve BQ at the point B (which will tend to coincide with P), and this is given by $\partial z / \partial y$.

Although this is not a rigorous proof, we may therefore conclude that, in the limit, the expression

$$\frac{\Delta z}{\Delta t} = m\frac{\Delta x}{\Delta t} + n\frac{\Delta y}{\Delta t}$$

will tend to

$$\frac{dz}{dt} = \frac{\partial z}{\partial x}\frac{dx}{dt} + \frac{\partial z}{\partial y}\frac{dy}{dt}$$

which is the expression that we used in the example.

We call dz/dt the *total* derivative of z with respect to t. It is "total" in the sense that it takes into account all the "partial" variations of z, and does not presuppose that anything is conveniently held constant.

It should be noted that if Δz, Δx and Δy are small, but not so small that they cannot be measured, then it is *approximately* true that

$$\Delta z = \frac{\partial z}{\partial x}\Delta x + \frac{\partial z}{\partial y}\Delta y$$

Although we have found it convenient to think of x and y as measuring prices, z demand and t time this is clearly not essential. If z is a function of x and y, which are in turn functions of t, then the above relationship is true. dz/dt measures the rate of increase of z with respect to t when both x and y are allowed to vary.

EXERCISE 19.1

1. If $z = x^2 - y^2$ where $x = e^t \cos t$ and $y = e^t \sin t$ find dz/dt
2. If $z = x^3y^2$ where $x = e^t$ and $y = (1+t)^2$ find dz/dt
3. If $z = x^3y^2$ where $x = (1+t)^2$ and $y = e^t$ find dz/dt
4. If $z = x \cos y$ where $x = e^t$ and $y = 1 + t^2$ find dz/dt
5. If $z = wx^2y^3$ where $w = t^2$, $x = e^t$ and $y = \sin t$ find dz/dt

$$\left(\text{use } \frac{dz}{dt} = \frac{\partial z}{\partial w}\frac{dw}{dt} + \frac{\partial z}{\partial x}\frac{dx}{dt} + \frac{\partial z}{\partial y}\frac{dy}{dt}\right)$$

An important special case arises. Suppose that z is a function of x and y, and that y is a function of x (e.g., demand for beef is a function of supply of mutton and price of beef; and price of beef is a function of the supply of mutton). Then this really means identifying t with x, and we have that

$$\frac{dz}{dt} = \frac{\partial z}{\partial x}\frac{dx}{dt} + \frac{\partial z}{\partial y}\frac{dy}{dt}$$

becomes

$$\frac{dz}{dx} = \frac{\partial z}{\partial x} + \frac{\partial z}{\partial y}\frac{dy}{dx} \quad \left(\text{since } \frac{dx}{dt} \text{ becomes } \frac{dt}{dt} = 1 \right)$$

which expresses the *total* derivative of z with respect to x in terms of the *partial* derivative of z with respect to x, and the *product* of the partial derivative of z with respect to y *by* the derivative of y with respect to x.

For example,

if $\qquad\qquad z = 2x + \sin y$

and $\qquad\qquad y = x + \cos x$

then $\qquad\qquad \dfrac{\partial z}{\partial x} = 2$

Also $\qquad\qquad \dfrac{\partial z}{\partial y} = \cos y$

and $\qquad\qquad \dfrac{dy}{dx} = 1 - \sin x$

and so $\qquad \dfrac{dz}{dx} = \dfrac{\partial z}{\partial x} + \dfrac{\partial z}{\partial y}\dfrac{dy}{dx} = 2 + (1 - \sin x)(\cos y)$

$$= 2 + (1 - \sin x)\cos(x + \cos x)$$

This is clearly much simpler than substituting $y = x + \cos x$ in the expression for z getting

$$z = 2x + \sin(x + \cos x)$$

and then differentiating.

EXERCISE 19.2

1. Find dz/dx in the following cases:

(i) $z = x^2 + y^3$ when $y = \sqrt{1 - x^2}$

(ii) $z = x^2 \sin y$ when $y = \sqrt{1 - x^2}$

(iii) $z = x^3 \sin y$ when $y = x \cos x$

(iv) $z = e^{x^2 + y^2}$ when $y = e^{x^2}$

2. The cost of producing x units of a good is $(100 + 0 \cdot 1x + 0 \cdot 01x^2)$, so that the net revenue obtained by producing x units and selling them at a price p is $R = xp - (100 + 0 \cdot 1x + 0 \cdot 01x^2)$. The demand equation shows that a quantity x will be sold if the price p is such that $p = \dfrac{x + 1}{(x + 2)^2}$, and this equation also indicates the price p that will prevail when production is x. Obtain the total rate of change of net revenue with respect to a change in output.

Before proceeding to consider some applications of these results the following may be noted, although there is no attempt to prove them.

✗   (1) If z is a function of x and y, and x and y are functions of t then (as we have shown)

$$\frac{dz}{dt} = \frac{\partial z}{\partial x}\frac{dx}{dt} + \frac{\partial z}{\partial y}\frac{dy}{dt}$$

and (as we have not shown)

$$\frac{d^2z}{dt^2} = \frac{\partial^2 z}{\partial x^2}\left(\frac{dx}{dt}\right)^2 + 2\frac{\partial^2 z}{\partial x\,\partial y}\frac{dx}{dt}\frac{dy}{dt} + \frac{\partial^2 z}{\partial y^2}\left(\frac{dy}{dt}\right)^2 + \frac{\partial z}{\partial x}\frac{d^2x}{dt^2} + \frac{\partial z}{\partial y}\frac{d^2y}{dt^2}$$

If we denote dx/dt by x_t, dy/dt by y_t, etc., and d^2x/dt^2 by x_{tt}, etc. then this can be more compactly written as

$$z_{tt} = z_{xx}x_t{}^2 + 2z_{xy}x_ty_t + z_{yy}y_t{}^2 + z_xx_{tt} + z_yy_{tt} \qquad ✗$$

Example :

If $\qquad z = x^2 + 3xy + 2y^2$

and $\qquad x = t + t^2$ and $y = \sin t$

then $\quad z_x = \dfrac{\partial z}{\partial x} = 2x + 3y$; $z_y = \dfrac{\partial z}{\partial y} = 3x + 4y$

$\qquad z_{xx} = 2 \qquad\quad$; $z_{yy} = \quad 4$

$\qquad\qquad z_{xy} = 3$

$\qquad x_t = 1 + 2t \qquad$; $y_t = \cos t$

$\qquad x_{tt} = \quad 2 \qquad\quad$; $y_{tt} = -\sin t$

and so

$$\begin{aligned}
z_{tt} &= 2(1+2t)^2 + 2\,.\,3(1+2t)(\cos t) + 4\cos^2 t \\
&\quad + (2x+3y)2 + (3x+4y)(-\sin t) \\
&= 2(1+2t)^2 + 6(1+2t)\cos t + 4\cos^2 t \\
&\quad + 2(2t+2t^2+3\sin t) - (3t+3t^2+4\sin t)\sin t \\
&= 2(1+6t+6t^2) + (6-3t-3t^2)\sin t \\
&\quad + 6(1+2t)\cos t + 4\cos^2 t - 4\sin^2 t
\end{aligned}$$

✗   (2) If z is a function of x and y, and x and y are themselves functions of u and v, then $\quad$ ✗

$$\frac{\partial z}{\partial u} = \frac{\partial z}{\partial x}\frac{\partial x}{\partial u} + \frac{\partial z}{\partial y}\frac{\partial y}{\partial u} \qquad\qquad z_u = z_xx_u + z_yy_u$$

and

$$\frac{\partial z}{\partial v} = \frac{\partial z}{\partial x}\frac{\partial x}{\partial v} + \frac{\partial z}{\partial y}\frac{\partial y}{\partial v} \qquad\qquad z_v = z_xx_v + z_yy_v$$

and
$$\frac{\partial^2 z}{\partial u^2} = \frac{\partial^2 z}{\partial x^2}\left(\frac{\partial x}{\partial u}\right)^2 + 2\frac{\partial^2 z}{\partial x\,\partial y}\frac{\partial x}{\partial u}\frac{\partial y}{\partial u} + \frac{\partial^2 z}{\partial y^2}\left(\frac{\partial y}{\partial u}\right)^2 + \frac{\partial z}{\partial x}\left(\frac{\partial^2 x}{\partial u^2}\right) + \frac{\partial z}{\partial y}\frac{\partial^2 y}{\partial u^2}$$

which is more neatly written as

$$z_{uu} = z_{xx}x_u{}^2 + 2z_{xy}x_u y_u + z_{yy}y_u{}^2 + z_x x_{uu} + z_y y_{uu}$$

and similarly

$$z_{vv} = z_{xx}x_v{}^2 + 2z_{xy}x_v y_v + z_{yy}y_v{}^2 + z_x x_{vv} + z_y y_{vv}$$

while

$$z_{uv} = z_{xx}x_u x_v + z_{xy}(x_u y_v + x_v y_u) + z_{yy}y_u y_v + z_x x_{uv} + z_y y_{uv}$$

Also

$$z_{xx} = z_{uu}u_x{}^2 + 2z_{uv}u_x v_x + z_{vv}v_x{}^2 + z_u u_{xx} + z_v v_{xx}$$

with similar expressions for z_{xy} and z_{yy}.

Example :

If $\qquad z = x^2 + 2xy + 3y^2$

and $\qquad x = u\sin v \qquad$ while $\qquad y = v\cos u$

$$\begin{aligned}
x_u &= \ \sin v & y_u &= -v\sin u \\
x_{uu} &= \quad 0 & y_{uu} &= -v\cos u \\
z_x &= 2x + 2y & z_y &= 2x + 6y \\
z_{xx} &= 2 & z_{yy} &= \quad 6 \\
& z_{xy} = 2
\end{aligned}$$

whence

$$\begin{aligned}
z_{uu} &= 2\sin^2 v + 2 \cdot 2\,(\sin v)(-v\sin u) + 6(-v\sin u)^2 \\
&\quad + 0 + (2x + 6y)(-v\cos u) \\
&= 2\sin^2 v - 4v\sin u\sin v + 6v^2\sin^2 u \\
&\quad + 2(u\sin v + 3v\cos u)(-v\cos u)
\end{aligned}$$

Finally we should note that although the above results are restricted to the case where z is a function of two variables x and y, they can be extended to the case where z is a function of n variables which we will denote by $x_1, x_2, x_3, \ldots, x_n$.

In such a case

$$\frac{dz}{dt} = \frac{\partial z}{\partial x_1}\frac{dx_1}{dt} + \frac{\partial z}{\partial x_2}\frac{dx_2}{dt} + \ldots + \frac{\partial z}{\partial x_n}\frac{dx_n}{dt}$$

which may conveniently be denoted by

$$\sum_{r=1}^{n}\frac{\partial z}{\partial x_r}\frac{dx_r}{dt}$$

or even by

$$\sum_{r=1}^{n} z_{x_r}(x_r)_t$$

K

The expression for the second derivative in such a case is somewhat more complicated. It is

$$\frac{d^2z}{dt^2} = \sum_{r=1}^{n} z_{x_r x_r}\left(\frac{dx_r}{dt}\right)^2 + 2 \sum_{\substack{r,s=1 \\ r\neq s}}^{n} z_{x_r x_s}\frac{dx_r}{dt}\frac{dx_s}{dt} + \sum_{r=1}^{n} z_{x_r}\frac{d^2x_r}{dt^2}$$

where $\displaystyle\sum_{r=1}^{n} z_{x_r x_r}\left(\frac{dx_r}{dt}\right)^2$ denotes the sum of all the products like

$$z_{x_1 x_1}\left(\frac{dx_1}{dt}\right)^2, \quad z_{x_2 x_2}\left(\frac{dx_2}{dt}\right)^2, \quad \text{etc., i.e., the}$$

products of the unmixed second derivatives and the square of the corresponding first derivative with respect to t;

where $\displaystyle\sum_{\substack{r,s=1 \\ r\neq s}}^{n} z_{x_r x_s}\frac{dx_r}{dt}\frac{dx_s}{dt}$ denotes the sum of all possible terms like

$$z_{x_1 x_2}\frac{dx_1}{dt}\frac{dx_2}{dt}\text{ i.e., the products of every mixed}$$

derivative with its two corresponding first derivatives;

and $\displaystyle\sum_{r=1}^{n} z_{x_r}\frac{d^2x_r}{dt^2}$ denotes the sum of all terms like $z_{xr}\dfrac{d^2x_r}{dt^2}$

i.e., the product of each first derivative with the corresponding second derivative with respect to t.

Similar expressions exist for the partial second derivatives. We shall not be using these results much.

CHAPTER XX

MAXIMA AND MINIMA OF FUNCTIONS OF TWO OR MORE VARIABLES

The statements was interesting, but tough.
MARK TWAIN : *The Adventures of Huckleberry Finn*

1. Functions of two variables

We now start to consider a problem which can become very complicated. There are complications in the theory, and complications in the practical application of the theory. Since this is intended to be a practical book, rather than a book on mathematical theory, we shall content ourselves in this chapter with stating the theoretical

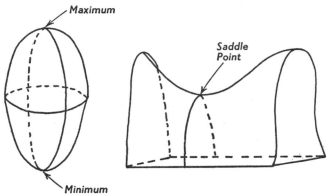

FIG. 20.1

results, and considering their application. No attempt at all will be made to justify these results.

We will begin by considering functions of two variables. If we have z as a function of x and y, then we define a maximum value of z to be the value which is such that a small movement along the surface from that point will result in a smaller value of z no matter in what direction the movement occurs. It is like the top of a mountain.

A minimum is defined similarly. It is a point which is such that a small displacement from it in any direction will result in an increased value of z.

275

We also define a saddle point which may be regarded as a point such that a movement from it parallel to some arbitrary axis will result in an increased value of z, but a movement parallel to some other axis will result in a decreased value of z. It is like the point at the top of a pass.

These are by no means rigorous definitions ; but they are adequate. They are illustrated in Diagram 20.1.

Just as in Chapter XI we found that when y was a function of x, its maximum and minimum values could be obtained by examining the first and second derivatives of y with respect to x, so, in this case, we have to examine the derivatives of z with respect to x and y. Without attempting to prove them, we now list the conditions under which maxima, minima and saddle points in the value of z may arise.

Maximum :

If (i) $z_x = z_y = 0$; $z_{xx} < 0$, $z_{yy} < 0$ and (ii) $z_{xx}z_{yy} > (z_{xy})^2$ then z *is a* maximum.

If, however, the first set of conditions holds but, instead of (ii), we have $z_{xx}z_{yy} = (z_{xy})^2$, there *may* be a maximum, but further investigation is necessary before we can be certain.

Minimum :

If (i) $z_x = z_y = 0$; $z_{xx} > 0$, $z_{yy} > 0$ and (ii) $z_{xx}z_{yy} > (z_{xy})^2$ then z *is a* minimum.

If, however, the first set of conditions holds but, instead of (ii), we have $z_{xx}z_{yy} = (z_{xy})^2$, there *may* be minimum.

Saddle Point :

If $z_x = z_y = 0$ and $z_{xx}z_{yy} < (z_{xy})^2$ then we have a saddle point.

Examples :

(i) Consider the surface $z = x^2 + y^2 - p$ where p is a constant. Here we have $z_x = 2x$; $z_y = 2y$; $z_{xx} = 2$; $z_{yy} = 2$; $z_{xy} = 0$. At the point $(0, 0)$ both z_x and z_y are zero. Also, $z_{xx}z_{yy} = 4 > (z_{xy})^2$ and both z_{xx} and z_{yy} are positive. Consequently z has a minimum value at $(0, 0)$, the value being $-p$.

(ii) Consider now the surface $z = p - x^2 - y^2$. Here we have that $z_x = -2x$; $z_y = -2y$; $z_{xx} = -2$; $z_{yy} = -2$; $z_{xy} = 0$. At the point $(0, 0)$, $z_x = z_y = 0$. The values of z_{xx} and z_{yy} are negative. Also, since $z_{xx}z_{yy} = 4 > (z_{xy})^2$, there is a maximum value to z when x and y are zero. This is the point $(0, 0, p)$.

(iii) The surface $z = x^2 - y^2 + p$ gives a similar result, but this time $z_{xx} = 2$ and $z_{yy} = -2$, leading to $z_{xx}z_{yy} < (z_{xy})^2$ which indicates a saddle point corresponding to $(0, 0, p)$.

(iv) The surface

$$z = 4x^3 - 24x^2 - 4xy + y^2 + 48x - 2y$$

gives first derivatives

$$z_x = 12x^2 - 48x - 4y + 48 \text{ and } z_y = -4x + 2y - 2$$

whence $\quad z_{xx} = 24x - 48$; $z_{yy} = 2$; $z_{xy} = -4$

In order for there to be a maximum, minimum or saddle point $z_x = z_y = 0$. This will be so when

$$12x^2 - 48x - 4y + 48 = 0$$
and $\qquad\qquad\quad -4x + 2y - 2 = 0$

The second of these may be written as $y = 1 + 2x$.

Solution of these two equations simultaneously yields the quadratic equation

$$3x^2 - 14x + 11 = 0$$

This has solutions $x = 1$ and $x = 3\frac{2}{3}$. In the former case, the condition for $z_x = 0$ shows us that $y = 1 + 2x = 3$, and in the latter case $y = 8\frac{1}{3}$. We must therefore consider the values of the second derivatives at the points on the surface where $x = 1$ and $y = 3$; and where $x = 3\frac{2}{3}$ and $y = 8\frac{1}{3}$.

At the point where $x = 1$ and $y = 3$, $z_{xx} = -24$, $z_{yy} = 2$, $z_{xy} = -4$. Consequently $z_{xx}z_{yy} < (z_{xy})^2$ and we have a saddle point, at which the value of z is 19.

At the point where $x = 3\frac{2}{3}$ and $y = 8\frac{1}{3}$, $z_{xx} = 40$, $z_{yy} = 2$, $z_{xy} = -4$. Consequently $z_{xx}z_{yy} > (z_{xy})^2$ and so, since the second derivatives are both positive, we have a minimum value of z at this point, it being $-18\frac{25}{27}$.

EXERCISE 20.1

1. Locate the maximum, minimum and saddle points of the following functions :

    (i) $z = x^3 + y^3 - p$ (ii) $z = x^4 + y^4 - 2(x-y)^2$
   (iii) $z = x^3 - y^3 - p$ (iv) $z = x^4 - y^4 + 2(x-y)^2$
    (v) $z = \cos x \sin y$ within the range $-\pi \leqslant x \leqslant \pi$, $-\pi \leqslant y \leqslant \pi$.

2. Functions of many variables ✓

The results we have just quoted are valid when z is a function of two variables. If z is a function of n variables then a *stationary value* of z will exist where all of the first derivatives are zero, i.e., where

$$z_{x_1} = z_{x_2} = z_{x_3} = \ldots z_{x_n} = 0$$

The point concerned will be a *minimum* if all of the determinants

$$z_{x_1 x_1}, \quad \begin{vmatrix} z_{x_1 x_1} & z_{x_1 x_2} \\ z_{x_1 x_2} & z_{x_2 x_2} \end{vmatrix}, \quad \begin{vmatrix} z_{x_1 x_1} & z_{x_1 x_2} & z_{x_1 x_3} \\ z_{x_1 x_2} & z_{x_2 x_2} & z_{x_2 x_3} \\ z_{x_1 x_3} & z_{x_2 x_3} & z_{x_3 x_3} \end{vmatrix} \quad \cdots \quad \begin{vmatrix} z_{x_1 x_1} & z_{x_1 x_2} & \cdots & z_{x_1 x_n} \\ z_{x_1 x_2} & z_{x_2 x_2} & \cdots & z_{x_2 x_n} \\ \cdots\cdots\cdots\cdots\cdots\cdots\cdots\cdots \\ \cdots\cdots\cdots\cdots\cdots\cdots\cdots\cdots \\ z_{x_1 x_n} & z_{x_2 x_n} & \cdots & z_{x_n x_n} \end{vmatrix}$$

are positive (see Chapter XXI).

It will be a *maximum* if the determinants of odd order are negative, but those of even order positive.

Example :

We may illustrate the application of this method by considering a manufacturer who is able to produce four different kinds of goods. We shall suppose that his net revenue is given by z (in thousands of pounds) which is related to his outputs of the four different goods by the equation

$$z = 2x_1 x_2 x_3 - (2x_1{}^2 + 4x_2{}^2 + 23x_3{}^2 + 2x_4{}^2) + 4x_2 + 90x_3 + 4x_4$$

where the x's represent the various outputs measured in hundreds of tons.

The manufacturer wishes to decide on the levels of production which will maximise his total net revenue. In other words, he wishes to maximise z subject to variations in the four x's.

Here we have that

$$
\begin{aligned}
z_{x_1} &= 2x_2 x_3 - 4x_1 \\
z_{x_2} &= 2x_1 x_3 \qquad\quad - 8x_2 \qquad\qquad + 4 \\
z_{x_3} &= 2x_2 x_1 \qquad\qquad\qquad - 46x_3 \qquad + 90 \\
z_{x_4} &= \qquad\qquad\qquad\qquad\qquad - 4x_4 + 4
\end{aligned}
$$

At a stationary value these have all to be zero. The fourth provides that $x_4 = 1$. The other three have now to be solved simultaneously.

From the first we have that

$$x_1 = \frac{x_2 x_3}{2}$$

Substitution of this in the second and third give

$$8x_2 = x_2 x_3{}^2 + 4$$

and

$$46x_3 = x_2{}^2 x_3 + 90$$

From the first of these new equations we get

$$x_2 = \frac{4}{8 - x_2{}^2}$$

and substitution of this in the second gives us

$$46x_3 = \frac{16x_3}{(8 - x_3{}^2)^2} + 90$$

This last equation has to be satisfied by the value of x at a turning point ; and so we have to solve it. To do this we write it as

$$(23x_3 - 45)(8 - x_3{}^2)^2 = 8x_3$$

and then expand it, getting the quintic equation

$$23x_3{}^5 - 45x_3{}^4 - 368x_3{}^3 + 720x_3{}^2 + 1464x_3 - 2880 = 0$$

We are now faced with the problem of solving this equation. It is good practice in such a case to begin by trying a few guesses with the help of the Remainder Theorem. If we do this we will find that two solutions are given by $x_3 = 2$ and $x_3 = 3$. This means that we may divide the above expression by $(x_3 - 2)(x_3 - 3) = (x_3{}^2 - 5x_3 + 6)$. If we perform this division we obtain the left-hand side of the above expression as

$$(x_3 - 2)(x_3 - 3)(23x_3{}^3 + 70x_3{}^2 - 156x_3 - 480)$$

and x_3 has to be such that this is zero. Two solutions are given by $x_3 = 2$ and $x_3 = 3$. Three other solutions are given by the roots of the cubic equation obtained by putting the last bracket equal to zero. Strictly speaking we should do this : but there is no easy solution to this equation, and as this is merely an artificial example there is no point in going to the immense amount of purely numerical labour that is involved in its solution. Let us just examine the two solutions we have found, keeping in mind the fact that the cubic equation will certainly provide one other solution (and possibly three).

We have found two values of x_3 that satisfy the four equations obtained by equating the first derivatives to zero ; we also know that the last of these equations is satisfied by $x_4 = 1$. By substitution of $x_3 = 2$ in the remaining equations we will find the values $x_1 = 1$ and $x_2 = 1$. This means that the point corresponding to productions of 1, 1, 2 and 1 will give a stationary value. If we do the same with $x = 3$ we obtain (− 6, − 4, 3, 1) as another set of production values.

But we have now to determine whether these stationary points are maxima, minima or points of inflexion. To do this we need to know the second derivatives which are

$$z_{x_1x_1} = -4\,;\; z_{x_2x_2} = -8\,;\; z_{x_3x_3} = -46\,;\; z_{x_4x_4} = -4\,;\; z_{x_1x_2} = 2x_3\,;$$
$$z_{x_1x_3} = 2x_2\,;\; z_{x_1x_4} = 0\quad;\; z_{x_2x_3} = 2x_1\quad;\; z_{x_2x_4} = 0\quad;\; z_{x_3x_4} = 0$$

Since all of the unmixed derivatives are negative, a minimum value of z cannot correspond to either of the solutions we are now examining.

We now examine the second, third and fourth order determinants in the two cases, taking the case of $x_1 = 1$, $x_2 = 1$, $x_3 = 2$ and $x_4 = 1$ first. The second order determinant gives

$$\begin{vmatrix} -4 & 2x_3 \\ 2x_3 & -8 \end{vmatrix} = \begin{vmatrix} -4 & 4 \\ 4 & -8 \end{vmatrix} = 16$$

The third order gives

$$\begin{vmatrix} -4 & 2x_3 & 2x_2 \\ 2x_3 & -8 & 2x_1 \\ 2x_2 & 2x_1 & -46 \end{vmatrix} = \begin{vmatrix} -4 & 4 & 2 \\ 4 & -8 & 2 \\ 2 & 2 & -46 \end{vmatrix} = 4 \begin{vmatrix} -2 & 2 & 2 \\ 2 & -4 & 2 \\ 1 & 1 & -46 \end{vmatrix}$$

$$= 4 \begin{vmatrix} -2 & 0 & 0 \\ 2 & -2 & 4 \\ 1 & 2 & -45 \end{vmatrix} = -8 \begin{vmatrix} -2 & 4 \\ 2 & -45 \end{vmatrix} = -8(90-8) < 0$$

Finally the fourth order determinant is

$$\begin{vmatrix} -4 & 2x_3 & 2x_2 & 0 \\ 2x_3 & -8 & 2x_1 & 0 \\ 2x_2 & 2x_1 & -46 & 0 \\ 0 & 0 & 0 & -4 \end{vmatrix}$$

which is clearly equal to the previous determinant multiplied by -4 and is, therefore, positive.

We, therefore, have that at the point corresponding to productions of 1, 1, 2 and 1 units the first derivatives are zero, the unmixed second derivatives are negative, and the second order determinant is positive, the third order negative and the fourth order positive. There is, therefore, a maximum at this point.

We now repeat this analysis with the values -6, -4, 3 and 1. We find that here the second order determinant is negative, which corresponds to a saddle point.

The manufacturer now knows that if he produces according to the scheme (1, 1, 2, 1) then his revenue will be greater than if he made any slight alteration to this scheme of production. He is clearly not interested in a saddle point that demands negative production of certain goods.

EXERCISE 20.2

1. A manufacturer of breakfast cereals has a monopoly in two competing products A and B. The cost of producing a packet of A is 7p., and that of a packet of B is 11p. The demands for the two products are given by

$$x_a = 200\,(p_b - p_a) - 500 \quad \text{and} \quad x_b = 25\,(p_a - 10p_b) + 3900$$

where p_a and p_b are the retail prices per packet and x_a and x_b the daily numbers of packets demanded. Show that the net joint revenue is a maximum when the prices charged are 10p. and 15p. respectively.

2. A manufacturer of ball-pens finds that it cost him 50p. to make a complete pen, and 5p. to make a refill. He has a monopoly of the two markets, and finds that the demand equation for the ball-pens is

$$x_b = 27{,}000 - 100\,(5p_b + 2p_r)$$

while that for the refills is

$$x_r = 13{,}200 - 40\,(5p_b + 6p_r)$$

Show that he maximises his net revenue by selling each refill at 5p. more than it costs him to make it, and each pen at 1p. less than it costs to make it.

3. A publisher sells a book (in hard covers) to libraries at a price p_h, and in soft covers to students at a price p_s.
The demand equations are

$$x_h = 5000 - 30p_h + 10p_s$$
$$x_s = 20{,}000 - 100p_s + 5p_h$$

The cost of producing one copy of the hard back is 50p. while for the paperback it is 45p. What prices should the publisher charge if he is to maximise his net revenue?

4. The above questions assume constant production costs. Rework them when the costs are simple variable linear functions, such as $c_a = 7 - 0{\cdot}01x_a$.

CHAPTER XXI

RESTRICTED MAXIMA AND MINIMA AND LAGRANGE MULTIPLIERS

Unbidden guests
Are often welcomest when they are gone
SHAKESPEARE : *King Henry VI,* part I

Often we are faced with a problem in which we wish to maximise a certain function subject to certain restrictions, or conditions. For example, we may wish to maximise utility subject to the restriction that our total spending does not exceed a certain amount ; or to maximise net revenue subject to the condition that the total labour force is kept constant. Problems of this kind can be solved by using the method of *Lagrange's Undetermined Multipliers.* Although we shall not prove the validity of this method, the principle of its application is quite simple. In this chapter we shall state the principle generally, and illustrate it with a reasonably simple numerical example, before using the principle to consider a problem in economic theory.

Suppose that we wish to maximise z, a function of the n variables $x_1, x_2, \ldots, x_n$, subject to the condition that $u = 0$ where u is another function of the same variables.

(For example, we may wish to maximise

$$z = w^2 + x^2 + y^2$$

subject to the condition that

$$u = w + x + y = 0)$$

To do so we consider a new function F which is defined to be

$$F = z + \lambda u$$

where λ is a new variable, independent of the x's and introduced for a reason which will become apparent later.

Now since $u = 0$, the values of x_r (i.e., the values of $x_1, x_2, \ldots, x_n$) that make F a maximum must also make z a maximum. We therefore proceed to find these values.

This involves two conditions. The first is that all of the first derivatives of F shall be zero. This will determine the values of the

n x's and of λ that give F a stationary value. The second condition will be discussed in a moment.

In order that the first derivatives of F (which is a function of the n x's and of λ) may all be zero, we have that

$$\frac{\partial F}{\partial x_r} = \frac{\partial z}{\partial x_r} + \lambda \frac{\partial u}{\partial x_r} = 0$$

for all values of r (giving n equations in all),

and that

$$\frac{\partial F}{\partial \lambda} = u = 0$$

We may write the first of these equations in the form

$$z_r + \lambda u_r = 0 \qquad (r = 1, \ldots, n)$$

where we understand that z_r is simply a convenient way of writing $\partial z / \partial x_r$. This equation

$$z_r + \lambda u_r = 0 \qquad (r = 1, \ldots, n)$$

is really, of course, n equations : one for each value of r. We have that

$$z_1 + \lambda u_1 = 0$$
$$z_2 + \lambda u_2 = 0$$

and so on.

We also have that

$$\frac{\partial F}{\partial \lambda} = 0$$

whence

$$u = 0$$

This means that we have $n + 1$ equations from which to determine the n values of x_r and the value of λ that make F a maximum ; and then these same values of x_r will make z a maximum subject to $u = 0$.

The first step in the determination of a maximum is therefore to solve the $n + 1$ equations summarised by

$$z_r + \lambda u_r = 0$$

and

$$u = 0$$

We refer to λ as an *undetermined multiplier*, since we multiply u by it before we have determined its value.

Before going any further, and deciding on the conditions to determine whether we have located a maximum or a minimum, we will illustrate the procedure so far with a numerical example.

Suppose that we wish to find the maxima and minima of

$$z = x^2 + y^2$$

subject to the condition

$$u = 2x + 3y - 4 = 0$$

To do so we take a new function

$$F = z + \lambda u = x^2 + y^2 + \lambda(2x + 3y - 4)$$

where λ is the undertermined multiplier. We now maximise F, treated as a function of x, y and λ. We have

$$F_x = 2x + 2\lambda$$
$$F_y = 2y + 3\lambda$$
$$F_\lambda = u = 2x + 3y - 4$$

For F (and therefore z) to be stationary

$$F_x = F_y = F_\lambda = 0$$

Therefore

$$2x + 2\lambda = 0$$
$$2y + 3\lambda = 0$$
$$2x + 3y - 4 = 0$$

Clearly $\qquad x = -\lambda \quad$ and $\quad y = -\dfrac{3\lambda}{2}$, whence

$$-2\lambda - 3\left(\frac{3\lambda}{2}\right) - 4 = 0$$

yielding $\qquad\qquad 13\lambda + 8 = 0$

$$\lambda = -\frac{8}{13}$$

whence $\qquad\qquad x = \dfrac{8}{13}, \quad y = \dfrac{12}{13}$

as the only solution. At these values of x and y there is a stationary value of F (and therefore of z) subject to $u = 0$. Whether it is a maximum, minimum or saddle pooint is a question to be deferred.

Now consider a more complicated example.

Suppose that we wish to maximise the function

$$z = x_1{}^2 x_2 x_3$$

subject to the condition that

$$u = x_1 + x_2 + x_3 - 20 = 0$$

To do so we take a new function

$$F = z + \lambda u$$

where λ is an undetermined multiplier. We proceed to maximise F. The first condition is that all the first derivatives of F shall be zero. i.e., since

$$F = x_1{}^2 x_2 x_3 + \lambda(x_1 + x_2 + x_3 - 20)$$

$$F_1 = \frac{\partial F}{\partial x_1} = 2x_1 x_2 x_3 + \lambda = 0 \tag{21.1}$$

$$F_2 = \frac{\partial F}{\partial x_2} = x_1{}^2 x_3 + \lambda = 0 \tag{21.2}$$

$$F_3 = \frac{\partial F}{\partial x_3} = x_1{}^2 x_2 + \lambda = 0 \tag{21.3}$$

We also have that

$$F_\lambda = \frac{\partial F}{\partial \lambda} = u = x_1 + x_2 + x_3 - 20 = 0 \tag{21.4}$$

We must now find the values of x_1, x_2, x_3 and λ that satisfy these four equations. As we shall now show, there are several alternative solutions.

From equations (21.2) and (21.3) we have that

$$x_1{}^2 x_3 = x_1{}^2 x_2$$

and so *either* $x_3 = x_2$ or $x_1{}^2 = 0$. Let us pursue the first alternative
(1) *If we put* $x_2 = x_3$ in equations (21.1) and (21.2) we have

$$2x_1 x_2{}^2 = -\lambda = x_1{}^2 x_2$$

and so *either* (i) $x_1 = 2x_2$ or $x_1 x_2 = 0$, in which case *either* (ii) $x_1 = 0$ *or* (iii) $x_2 = 0$. We consider these three alternatives in turn :
(i) $x_1 = 2x_2$ *with* $x_2 = x_3$
We see, with the help of equation (21.4), that

$$4x_2 - 20 = 0$$

which yields

$$x_2 = 5$$

and so $x_1 = 10$, $x_2 = 5$, $x_3 = 5$ and (from (21.2)) $\lambda = -500$. On the other hand, if we pursue the second alternative we have

(ii) $x_1 = 0$ *with* $x_2 = x_3$

We have from equation (21.1) that

$$\lambda = 0$$

and from equation (21.4) that $x_2 + x_3 = 20$

which, since $\qquad\qquad x_2 = x_3$

gives us the solution

$$x_1 = 0,\ x_2 = 10,\ x_3 = 10,\ \lambda = 0$$

(iii) $x_2 = 0$ *with* $x_2 = x_3$

We now find that

$$x_3 = 0$$

and so $x_1 = 20$ and $\lambda = 0$ giving us the solution

$$x_1 = 20,\ x_2 = 0,\ x_3 = 0,\ \lambda = 0$$

We thus have that if we accept the set of solutions involving $x_2 = x_3$ we have the alternative solutions

(i) $x_1 = 10$ $\quad x_2 = 5$ $\quad x_3 = 5$ $\quad \lambda = -500$
(ii) $x_1 = 0$ $\quad x_2 = 10$ $\quad x_3 = 10$ $\quad \lambda = 0$
(iii) $x_1 = 20$ $\quad x_2 = 0$ $\quad x_3 = 0$ $\quad \lambda = 0$

(2) We can now consider the solutions that arise out of taking

$$x_1{}^2 = 0$$

It is obvious that in this case, $x_1 = 0$, $\lambda = 0$ and $x_2 + x_3 = 20$. We can say no more. The condition for a stationary value is simply that x_1, x_2, x_3 and λ satisfy the conditions $x_1 = 0$, $\lambda = 0$ and $x_2 + x_3 = 20$. This condition includes the solution (ii) above.

The solutions to the problem of finding a stationary value of z therefore fall into two categories. There are the solutions which involve precise values for all the x's, viz.,

$$x_1 = 10 \qquad x_2 = 5 \qquad x_3 = 5 \qquad \lambda = -500$$

and $\qquad x_1 = 20 \qquad x_2 = 0 \qquad x_3 = 0 \qquad \lambda = 0$

and the whole set of solutions for which

$$x_1 = 0 \quad \text{and} \quad x_2 + x_3 = 20, \quad \lambda = 0$$

In order to determine whether any of these stationary values are maxima we have to consider a certain determinant, which will show us the importance of λ.

The student who wishes to read this chapter before acquiring an understanding of determinants may obtain an adequate quick insight from Appendix 6. But a better understanding of them may be obtained from Chapter XXIX.

It can be shown that if we have to determine the stationary values of z, a function of n variables, subject to the condition that $u = 0$, then the conditions for a maximum or minimum depend on the values of the determinant formed from the second derivatives of

$$F = z + \lambda u$$

and the first derivatives of u, thus

$$\Delta_1 = \begin{vmatrix} F_{11} & F_{12} & F_{13} \dots F_{1n} & u_1 \\ F_{21} & F_{22} & F_{23} \dots F_{2n} & u_2 \\ \hdotsfor{4} \\ \hdotsfor{4} \\ F_{n1} & F_{n2} & F_{n3} \dots F_{nn} & u_n \\ u_1 & u_2 & u_3 \dots u_n & 0 \end{vmatrix}$$

where $F_{11} = \dfrac{\partial^2 F}{\partial x_1 \, \partial x_1} = \dfrac{\partial^2 F}{\partial x_1{}^2}$

$F_{23} = \dfrac{\partial^2 F}{\partial x_2 \, \partial x_3}$

and $u_1 = \dfrac{\partial u}{\partial x_1}$

and the values of the set of principal minors

$$\Delta_2 = \begin{vmatrix} F_{22} F_{23} \dots u_2 \\ F_{23} F_{33} \dots u_3 \\ \dotfill \\ \dotfill \\ u_2 \; u_3 \dots 0 \end{vmatrix}, \; \Delta_3 = \begin{vmatrix} F_{33} F_{34} \dots u_3 \\ F_{34} F_{44} \dots u_4 \\ \dotfill \\ u_3 \; u_4 \dots 0 \end{vmatrix}, \dots \Delta_{n-1} = \begin{vmatrix} F_{n-1,\,n-1} & F_{n-1,\,n} & u_{n-1} \\ F_{n-1,\,n} & F_{n,n} & u_n \\ u_{n-1} & u_n & 0 \end{vmatrix}$$

where all the partial derivatives involved have the values that correspond to the situation under consideration. We shall illustrate this shortly.

For z to be a maximum, when there is one restriction, we must have that $\Delta_1 < 0$, $\Delta_2 > 0$, $\Delta_3 < 0$, and so on, if n is odd; but $\Delta_1 > 0$, $\Delta_2 < 0$, $\Delta_3 > 0$, ... , if n is even.

For a minimum we must have that Δ_1 and all its principal minors within the range listed above must be negative.

We shall now apply this test to the results we have found. First, however, let us notice that in general the second derivatives of F

may well contain λ, and it is in this respect that λ becomes important, since it may affect the value of the above determinant. It happens that, in the example we have chosen, λ disappears from the second derivatives; but that is only because the condition $u = 0$ happens to be linear in the variables x_r.

In our first, simpler, example

$$F_{11} = F_{xx} = 2 \qquad F_{12} = F_{xy} = 0$$
$$F_{21} = F_{yx} = 0 \qquad F_{22} = F_{yy} = 2$$

Also $\qquad u_1 = u_x = 2 \; ; \; u_2 = u_y = 3$

Our determinant becomes

$$\Delta_1 = \begin{vmatrix} 2 & 0 & 2 \\ 0 & 2 & 3 \\ 2 & 3 & 0 \end{vmatrix} = 2 \begin{vmatrix} 2 & 3 \\ 3 & 0 \end{vmatrix} + 2 \begin{vmatrix} 0 & 2 \\ 2 & 3 \end{vmatrix} = -18 - 8$$

$$= -26$$

and its principal minor is

$$\Delta_2 = \begin{vmatrix} 2 & 3 \\ 3 & 0 \end{vmatrix} = -9$$

For z to be a maximum where there is only one restriction and two independent variables we must have that

$$\Delta_1 < 0 \quad \text{and} \quad \Delta_2 > 0$$

For z to be a minimum,

$$\Delta_1 < 0 \quad \text{and} \quad \Delta_2 < 0$$

In our case $\Delta_1 < 0$, $\Delta_2 < 0$ and so z is a minimum.
Let us now turn to our second example.
Returning to equations (21.1), (21.2) and (21.3) we have that

$$F_{11} = 2x_2 x_3 \; ; \; F_{12} = 2x_1 x_3 \; ; \; F_{13} = 2x_1 x_2$$
$$F_{22} = 0 \qquad ; \; F_{23} = x_1^2 \quad ; \; F_{33} = 0$$

Also $\qquad u_1 = u_2 = u_3 = 1$

We thus have that the determinant is

$$\begin{vmatrix} 2x_2x_3 & 2x_1x_3 & 2x_1x_2 & 1 \\ 2x_1x_3 & 0 & x_1{}^2 & 1 \\ 2x_1x_2 & x_1{}^2 & 0 & 1 \\ 1 & 1 & 1 & 0 \end{vmatrix}$$

and we have to consider the sign of this and its minors under the various conditions dictated by the three different solutions.

Let us begin with the solution $x_1 = 10$, $x_2 = 5$ and $x_3 = 5$. We have that

$$\begin{vmatrix} 50 & 100 & 100 & 1 \\ 100 & 0 & 100 & 1 \\ 100 & 100 & 0 & 1 \\ 1 & 1 & 1 & 0 \end{vmatrix} = \begin{vmatrix} -50 & 0 & 100 & 0 \\ 0 & -100 & 100 & 0 \\ 100 & 100 & 0 & 1 \\ 1 & 1 & 1 & 0 \end{vmatrix}$$ (by subtracting row 3 from row 1 and row 2 in turn)

$$= 5000 \begin{vmatrix} -1 & 0 & 2 & 0 \\ 0 & -1 & 1 & 0 \\ 100 & 100 & 0 & 1 \\ 1 & 1 & 1 & 0 \end{vmatrix} = 5000 \begin{vmatrix} -1 & 0 & 0 & 0 \\ 0 & -1 & 1 & 0 \\ 100 & 100 & 200 & 1 \\ 1 & 1 & 3 & 0 \end{vmatrix}$$

(by first removing 50×100 as a factor, and then adding twice column (1) to column (3) to obtain a new column (3)).

$$= -5000 \begin{vmatrix} -1 & 1 & 0 \\ 100 & 200 & 1 \\ 1 & 3 & 0 \end{vmatrix} = -5000 \begin{vmatrix} -1 & 0 & 0 \\ 100 & 300 & 1 \\ 1 & 4 & 0 \end{vmatrix}$$

$$= +5000 \begin{vmatrix} 300 & 1 \\ 4 & 0 \end{vmatrix}$$

which is $-20,000$. We thus have that $\Delta_1 < 0$, showing that the stationary point *can* be a maximum or minimum. If Δ_1 had been > 0 then this would not have been possible.

We must now examine the sign of

$$\Delta_2 = \begin{vmatrix} 0 & 100 & 1 \\ 100 & 0 & 1 \\ 1 & 1 & 0 \end{vmatrix}$$

The value of this determinant is easily seen to be $+200$.

Since n, the number of variables, is 3, and we need only consider minors as far as $n - 1$, we need proceed no further. The condition for a maximum is clearly satisfied at $x_1 = 10$, $x_2 = 5$, $x_3 = 5$.

In the case of the solution $x_1 = 20$, $x_2 = 0$ and $x_3 = 0$ the determinant becomes

$$\Delta_1 = \begin{vmatrix} 0 & 0 & 0 & 1 \\ 0 & 0 & 400 & 1 \\ 0 & 400 & 0 & 1 \\ 1 & 1 & 1 & 0 \end{vmatrix} = -\begin{vmatrix} 0 & 0 & 400 \\ 0 & 400 & 0 \\ 1 & 1 & 1 \end{vmatrix} = 160{,}000$$

and

$$\Delta_2 = \begin{vmatrix} 0 & 400 & 1 \\ 400 & 0 & 1 \\ 1 & 1 & 0 \end{vmatrix} = 800$$

and so, since both Δ_1 and Δ_2 are positive, this solution corresponds to neither a maximum nor a minimum : it is a saddle point.

Finally we come to the third solution which is rather more complicated. It is $x_1 = 0$ and $x_2 + x_3 = 20$, so that $x_2 x_3 = x_2(20 - x_2)$. In this case the determinant becomes

$$\Delta_1 = \begin{vmatrix} 2x_2(20 - x_2) & 0 & 0 & 1 \\ 0 & 0 & 0 & 1 \\ 0 & 0 & 0 & 1 \\ 1 & 1 & 1 & 0 \end{vmatrix}$$

Since two rows of this are identical its value is zero. The minor Δ_2 is also zero. We are therefore unable to decide whether or not there is a maximum or a minimum in this position. It would be possible to reach a decision if we knew a little more mathematics : but readers of this book must not expect to run.

EXERCISE 21.1

1. Find the maximum and minimum values of $z = x^2 + y^2$ when $\dfrac{x^2}{16} + \dfrac{y^2}{9} = 1$

2. Examine the function $z = xy^2$ subject to $x^2 + y^2 = 9$

3. Examine the function $z = x_1 x_2 x_3$ subject to $x_1 + x_2 + x_3 = a$

2. The case of more than one restriction

We have now seen how to maximise (or minimise) the function z subject to the restriction that $u = 0$. It is possible, however, that there is more than one restriction. Let us suppose that there are two, $u = 0$ and $v = 0$. In this case we maximise

$$F = z + \lambda u + \mu v$$

obtaining the n equations

$$z_r + \lambda u_r + \mu v_r = 0 \qquad\qquad (r = 1, \dots, n)$$

which, with the two conditions $u=0$ and $v=0$, enable us to find the n values of x_r and the values of λ and μ which maximise (or minimise) F, and therefore z. If we had three restrictions, the third being $w=0$, then the equations would be of the form

$$z_r + \lambda u_r + \mu v_r + \nu w_r = 0$$

and so on.

To determine whether the stationary values so found are maxima or minima we consider the determinant

$$\Delta_1 = \begin{vmatrix} F_{11} & F_{12} & F_{13} \cdots F_{1n} & a_1 & b_1 & c_1 \cdots v_1 & w_1 \\ F_{21} & F_{22} & F_{23} \cdots F_{2n} & a_2 & b_2 & c_2 \cdots v_2 & w_2 \\ \cdots & & & & & & \\ \cdots & & & & & & \\ F_{n1} & F_{n2} & F_{n3} \cdots F_{nn} & a_n & b_n & c_n \cdots v_n & w_n \\ a_1 & a_2 & a_3 \cdots a_n & 0 & 0 & 0 \cdots 0 & 0 \\ b_1 & b_2 & b_3 \cdots b_n & 0 & 0 & 0 \cdots 0 & 0 \\ c_1 & c_2 & c_3 \cdots c_n & 0 & 0 & 0 \cdots 0 & 0 \\ \cdots & & & & & & \\ \cdots & & & & & & \\ v_1 & v_2 & v_3 \cdots v_n & 0 & 0 & 0 \cdots 0 & 0 \\ w_1 & w_2 & w_3 \cdots w_n & 0 & 0 & 0 \cdots 0 & 0 \end{vmatrix}$$

where there are m restrictions denoted by $a=0$, $b=0 \dots w=0$. There will also be the minors obtained by removing successive rows and columns, starting at the top and left, as in the case of one restriction. We denote these by Δ_2, Δ_3, etc. If there are m restrictions, and F is a function of n variables, we need the minors from Δ_2 to Δ_{n-m}.

The condition for a maximum is that

(1) If n, the number of variables, is even, then $\Delta_1 > 0$, $\Delta_2 < 0$, $\Delta_3 > 0$, ... , etc., ending with Δ_{n-m} which must be <0 if $n-m$ is even, and >0 if $n-m$ is odd.

(2) If n, the number of variables, is odd, then $\Delta_1 < 0$, $\Delta_2 > 0$, ... , etc., ending with Δ_{n-m} which must be >0 if $n-m$ is even and <0 if $n-m$ is odd.

The condition for a minimum is that

(1) If m, the number of restrictions, is even $\Delta_1 > 0$, $\Delta_2 > 0$, $\Delta_3 > 0$ down to $\Delta_{n-m} > 0$.

(2) If m, the number of restrictions, is odd, then the above minors must all be negative.

These results are, of course, sufficient conditions. If some of the minors are zero, there may still be a maximum or a minimum : but if any of the above conditions are violated for some reason other than a zero minor, then the point is a saddle point.

Before leaving this subject we should draw attention to the fact that the chapter on differentials has a little more to say about it.

We now consider the application of these ideas to a problem in economics.

Example :

Suppose that a person has a fixed income M and is able to spend it upon two goods, whose fixed prices are p_1 and p_2.

From given quantities of these goods he derives a certain degree of satisfaction, and the utility function expressing this is

$$U = x_1{}^{\alpha} x_2{}^{\beta}$$

where x_1 and x_2 are the quantities that he buys, and α and β are positive constants.*

His aim is to maximise U subject to the restriction that his total expenditure is equal to his total income, i.e., that

$$p_1 x_1 + p_2 x_2 = M$$

which is more usefully written as

$$u \equiv p_1 x_1 + p_2 x_2 - M = 0$$

To solve the problem of how much to spend on one good and how much on another, we consider

$$F = x_1{}^{\alpha} x_2{}^{\beta} + \lambda (p_1 x_1 + p_2 x_2 - M)$$

where λ is an undetermined multiplier.

For F to be a maximum, certain conditions must hold. The first of these are that F_1, F_2 and F_λ shall be zero.

i.e.,
$$F_1 = \frac{\partial F}{\partial x_1} = \alpha x_1{}^{\alpha-1} x_2{}^{\beta} + \lambda p_1 = 0$$

$$F_2 = \frac{\partial F}{\partial x_2} = \beta x_1{}^{\alpha} x_2{}^{\beta-1} + \lambda p_2 = 0$$

$$F_\lambda = \frac{\partial F}{\partial \lambda} = p_1 x_1 + p_2 x_2 - M = 0$$

We now have to find values of x_1, x_2 and λ that satisfy these conditions.

---

* This example is based on an Exercise in R. G. D. Allen's *Mathematical Analysis for Economists*.

The first condition gives $\quad p_1 = -\dfrac{\alpha x_1^{\alpha-1} x_2^{\beta}}{\lambda}$

and the second gives $\quad p_2 = -\dfrac{\beta x_1^{\alpha} x_2^{\beta-1}}{\lambda}$

Substitution of these values in the third condition gives

$$-\frac{\alpha x_1^{\alpha} x_2^{\beta}}{\lambda} - \frac{\beta x_1^{\alpha} x_2^{\beta}}{\lambda} - M = 0$$

which yields

$$x_1^{\alpha} x_2^{\beta} = -\frac{\lambda M}{\alpha + \beta}$$

It follows that

$$p_1 x_1 = \left(-\frac{\alpha x_1^{\alpha-1} x_2^{\beta}}{\lambda}\right) x_1$$

$$= -\frac{\alpha x_1^{\alpha} x_2^{\beta}}{\lambda}$$

$$= \frac{\alpha M}{\alpha + \beta}$$

and, similarly, that $\quad p_2 x_2 = \dfrac{\beta M}{\alpha + \beta}$

The values of x_1 and x_2 that make $F_1 = F_2 = F_\lambda = 0$ are therefore

$$x_1 = \frac{\alpha M}{(\alpha + \beta) p_1}$$

and

$$x_2 = \frac{\beta M}{(\alpha + \beta) p_2}$$

It follows that the associated value of λ is given by

$$\lambda = -\frac{\alpha + \beta}{M} x_1^{\alpha} x_2^{\beta}$$

$$= -\frac{\alpha + \beta}{M} \left(\frac{\alpha M}{(\alpha + \beta) p_1}\right)^{\alpha} \left(\frac{\beta M}{(\alpha + \beta) p_2}\right)^{\beta}$$

$$= -\frac{(\alpha + \beta)}{M} \frac{\alpha^{\alpha} \beta^{\beta} M^{\alpha+\beta}}{p_1^{\alpha} p_2^{\beta} (\alpha + \beta)^{\alpha+\beta}}$$

$$= -\frac{M^{\alpha+\beta-1}}{(\alpha + \beta)^{\alpha+\beta-1}} \left(\frac{\alpha}{p_1}\right)^{\alpha} \left(\frac{\beta}{p_2}\right)^{\beta}$$

$$= -\left(\frac{M}{\alpha + \beta}\right)^{\alpha+\beta-1} \left(\frac{\alpha}{p_1}\right)^{\alpha} \left(\frac{\beta}{p_2}\right)^{\beta}$$

Since α, β, p_1, p_2 and M are all positive, this quantity is negative.

We have now to consider whether these values lead to a maximum U, or, possibly to a minimum or saddle point. To do this we have to consider the second derivatives which are

$$F_{11} = \frac{\partial^2 F}{\partial x_1{}^2} = \alpha(\alpha - 1)x_1{}^{\alpha-2}x_2{}^{\beta}$$

$$F_{22} = \frac{\partial^2 F}{\partial x_2{}^2} = \beta(\beta - 1)x_1{}^{\alpha}x_2{}^{\beta-2}$$

$$F_{12} = \frac{\partial^2 F}{\partial x_1\,\partial x_2} = \alpha\beta x_1{}^{\alpha-1}x_2{}^{\beta-1}$$

When x_1 and x_2 have the values just derived these expressions take the values

$$F_{11} = \frac{\alpha(\alpha-1)}{x_1{}^2}\left(-\frac{\lambda M}{\alpha+\beta}\right)$$

$$= \alpha(\alpha-1)\left(-\frac{\lambda M}{\alpha+\beta}\right)\frac{(\alpha+\beta)^2 p_1{}^2}{\alpha^2 M^2}$$

$$= -\frac{\lambda(\alpha-1)(\alpha+\beta)p_1{}^2}{\alpha M}$$

$$F_{22} = -\frac{\lambda(\beta-1)(\alpha+\beta)p_2{}^2}{\beta M} \qquad \text{(similarly)}$$

and

$$F_{12} = \frac{\alpha\beta}{x_1 x_2}\left(-\frac{\lambda M}{\alpha+\beta}\right)$$

$$= -\frac{\lambda(\alpha+\beta)}{M}p_1 p_2$$

We also have the first derivatives of the restriction

$$u \equiv p_1 x_1 + p_2 x_2 - M = 0$$

which are $u_1 = p_1$ and $u_2 = p_2$

The criteria for distinguishing between the critical points are based on the value of the following determinant and its principal minors.

$$\begin{vmatrix} F_{11} & F_{12} & u_1 \\ F_{12} & F_{22} & u_2 \\ u_1 & u_2 & 0 \end{vmatrix}$$

In our case, this has the value

$$\Delta_1 = \begin{vmatrix} -\dfrac{\lambda(\alpha-1)(\alpha+\beta)}{\alpha M}p_1{}^2 & -\dfrac{\lambda(\alpha+\beta)}{M}p_1 p_2 & p_1 \\[2mm] -\dfrac{\lambda(\alpha+\beta)}{M}p_1 p_2 & -\dfrac{\lambda(\beta-1)(\alpha+\beta)}{\beta M}p_2{}^2 & p_2 \\[2mm] p_1 & p_2 & 0 \end{vmatrix}$$

$$= p_1 p_2 \begin{vmatrix} -\dfrac{\lambda(\alpha-1)(\alpha+\beta)}{\alpha M}p_1 & -\dfrac{\lambda(\alpha+\beta)}{M}p_2 & 1 \\[2mm] -\dfrac{\lambda(\alpha+\beta)}{M}p_1 & -\dfrac{\lambda(\beta-1)(\alpha+\beta)}{\beta M}p_2 & 1 \\[2mm] p_1 & p_2 & 0 \end{vmatrix}$$

$$= p_1{}^2 p_2{}^2 \begin{vmatrix} -\dfrac{\lambda(\alpha-1)(\alpha+\beta)}{\alpha M} & -\dfrac{\lambda(\alpha+\beta)}{M} & 1 \\[2mm] -\dfrac{\lambda(\alpha+\beta)}{M} & -\dfrac{\lambda(\beta-1)(\alpha+\beta)}{\beta M} & 1 \\[2mm] 1 & 1 & 0 \end{vmatrix}$$

The student who has read Appendix 6 should be able to show in a few lines that this determinant has the value

$$\Delta_1 = -\frac{(\alpha+\beta)^2}{\alpha\beta}\frac{\lambda}{M}p_1{}^2 p_2{}^2$$

The first minor is

$$\Delta_2 = \begin{vmatrix} -\dfrac{\lambda(\beta-1)(\alpha+\beta)p_2{}^2}{\beta M} & p_2 \\[2mm] p_2 & 0 \end{vmatrix}$$

which is clearly $-p_2{}^2$.

We know that if we have 2 independent variables the conditions for a maximum are

(1) that $z_x = z_y = 0$

and   (2) that $\Delta_1 > 0$, $\Delta_2 < 0$.

In the case we are considering the first condition clearly holds. Furthermore

$$\Delta_1 = -\frac{(\alpha+\beta)^2}{\alpha\beta}\frac{\lambda}{M}p_1{}^2 p_2{}^2$$

must be positive, because $\dfrac{(\alpha+\beta)^2}{\alpha\beta M}$ is positive (since α, β and M

are positive) while λ has just been shown to be negative.   We have seen that also $\Delta_2 < 0$.   It follows that when

$$x_1 = \frac{\alpha M}{(\alpha + \beta)p_1} \quad \text{and} \quad x_2 = \frac{\beta M}{(\alpha + \beta)p_2}$$

the utility function U is at a maximum.   It is *these* quantities which the man should buy.

CHAPTER XXII

HOMOGENEOUS FUNCTIONS AND EULER'S THEOREM

This too is an experience of the soul.

KATHLEEN RAINE : *Isis Wanderer*

1. Homogeneous functions

Often in a study of economics we come across the idea of " constant returns to scale ". We may have, for example, that three men and ten acres will produce a certain amount of wheat, while six men and twenty acres will produce double that amount, nine men and thirty acres treble that amount and so on. This is just one simple example of *linear homogeneous function*. We now define these functions more precisely, and then consider a few of their properties.

Definition :

A function $f(x_1, x_2, x_3, x_4, \ldots, x_n)$ is said to be *homogeneous* of degree n if

$$f(tx_1, tx_2, tx_3, tx_4, \ldots, tx_n) = t^n f(x_1, x_2, x_3, x_4, \ldots, x_n)$$

for all values of t.

For example, if our function is

$$f(x, y, z) = x^2 + y^2 + z^2$$

then we have that

$$\begin{aligned}
f(tx, ty, tz) &= (tx)^2 + (ty)^2 + (tz)^2 \\
&= t^2 x^2 + t^2 y^2 + t^2 z^2 \\
&= t^2 (x^2 + y^2 + z^2) \\
&= t^2 f(x, y, z)
\end{aligned}$$

and so $(x^2 + y^2 + z^2)$ is homogeneous of the second degree.

297

Similarly it can be shown that

$\dfrac{x+y}{x^4-y^4}$ is homogeneous of degree -3 ; $f(tx, ty)=\dfrac{1}{t^3}f(x, y)$

$x+y$ is homogeneous of degree 1 ; $f(tx, ty)=tf(x, y)$

and $\dfrac{x}{y}$ is homogeneous of degree 0 ; $f(tx, ty)=f(x, y)$, t^0 being unity.

EXERCISE 22.1

1. Consider whether the following functions are homogeneous, evaluating the degree of homogeneity where appropriate.

(i) $x^3+y^3+3z^3$ (ii) $\sin x+\sin y$

(iii) x^2+xy+y^2 (iv) $x^2+2xy+y^2$

(v) $x^2+2xy+7y^2$ (vi) $\dfrac{x}{y}+\dfrac{4x^2}{y^2}$

(vii) x^2+xy+y^2+1 (viii) $\dfrac{x}{y}+\dfrac{4x^2}{y^2}+1$

(ix) $\dfrac{x^2}{y}+\dfrac{4x^3}{y^2}$ (x) $\dfrac{x^2}{y}+\dfrac{4x^3}{y^2}+1$

2. Euler's theorem

Euler's Theorem states that if f is a function of the variables x_1, x_2, x_3, ... , x_m, and is homogeneous of degree n in these variables, then

$$x_1 f_{x_1}+x_2 f_{x_2}+x_3 f_{x_3}+\dots+x_m f_{x_m}=nf$$

We may illustrate this by referring to the examples just cited.

(i) If $f=x^2+y^2+z^2$ (homogeneous of degree 2)

then $f_x=2x,\quad f_y=2y$ and $f_z=2z$

Therefore, $xf_x+yf_y+zf_z=2x^2+2y^2+2z^2=2f$

(ii) If $f=\dfrac{x+y}{x^4-y^4}$ (homogeneous of degree -3)

we have by differentiation of the quotient that

$$f_x=\dfrac{(x^4-y^4)-4x^3(x+y)}{(x^4-y^4)^2}$$

and $f_y=\dfrac{(x^4-y^4)+4y^3(x+y)}{(x^4-y^4)^2}$

whence $$xf_x + yf_y = \frac{x^5 - xy^4 - 4x^4(x+y) + x^4y - y^5 + 4y^4(x+y)}{(x^4 - y^4)^2}$$

$$= \frac{(x+y)(x^4 - y^4 - 4[x^4 - y^4])}{(x^4 - y^4)^2}$$

$$= -\frac{3(x+y)}{x^4 - y^4}$$

$$= -3f$$

(iii) If $\qquad f = x + y \qquad$ (homogeneous of degree 1)

then $\qquad f_x = 1, \quad f_y = 1$

and so $\qquad xf_x + yf_y = x + y = f$

(iv) If $\qquad f = \dfrac{x}{y} \qquad$ (homogeneous of degree 0)

then $\qquad f_x = \dfrac{1}{y}, \quad f_y = -\dfrac{x}{y^2}$

and so $\qquad xf_x + yf_y = \dfrac{x}{y} - \dfrac{x}{y} = 0 = 0 \times f$

EXERCISE 22.2

Verify Euler's Theorem for those functions in Exercise 22.1 which are homogeneous.

Also consider the validity of the theorem in some of the other cases, by evaluating $xf_x + yf_y$, and considering whether there is a number m such that $xf_x + yf_y = mf$.

3. Some properties of homogeneous functions

Further properties of homogeneous functions are :

(1) If f is a homogeneous function of the variables $x_1, x_2, x_3, \ldots, x_m$ then it can be written as

$$f = x_1{}^n \phi \left(\frac{x_2}{x_1}, \frac{x_3}{x_1}, \frac{x_4}{x_1}, \ldots, \frac{x_m}{x_1} \right)$$

where n is the degree of homogeneity and ϕ is a new function. For example,

if $\qquad f = x_1{}^2 + x_2{}^2 + x_3{}^2$

then it can be written as

$$f = x_1{}^2 \left[1 + \left(\frac{x_2}{x_1} \right)^2 + \left(\frac{x_3}{x_1} \right)^2 \right] = x_1{}^2 (1 + u^2 + v^2)$$

$$= x_1{}^2 \phi(u, v) \qquad \text{where } u = x_2/x_1 \text{ and } v = x_3/x_1$$

(2) If f is a homogeneous function of degree n then

$$x_1^2 f_{x_1 x_1} + x_2^2 f_{x_2 x_2} + \ldots + x_m^2 f_{x_m x_m}$$
$$+ 2 \left(x_1 x_2 f_{x_1 x_2} + x_1 x_3 f_{x_1 x_3} + \ldots + x_1 x_m f_{x_1 x_m} + x_2 x_3 f_{x_2 x_3} \right.$$
$$\left. + x_2 x_4 f_{x_2 x_4} + \ldots + x_2 x_m f_{x_2 x_m} + \ldots + \ldots \right) = n(n-1)f$$

where the bracket contains all possible mixed derivatives of second order.   An example of this is shown in the next paragraph.

(3) To avoid expressions that look a great deal more complicated than they really are, we shall now state a result that applies only to functions of two variables.   An extension of it to the case of m variables may be made ; and the preceding paragraph contains a special case of this extension.   The result is that the pth derivatives are related, in the case of an nth degree homogeneity, by

$$x^p \frac{\partial^p f}{\partial x^p} + {}_pC_1 x^{p-1} y \frac{\partial^p f}{\partial x^{p-1} \partial y} + {}_pC_2 x^{p-2} y^2 \frac{\partial^p f}{\partial x^{p-2} \partial y^2} + \ldots + y^p \frac{\partial^p f}{\partial y^p}$$
$$= n(n-1) \ldots (n-p+1)f$$

It can be seen from this that the second derivatives are related by

$$x^2 f_{xx} + 2xy f_{xy} + y^2 f_{yy} = n(n-1)f$$

and the expression in (2) above reduces to this if there are only two variables.

We may consider

$$f = x^2 + y^2 + xy$$

which is homogeneous of degree 2.   The derivatives are

$$f_x = 2x + y, \quad f_y = 2y + x, \quad f_{xx} = 2, \quad f_{yy} = 2, \quad f_{xy} = 1$$

Consequently

$$x f_x + y f_y = 2x^2 + xy + 2y^2 + xy = 2f$$

verifying Euler's Theorem for the first derivatives.   Also

$$x^2 f_{xx} + 2xy f_{xy} + y^2 f_{yy} = 2x^2 + 2xy + 2y^2 = 2(2-1)f$$

verifying the theorem for the second derivative.

This particular example demonstrates an interesting consequence of some of the above results, namely that for a homogeneous function of the second degree

$$x_1 f_{x_1} + x_2 f_{x_2} + \ldots = 2f = 2(2-1)f$$
$$= x_1^2 f_{x_1 x_1} + x_2^2 f_{x_2 x_2} + \ldots + 2 \left(x_1 x_2 f_{x_1 x_2} + x_1 x_3 f_{x_1 x_3} + \ldots + \ldots \right)$$

4. Some economic examples ⊲

We may now consider some of the applications of these results. Many of the applications occur in some piece of mathematical theory, often mathematical economic theory. For the moment we shall take a simple case, the one in fact with which we started this chapter.

(i) Suppose that if we employ x men and y acres then the amount of wheat produced is z units. It is reasonable to suppose that, provided the men are all born equal and that the land is of uniform quality, then doubling or trebling the quantities of these factors of production will double or treble the volume of production. This entitles us to assume that the production function

$$z = f(x, y)$$

is linear and homogeneous (i.e., homogeneous of the first degree).

Now the marginal product of men is given by

$$\frac{\partial z}{\partial x} \quad \text{which is} \quad z_x$$

and that of land by

$$\frac{\partial z}{\partial y} \quad \text{which is} \quad z_y$$

and Euler's Theorem states that $xz_x + yz_y = z$, i.e., the number of men multiplied by their marginal product *plus* the amount of land multiplied by its marginal product is equal to the total product.

Suppose now that we wish to vary the number of men (x) so that we obtain a maximum total product for a given quantity of land (y). This means that we have to choose a value of x that will make z_x zero. Since

$$xz_x + yz_y = z$$

if we make $z_x = 0$ then

$$z_y = \frac{z}{y}$$

which shows that marginal product of land (z_y) is equal to the average product of land $\left(\dfrac{z}{y} = \dfrac{\text{total product}}{\text{total area}} \right)$. In other words, if the production function for the two variables is homogeneous and linear, and the supply of one factor is fixed, the production is at a stationary point when the marginal product of the variable factor is zero and the marginal product of the fixed factor is equal to the average product of that factor.

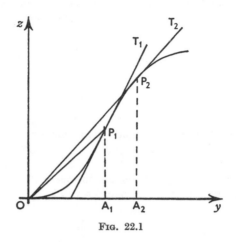

FIG. 22.1

We may note that, since the average product is given by $z/y = PA/OA$ (in Diagram 22.1) it measures the slope of OP and the marginal product is given by the slope of the tangent at P. The average and marginal products can be equal only when the tangent at P coincides with OP and passes through the origin, as at P_2. At P_1 the average and marginal products must differ.

(ii) Another consequence of Euler's Theorem is that if we have a supply function relating the supply x of a good to the prices p_1, p_2, p_3, ... , of several goods, possibly including itself, then if that supply function is homogeneous of degree n the sum of the partial price elasticities of that good must total n. Thus we have

$$x = f(p_1, p_2, p_3, \ldots) \quad \text{(homogeneous of degree } n\text{)}$$

Therefore,

$$p_1 \frac{\partial x}{\partial p_1} + p_2 \frac{\partial x}{\partial p_2} + \ldots = nx$$

and so

$$\frac{p_1}{x} \frac{\partial x}{\partial p_1} + \frac{p_2}{x} \frac{\partial x}{\partial p_2} + \ldots = n$$

and each term on the left is a partial price elasticity of supply.

For example, suppose that we have a supply function.

$$z = a\sqrt{x-y}$$

where z is the quantity of butter that is supplied when its price is x and the price of milk is y. Then this is homogeneous of degree one-half, for

$$
\begin{aligned}
z(tx,\, ty) &= a\sqrt{tx-ty}\\
&= t^{\frac{1}{2}}a\sqrt{x-y}\\
&= t^{\frac{1}{2}}z(x,\, y)
\end{aligned}
$$

The above result states that the elasticity of supply of butter with respect to its own price, *plus* the elasticity of supply of butter with respect to the price of milk is one-half. To check this we have

$$z_x = \frac{\partial z}{\partial x} = \frac{a}{2\sqrt{x-y}};\ \ z_y = \frac{\partial z}{\partial y} = -\frac{a}{2\sqrt{x-y}}$$

and so the sum of the elasticities is

$$
\frac{x}{z}z_x + \frac{y}{z}z_y = \frac{ax-ay}{2z\sqrt{x-y}} = \frac{a\sqrt{x-y}}{2z}
$$
$$
= \tfrac{1}{2}\ \ \text{since}\ z = a\sqrt{x-y}
$$

(iii) Sometimes Euler's Theorem reveals that seemingly independent assumptions are related to each other.

For example if we assume that

(*a*) every factor of production is rewarded with its marginal product, and

(*b*) the total output is divided between the factors of production

then the first assumption means that if we employ x_n units of factor n, the total reward to this factor is

$$x_n \frac{\partial Y}{\partial x_n}$$

The second assumption means that the sun of these rewards, added over all factors, equals the total output Y, i.e.,

$$x_1 \frac{\partial Y}{\partial x_1} + x_2 \frac{\partial Y}{\partial x_2} + \ldots = Y$$

From the converse of Euler's Theorem, this implies that Y is a homogeneous function of the first degree.

Consequently, the two assumptions which we have just made can be simultaneously valid only if there are constant returns to scale.

If we consider an economy which has increasing returns to scale, and make assumption (*a*) then the sum of the rewards to factors of production will not absorb the whole output. It could, however, be absorbed in some form of tax on output, or in some other way.

CHAPTER XXIII

MULTIPLE, LINEAR AND SURFACE INTEGRATION

Shall lure it back to cancel half a line.

E. FitzGerald

1. Introduction

When we introduced the idea of integration we saw that it is the reverse of differentiation. Now that we have extended the idea of differentiation so that it deals with functions of many variables, we may enquire whether there is here, too, a reverse process which enables us to integrate when several independent variables exist.

The brief answer is that there is such a process; and that just as definite integration of a function of one variable yields the area under a curve, so the definite integration of a function of two variables will yield the volume under a surface. For more than two variables the geometrical interpretation becomes more complicated.

This process of multiple integration is used more in the development of statistical and econometric theory than in pure economics. We shall show that it is closely related to linear integration. This relationship is the basis of a great deal of advanced work, especially in theory involving the complex variable, and many important definite integrals, such as, for example,

$$\int_0^\infty e^{-x^2}\, dx$$

are most easily evaluated by using techniques which are based upon it. We shall not go into these techniques here, but we present the basic ideas of multiple and linear integration so that the reader may have some idea about them when he comes across them. They are best developed in terms of complex functions, if more advanced work is contemplated, but the following approach is simpler. For a more sophisticated treatment, including the evaluation of the above integral, one should consult a work on complex functions.

2. An economic example

Suppose that we have a detailed breakdown of people's consumer expenditures according to their incomes y and their ages a, and that the information is adequately represented by

$$C = (\alpha a - \beta a^2)(\gamma + \delta y)$$

where C is the total consumer expenditure of all people who have a given age a and income y.

We know that a varies between 10 and 70 and that y ranges from 20 to 100.

We wish to find the total consumer expenditure of all people.

The problem can be solved in the following way. The rectangle outlined in the (a, y) plane of Diagram 23.1 contains all possible combinations of age and income. To each point in it there corresponds a value of C, which may be represented by a vertical line of appropriate length.

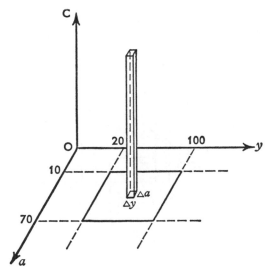

FIG. 23.1

If we imagine a point in this rectangle, having coordinates (a_1, y_1) to be in the centre of a very small rectangle of dimensions $\Delta a, \Delta y$, then we can erect a column upon it, of height

$$C_1 = (\alpha a_1 - \beta a_1^2)(\gamma + \delta y_1)$$

The volume of this column will be

$$C_1 \, \Delta a \, \Delta y$$

L

By an argument similar to that used in discussing the area under a curve, we can see that this volume approximates to the total expenditure of all people whose ages and incomes correspond to points in this small rectangle of area $\Delta a\, \Delta y$.

As a and y vary, so does C, and consequently the height of the column. If they vary over their complete ranges we obtain a set of tightly packed slender columns whose upper extremities define the three-dimensional surface

$$C = (\alpha a - \beta a^2)(\gamma + \delta y)$$

The sum of their volumes (which will tend to the volume under this surface and between the extreme values of a and y) will tend to the sum of the expenditures of all people.

The question, now, is how to find this volume.

Consider the point (a_1, y_1).

Draw a thin strip on the (a, y) plane, containing this point, of thickness Δa, from $y = 20$ to $y = 100$.

Use this strip to define the base of a thin slice of the solid, as shown in Diagram 23.2.

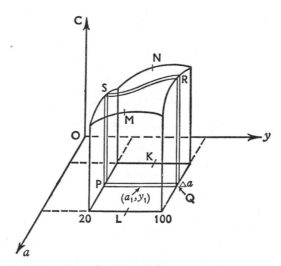

Fig. 23.2

If Δa is very small then the volume of this slice tends to

$$\Delta V = (\text{Area of section } PQRS) \times \Delta a$$

But the area of the section $PQRS$ is

$$\int_{y=20}^{y=100} C \, dy$$

where $C = (\alpha a_1 - \beta a_1^2)(\gamma + \delta y)$, for along the narrow strip, a is constant at a_1, but y varies.

Thus the element of volume above the strip containing (a_1, y_1) is therefore

$$\Delta V = \int_{y=20}^{y=100} (\alpha a_1 - \beta a_1^2)(\gamma + \delta y) \, \Delta a \, dy$$

$$= (\alpha a_1 - \beta a_1^2)\Delta a \int_{y=20}^{y=100} (\gamma + \delta y) \, dy$$

since the bracket containing a_1 has a value which does not depend upon y.

This can be evaluated; but let us defer that stage.

The total volume is the sum of the volumes of separate slices, as a varies from 10 to 70.

Thus

$$V = \int \Delta V = \int_{a=10}^{70} (\alpha a - \beta a^2) \left[\int_{y=20}^{y=100} (\gamma + \delta y) \, dy \right] da$$

$$= \int_{a=10}^{70} (\alpha a - \beta a^2) \left[\gamma y + \frac{\delta y^2}{2} \right]_{20}^{100} da$$

$$= \int_{a=10}^{70} (\alpha a - \beta a^2) \left[80\gamma + 4800\delta \right] da$$

$$= (80\gamma + 4800\delta) \int_{a=10}^{70} (\alpha a - \beta a^2) \, da$$

$$= (80\gamma + 4800\delta) \left[\frac{\alpha a^2}{2} - \frac{\beta a^3}{3} \right]_{10}^{70}$$

$$= (80\gamma + 4800\delta) \left[2400\alpha - 114000\beta \right]$$

which represents the total consumer expenditure.

3. A geometric example

We have chosen to evaluate the integral in this way, rather than at a slightly earlier stage, because it points the way to the solution of more complicated problems.

Suppose, for example, that the boundary in the (a, y) plane was not a rectangle but a circle of equation

$$(a - 20)^2 + (y - 40)^2 = 400$$

having its centre at (20, 40) and a radius of 20.

Now the strip containing the point (a_1, y_1) has a length which depends on its position. The geometry of Diagram 23.3 shows

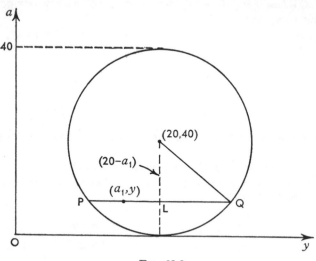

FIG. 23.3

that the chord PQ has a length which is double QL, and that

$$QL^2 = 20^2 - (20 - a_1)^2$$
$$= 40a_1 - a_1{}^2$$

Thus

$$PQ = 2\sqrt{40a_1 - a_1{}^2}$$

The value of y at P is therefore $40 - \sqrt{40a_1 - a_1{}^2}$

and at Q it is $40 + \sqrt{40a_1 - a_1{}^2}$

To find the element of volume ΔV above the thin strip containing (a_1, y) we therefore have to integrate C with respect to y between these two values, and then multiply by Δa.

We have

$$\Delta V = \int_{y=40-\sqrt{40a_1-a_1{}^2}}^{40+\sqrt{40a_1-a_1{}^2}} (\alpha a_1 - \beta a_1{}^2)(\gamma + \delta y)\, \Delta a\, dy$$

$$= (\alpha a_1 - \beta a_1{}^2) \int_{y=40-\sqrt{40a_1-a_1{}^2}}^{40+\sqrt{40a_1-a_1{}^2}} (\gamma + \delta y)\, dy\, \Delta a$$

The integral can be evaluated to yield an answer in terms of a_1. Call it $f(a_1)$. Then

$$\Delta V = (\alpha a_1 - \beta a_1{}^2) f(a_1) \, \Delta a$$

and so the complete volume is obtained by summing these elementary volumes as a moves from 0 to 40. We get

$$V = \int \Delta V = \int_{a=0}^{40} (\alpha a - \beta a^2) f(a) \, da$$

Let us now rewrite this more fully.

$$V = \int_{a=0}^{40} (\alpha a - \beta a^2) \int_{y=40-\sqrt{40a-a^2}}^{40+\sqrt{40a-a^2}} (\gamma + \delta y) \, dy \, da$$

which is often written as

$$V = \int_{a=0}^{40} \int_{y=40-\sqrt{40a-a^2}}^{40+\sqrt{40a-a^2}} (\alpha a - \beta a^2)(\gamma + \delta y) \, dy \, da$$

t being understood that the innermost integral sign is used first, with everything except the variable specified as its subscript (y) being treated as constant.

To evaluate it we may most conveniently go back to the form

$$\begin{aligned}
V &= \int_{a=10}^{40} (\alpha a - \beta a^2) \int_{y=40-\sqrt{40a-a^2}}^{40+\sqrt{40a-a^2}} (\gamma + \delta y) \, dy \, da \\
&= \int_{a=0}^{40} (\alpha a - \beta a^2) \left[\gamma y + \frac{\delta y^2}{2} \right]_{40-\sqrt{40a-a^2}}^{40+\sqrt{40a-a^2}} da \\
&= \int_{a-0}^{40} (\alpha a - \beta a^2) \left[2\gamma \sqrt{40a - a^2} + 80\delta \sqrt{40a - a^2} \right] da \\
&= \int_{a=0}^{40} (\alpha a - \beta a^2) \sqrt{40a - a^2} (2\gamma + 80\delta) \, da \\
&= (2\gamma + 80\delta) \int_{a=0}^{40} \left[\alpha a \sqrt{40a - a^2} - \beta a^2 \sqrt{40a - a^2} \right] da
\end{aligned}$$

We have thus reduced the problem to that of evaluating

$$I_1 = \int a \sqrt{40a - a^2} \, da$$

and

$$I_2 = \int a^2 \sqrt{40a - a^2} \, da$$

Most integration is trickery. Here the trick is to write

$$a = 20 - x$$

$$da = \quad - dx$$

$$I_1 = -\int (20 - x)\sqrt{40(20 - x) - (20 - x)^2}\, dx$$

$$= -\int (20 - x)\sqrt{800 - 40x - 400 + 40x - x^2}\, dx$$

$$= -\int (20 - x)\sqrt{400 - x^2}\, dx$$

$$= -20 \int \sqrt{400 - x^2}\, dx + \int x\sqrt{400 - x^2}\, dx$$

$$= -20 \left[\frac{x}{2}\sqrt{400 - x^2} + \frac{400}{2}\, \sin^{-1} \frac{x}{20} \right]$$

$$- \tfrac{1}{3}(400 - x^2)^{3/2} + C$$

where the solution of the first integral comes from Chapter XVI, Section 3, example (iii).

To evaluate $I_2 = \int a^2 \sqrt{40a - a^2}\, da$ we can use the same substitution, giving

$$I_2 = -\int (20 - x)^2 \sqrt{400 - x^2}\, dx$$

$$= -400 \int \sqrt{400 - x^2}\, dx$$

$$+ 40 \int x\sqrt{400 - x^2}\, dx$$

$$- \int x^2 \sqrt{400 - x^2}\, dx$$

The first two of these integrals were evaluated in finding I_1.

An instructive method for the evaluation of the third integral is to write

$$x = 20 \sin \theta \qquad\qquad dx = 20 \cos \theta\, d\theta$$

Then
$$\int x^2 \sqrt{400 - x^2}\, dx$$

$$= \int 400 \sin^2 \theta \sqrt{400 - 400 \sin^2 \theta}\; 20 \cos \theta\, d\theta$$

$$= \int 8000 \sin^2 \theta \sqrt{400 \cos^2 \theta}\; 20 \cos \theta\, d\theta$$

$$= \int 8000 \sin^2 \theta \cdot 400 \cos^2 \theta \, d\theta$$

$$= \int 3{,}200{,}000 \sin^2 \theta \cos^2 \theta \, d\theta$$

which can be evaluated by integration by parts, or by the use of a reduction formula derived in Chapter XVI.

Finally, in these solutions, we must rewrite them in terms of the original variable a, using

$$a = 20 - x$$
$$= 20 - 20 \sin \theta$$

so that

$$x = 20 - a$$

and

$$\sin \theta = \frac{20 - a}{20}$$

and then insert the extreme values of a, namely 0 and 40, between which integration occurs.

4. The order of integration

But let us go back to

$$V = \int_{a=0}^{40} \int_{y=40-\sqrt{40a-a^2}}^{40+\sqrt{40a-a^2}} (\alpha a - \beta a^2)(\gamma + \delta y) \, dy \, da$$

Compare this, part by part, with

$$V = \int_{a=a_1}^{a_2} \int_{y=f_1(a)}^{f_2(a)} \phi(a, y) \, dy \, da,$$

which is simply a more general form of the equation. We have $\phi(a, y)$ a function of a and y, which is integrated with respect to y between two limits. The lower limit is a certain function of a, denoted by $f_1(a)$ (which in our example denoted the value of y at the point P). The upper limit is a different function of a, denoted by $f_2(a)$ (which showed the value of y at the point Q). This integration with respect to y between these two limits yields a function of a, whose value will be

$$F(a) = \left[\int_{y=f_1(a)}^{f_2(a)} \phi(a, y) \, dy \right]$$

and it is this which is now integrated with respect to a between limits a_1 and a_2.

Thus

$$V = \int_{a=a_1}^{a_2} \int_{y=f_1(a)}^{f_2(a)} \phi(a, y)\, dy\, da$$

$$= \int_{a=a_1}^{a_2} F(a)\, da$$

where $F(a)$ has the above value.

If one considers the argument based on Diagram 23.3, in which we put

$$\Delta V = (\text{Area } PQRS) \times \Delta a$$

we will see that we could equally well have sliced the solid parallel to the a-axis, through (say) $KLMN$, and obtained

$$\Delta V = (\text{Area } KLMN) \times \Delta y$$

Proceeding in this way we would obtain

$$V = \int_{y=y_1}^{y_2} \left[\int_{a=\psi_1(y)}^{\psi_2(y)} \phi(a, y)\, da \right] dy$$

where $\psi_1(y)$ and $\psi_2(y)$ would denote the values of a, for varying values of y, at the two points K and L, corresponding to P and Q in the original argument.

More usually we would employ x and y as variables, and think of a typical *double integral* as

$$V = \int_{x_1}^{x_2} \int_{y=f_1(x)}^{y=f_2(x)} \phi(x, y)\, dy\, dx$$

We have just argued that

$$V = \int_{x_1}^{x_2} \int_{y=f_1(x)}^{f_2(x)} \phi(x, y)\, dy\, dx$$

$$= \int_{y_1}^{y_2} \int_{x=\psi_1(y)}^{\psi_2(y)} \phi(x, y)\, dx\, dy$$

The order of integration does not matter provided that one appropriately alters the limits. Here, great care is essential. Suppose, for example, that we wish to evaluate

$$\iint x^2 y\, dy\, dx$$

over the area shaded in Diagram 23.4.

If we consider this divided into vertical strips then the upper and lower limits are set by

$$y = 2x \quad \text{and} \quad y = x$$

with x ranging from 0 to 2.

If we divide it horizontally then between $y = 0$ and $y = 2$ the limits of x are defined by

$$x = \frac{y}{2} \quad \text{and} \quad x = y$$

but between $y = 2$ and $y = 4$ the limits of x are

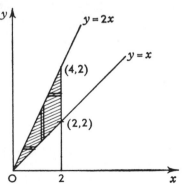

Fig. 23.4

$$x = \frac{y}{2} \quad \text{and} \quad x = 2$$

Let us evaluate the integral of the " vertical " approach:

$$\int_{x=0}^{2} \int_{y=x}^{y=2x} x^2 y \, dy \, dx$$

and that of the " horizontal " approach:

$$\int_{y=0}^{2} \int_{x=y/2}^{x=y} x^2 y \, dx \, dy + \int_{y=2}^{4} \int_{x=y/2}^{2} x^2 y \, dx \, dy$$

The former evaluation gives us

$$\int_{x=0}^{2} \left[\frac{x^2 y^2}{2} \right]_{x}^{2x} dx$$

$$= \int_{0}^{2} \frac{4x^4 - x^4}{2} \, dx = \int_{0}^{2} \frac{3x^4}{2} \, dx$$

$$= \left[\frac{3x^5}{10} \right]_{0}^{2} \qquad = \frac{96}{10}$$

The latter gives

$$\int_{y=0}^{2} \left[\frac{x^3 y}{3} \right]_{y/2}^{y} dy + \int_{2}^{4} \left[\frac{x^3 y}{3} \right]_{y/2}^{2} dy$$

$$= \int_{y=0}^{2} \left[\frac{y^4}{3} - \frac{y^4}{24} \right] dy + \int_{2}^{4} \left[\frac{8y}{3} - \frac{y^4}{24} \right] dy$$

$$= \int_{y=0}^{2} \frac{7y^4}{24} \, dy + \int_{2}^{4} \frac{8y}{3} \, dy - \int_{2}^{4} \frac{y^4}{24} \, dy$$

$$= \left[\frac{7y^5}{120}\right]_0^2 + \left[\frac{4y^2}{3}\right]_2^4 - \left[\frac{y^5}{120}\right]_2^4$$

$$= \frac{224}{120} + \frac{64}{3} - \frac{16}{3} - \frac{1024}{120} + \frac{32}{120}$$

$$= \frac{48}{3} - \frac{768}{120} = \frac{1152}{120} = \frac{96}{10}$$

as before. The two approaches give identical results. The example shows that when we change the order of integration it may be necessary to split the integral into two or more integrals, one for each part of the area.

5. Multiple integration

Just as it is possible to extend differentiation to functions of more than two variables, so can we have multiple integration. For example, a triple integral might be

$$\iiint_A f(x, y, z)\, dx\, dy\, dz$$

where A denotes that the integration is performed for all points in the three-dimensional space A.

If the boundaries are indicated by

$$z = \psi_1(x, y) \quad \text{and} \quad \psi_2(x, y)$$
$$y = \phi_1(x) \quad \text{and} \quad \phi_2(x)$$
and
$$x = a \quad \text{and} \quad b$$

then we may write

$$I = \int_a^b \left[\int_{y=\phi_1(x)}^{\phi_2(x)} \left\{ \int_{z=\psi_1(x,y)}^{\psi_2(x,y)} f(x, y, z)\, dz \right\} dy \right] dx$$

$$= \int_a^b \int_{\phi_1(x)}^{\phi_2(x)} \int_{\psi_1(x,y)}^{\psi_2(x,y)} f(x, y, z)\, dz\, dy\, dx$$

which is sometimes written as

$$\int_a^b dx \int_{\phi_1(x)}^{\phi_2(x)} dy \int_{\psi_1(x,y)}^{\psi_2(x,y)} f(x, y, z)\, dz$$

6. Change of variable

It is sometimes useful, or even necessary, to re-write a problem in terms of new variables.

Suppose that we wish to evaluate the double integral

$$\iint_A f(x, y)\, dx\, dy$$

over some area A in the (x, y) plane. Let x and y be expressed in terms of new variables u and v, such that

$$x = \phi(u, v) \quad \text{and} \quad y = \psi(u, v)$$

such that for every point in the (u, v) plane there is a corresponding point in the (x, y) plane, and vice versa. For example, possibly

$$x = u \cos v \quad \text{and} \quad y = u \sin v$$

This is clearly nothing other than a transformation from Cartesian to polar co-ordinates, and every (x, y) point corresponds to a (u, v) point.

It is likely to be easy enough for us to rewrite $f(x, y)$ in terms of u and v. The problem is how to cope with the $dx\, dy$ at the end of the integration expression. Simply to replace it by $du\, dv$ would be wrong, because the small element of area $\Delta x \Delta y$ will probably not be the same as the small element $\Delta u \Delta v$.

It is shown in all of the standard works (such as, for example, Gillespie's *Integration* in the Oliver & Boyd series) that we have to replace the integral by another integral which contains partial derivatives. Let us write

$$f(x, y) = F(u, v)$$

where $x = \phi(u, v)$ and $y = \psi(u, v)$. Then it can be shown that

$$\iint_A f(x, y)\, dx\, dy = \iint_{A'} F(u, v)\left[\frac{\partial x}{\partial u}\frac{\partial y}{\partial v} - \frac{\partial x}{\partial v}\frac{\partial y}{\partial u}\right] du\, dv$$

where A' is the area in the (u, v) plane corresponding to A in the (x, y) plane.

Readers who have a slight familiarity with determinants will see that the expression in square brackets is the value of the determinant

$$\begin{vmatrix} \dfrac{\partial x}{\partial u} & \dfrac{\partial x}{\partial v} \\ \dfrac{\partial y}{\partial u} & \dfrac{\partial y}{\partial v} \end{vmatrix}$$

which is frequently written in the abbreviated form

$$\begin{vmatrix} \dfrac{\partial(x, y)}{\partial(u, v)} \end{vmatrix}$$

It is called the Jacobian of (x, y) with respect to (u, v).

In the case of a triple integral we would need to use the Jacobian

$$\left| \frac{\partial(x, y, z)}{\partial(u, v, w)} \right| = \left| \begin{array}{ccc} \dfrac{\partial x}{\partial u} & \dfrac{\partial x}{\partial v} & \dfrac{\partial x}{\partial w} \\[8pt] \dfrac{\partial y}{\partial u} & \dfrac{\partial y}{\partial v} & \dfrac{\partial y}{\partial w} \\[8pt] \dfrac{\partial z}{\partial u} & \dfrac{\partial z}{\partial v} & \dfrac{\partial z}{\partial w} \end{array} \right|.$$

and then we would have that

$$\iiint_A f(x, y, z)\, dx\, dy\, dz = \iiint_{A'} F(u, v, w) \left| \frac{\partial(x, y, z)}{\partial(u, v, w)} \right| du\, dv\, dw$$

Example:

If

$$x = u \cos v \quad \text{and} \quad y = u \sin v$$

then

$$\partial(x, y)/\partial(u, v) = \left| \begin{array}{cc} \cos v & \sin v \\ -u \sin v & u \cos v \end{array} \right| = u$$

and so

$$\iint_A f(x, y)\, dx\, dy = \iint_{A'} f(u \cos v,\, u \sin v)\, u\, du\, dv$$

where the right-hand expression incorporates a u, which is the value of the Jacobian.

EXERCISE 23.1

1. Show that the area of the triangle whose vertices are at $(0, 0)$, $(3, 0)$ and $(3, 6)$ is given by both

$$\int_{y=0}^{6} \int_{x=y/2}^{3} f(x, y)\, dx\, dy \quad \text{and} \quad \int_{x=0}^{3} \int_{y=0}^{2x} f(x, y)\, dy\, dx$$

provided that $f(x, y) = 1$. Evaluate these integrals and check that they are equal to the value of the area as determined by elcmentary geometry.

2. Evaluate the integrals of question 1, when $f(x, y) = x$. Check that they are equal.

7. Line integration

Ordinary integration of the kind

$$\int_a^b f(x)\, dx$$

is integration of $f(x)$ as x proceeds along the straight x-axis from a to b.

We have also looked at double integration over an area in the (x, y) plane.

A question we may now consider is how to integrate over a curved line in the (x, y) plane. This could arise in several ways. One could, for example, twist the axis Ox. Or one could think of an area such as that shown in Diagram 23.5 being slowly squeezed up until it becomes the line ABC.

A physical example of such a problem is easily given.

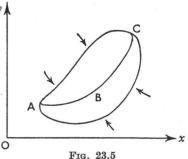

FIG. 23.5

Suppose we have a circular line of equation

$$x^2 + y^2 = 9$$

centred on the origin. Let us build a cylinder upon it, and on this cylinder draw a continuous line, starting where $x = 0$ and $y = 3$, and going around it until we are back to the same position, at a variable height

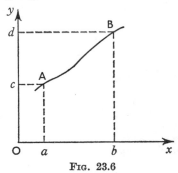

FIG. 23.6

$$z = x^2 y$$

What is the area on the surface of the cylinder below this line?

Consider a function $F(x, y)$, and let $y = f(x)$ be a single-valued function of x, in the sense that for each value of x there is only one value of y. Let AB in Diagram 23.6 be a section of the curve $y = f(x)$.

The integral

$$\int_a^b F[x, f(x)]\, dx$$

is called the *line integral* along the curve $y = f(x)$ from A to B. It may be written

$$\int_{AB} F[x, f(x)]dx \quad \text{or} \quad \int_{AB} F(x, y)\, dx$$

and is sometimes called a *curvilinear integral*.

If, in addition, $y = f(x)$ is such that it implies $x = \phi(y)$ where this, too, is a single-valued function, then we can likewise define

$$\int_{AB} F[\phi(y),\, y]\, dy$$

$$= \int_{c}^{d} F[\phi(y),\, y]\, dy$$

as another line integral along the curve AB. In general, its value will differ from that of the line integral with respect to x.

For example, consider the curve

$$y = 2x + 3$$

between the points $(0,\, 3)$ and $(2,\, 7)$.

Let
$$F(x,\, y) = x + 3y$$
$$= 7x + 9$$

Then
$$\int_{0}^{2} F(x,\, y)\, dx = \int_{0}^{2} (7x + 9)\, dx$$
$$= \left[\frac{7x^2}{2} + 9x \right]_{0}^{2}$$
$$= 32$$

But we can also write $F(x,\, y)$ as

$$F(x,\, y) = x + 3y = \frac{y-3}{2} + 3y$$

$$= \frac{7y-3}{2}$$

And
$$\int_{3}^{7} F(x,\, y)\, dy = \int_{3}^{7} \frac{7y-3}{2}\, dy$$
$$= \left[\frac{7y^2}{4} - \frac{3y}{2} \right]_{3}^{7}$$
$$= \frac{343 - 63}{4} - \frac{(21 - 9)}{2}$$
$$= 70 - 6$$
$$= 64$$

which means that in this case

$$\int_{AB} F[\phi(y),\, y]\, dy = 2 \int_{AB} F[x,\, f(x)]\, dx$$

With these definitions before us a few important results quickly emerge.

Because
$$\int_{x_1}^{x_2} f(x)\, dx = -\int_{x_2}^{x_1} f(x)\, dx$$

it is easy to show that the *direction of movement along the curve matters*, and that

$$\int_{AB} F(x, y)\, dx = -\int_{BA} F(x, y)\, dx$$

Example:

Integrate $F = xy$ along the curve $y = x^3 + 3$ between $x = 1$ and $x = 2$. We have

$$\int_1^2 F(x, y)\, dx = \int_1^2 x(x^3 + 3)\, dx$$

$$= \int_1^2 (x^4 + x)\, dx = \left[\frac{x^5}{5} + \frac{x^2}{2}\right]_1^2$$

$$= \frac{31}{5} + \frac{3}{2} = 7 \cdot 7$$

If we have a closed curve then we can divide it into two or more curves corresponding to single-valued functions, and integrate along each separately. But now we must be consistent in our direction. If there is a closed curve we define the positive direction to be that which is such that a person walking along the curve will always have the enclosed area on his left. It is the counter-clockwise direction.

Example:

Evaluate $I = \int y\, dx$ round the circle $x^2 + y^2 = 9$. We divide the circle into halves as shown in Diagram 23.7.

For the top half we evaluate

$$\int_{x=3}^{-3} y\, dx = \int_3^{-3} \sqrt{9 - x^2}\, dx$$

while for the bottom half we take

$$\int_{-3}^3 -\sqrt{9 - x^2}\, dx$$

since here,
$$y = -\sqrt{9 - x^2}$$

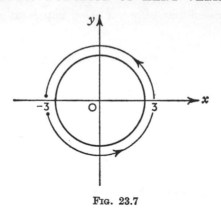

FIG. 23.7

These two integrals yield, on addition

$$I = \int_{x=3}^{-3} \sqrt{9-x^2}\, dx - \int_{-3}^{3} \sqrt{9-x^2}\, dx$$

$$= 2 \int_{3}^{-3} \sqrt{9-x^2}\, dx$$

$$= 2 \left[\frac{x}{3}\sqrt{9-x^2} + \frac{9}{2}\sin^{-1}\frac{x}{3} \right]_{3}^{-3}$$

$$= 2 \left[\left(-\frac{3}{3}\times 0 + \frac{9}{2}\sin^{-1}(-1) \right) - \left(0 + \frac{9}{2}\sin^{-1} 1 \right) \right]$$

$$= 9 \left[\sin^{-1}(-1) - \sin^{-1} 1 \right]$$

$$= 9 \left[-\frac{\pi}{2} - \frac{\pi}{2} \right] = -9\pi$$

8. Green's (or Gauss's) theorem

There is an important theorem linking line integrals and double integrals.

Suppose that $P(x, y)$ and $Q(x, y)$ are two functions defined at all points inside and on the boundary of a closed area A, bounded by a curve C as in Diagram 23.8.

Let C be such that no line parallel to Ox or to Oy can cut it in more than two points.

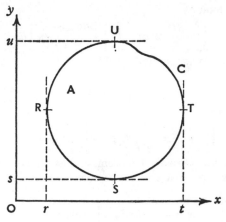

FIG. 23.8

Consider

$$\iint_A \frac{\partial Q}{\partial x}\, dx\, dy$$

$$= \int_s^u \left[\int_{RST}^{RUT} \frac{\partial Q}{\partial x}\, dx \right] dy$$

$$= \int_s^u Q\left[\psi_2(y),\, y \right] dy - \int_s^u Q\left[\psi_1(y),\, y \right] dy$$

where $Q[\psi_2(y),\, y]$ is the result of evaluating

$$\int \frac{\partial Q}{\partial x}\, dx$$

and replacing x by the functional relationship $\psi_2(y)$ which relates x to y along the curve RUT, while the second integral is similarly obtained, but for the curve RST.

From our definition of a line integral, we can now write our result immediately as

$$\int_{STU} Q(x,\, y)\, dy - \int_{SRU} Q(x,\, y)\, dy$$

$$= \int_{STU} Q(x,\, y)\, dy + \int_{URS} Q(x,\, y)\, dy$$

$$= \int_C Q(x,\, y)\, dy$$

taken in the positive direction around the curve.

Now let us look at

$$\iint_A \frac{\partial P}{\partial y}\,dx\,dy = \int_\tau^t \left[\int_{STU}^{SRU} \frac{\partial P}{\partial y}\,dy\right]dx$$

which, by a similar argument, yields

$$-\int P(x,\,y)\,dx$$

It follows that

$$\iint_A \left(\frac{\partial Q}{\partial x} - \frac{\partial P}{\partial y}\right)dx\,dy = \int_C (P\,dx + Q\,dy)$$

which is known as Green's Theorem, and as Gauss's Theorem. For example, over any area and curve such as that drawn in Diagram 23.8.

$$\iint_A (\cos x - x^2)\,dx\,dy = \int_C (x^2 y\,dx + \sin x\,dy)$$

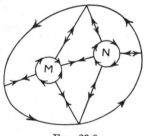

FIG. 23.9

The reader could verify this, and similar results, for simple shapes, such as rectangles.

This theorem holds in a modified form if the area has holes in it, as in Diagram 23.9. This can be shown by dividing into as many contiguous areas as may be necessary, each conforming to the condition that no line parallel to an axis can cut it more than twice.

Repeated application of the theorem will show that the integration over each sub-area is equal to a line integral. But as all the line integrals are in counter-clockwise direction substantial parts cancel out, and in the end we are left with the counter-clockwise integral around the outer curve *plus* the *clockwise* integral around the holes M and N.

9. Area

Now consider

$$\iint_A f(x,\,y)\,dx\,dy \quad \text{where} \quad f(x,\,y) = 1$$

It is easy to see from the definition of a double integral that this gives the volume of a solid of unit height built on the area A, which is numerically equal to the area of area A.

Now define $Q(x,\,y)$ and $P(x,\,y)$ such that $\dfrac{\partial Q}{\partial x} = 1$ and $\dfrac{\partial P}{\partial y} = 0$

Then

$$\iint_A \left(\frac{\partial Q}{\partial x} - \frac{\partial P}{\partial y}\right) dx\, dy = \iint_A dx\, dy$$

But also this double integral is, by Gauss's Theorem, equal to

$$\int_C (P\, dx + Q\, dy)$$

To satisfy our requirements for P and Q, put $P = 0$ and $Q = x$ (which we have not shown to be necessary, but these are values which satisfy our needs).

Then

$$\int_C (P\, dx + Q\, dy) = \int_C x\, dy$$

Thus

$$\iint_A dx\, dy = \int_C x\, dy$$

are two alternative expressions for the area of A which is bounded by C. Instead of putting $\frac{\partial Q}{\partial x} = 1$ and $\frac{\partial P}{\partial y} = 0$ we could have chosen

$$P = y, \qquad Q = 0$$

$$\frac{\partial P}{\partial y} = 1, \qquad \frac{\partial Q}{\partial x} = 0$$

whence

$$\iint_A \left(\frac{\partial Q}{\partial x} - \frac{\partial P}{\partial y}\right) dx\, dy = -\iint_A dx\, dy$$

and so we find that

$$\iint_A dx\, dy = -\int_C y\, dx$$

It follows, by adding these two results, that the area of A, being the region enclosed by the curve C, is given by any of the four integrals

$$\iint_A dx\, dy, \quad \int_C x\, dy, \quad -\int_C y\, dx, \quad \tfrac{1}{2}\int_C (x\, dy - y\, dx)$$

Example:

To evaluate the area of the circle of radius a, centred at the origin, whose equation is

$$x^2 + y^2 = a^2$$

we may proceed in any of the following ways.

(i) $$\int_{x=-a}^{a} \int_{y=-\sqrt{a^2-x^2}}^{+\sqrt{a^2-x^2}} dy\, dx$$

$$= \int_{x=-a}^{a} \left[y \right]_{-\sqrt{a^2-x^2}}^{\sqrt{a^2-x^2}} dx$$

$$= 2 \int_{x=-a}^{a} \sqrt{a^2 - x^2}\, dx = 2 \left[\frac{x}{2}\sqrt{a^2 - x^2} + \frac{a^2}{2} \sin^{-1}\frac{x}{a} \right]_{-a}^{a}$$

$$= 2 \left[\frac{a^2}{2} \sin^{-1} 1 + \frac{a^2}{2} \sin^{-1}(-1) \right] = 2 \left[a^2 \sin^{-1} 1 \right]$$

$$= 2a^2 \frac{\pi}{2}$$

$$= \pi a^2$$

(ii) $$\int_C x\, dy$$

where this integral has to be done positively around the circle, which has to be divided into parts. We can start anywhere, but we must keep in mind that to the right of the y-axis, $x = +\sqrt{a^2 - y^2}$, while to its left, $x = -\sqrt{a^2 - y^2}$. It is therefore convenient to start at $(x=0, y=-a)$ proceeding counter-clockwise to $(x=0, y=+a)$ and then, in the same direction, around to our starting point.

We get

$$\int_C x\, dy = \int_{-a}^{+a} \sqrt{a^2 - y^2}\, dy + \int_{+a}^{-a} -\sqrt{a^2 - y^2}\, dy$$

$$= 2 \int_{-a}^{+a} \sqrt{a^2 - y^2}\, dy$$

which, compared with our integral in case (i), clearly yields

$$\pi a^2.$$

(iii) Here we look at $\int_C y\, dx$. Now the positive value of y is above the x-axis and the negative value below it. We consequently evaluate

$$\int_a^{-a} \sqrt{a^2 - x^2}\, dx + \int_{-a}^{a} -\sqrt{a^2 - x^2}\, dx$$

which yields not πa^2 but $-\pi a^2$, showing that the area is

$$-\int_C y\, dx.$$

(iv) The fourth result follows from combining (ii) and (iii).

10. Integration along a curve

Sometimes one comes across the curvilinear integral in a slightly different form, usually writen as

$$\int_{AB} F(x, y)\, ds$$

Let P and Q be two close points on the curve, with coordinates (x, y) and $(x + \Delta x, y + \Delta y)$. Let the length of the curve joining P to Q be Δs and let the length of the straight line joining them be Δl. Then

$$(\Delta l)^2 = (\Delta x)^2 + (\Delta y)^2$$

and

$$(\Delta s)^2 = \left(\frac{\Delta s}{\Delta l}\right)^2 (\Delta l)^2$$

$$= \left(\frac{\Delta s}{\Delta l}\right)^2 \left[(\Delta x)^2 + (\Delta y)^2\right]$$

whence

$$\left(\frac{\Delta s}{\Delta x}\right)^2 = \left(\frac{\Delta s}{\Delta l}\right)^2 \left[1 + \left(\frac{\Delta y}{\Delta x}\right)^2\right]$$

When $\Delta x \to 0$ then $\dfrac{\Delta s}{\Delta l} \to 1$, $\dfrac{\Delta s}{\Delta x} \to \dfrac{ds}{dx}$ and $\dfrac{\Delta y}{\Delta x} \to \dfrac{dy}{dx}$

whence

$$\left(\frac{ds}{dx}\right)^2 = 1 + \left(\frac{dy}{dx}\right)^2$$

$$\frac{ds}{dx} = \sqrt{1 + \left(\frac{dy}{dx}\right)^2}$$

Thus the length of the arc of the curve from $x = a$ to $x = b$ is

$$\int_{x=a}^{x=b} ds = \int_{x=a}^{b} \frac{ds}{dx}\, dx$$

$$= \int_a^b \sqrt{1 + \left(\frac{dy}{dx}\right)^2}\, dx$$

For example, if $y = x^2$ is the curve then its length from $x = a$ to $x = b$ is

$$\int_a^b \sqrt{1 + \left(\frac{dy}{dx}\right)^2}\, dx = \int_a^b \sqrt{1 + 2x}\, dx$$

$$= \left[\frac{(1 + 2x)^{3/2}}{3}\right]_a^b$$

$$= \frac{1}{3}\left[(1 + 2b)^{3/2} - (1 + 2a)^{3/2}\right]$$

In our treatment of line integrals we have considered integration with respect to x and with respect to y—i.e. in directions parallel to the axes. Now we consider it in the direction of the curve itself, writing

$$I = \int_{AB} f(x, y)\, ds$$

$$= \int_{AB} f(x, y)\sqrt{1 + \left(\frac{dy}{dx}\right)^2}\, dx$$

$$= \int_{AB} f(x, y)\sqrt{1 + \left(\frac{dx}{dy}\right)^2}\, dy$$

We may notice that if

$$\frac{dy}{dx} = \lambda$$

where λ is constant, then

$$\frac{dx}{dy} = \frac{1}{\lambda}$$

and we have

$$I = \int_{AB} f(x, y)\sqrt{1 + \lambda^2}\, dx = \int_{AB} f(x, y)\sqrt{1 + \frac{1}{\lambda^2}}\, dy$$

whence

$$\sqrt{1 + \lambda^2} \int_{AB} f(x, y)\, dx = \sqrt{1 + \frac{1}{\lambda^2}} \int_{AB} f(x, y)\, dy$$

whence

$$\int_{AB} f(x, y)\, dx \Big/ \int_{AB} f(x, y)\, dy = \sqrt{1 + \frac{1}{\lambda^2}} \Big/ \sqrt{1 + \lambda^2}$$

$$= \frac{1}{\lambda}$$

It is for this reason that when we considered the line integration along

$$y = 2x + 3$$

of $F(x, y)$ we found that one integral was half the other—for here $\frac{dy}{dx} = 2$.

Now let us consider the positive quadrant of the circle $x^2 + y^2 = a^2$ —i.e. the portion between $A = (x = a, y = 0)$ and $B = (x = 0, y = a)$.

Also let

$$f(x, y) = 1$$

Consider

$$\int_{AB} f(x, y)\, dx = \int_a^0 dx = -a$$

and

$$\int_{AB} f(x, y)\, dy = \int_0^a dy = \quad a$$

and

$$\int_{AB} f(x, y)\, ds = \int_a^0 \sqrt{1 + \left(\frac{dy}{dx}\right)^2}\, dx$$

where

$$\frac{dy}{dx} = \frac{d}{dx}\sqrt{a^2 - x^2}$$

$$= -\frac{x}{\sqrt{a^2 - x^2}}$$

and so

$$\int_{AB} f(x, y)\, ds = \int_a^0 \sqrt{1 + \frac{x^2}{a^2 - x^2}}\, dx$$

$$= \int_a^0 \sqrt{\frac{a^2}{a^2 - x^2}}\, dx$$

$$= a\int_a^0 \frac{dx}{\sqrt{a^2 - x^2}}$$

$$= a\left[\sin^{-1}\frac{x}{a}\right]_a^0$$

$$= a\left[\sin^{-1} 0 - \sin^{-1} 1\right]$$

$$= a \left[0 - \frac{\pi}{2} \right]$$

$$= -\frac{\pi a}{2}$$

where the negative sign arises simply out of the direction of integration. We have that the length of the arc from AB, being a quarter-circle of radius a, is

$$\frac{\pi a}{2}$$

This is also the area of a surface of unit height built perpendicularly on the arc. If the surface is not of unit height, but is such that its height at (x, y) is $f(x, y)$, then the integral

$$\int_{AB} f(x, y) \, ds$$

determines its area.

EXERCISE 23.2

1. Show that the area of the triangle defined in Exercise 23.1, question 1, is also given by the line integral

$$\int_C x \, dy$$

over its perimeter. Show that this has the value

$$\int_{y=0}^{6} 3 \, dy + \int_{y=6}^{0} \frac{y}{2} \, dy$$

and confirm that this yields the same result as in Exercise 23.1, question 1.

2. Show similarly that the area is also equal to

$$-\int_{3}^{0} 2x \, dx$$

3. Invent other problems of the same kind, checking that the answers obtained by the various methods agree with each other.

11. Surface integration

A surface integral is similar to a line integral but involves more dimensions. To introduce it we must first make a few remarks about surfaces.

Suppose that

$$z = f(x, y)$$

defines a surface such that for any point (x, y) in the (x, y) plane there is a single associated value of z, and a corresponding point on the surface, having co-ordinates (x, y, z).

Let us draw some closed curve on this surface, so that a portion S of the surface is contained within it.

From all points on this curve draw lines parallel to the z-axis so that they end in points on the (x, y) plane. The curve connecting these points contains an area A in the (x, y) plane. We call it the *projection* of S onto the (x, y) plane.

Let us also think of any point on the surface S and consider a plane which is tangential to the surface at that point. A straight line through the point and perpendicular to the tangential plane is the *normal* at that point.

Now consider the surface and choose one side of it. From a point on the surface draw a normal outwards into the space on that side. We call this the " outward drawn normal " to that particular side of the surface at that point.

Suppose that the outward drawn normal makes an angle α with the z-axis.

If $\cos \alpha > 0$ we say that that side of the surface is the " upper " side. If $\cos \alpha < 0$ then it is the " lower " side.

Consider now the integral denoted by

$$\iint_S F(x, y, z) \, dx \, dy$$

which we define as follows.

For the portion S of the side of the surface $z = f(x, y)$ for which $\cos \alpha > 0$, we define the value of the above integral to be

$$+ \iint_A F(x, y, f(x, y)) \, dx \, dy$$

where A is the projection of S onto the (x, y) plane. For the portion for which $\cos \alpha < 0$ we define the value to be the negative of this integral. Thus the integrals over opposite sides of the same surface are equal in magnitude but of opposite sign.

We can similarly define surface integrals which are evaluated over projection of the surface onto the other coordinate planes, these being

$$\iint_S F(x, y, z) \, dy \, dz \quad \text{and} \quad \iint_S F(x, y, z) \, dz \, dx$$

In general they will have different values.

Just as there is an integral along a curve, so there is an integral over a surface, which will have a value independent of the side over which we integrate it. This is

$$\iint_A F(x, y, f(x, y)) \frac{dx\,dy}{|\cos \alpha|}$$

There is also a version of Gauss's Theorem for surface integrals. It is that if $P(x, y, z)$, $Q(x, y, z)$ and $R(x, y, z)$ are functions defined at all points in and on the boundary of the solid K which is bounded by the surface S then, provided the surface integral is evaluated over the *outside* surface,

$$\iiint_K \left(\frac{\partial P}{\partial x} + \frac{\partial Q}{\partial y} + \frac{\partial R}{\partial z} \right) dx\,dy\,dz = \iint_S (P\,dy\,dz + Q\,dz\,dx + R\,dx\,dy)$$

The reader should go over this brief note and compare it step by step with the longer account of line integrals.

CHAPTER XXIV

DIFFERENTIALS

did appear
A very little thing.
THOMAS TRAHERNE : *News*

1. Basic ideas and definitions

The reader who has looked at many economics books containing much mathematics will almost certainly have come across dx and dy used by themselves, and may possibly have been puzzled by this usage, especially after reading Chapter IX in which it is stressed that, as there defined, dx and dy have no separate existence. This chapter indicates the solution to this apparent difference of opinion.

In Chapter IX we considered a function y of the variable x. We let the independent variable x increase by a small, but measurable, amount Δx, and defined the amount Δy to be the corresponding change in y. We then took the ratio $\Delta y/\Delta x$ and said that this would tend to a limit as Δx tended to zero. We denoted that limit by dy/dx, using this symbol to remind us that the limit has its origin in $\Delta y/\Delta x$. Clearly, this defines the symbol dy/dx quite precisely. But it does not imply the separate existence of dy or of dx. To remind ourselves of this, we shall print it in bold type in the next few pages, thus $\frac{\mathbf{dy}}{\mathbf{dx}}$.

Now let us, at the risk of seeming to split hairs, but in the interests of knowing the validity of what we are doing, introduce a *new variable dx* which will have an existence of its own. Let us note first of all, that we are considering y to be a function of x so that x is the *independent* variable and y the *dependent* variable. We decide to introduce a *new variable dx* which will be quite independent of x and may take any arbitrary values. We do not even define dx to be small. All we say is that, in some way or other, it is associated with x, but is independent of it. We will call this new variable the *differential* of x ; and we must note that this definition is the definition of the differential of the *independent* variable.

Now y is a function of x ; it is a *dependent* variable. We *define* the differential of y to be the differential of x (i.e., dx) multiplied by $\frac{\mathbf{dy}}{\mathbf{dx}}$ as defined in Chapter IX.

In other words, we first decide to denote a certain limit concerning small quantities capable of tending to zero (Δx and Δy) by $\dfrac{dy}{dx}$. We then introduce the differential dx as a new variable; and we finally

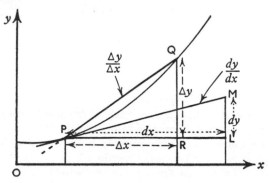

FIG. 24.1

define the differential of the dependent variable to be the product of these two. The point is a tricky one and is illustrated by Diagram 24.1. Here we have a curve whose equation is $y = f(x)$. Let us take two fairly close points P and Q whose co-ordinates are (x, y) and $(x + \Delta x, y + \Delta y)$. Then

$$\Delta x = PR$$

and

$$\Delta y = RQ$$

The slope of the chord PQ is $\dfrac{RQ}{PR} = \dfrac{\Delta y}{\Delta x}$

The slope of the tangent at P is the limit to which $\Delta y/\Delta x$ tends as Q approaches P. We have agreed to denote it by

$$\frac{dy}{dx} = \underset{\Delta x \to 0}{\text{Limit}} \frac{\Delta y}{\Delta x}$$

Let us now take any point L on the horizontal through P. Let us agree to denote the distance PL by dx. Let the vertical through L meet the tangent at P in M. Then we will denote the distance ML by dy. We will call dx the differential of x, and dy the differential of y. We may note that

$$\frac{dy}{dx} = \frac{\text{differential of } y}{\text{differential of } x} = \frac{ML}{LP} = \text{slope of tangent}$$

$$= \frac{dy}{dx} = \underset{\Delta x \to 0}{\text{Limit}} \frac{\Delta y}{\Delta x}$$

where dy/dx is the ratio of separate entities dy and dx while $\mathbf{dy/dx}$ has a meaning only when regarded as a whole.

Clearly the definition of the differential of y enables us to write

$$dy = \frac{\mathbf{dy}}{\mathbf{dx}} \times dx$$

where dy is the differential of y defined to be the product of the rate of increase of y with respect to x (i.e., $\mathbf{dy/dx}$) and the differential of x (defined to be a new variable dx associated with x).

We cannot emphasise too much that the differentials here introduced are *not* the limits to which Δx and Δy tend . . . for those limits are each zero. If we divide each side of the above definition by the differential dx we will have

$$\frac{dy}{dx} = \frac{\mathbf{dy}}{\mathbf{dx}}$$

where the left-hand side is the ratio of two finite differentials, and the right-hand side is the limit to which the ratio $\Delta y/\Delta x$ of two small increments tends as these increments tend to zero.

If the reader is puzzled by these remarks he should not be too perturbed. When the same symbol appears to be used in two different ways, or a couple of pages appear to be devoted to proving that something is equal to itself, then some confusion is bound to arise : but a careful reading of these few remarks should remove it.

The important part of all this is that if we introduce the differentials of x and of y as above, then they have separate existences and are such that whatever is true of the *limit* $\mathbf{dy/dx}$ is also true of the *ratio* dy/dx. What is true of the right-hand side of the above equations is true of the left. Also, if the differential dx is put to *represent a small increment* of x (i.e., if $dx = \Delta x$) then the differential dy (defined as above) will *approximate* to the increment Δy, for

$$dy = \frac{\mathbf{dy}}{\mathbf{dx}} dx = \left(\underset{\Delta x \to 0}{\text{Limit}} \frac{\Delta y}{\Delta x} \right) dx = \left(\underset{\Delta x \to 0}{\text{Limit}} \frac{\Delta y}{\Delta x} \right) \Delta x$$

when dx is defined to be equal to Δx.

A few examples will illustrate these points.

Suppose that

$$y = 4x^2$$

Let us define the differential of x to be a new variable dx. Then the differential of y is defined to be

$$\frac{dy}{dx} dx$$
$$= 8x \, dx$$

If dx be used to denote a small increment in x, then

$$dx = \Delta x$$

therefore, $\qquad dy = 8x\, \Delta x \text{ (exactly)}$

and so the corresponding increment in y is *approximately* equal to this same quantity ; and it is therefore true to say that

$$\Delta y = 8x\, \Delta x \text{ (approximately)}$$

This is shown in Diagram 24.2 where we have put $dx = \Delta x$. We can see that $MR = dy = 8x\, \Delta x$ exactly, while Δy is QR which is approx-

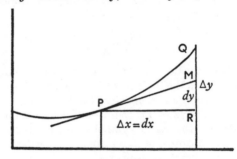

FIG. 24.2

imately equal to MR provided that Δx is small.

In the same way we may define the differential of a function which is the product of two functions u and v (each being a function of x) as being given by $dy = u\, dv + v\, du$

where $\qquad du = \dfrac{du}{dx} dx \quad \text{and} \quad dv = \dfrac{dv}{dx} dx$

If $\qquad y = f(u,\, v) = uv$

then $\qquad dy = \dfrac{df}{dx}(u,\, v)\, dx$

$$= \left(u\frac{dv}{dx} + v\frac{du}{dx} \right) dx$$

$$= u\, dv + v\, du$$

For example, if

$$y = (x^3 + 3)(x^2 - 2x)$$
$$dy = (x^3 + 3)d(x^2 - 2x) + (x^2 - 2x)d(x^3 + 3)$$
$$= (x^3 + 3)(2x - 2)dx + (x^2 - 2x)(3x^2)dx$$

which may be compared with

$$\frac{dy}{dx} = (x^3 + 3)(2x - 2) + (x^2 - 2x)(3x^2)$$

EXERCISE 24.1

1. Use the above method to obtain the differential dy in the following cases :

(i) $y = (x^2 + 3)(x - 4)$ (ii) $y = (x^3 + 4x)(x^2 - 1)$

(iii) $y = (x^3 + x)(\cos x)$ (iv) $y = (x^3 - 3x^2)(\cos x + \sin x)$

2. Complete and partial differentials

We may also define a differential of a function of more than one variable. We let dx, dy be the differentials of the independent variables x and y, each defined to be a new variable associated with x or y as the case may be. Then we *define* the differential of the dependent variable z to be such that

$$dz = \frac{\partial z}{\partial x} dx + \frac{\partial z}{\partial y} dy = z_x \, dx + z_y \, dy$$

is *exactly* true. It follows that dz is a function of *four* independent variables x, y, dx and dy, and is linear and homogeneous in dx and dy. We call dz the *complete* or *exact differential* of z, while $z_x \, dx$ and $z_y \, dy$ are the *partial differentials*.

For example, if

$$z = x^2 - xy + y^2$$

then if we denote the differentials of x and y by dx and dy we will have that

$$dz = (2x - y)dx + (-x + 2y)dy$$

3. Implicit differentiation

Now an equation of the form $z = f(x, y)$ represents a three-dimensional surface. If we put $z = 0$ so that $f(x, y) = 0$ then we have

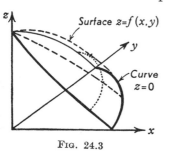

FIG. 24.3

the equation of the curve in which this surface is intersected by the plane $z = 0$. The equation $dz = z_x \, dy + z_y \, dy$ tells us something about the surface for any values of x, y and z. If we stipulate that z must be kept at zero, so that dz is also zero, then we learn something about the curve $f(x, y) = 0$. We have, in fact, that

$$0 = z_x \, dx + z_y \, dy$$

where dx is the differential of x, and dy is a differential of y chosen to satisfy the condition that $z = dz = 0$. If we divide down by dx we have that

$$0 = z_x + z_y \frac{dy}{dx}$$

and a moment's consideration will show that this ratio of dy to dx may be interpreted for all practical purposes as the derivative of y with respect to x. We can, therefore, write that

$$\frac{dy}{dx} = -\frac{z_x}{z_y}$$

where

$$z = f(x, y) = 0$$

Let us now consider the equation

$$x^2 - xy + y^2 = 0$$

Suppose that we want to find dy/dx. First of all, to solve this quadratic equation in y, and then to differentiate, would be tedious and rather complicated. We can, however, write the left-hand side as z and obtain

$$z_x = 2x - y \; ; \; z_y = -x + 2y$$

and so

$$\frac{dy}{dx} = -\frac{z_x}{z_y} = -\left[\frac{2x - y}{-x + 2y}\right]$$

where, if the result is really required entirely in terms of x, the quadratic may be solved and the appropriate values of y inserted. The same result is obtained if we adopt a very slightly different method, obtaining the differential of each term as follows :

$$x^2 - xy + y^2 = 0$$

therefore,

$$d(x^2) - d(xy) + d(y^2) = d(0) = 0$$

i.e.,

$$2x\,dx - (x\,dy + y\,dx) + 2y\,dy = 0$$

whence

$$2x - \left(x\frac{dy}{dx} + y\right) + 2y\frac{dy}{dx} = 0$$

which gives the same result as before. A natural extension of this method is to proceed

$$x^2 - xy + y^2 = 0$$

therefore

$$\frac{d}{dx}\left(x^2\right) - \frac{d}{dx}\left(xy\right) + \frac{d}{dx}\left(y^2\right) = 0$$

giving

$$2x - \left(x\frac{dy}{dx} + y\right) + 2y\frac{dy}{dx} = 0$$

the justification for this being found by comparison with the previous method.

These methods are examples of *implicit differentiation*, i.e., the differentiation of a function in which y is not given explicitly (as in $y = 4x^2 + 3$) but implicitly, as above.

EXERCISE 24.2

1. Differentiate the following functions implicitly :

(i) $x^2 + xy + y^2 = 0$	(ii) $x^3 + x^2y + xy^2 + y^3 = 0$
(iii) $x^2 \sin y + y^2 \sin x = 0$	(iv) $\sin x \sin y = 0$
(v) $x^3 + y^3 - 3xy = 0$	(vi) $x \tan y + y \tan x = 0$
(vii) $y \log_e x + x \log_e y = 0$	(viii) $x \cosh y + y \cosh x = 0$
(ix) $x \sinh y + y \tanh x = 0$	(x) $x^2 \tanh y + y^2 \cosh x = 0$

4. Second differentials

We can obtain *second differentials* in a similar way, defining d^2z to be $d(dz)$, which is the differential of the differential. In this case a certain amount of care is necessary when considering the second (and higher) differentials of two or more variables. There are two cases.

(1) If x and y are *independent of each other* then

$$d^2z = d(z_x\,dx + z_y\,dy) = d(z_x)dx + d(z_y)dy$$
$$= (z_{xx}\,dx + z_{xy}\,dy)dx + (z_{xy}\,dx + z_{yy}\,dy)dy$$
$$= z_{xx}\,dx^2 + 2z_{xy}\,dx\,dy + z_{yy}\,dy^2$$

where dx and dy have, in the process of differentiation, been treated as constants.

For example, if

$$z = x^2 - xy + y^2$$
$$z_{xx} = 2, \quad z_{xy} = -1 \quad \text{and} \quad z_{yy} = 2$$

and so

$$d^2z = 2\,dx^2 - 2\,dx\,dy + 2\,dy^2$$

This result will also be obtained if we proceed,

$$z = x^2 - xy + y^2$$
$$dz = (2x - y)dx + (-x + 2y)dy$$
$$d^2z = d(2x - y)dx + d(-x + 2y)dy$$
$$= (2\,dx - dy)dx + (-dx + 2\,dy)dy$$
$$= 2\,dx^2 - 2\,dx\,dy + 2\,dy^2$$

M

(2) If, however, *x and y are functions of some other variable* or variables, and therefore related to each other, then it can be shown that

$$d^2z = z_{xx}\,dx^2 + 2z_{xy}\,dx\,dy + z_{yy}\,dy^2 + z_x\,d^2x + z_y\,d^2y$$

A consequence of this result is that

$$\frac{d^2z}{dx^2} = z_{xx} + 2z_{xy}\frac{dy}{dx} + z_{yy}\left(\frac{dy}{dx}\right)^2 + z_y\frac{d^2y}{dx^2}$$

gives the second total derivative of z with respect to x when $z = f(x, y)$ and y is a function of x.

EXERCISE 24.3

1. Given that x and y are independent of each other, obtain the second differential of z in the following cases :

(i) $z = x^2 + xy + y^2$ (ii) $z = x^3 + x^2y + y^3$
(iii) $z = x^2 \sin y$ (iv) $z = x^2 \sin y + y^2 \sin x$
(v) $z = x^2 \sin y + y^2 \cos x$ (vi) $z = x \tan y + y \tan x$
(vii) $z = x \log_e y + y \log_e x$ (viii) $z = x^3 \sin^2 y$

2. Given that x and y are not independent, obtain the second differential of z in the cases (i)–(iv) above.

5. Maxima and minima

Differentials are often used when considering maxima and minima. The results obtained are, of course, the same as those obtained by the methods we have already discussed. But it is sometimes more convenient to use differentials than to use the ordinary process of differentiation.

It can be shown that a function y given in terms of several variables $x_1, x_2, x_3, \ldots, x_n$ is subject to the following criteria :

(1) $y = f(x_1, x_2, \ldots, x_n)$ has an extreme value at the point $x_1 = a_1$, $x_2 = a_2$, etc., if $dy = 0$ for all changes of the variables from these constant values.

(2) If, in addition, there are some variations of the variables from these constant values which result in a non-zero value of d^2y, then y has a stationary value, which is a maximum if $d^2y < 0$ for all variations, a minimum if $d^2y > 0$ for all variations, and a saddle point if d^2y is positive for some variations and negative for others. If $d^2y = 0$ for variations from these constant values then a further examination is necessary.

These rules result in criteria identical with those mentioned in Chapter XX.

In practice the method is employed as follows:
Suppose that we wish to determine the stationary values of

$$y = x_1^2 + x_2^2 + x_1 x_2 - x_3^2 + x_1 - x_3$$

then we have that

$$dy = 2x_1\,dx_1 + 2x_2\,dx_2 + (x_1\,dx_2 + x_2\,dx_1) - 2x_3\,dx_3 + dx_1 - dx_3$$
$$= (2x_1 + x_2 + 1)dx_1 + (2x_2 + x_1)dx_2 - (2x_3 + 1)dx_3$$

Now the first conditions hold if we make the coefficients of dx_1, dx_2 and dx_3 zero. This gives us three equations to solve simultaneously, and these lead to $x_1 = -\frac{2}{3}$, $x_2 = \frac{1}{3}$ and $x_3 = -\frac{1}{2}$.
Furthermore

$$d^2y = (2\,dx_1 + dx_2)dx_1 + (2\,dx_2 + dx_1)dx_2 - 2\,dx_3\,dx_3$$
$$= 2(dx_1^2 + dx_2^2 - dx_3^2 + dx_1\,dx_2)$$

Now we may compare this with the result

$$d^2y = \frac{\partial^2 y}{\partial x_1^2}dx_1^2 + \frac{\partial^2 y}{\partial x_2^2}dx_2^2 + \frac{\partial^2 y}{\partial x_3^2}dx_3^2$$
$$+ 2\left(\frac{\partial^2 y}{\partial x_1\,\partial x_2}dx_1\,dx_2 + \frac{\partial^2 y}{\partial x_2\,\partial x_3}dx_2\,dx_3 + \frac{\partial^2 y}{\partial x_3\,\partial x_1}dx_3\,dx_1\right)$$

which is analogous to the equations listed on pages 272–4.
We see at once that

$$\frac{\partial^2 y}{\partial x_1^2} = 2\,;\ \frac{\partial^2 y}{\partial x_2^2} = 2\,;\ \frac{\partial^2 y}{\partial x_3^2} = -2\,;\ \frac{\partial^2 y}{\partial x_1\,\partial x_2} = 1\,;\ \frac{\partial^2 y}{\partial x_1\,\partial x_2} = 0 \quad \text{and}$$
$$\frac{\partial^2 y}{\partial x_2\,\partial x_3} = 0$$

It follows that the determinant involved in deciding between a maximum and a minimum is

$$\begin{vmatrix} 2 & 1 & 0 \\ 1 & 2 & 0 \\ 0 & 0 & -2 \end{vmatrix}$$

and consideration of the sign of this and its minors enables us to decide on the nature of the stationary point $(-\frac{2}{3}, \frac{1}{3}, -\frac{1}{2})$.
It should be pointed out that if the variables x_1, x_2 and x_3 are not independent of each other then the above process for finding the second derivative must be replaced by the process indicated on page 338.

EXERCISE 24.4

1. Obtain the stationary values of the following functions :

    (i) $y = x_1^2 + 2x_2^2 - x_3^2 + x_1$

    (ii) $y = x_1^2 - x_2^2 + x_1 x_2 - x_3$

    (iii) $y = x_1^2 - x_2^2 + x_3^2 - x_1 - x_2 - x_3$

    (iv) $y = x_1^3 + x_2^2 + x_3^2 - x_1 x_2$

SECTION V

MATHEMATICS OF FLUCTUATIONS
AND GROWTH

The first chapter introduces imaginary and complex numbers and develops the more important elementary ideas that are necessary for an understanding of the solution of differential and difference equations. These equations form the subjects of the two remaining chapters which are more practical than theoretical.

COMPLEX NUMBERS

Find the way now, blind Samson, with your fingers
Feel at the latch. The door will open,
And you will let the sky in, your wide sockets
Open these blind temples to the sun!

KATHLEEN RAINE : *Samson*

1. Introduction

In our first chapter we saw that the quadratic equation

$$2x^2 - 2x + 1 = 0$$

had solutions

$$x = \frac{2 \pm \sqrt{-4}}{4}$$

Because these solutions contained the square root of a negative quantity we deferred consideration of them. We now return to this problem. It had long been apparent that quite innocent looking equations were sometimes likely to result in solutions which involved the square root of a negative number, and there was a very understandable tendency to avoid these solutions. During the sixteenth century some mathematicians were a little braver, and examined solutions of this kind, but it was not until 1629 that Girard pointed to three reasons for considering solutions of this kind. One reason was that they helped to establish general rules for the solution of equations. Another was for establishing, in certain cases, the lack of other solutions. The third reason was that solutions of this kind had their own usefulness. Fifty years afterwards, Wallis had claimed that the idea of such an " imaginary quantity " was no more unreasonable than that of a negative one. We now consider the matter for ourselves. It is a vast subject, which can become very difficult : but the basic ideas presented in this chapter are quite simple.

2. Some properties of numbers

Rational numbers

All the numbers we have come across fall into one of two classes. Some have been in the class consisting of whole numbers and of fractions that can be obtained by dividing one of these numbers by another (e.g., 1, 2, 3, 4, ... , $\frac{1}{2}$, $\frac{1}{4}$, $\frac{3}{4}$, ... , etc.). All of these numbers have the property that they can be represented by distances on a straight line, as shown in Diagram 25.1. Furthermore, it is possible to represent negative numbers in the same way. The numbers of

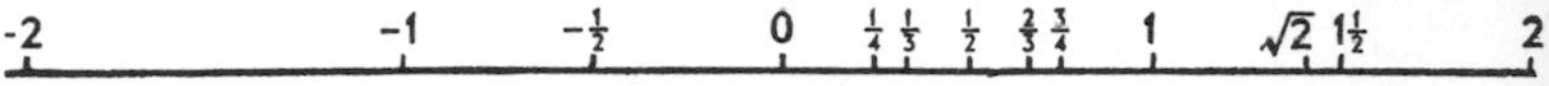

FIG. 25.1

this class have the property that any one of them can be built up in an infinite number of ways from two or more other members of the class. For instance

$$\tfrac{15}{8} = \tfrac{5}{8} + \tfrac{10}{8} = 2 - \tfrac{1}{8} = \tfrac{3}{7} + \tfrac{81}{56} = \text{etc.}$$

We call this class of numbers " rational numbers ". We should note that to every positive rational number, there corresponds a negative rational number. If we consider positive rational numbers to be represented by distances measured along the axis Ox, then we can think of negative rational numbers as being represented by distances measured along the negative axis Ox'.

Irrational numbers

There are other numbers which we have used that cannot be expressed as fractions or rational numbers. For instance it is impossible to represent $\sqrt{2}$ exactly as a fraction. On the other hand it is quite easy, theoretically, to represent it exactly in Diagram 25.1. If we consider a square of side one inch, we see, from Pythagoras' Theorem, that the diagonal will have a length of exactly $\sqrt{2}$ inches ; and so, in order to locate a point on the above line that corresponds to $\sqrt{2}$ we need only draw a square of side one inch and take an accurate measurement of its diagonal. Only practical considerations limit our accuracy. Numbers which, like $\sqrt{2}$, can be represented on this line, but which cannot be represented exactly by a rational number, are called *irrational numbers*. Examples are $\sqrt{3}$, $\sqrt{5}$, $\sqrt{6}$, $\sqrt{7}$, $\sqrt{8}$, $\sqrt{10}$, $\sqrt[3]{2}$, $\sqrt[3]{3}$, $\sqrt[3]{4}$, $\sqrt[3]{5}$, $\sqrt[3]{6}$, $\sqrt[3]{7}$, $\sqrt[3]{9}$, ... , etc. But $\sqrt{4}$ and $\sqrt[3]{8}$ are rational, since they can be represented exactly by 2 and 2. All exact decimals are rational, since, for example, 1·23456876 can be represented as $\frac{123456876}{100000000}$ exactly.

Surds

We should note that sometimes we come across combinations of rational and irrational numbers, such as $3 + \sqrt{5}$. Since we can represent both parts of this by distances on the line Ox then we can represent the whole of it by such a distance. We call such a number a *surd*.

There is one very important property concerning surds. If two surds are equal, then the rational part of the one must equal the rational part of the other, and the irrational part of the one must equal the irrational part of the other. If, for example,

$$a + \sqrt{b} = c + \sqrt{d}$$

then, unless b and d are perfect squares (in which case $\sqrt{b}$ and $\sqrt{d}$ are rational and we have completely rational numbers on both sides of the equation) then

$$a = c$$

and $$b = d$$

If, for example,

$$3 + \sqrt{7} = x + \sqrt{y}$$

then there can be no solution except $x = 3$ and $y = 7$. But if we have

$$3 + \sqrt{4} = x + \sqrt{y}$$

then we can have any number of solutions, since the left-hand side is equal to 5, and the equation is simply $x + \sqrt{y} = 5$ which has as many solutions as we please.

3. The rotation of a line

During the course of this book we have become used to performing various operations on numbers. We have differentiated functions, including some constants. We have integrated them. In doing this we have introduced new symbols, and given them meanings. We came to regard d/dx as an " operator " which, when placed before y, operated on it in the sense that it differentiated it. We can regard $\int dx$ as an operator, in the sense that when placed around y it integrates it. We now introduce a new operator.

First of all we may note that if we let a rational number be represented by a straight line of a given length OA in the direction of Ox (as in Diagram 25.2) then we may regard the

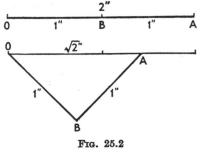

FIG. 25.2

square root sign as performing an operation on that line. The operation is to distort the line into the shape shown, so that the new distance between O and A in the direction of Ox becomes the square root of the original distance.

Alternatively we may think of the square root as operating on the line OA in such a way as to alter its length (from $2''$ to $\sqrt{2}''$) without altering its direction (along Ox).

Suppose now that we are interested not only in lines drawn along Ox, but also in lines drawn in other directions, and, in particular, in lines drawn along Oy. It would be useful if these, too, could be represented by numbers. We have just seen that the square root is an operator that changes the magnitude of a line but leaves the direction unchanged. Why can we not have an operator that will change the direction but leave the magnitude unchanged?

There is no reason at all why we should not. Let us, in fact, introduce one at once. Let us use the symbol i for this purpose, and give it the following property.

If a denotes a length of a units in the direction of Ox then we will let ia denote a length of a units in the direction of Oy. Diagram 25.3 illustrates this.

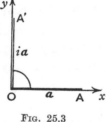

FIG. 25.3

We will, in fact, be a little more generous, and give i a more general property. If b denotes a distance of b units in any direction whatsoever, then we will let ib denote the same distance turned through a right angle in the counter-clockwise direction. This is shown in Diagram 25.4.

Now let us consider the meaning of this definition. The line OA in Diagram 25.5 may be denoted by a. The line OA' may be denoted by ia. Let us now turn OA' through a right angle putting it into a position OA''. Then we have that this line OA'' may clearly be represented by $i(i$a$)$ which may conveniently be written as i^2a.

But OA'' is the same length as OA and is in the opposite direction. In fact $OA'' = -OA$.

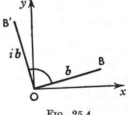

FIG. 25.4

(Instead of going three miles to the East we go three miles to the West, which is the same as going three miles *away from* the East.) We therefore have that

$$-\mathbf{a} = -OA = OA'' = i^2\mathbf{a}$$

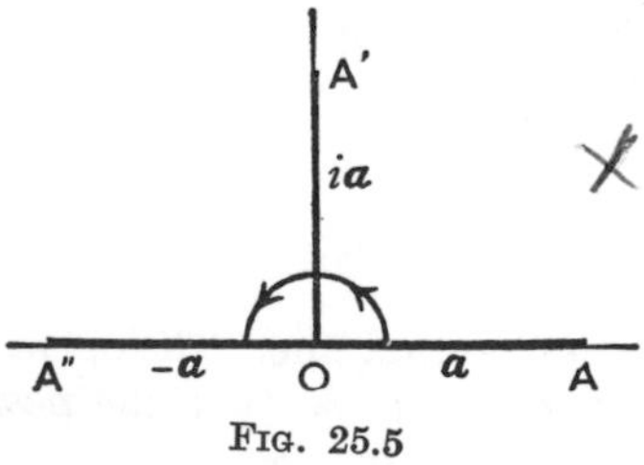

and so we can see that i is an operator with the following properties.

FIG. 25.5

(1) When i operates on a line of given direction and length it turns it through a right angle in the counter-clockwise direction without altering its length.

(2) If we operate *twice* with i we obtain the same result as if we had operated with -1, i.e., we reverse the direction of the line, without altering its length. Formally we may indicate this by identifying the double operation i^2 with -1, thus

$$ii = i^2 = -1$$

We can now interpret $\sqrt{-1}$ as being equivalent to the operator i.

We call all numbers in the class that we can represent along the x-axis *real*. The class includes positive and negative numbers, both rational and irrational. It also includes surds.

Numbers obtained from real numbers by operating on them with i are called *imaginary* numbers. All imaginary numbers can be represented along the y-axis. To every real number there corresponds an imaginary number, obtained from it by operating on it with i.

The terms *real* and *imaginary* are due to Descartes, who introduced them in 1637. The use of the symbol i was recommended by Euler in 1748.

4. Complex numbers

Graphical representation

At the beginning of this chapter we referred to an equation which had solutions

$$x = \frac{2 \pm \sqrt{-4}}{4}$$

which may be put as

$$x = \frac{2 \pm \sqrt{4(-1)}}{4}$$
$$= \frac{2 \pm 2\sqrt{-1}}{4}$$
$$= \tfrac{1}{2} \pm \tfrac{1}{2}\sqrt{-1}$$
$$= \tfrac{1}{2} \pm \frac{i}{2}$$

This number is neither real nor imaginary. It is, in fact, made up of two parts, the first being real and the second imaginary. We call such numbers *complex*.

It was shown by Wessel in 1797 that numbers of this kind can be represented graphically, once it is realised that they consist of two parts.

Let us consider a point P in Diagram 25.6 whose co-ordinates are

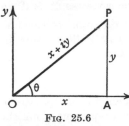

FIG. 25.6

(x, y). We may join it to the origin O by the line OP. This line, like all others, has two properties : it has length and direction.

Now it is posssible to reach the point P from O in an infinite number of ways. Two of them are very important. The more obvious is to follow the straight path along the line OP. The other is to proceed along the x-axis as far as A, and then in a direction parallel to the y-axis as far as P. In either case, we arrive at P.

We have seen, however, that the distance OA may be represented by a real number of magnitude equal to the length of OA, which we will denote by x. The distance AP may also be represented by the imaginary number of magnitude equal to the length of AP, which we will denote by iy.

If we are concerned not with the path followed but with the location of the end of the journey relative to the location of its commencement, we can say that the journey represented by OP is equivalent to the sum of the journeys represented by OA and AP. This means that we may write

$$OP \equiv OA + AP$$

where we interpret the sign $\equiv$ in the way just indicated. But AO may be represented by x and AP by iy, and so we have that the journey OP (or, more formally, the line OP in both magnitude and direction), is equivalent to

$$x + iy$$

We thus have the very important result that the line joining the point (x, y) to the origin may be represented by the complex number

$$x + iy$$

It may be noted that the length of this line is the positive root

$$\sqrt{x^2 + y^2}$$

and the direction it makes with the x-axis is such that

$$\tan \theta = \frac{y}{x}$$

The length of the line is sometimes called the *modulus*, and the angle θ the *argument* of the *complex* number. We thus have that

$$\mod (x + iy) = \sqrt{x^2 + y^2}$$

$$\arg (x + iy) = \theta = \tan^{-1} \frac{y}{x}$$

The complex number itself is often denoted by z, in order to save space and to facilitate manipulation. We also write the modulus as $|z|$ and so we have that

$$|z| = \sqrt{x^2 + y^2}$$
$$\arg z = \theta$$

We should be careful to notice that just as any given point P can have only one pair of co-ordinates (x, y) so can any given point P correspond to only one complex number. It follows that if two complex numbers are equal to each other, and therefore represent the same line OP, then the real part of the one must equal the real part of the other, and the imaginary part of the one must equal the imaginary part of the other. If, for example,

$$3 + i4 = x + iy$$

then $x = 3$ and $y = 4$ are the only possible solutions. More generally, if

$$a + ib = c + id$$

then $a = c$ and $b = d$.

EXERCISE 25.1

1. Solve the equations :

    (i) $z^2 + 2z + 5 = 0$ (ii) $2z^2 + 2z + 5 = 0$
    (iii) $3z^2 - z + 3 = 0$ (iv) $25z^2 + 6z + 1 = 0$

2. Write down the moduli and arguments of the solutions to question 1 above.

3. Given that x and y are real, determine their values if :

    (i) $(x + 3y) + i(2x - y) = 7$
    (ii) $(x + 3y) + i(2x - y) = 5 + 3i$
    (iii) $(x + 3y) + i(2x - y) = 7i$

5. A trigonometric interpretation

    Suppose that we have a line OQ of length r in the direction of Ox and we wish to turn it through an angle θ. As Diagram 25.7 shows,

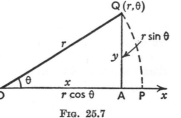

FIG. 25.7

the line will then be represented by the complex number of which the real part (OA) will be $x = r \cos \theta$, and the imaginary part (AQ)

will be $iy = ir \sin \theta$. This means that we may represent the direction and length of the line by

$$z = x + iy$$
$$= r \cos \theta + ir \sin \theta$$
$$= (\cos \theta + i \sin \theta)r$$

An alternative way of saying this is that we may think of the complex number $(\cos \theta + i \sin \theta)$ as being an operator whose effect is to turn the line OP of length r through an angle θ in the counter-clockwise direction, so that it occupies the position OQ but is of unchanged length.

It may be seen also that the complex number

$$z = x + iy$$

has a modulus of

$$|z| = \sqrt{x^2 + y^2}$$
$$= r$$

and an argument of

$$\arg z = \tan^{-1} \frac{y}{x} = \theta ;$$

and that the number

$$z = (\cos \theta + i \sin \theta)r$$

has the same modulus r and argument θ.

✕ EXERCISE 25.2 ✕

1. Express the following complex numbers in the form
$$r(\cos \theta + i \sin \theta)$$

(i) $3 + 4i$ (ii) $3 - 4i$ (iii) $4 + 3i$

(iv) $4 - 3i$ (v) $-3 + 4i$ (vi) $-3 - 4i$

The result we have just proved is of great importance, even as it stands. It is, however, of even greater importance because of a result which was first proved by Cotes (1682–1716), who succeeded in proving one of the most important results in mathematics even before an adequate notation had been devised for work of this kind.

We have already seen that, provided that θ is measured in radians, then $\cos \theta$ and $\sin \theta$ may be represented by the converging series

$$\cos \theta \equiv 1 - \frac{\theta^2}{2!} + \frac{\theta^4}{4!} - \ldots$$
$$\sin \theta \equiv \theta - \frac{\theta^3}{3!} + \frac{\theta^5}{5!} - \ldots$$

If we now add these, keeping the terms in the right order, and multiplying the terms arising out of sin θ by i we shall have that

$$\cos\theta + i\sin\theta \equiv 1 + i\theta - \frac{\theta^2}{2!} - \frac{i\theta^3}{3!} + \ldots$$

But since $i = \sqrt{-1}$ we have that $i^2 = -1$ and so $i^4 = 1$. Also, $i^3 = -i$, etc. In the series we have just written down for $\cos\theta + i\sin\theta$ we may rewrite $-\dfrac{\theta^2}{2!}$ as $+\dfrac{(i\theta)^2}{2!}$, $-\dfrac{i\theta^3}{3!}$ as $+\dfrac{(i\theta)^3}{3!}$, and so on, thus obtaining the series

$$\cos\theta + i\sin\theta \equiv 1 + i\theta + \frac{(i\theta)^2}{2!} + \frac{(i\theta)^3}{3!} + \ldots$$

which is, of course, $e^{i\theta}$ as we may see from Chapter XIV.

We thus have the very important result that

$$e^{i\theta} \equiv \cos\theta + i\sin\theta$$

for all values of θ.

We also have that $e^{i\theta}$ can be interpreted as an operator which turns a line through an angle θ, leaving the magnitude unchanged. For example, if we represent a line in the direction Ox of magnitude two inches by 2, then $e^{i\theta}2$ will represent a line of the same length at an angle of θ with the x-axis. We attribute the same meaning to $2e^{i\theta}$.

Furthermore, if we let a line of any length and any direction be represented by the complex number z, then $e^{i\theta}z$ will represent a line of the same magnitude at an angle of θ to the original line.

It can also be shown that

$$e^{in\theta} \equiv (\cos n\theta + i\sin n\theta)$$

which may be proved by induction, and that

$$e^{-i\theta n} \equiv (\cos n\theta - i\sin n\theta)$$

From these results it is possible to deduce that

$$\{r(\cos\theta \pm i\sin\theta)\}^n \equiv r^n(\cos n\theta \pm i\sin n\theta)$$

This highly important result is known as *De Moivre's Theorem*.

EXERCISE 25.3

1. By expressing $e^{in\theta}$ in the form $(\cos n\theta + i\sin n\theta)$, obtain the values of θ satisfying the following equations :

(i) $e^{2i\theta} = 1$ (ii) $e^{2i\theta} = i$ (iii) $e^{2i\theta} = -i$

(iv) $e^{2i\theta} = \frac{1}{2} + \frac{\sqrt{-3}}{2}$ (v) $e^{3i\theta} = \frac{1}{2} - \frac{\sqrt{-3}}{2}$ (vi) $e^{3i\theta} = -\frac{1}{2} + \frac{\sqrt{-3}}{2}$

6. Addition and multiplication

Having introduced complex numbers we must decide on how to manipulate them. Since they arise quite naturally in the solution of certain quadratic (and other) equations when these involve only real terms, it will be useful if we can find ways of operating on complex numbers which are compatible with the operations used on real numbers.

Let us define the process of adding complex numbers, and then consider its meaning.

✕ *Definition:* ✕

We define the sum of two complex numbers $z_1 = x_1 + iy_1$ and $z_2 = x_2 + iy_2$ to be the complex number

$$z_3 = z_1 + z_2 = (x_1 + x_2) + i(y_1 + y_2)$$

which clearly equals $\quad (x_2 + x_1) + i(y_2 + y_1)$

$$= z_2 + z_1$$

For example, if $z_1 = 2 + 4i$ and $z_2 = 3 - 2i$ then $z_1 + z_2 = 5 + 2i$. The graphical interpretation of this definition is of some interest.

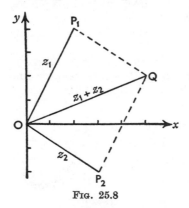

FIG. 25.8

If we represent the number z_1 by OP_1 of length $\sqrt{2^2 + 4^2}$ and direction $\tan^{-1} \frac{4}{2}$ as in Diagram 25.8, and z_2 by OP_2 of length $\sqrt{3^2 + 2^2}$ and direction $\tan^{-1} - \frac{2}{3}$, then the sum of these complex numbers is represented by OQ which is the diagonal of the parallelogram which has OP_1 and OP_2 as adjacent sides. Since P_1Q is equal and parallel to OP_2 the co-ordinates of the point Q may easily be found, and a study of the diagram clearly shows that they are $(2 + 3, 4 + (-2))$.

We now consider multiplication.

✕ *Definition:* ✕

We define the product of z_1 by z_2 to be

$$z_3 = z_1 z_2 = (x_1 + iy_1)(x_2 + iy_2)$$

where the multiplication proceeds as for real numbers, with $i^2 = -1$, yielding the complex number

$$x_1x_2 + i(x_1y_2 + x_2y_1) + i^2y_1y_2$$
$$= (x_1x_2 - y_1y_2) + i(x_1y_2 + x_2y_1).$$

For example,

$$(3+4i)(2+3i) = 6 + i(8+9) + i^2 12$$
$$= 6 + 17i - 12 = -6 + 17i$$

It is useful to consider this definition in trigonometric terms. We have

$$z_1 = r_1 e^{i\theta_1} \quad \text{and} \quad z_2 = r_2 e^{i\theta_2}$$

and so

$$z_1 z_2 = r_1 r_2 e^{i\theta_1} \cdot e^{i\theta_2}$$
$$= r_1 r_2 e^{i(\theta_1 + \theta_2)}$$

which denotes a complex number of modulus $r_1 r_2$ and in a direction $(\theta_1 + \theta_2)$. This means that the product is a complex number whose modulus is the product of the individual moduli, and whose argument is the sum of the individual arguments. If we think of $z_1 = r_1 e^{i\theta_1}$ as an operation which multiplies the length of a line by r_1, and turns it through an angle θ_1 in a counter-clockwise direction, and of $z_2 = r_2 e^{i\theta_2}$ as an operation which multiplies the length by r_2 and turns the line through an angle θ_2, then clearly the product $z_1 z_2$ may be interpreted as one of these operations followed by the other. The result will be that the length has been multiplied by $r_1 r_2$ and the line itself has been turned through an angle of $(\theta_1 + \theta_2)$.

We may rewrite this result as

$$z_1 z_2 = r_1 r_2 e^{i(\theta_1 + \theta_2)} = r_1 r_2 [\cos(\theta_1 + \theta_2) + i \sin(\theta_1 + \theta_2)]$$

$$= r_1 r_2 [(\cos\theta_1 \cos\theta_2 - \sin\theta_1 \sin\theta_2) + i(\sin\theta_1 \cos\theta_2 + \cos\theta_1 \sin\theta_2)]$$

$$= \sqrt{x_1^2 + y_1^2}\sqrt{x_2^2 + y_2^2}\left\{\left[\frac{x_1}{\sqrt{x_1^2 + y_1^2}}\frac{x_2}{\sqrt{x_2^2 + y_2^2}} - \frac{y_1}{\sqrt{x_1^2 + y_1^2}}\frac{y_2}{\sqrt{x_2^2 + y_2^2}}\right]\right.$$

$$\left. + i\left[\frac{y_1}{\sqrt{x_1^2 + y_1^2}}\frac{x_2}{\sqrt{x_2^2 + y_2^2}} + \frac{x_1}{\sqrt{x_1^2 + y_1^2}}\frac{y_2}{\sqrt{x_2^2 + y_2^2}}\right]\right\}$$

$$= x_1 x_2 - y_1 y_2 + i(x_1 y_2 + x_2 y_1)$$

which is the result previously obtained.

EXERCISE 25.4

1. Evaluate the following products:

    (i) $(3+5i)(2+4i)$ (ii) $(3+5i)(2-4i)$

  (iii) $(3+5i)(-2+4i)$ (iv) $(3+5i)(-2-4i)$

    (v) $(3-5i)(2-4i)$ (vi) $(3-5i)(-2-4i)$

  (vii) $(-3-5i)(-2-4i)$ (viii) $(3+5i)(3-5i)$

  (ix) $(3+5i)^2$ (x) $(3+5i)(2+4i)(2+3i)$

  (xi) $(3-5i)(3+4i)(2-3i)$ (xii) $(3-5i)(3-4i)(2-3i)$

2. Express the following products in the form $r(\cos\theta + i\sin\theta)$:

    (i) $2e^{i\pi/4} \times 3e^{i\pi/3}$ (ii) $2e^{-i\pi/4} \times 3e^{i\pi/3}$

  (iii) $4e^{i\pi/4} \times 3e^{i\pi/4}$ (iv) $4e^{-i\pi/4} \times 3e^{-i\pi/3}$

7. Subtraction and division

The subtraction of complex numbers proceeds simply as the opposite of addition, according to the ordinary laws of algebra.   For example,

$$(5+3i)-(2+4i)=3-i$$

or, more generally,

$$(x_1+iy_1)-(x_2+iy_2)=(x_1-x_2)+i\,(y_1-y_2)$$

Division is easily performed in the polar notation.   If we interpret multiplication by $r_1 e^{i\theta_1}$ as an operation which multiplies the length by r_1 and then rotates the line through an angle of θ_1 in the counterclockwise direction, then we may interpret division by $r_2 e^{i\theta_2}$ as the inverse operation, i.e., as dividing the length by r_2 and turning the line through an angle θ_2 in the clockwise direction.   It follows that

$$\frac{z_1}{z_2}=\frac{r_1 e^{i\theta_1}}{r_2 e^{i\theta_2}}=\frac{r_1}{r_2}e^{i(\theta_1-\theta_2)}$$

Example

$$z_1=4e^{i\pi} \qquad z_2=3e^{i\pi/3}$$

$$\frac{z_1}{z_2}=\tfrac{4}{3}e^{i[\pi-(\pi/3)]}=\tfrac{4}{3}e^{i2\pi/3}$$

$$=\tfrac{4}{3}e^{2i\pi/3}$$

If we translate this result into the other notation we obtain

$$\frac{z_1}{z_2}=\frac{r_1}{r_2}\,e^{i(\theta_1-\theta_2)}$$

$$=\frac{\sqrt{x_1{}^2+y_1{}^2}}{\sqrt{x_2{}^2+y_2{}^2}}\left[\,\cos\,(\theta_1-\theta_2)+i\sin\,(\theta_1-\theta_2)\,\right]$$

$$=\frac{\sqrt{x_1{}^2+y_1{}^2}}{\sqrt{x_2{}^2+y_2{}^2}}\Big[\,(\cos\theta_1\cos\theta_2+\sin\theta_1\sin\theta_2)$$

$$+i\,(\sin\theta_1\cos\theta_2-\cos\theta_1\sin\theta_2)\,\Big]$$

$$=\frac{\sqrt{x_1{}^2+y_1{}^2}}{\sqrt{x_2{}^2+y_2{}^2}}\Big[\,\frac{x_1}{\sqrt{x_1{}^2+y_1{}^2}}\frac{x_2}{\sqrt{x_2{}^2+y_2{}^2}}+\frac{y_1}{\sqrt{x_1{}^2+y_1{}^2}}\frac{y_2}{\sqrt{x_2{}^2+y_2{}^2}}$$

$$+i\left(\frac{y_1}{\sqrt{x_1{}^2+y_1{}^2}}\frac{x_2}{\sqrt{x_2{}^2+y_2{}^2}}-\frac{x_1}{\sqrt{x_1{}^2+y_1{}^2}}\frac{y_2}{\sqrt{x_2{}^2+y_2{}^2}}\right)\Big]$$

$$=\frac{1}{x_2{}^2+y_2{}^2}\Big[\,(x_1x_2+y_1y_2)+i\,(x_2y_1-y_2x_1)\,\Big]$$

The same result may be obtained if we proceed

$$\frac{z_1}{z_2} = \frac{x_1 + iy_1}{x_2 + iy_2} = \frac{(x_1 + iy_1)(x_2 - iy_2)}{(x_2 + iy_2)(x_2 - iy_2)}$$
$$= \frac{(x_1x_2 + y_1y_2) + i(x_2y_1 - x_1y_2)}{x_2^2 + y_2^2}$$

There is little point in trying to remember this result. It is much better to perform each division separately as in the example below. The trick is to multiply the numerator and the denominator by a new complex number, which is derived from the denominator by reversing the sign of the imaginary part. Thus, if we have a denominator of $2 + 5i$, then we multiply throughout by $2 - 5i$; in the above derivation of the result we had a denominator of $x_2 + iy_2$ and multiplied by $x_2 - iy_2$. $x + iy$ is called the *conjugate* of $x - iy$. As examples of conjugates we have that

$$4 + 3i \text{ has a conjugate } \quad 4 - 3i$$
$$4 - 3i \text{ has a conjugate } \quad 4 + 3i$$
$$-3 + 5i \text{ has a conjugate } \quad -3 - 5i$$
$$-3 - 5i \text{ has a conjugate } \quad -3 + 5i$$

It will be noticed that the product of two conjugates is always a real number.

Example

$$\frac{3 + 4i}{2 + 5i} = \frac{(3 + 4i)(2 - 5i)}{(2 + 5i)(2 - 5i)}$$
$$= \frac{(6 + 20) + i(8 - 15)}{2^2 + 5^2}$$
$$= \frac{26 - 7i}{29}$$

EXERCISE 25.5

1. Evaluate the following :

(i) $\dfrac{3 + 5i}{2 + 4i}$ (ii) $\dfrac{3 + 5i}{2 - 4i}$ (iii) $\dfrac{3 - 5i}{2 - 4i}$

(iv) $\dfrac{(3 + 4i)(3 + 5i)}{2 + 4i}$ (v) $\dfrac{(3 + 4i)(3 + 5i)}{(2 + 4i)(2 + 5i)}$ (vi) $\dfrac{(3 + 4i)^2}{(2 - 3i)^2}$

2. Solve the following for x and y:

(i) $\dfrac{3 + 5i}{2 + 4i} = 2x + 3iy$ (ii) $\dfrac{3 - 5i}{2 - 4i} = 3x - 2iy$

8. Complex roots of an equation

There is a great deal more that could be said about the elementary properties of complex numbers, but we now know enough to enable us to say something about the use of complex numbers in the solution of equations. To anticipate a little, we shall see in the next two chapters that very many equations involve complex roots, and that these are often the most interesting cases. Complex roots are of particular importance in the study of fluctuations.

We saw in Chapter I that sometimes it is impossible to find the n roots of an equation. A method was given for locating all the *real* roots of a polynomial equation of degree n, such as

$$a_0 p^n + a_1 p^{n-1} + \ldots + a_{n-1} p + a_n = 0$$

Let us now consider

$$9p^4 + 9p^3 - 73p^2 - 55p - 150 = 0$$

which is of the fourth degree.

The graph of this is as shown approximately in Diagram 25.9 and

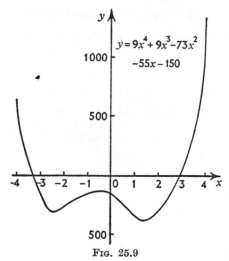

$$y = 9x^4 + 9x^3 - 73x^2 - 55x - 150$$

FIG. 25.9

it can be seen that the left-hand side is equal to zero for only two values. These can be found by using the Remainder Theorem, and are

$$p = 3 \quad \text{and} \quad p = -\frac{10}{3}$$

If we divide the left-hand side by the product

$$(p - 3)(3p + 10)$$

then we will obtain

$$3p^2 + 2p + 5$$

and so the equation may be rewritten

$$(p-3)(3p+10)(3p^2+2p+5) = 0$$

There is one root corresponding to the first bracket, one to the second, and two to the third (since the third is of the second degree). If, however, we attempt to locate the roots corresponding to the third bracket we will obtain (by use of the formula for the solution of a quadratic equation) the two complex (and conjugate) roots

$$p = \frac{-2 + \sqrt{4 - 60}}{6} \quad \text{and} \quad p = \frac{-2 - \sqrt{4 - 60}}{6}$$

i.e.,

$$p = \frac{-2 + \sqrt{-56}}{6} \quad \text{and} \quad p = \frac{-2 - \sqrt{-56}}{6}$$

$$= \frac{-1 + \sqrt{-14}}{3} \qquad\qquad = \frac{-1 - \sqrt{-14}}{3}$$

$$= -\frac{1}{3} + i\frac{\sqrt{14}}{3} \qquad\qquad = -\frac{1}{3} - i\frac{\sqrt{14}}{3}$$

The original equation, therefore, has the four roots

$$p_1 = 3$$

$$p_2 = -\frac{10}{3}$$

$$p_3 = -\frac{1}{3} + i\frac{\sqrt{14}}{3}$$

$$p_4 = -\frac{1}{3} - i\frac{\sqrt{14}}{3}$$

We may now ask what meaning is to be attached to these roots. The answer depends on the problem that has given rise to the equation. If, for example, p is a price at which a revenue is maximised, the only meaningful solution is $p = 3$. Neither negative nor complex prices will interest us. If, on the other hand, p denotes the change that must take place in production in order that certain conditions may hold, then the solutions $p = 3$ and $p = -\frac{10}{3}$ mean something to us—the former indicating an increase in production and the latter a decrease. The complex roots will be of importance if p is used to denote the position of a factory in order that certain location problems may be solved. In this case there will be four possible positions :

$$(3, 0) \qquad \text{corresponding to} \qquad p_1 = 3$$

$$\left(-\frac{10}{3}, 0\right) \qquad \text{corresponding to} \qquad p_2 = -\frac{10}{3}$$

$$\left(-\frac{1}{3}, \frac{\sqrt{14}}{3}\right) \qquad \text{corresponding to} \qquad p_3 = -\frac{1}{3} + i\frac{\sqrt{14}}{3}$$

$$\left(-\frac{1}{3}, -\frac{\sqrt{14}}{3}\right) \qquad \text{corresponding to} \qquad p_4 = -\frac{1}{3} - i\frac{\sqrt{14}}{3}$$

Later we shall see other examples of meaningful complex roots. At this stage the important point to notice is that every equation of degree n with real coefficients has n roots. Some of these may coincide. Some even number of them may be complex—for complex roots occur in conjugate pairs. Whether any one of them is meaningful depends on the problem under consideration.

CHAPTER XXVI

DIFFERENTIAL EQUATIONS

Mark how the curl'd Waves work and wind,
All hating to be left behind.
Each bigge with business thrusts the other,
And seems to say, Make haste, my Brother.

RICHARD CRASHAW: *A Letter to the Countess of Denbigh*

1. Differential equations and difference equations

The reader of modern economic theory is unlikely to proceed far without coming across either a differential equation or a difference equation. On these subjects many books, some of them very difficult, have been written. To attempt anything like a comprehensive summary of even the more elementary points in a book like this would be futile : but this and the next chapter should enable the reader to see what these equations are about, and to appreciate the broad principles that appear in their manipulation and solution.

Let us first of all make the distinction between these two kinds of equations clear. We will do so by considering a simple relationship between investment, consumption and income.

In the behaviour of any society in reasonably normal times the bulk of spending on consumption goods is out of money that is already, so to speak, in the pocket. If my income this week is £10, then I consider that I can spend all, or some, of this next week. Occasionally I might be inclined to spend rather more than I have : but while this is common enough in the case of purchasing a house— which is an investment—it is infrequent in the case of purchasing some consumption goods. Of course, if some period occurred during which I had no income, I should still spend something on consumption ; but generally the level of my spending would be determined by the level of my recent income. Let us sum this up mathematically by writing

$$C_t = \alpha Y_{t-1} + A$$

where C_t is my spending this week on consumption goods, Y_{t-1} is the income I received last week, α is a constant smaller than unity, and A is a constant denoting the amount I would spend if I had no income. α is the marginal propensity to consume, since it represents the additional consumption that will occur as the result of a rise in income.

Now in any society all the individuals have equations of this kind, and a similar equation may be written for the society as a whole, relating the total consumption expenditure in the society in any one week to the total of all the incomes in the previous week, α now being the marginal propensity to consume for the whole society.

In order to simplify things, and to avoid the difficulties that surround definitions of savings and investment, let us now suppose that all the income that is not spent on consumption goods is immediately devoted to investment, so that if the investment in period t is I_t then, by definition,

$$Y_t \equiv C_t + I_t$$

but we have already seen that

$$C_t = \alpha Y_{t-1} + A$$

and so

$$Y_t = \alpha Y_{t-1} + I_t + A \qquad (1)$$

Here we have the income of one period (Y_t) expressed in terms of the income in a previous period (Y_{t-1}). The periods concerned are, in this example, weeks. The important point is that they are *finite* periods. We call an equation of this kind, relating the value of a variable in one period to the value in some other period a *difference equation*. We could write

$$Y_t \equiv Y_{t-1} + \Delta Y_t$$

and then the equation would be

$$\Delta Y_t = (\alpha - 1) Y_{t-1} + I_t + A$$

which emphasises its nature, it being borne in mind that ΔY_t is the difference between the income in one finite period and the income in the next finite period.

If we further assume that investment I_t is related to the increase in consumption since the last period by the equation

$$I_t = \beta(C_t - C_{t-1})$$
$$= \alpha\beta(Y_{t-1} - Y_{t-2})$$

then substitution in equation (1) will give

$$Y_t = \alpha Y_{t-1} + \alpha\beta(Y_{t-1} - Y_{t-2}) + A$$
$$= \alpha(1 + \beta) Y_{t-1} - \alpha\beta Y_{t-2} + A$$

which relates the income of any period directly to the incomes of the two preceding periods. This is another example of a difference equation, since it relates the value of a variable in one period to the values in previous periods.

Let us now consider a different kind of equation. We might well, for example, have an equation relating income to some rate of change, such as the rate of change of population. For example, we could have

$$Y = Y_0\left(1 + a\frac{dP}{dt}\right)$$

where dP/dt is the rate of growth of population. And if it happens that the rate of growth of population is in turn related to the rate of growth of income by an equation such as

$$\frac{dP}{dt} = b + c\frac{dY}{dt}$$

then substitution will lead to

$$Y = Y_0\left(1 + ab + ac\frac{dY}{dt}\right)$$

which relates the income at any moment to the rate of growth of income at that moment. Such an equation, involving a derivative, is an example of a *differential equation*.

We may sum up this discussion, which has been illustrative rather than exhaustive, with the following definitions.

Definition :

> If y is a function of x, and x is allowed to increase by equal finitely large amounts, then the equation relating the value of y corresponding to one value of x to the value(s) of y corresponding to some other value(s) of x is called an *ordinary difference equation*.

In the discussion above, our x has been time. Income (y) has been a function of time, and we have allowed it to increase each week.

Definition :

> If y is a function of x, then an equation involving the derivatives of y with respect to x is called an *ordinary differential equation*.

We should notice that there are difference equations and differential equations which are not " ordinary " and which have more general definitions than those just given. We shall not deal with these.

2. The solution of a differential equation

We now consider the differential equation.

We call an equation of the form $f(x, y, dy/dx) = 0$ an *ordinary* differential equation, as opposed to a *partial* differential equation which would involve partial derivatives. Examples will appear in a moment.

The *order* of a differential equation is the order of the highest derivative in it. The equation just given is strictly an ordinary differential equation of the first order. An equation of the second order would involve the second derivative, and might be, for example,

$$x\frac{d^2y}{dx^2} + 4\frac{dy}{dx} + xy = 3$$

The *solution* of a differential equation is the equation (or, as we shall see in a moment, the set of equations) derived from the differential equation but free from all derivatives (or differentials).

We have come across simple differential equations when we have dealt with integrals. For example, we have said that if

$$\frac{dy}{dx} = 4x$$

then $$y = \int 4x\, dx = 2x^2 + C$$

where C is an arbitrary constant. Now the equation with which we began, and which could well be written as

$$4x - \frac{dy}{dx} = 0$$

or even as $$4x\, dx - dy = 0$$

is a differential equation. And clearly *one* solution to it is given by

$$y = 2x^2 + C \tag{2}$$

since this equation is derived from the differential equation but does not contain any differentials itself. But in the chapter on integration we saw that C could take any value; and so this equation represents an *infinite number* of solutions, one to each value we care to give C. We call equation (2) the *General Integral* of the differential equation, since it contains an infinite number of solutions grouped under one formula or equation. The solution obtained by giving C a particular value is called a *Particular Solution*. Any solution that is not contained within the general integral is a *Singular Solution*. Let us consider a simple example.

Example

Consider the equation

$$\left(\frac{dy}{dx}\right)^2 - x\frac{dy}{dx} + y = 0$$

It can be shown (and is easily verified by differentiation) that the *general integral* is

$$y = Cx - C^2$$

where C is a constant which may take any value. If we put $C = 1$ (or anything else) then we will obtain a *particular solution*, namely (in this case), $y = x - 1$. But the equation is also true if

$$y = \frac{x^2}{4}$$

and it is impossible to obtain this from the general integral; nevertheless it is a solution. It is a *singular solution*.

It can be shown that a differential equation of the first order has one and only one general integral.

Let us demonstrate the lines of proof.

Suppose

$$\left(\frac{dy}{dx}\right)^2 - x\frac{dy}{dx} + y = 0$$

has a general integral of the form

$$y = Af(x) + B$$

where A or B is able to take arbitrary values.

Then $\qquad (Af')^2 - xAf' + Af + B = 0$

Let there be another solution of the form

$$y = A\phi(x) + B$$

where, again, A or B is able to take arbitrary values, which we can therefore stipulate to be the same as those chosen for the other solution. Then

$$(A\phi')^2 - xA\phi' + A\phi + B = 0$$

These two results must hold for all values of A, B and x. They can be combined to yield

$$A^2(f'^2 - \phi'^2) - xA(f' - \phi') + A(f - \phi) = 0$$

which has to be true for all values of A and x.

It is fairly easy to see that this can be so only if

$$f = \phi.$$

We now mention a few of the simpler types of differential equation and indicate how they may be solved. We begin by considering equations of the first order and degree, i.e., involving only the first derivative, and this to only the first power. (The equation we have just considered is of the first order and second degree.)

3. Differential equations of the first order and first degree

These may all be written in the form

$$P(x, y)\, dx + Q(x, y)\, dy = 0$$

where P and Q are functions of x and y. Some of the more important kinds are listed below.

(a) *Separable:*

Here the equation is capable of being written as

$$P(x)\, dx + Q(y)\, dy = 0$$

where P is a function of x only, and Q of y only. In this case the general integral is obtained immediately by integrating the two parts of the equation separately, thus:

$$\frac{x^2}{y+1} - 4x\frac{dy}{dx} = 0$$

may be written as

$$\frac{x}{4}\, dx - (y+1)\, dy = 0$$

which shows that it is separable in form. The General Integral is given by

$$\int \frac{x}{4}\, dx - \int (y+1)\, dy = 0$$

i.e.,

$$\frac{x^2}{8} - \left(\frac{y^2}{2} + y\right) + C = 0$$

where C is any constant. Particular Solutions may be obtained by giving different values to C.

(b) *Homogeneous:*

If P and Q are functions homogeneous in x and y to the *same degree*, then the equation can be reduced to the separable form by writing

$$y = vx$$

For example,

$$(x^2 - y^2)\, dx + 2xy\, dy = 0$$

is such that the coefficients of dx and dy are homogeneous and of the second degree in x and y. Writing $y = vx$ we have that

$$dy = v\, dx + x\, dv$$

and so $$(x^2 - v^2 x^2)\, dx + 2vx^2(v\, dx + x\, dv) = 0$$

i.e., $$x^2\, dx + x^2 v^2\, dx + 2vx^3\, dv = 0$$

which gives, when x^2 cancels out,

$$(1 + v^2)\, dx + 2vx\, dv = 0$$

This is separable, giving

$$\frac{dx}{x} + \frac{2v}{1 + v^2}\, dv = 0$$

which, on integration, yields

$$\log x + \log (1 + v^2) = \log C$$

which may be written as

$$x(1 + v^2) = C$$

where $$v = \frac{y}{x}$$

and so $$x^2 + y^2 = Cx$$

which is the general integral. Particular integrals may be obtained by giving C particular values.

(c) *Linear coefficients*:

An equation of the form

$$(ax + by + c)\, dx + (a'x + b'y + c')\, dy = 0$$
$$[a,\, b,\, c,\, a',\, b',\, c',\ \text{constants}]$$

is not homogeneous as it stands, but it can be reduced to the homogeneous form by writing the first bracket as X and the second as Y when it becomes

$$X\, dx + Y\, dy = 0$$

from which we have

$$\frac{dy}{dx} = -\frac{X}{Y}$$

Now as $\qquad X = ax + by + c$

and $\qquad Y = a'x + b'y + c'$

$$\frac{dX}{dx} = a + b\frac{dy}{dx}$$

and $\qquad \dfrac{dY}{dx} = a' + b'\dfrac{dy}{dx}$

Furthermore $\qquad \dfrac{dY}{dX} = \dfrac{dY}{dx} \bigg/ \dfrac{dX}{dx}$

$$= \frac{a' + b'\dfrac{dy}{dx}}{a + b\dfrac{dy}{dx}}$$

$$= \frac{a' - b'\dfrac{X}{Y}}{a - b\dfrac{X}{Y}}$$

$$= \frac{a'Y - b'X}{aY - bX}$$

This puts the equation into the form

$$\frac{dY}{dX} = \frac{a'Y - b'X}{aY - bX}$$

which is homogeneous and may be solved as in (b), giving Y in terms of X and therefore y in terms of x.

For example, suppose we have

$$(x + 2y + 3)\,dx + (2x + y + 4)\,dy = 0$$

To solve this we put

$$x + 2y + 3 = X$$

$$2x + y + 4 = Y$$

We then have $\qquad \dfrac{dX}{dx} = 1 + 2\dfrac{dy}{dx}$

$$\frac{dY}{dx} = 2 + \frac{dy}{dx}$$

therefore,

$$\frac{dY}{dX} = \frac{2 + \dfrac{dy}{dx}}{1 + 2\dfrac{dy}{dx}}$$

$$= \frac{2 - \dfrac{X}{Y}}{1 - 2\dfrac{X}{Y}}$$

i.e.,
$$\frac{dY}{dX} = \frac{X - 2Y}{2X - Y}$$

Now put $Y = VX$ and we have

$$\frac{dY}{dX} = V\frac{dX}{dX} + X\frac{dV}{dX} = V + X\frac{dV}{dX}$$

while $\dfrac{X - 2Y}{2X - Y}$ becomes $\dfrac{X(1 - 2V))}{X(2 - V}$

Consequently

$$V + X\frac{dV}{dX} = \frac{1 - 2V}{2 - V}$$

$$X\frac{dV}{dX} = \frac{1 - 2V}{2 - V} - V = \frac{1 - 4V + V^2}{2 - V}$$

and so
$$\frac{dX}{X} = \frac{2 - V}{1 - 4V + V^2}\,dV$$

i.e.,
$$\int\frac{dX}{X} = \int\frac{2 - V}{1 - 4V + V^2}\,dV$$

$$\log X = -\tfrac{1}{2}\log(1 - 4V + V^2) + \log A$$

i.e.,
$$\log X = \log\frac{A}{\sqrt{1 - 4V + V^2}}$$

and so
$$X = \frac{A}{\sqrt{1 - 4V + V^2}}$$

where A is an arbitrary constant.

Since
$$V = \frac{Y}{X}$$

this gives
$$X = \frac{A}{\sqrt{1 - \dfrac{4Y}{X} + \dfrac{Y^2}{X^2}}}$$

i.e.,
$$X = \frac{AX}{\sqrt{Y^2 - 4XY + X^2}}$$

where
$$X = x + 2y + 3$$

and
$$Y = 2x + y + 4$$

Substitution of these values yields the final result in terms of x, y and A.

(d) *Exact:*

It sometimes happens that the left-hand side may be written as an exact differential. For example,

$$y \, dx + x \, dy = 0$$

may be written as
$$d(xy) = 0$$

immediately giving a solution

$$xy = C$$

Again, if we consider

$$\tan y \, dx + \tan x \, dy = 0$$

we will find that, although this is not " exact " as it stands, we can multiply it throughout by $\cos x \cos y$, and get

$$\sin y \cos x \, dx + \sin x \cos y \, dy = 0$$

which is
$$d(\sin y \sin x) = 0$$

whence
$$\sin y \sin x = C$$

An equation of the form $P(x, y) \, dx + Q(x, y) \, dy = 0$ is exact if

$$\frac{\partial P}{\partial y} = \frac{\partial Q}{\partial x}$$

In the case we have just considered we have turned a non-exact equation into an exact equation by multiplying throughout by $\cos x \cos y$. In fact, this factor $\cos x \cos y$ has enabled us to integrate the equation, and it is therefore called an *integrating factor*. It can be shown that every non-exact equation $P \, dx + Q \, dy = 0$ may be converted into an exact equation $\mu P \, dx + \mu Q \, dy = 0$. It is sometimes very difficult to find this integrating factor μ; but it always exists.

EXERCISE 26.1

Find the General Integrals of:

(i) $x(y+2) = 3\dfrac{dy}{dx}$
(ii) $x^2(y+2) + (y+3)^2\dfrac{dy}{dx} = 0$

(iii) $x^2 + y^2 = 2xy\dfrac{dy}{dx}$
(iv) $x^2y\,dx - (x^3 + y^3)\,dy = 0$

(v) $\dfrac{dy}{dx} = \dfrac{3x + 2y + 1}{x + y + 1}$
(vi) $x\cos y\,dy + \sin y\,dx = 0$

4. The arbitrary constant

Before we go any further we should say a word about the importance of the arbitrary constant.

We introduced the subject of differential equations by considering a hypothetical relationship between population and income, deriving the equation

$$Y = Y_0\left(1 + ab + ac\dfrac{dY}{dt}\right)$$

which may be written as

$$Y = Y_0(1 + ab) + acY_0\dfrac{dY}{dt} = P + Q\dfrac{dY}{dt}$$

where P and Q are constants. The solution proceeds

$$Y = P + Q\dfrac{dY}{dt}$$

i.e., $\qquad (Y - P)\,dt = Q\,dY$

therefore, $\qquad dt = \dfrac{Q}{Y - P}\,dY$

$$\int dt = \int \dfrac{Q}{Y - P}\,dY$$

$$t = Q\log(Y - P) + C \qquad \text{where } C = \text{constant}$$
$$ = Q\log A(Y - P) \qquad \text{where } \log A = C$$

$$\dfrac{t}{Q} = \log A\,(Y - P)$$

therefore, $\quad A(Y - P) = e^{t/Q}$

$$Y - P = \dfrac{1}{A}e^{t/Q}$$

whence $\qquad Y = Y_0(1 + ab) + \dfrac{1}{A}e^{t/acY_0}$

N

This, however, is not very useful as it stands. If we want to know the level of income at any particular time we are immediately confronted with the fact that we do not know what value is to be given to the arbitrary constant A. We can, however, determine this value if we are given a little more information about the conditions which are summarised by the equation. Suppose, for example, that we are told that the level of income at time $t = 0$ is γY_0, where $\gamma > 0$.

Then we have

$$\gamma Y_0 = Y_0(1 + ab) + \frac{1}{A} e^0$$

i.e.

$$\gamma Y_0 = Y_0(1 + ab) + \frac{1}{A}$$

whence

$$A = \frac{1}{(\gamma - 1 - ab) Y_0}$$

unless $\gamma = 1 + ab$, when A will be infinitely large. We thus have the solution

$$Y = Y_0(1 + ab) + \frac{Y_0 e^{t/acY_0}}{\gamma - 1 - ab} \qquad\qquad \gamma \neq 1 + ab$$

If $\gamma = 1 + ab$ then, because A is infinitely large the term containing $1/A$ disappears and we obtain

$$Y = Y_0(1 + ab) \qquad\qquad \gamma = 1 + ab$$

This solution satisfies our assumptions which produced the original differential equation. It also satisfies our *initial condition* which is that at time $t = 0$, $Y = \gamma Y_0$. By giving t various values we can trace the growth of Y.

There are three cases for us to consider.

If $\gamma > 1 + ab$ then the denominator in the solution is positive and Y grows as t increases.

If $\gamma = 1 + ab$ then Y is constant.

If $\gamma < 1 + ab$ then Y declines as t increases.

This means that if our two basic assumptions are stated with specified values of a and b, and γ is also given, then we will already have determined whether Y is to grow, to remain constant, or to decline. They also tell us what is happening to P. It is an example not only of the ways in which our assumptions can imply unsuspected properties of the system, but also of the way in which the initial conditions, by defining the arbitrary constant, can completely alter the character of the solution. Thus, the assumptions

$$Y = Y_0\left(1 + \frac{dP}{dt}\right) \quad \text{(implying } a = 1\text{)}$$

and $\quad \dfrac{dP}{dt} = 1 + 0.5\dfrac{dY}{dt} \quad \text{(implying } b = 1, \ c = 0.5\text{)}$

will lead to growing income and population provided that at time $t = 0$, $Y = \gamma Y_0$ where $\gamma > 2$. But if we had said that at time 0, $Y = 2Y_0$, then there would have been no population or income growth.

Whenever any real problem results in a differential equation which has n arbitrary constants in its solution it is necessary to find conditions which the solution must satisfy, in order that the arbitrary constants may be given definite values. Until this is done, the solution is too imprecise to be of much use. These conditions very often relate to the value of the dependent variable and some of its derivatives for the initial value of the independent variable, and are called *initial conditions*, as we have already illustrated.

There is, of course, a great deal more to be said on this subject, but the reader who wishes to obtain a greater knowledge should consult a book such as Piaggio's *Differential Equations* (Bell) or Ince's *Integration of Ordinary Differential Equations* (Oliver and Boyd). In this chapter we shall, having introduced the subject, content ourselves with considering in more detail a special kind of differential equation, namely the linear differential equation.

5. The general linear differential equation

The general linear equation of order n is defined to be of the form

$$p_0(x)\frac{d^n y}{dx^n} + p_1(x)\frac{d^{n-1}y}{dx^{n-1}} + \ldots + p_{n-1}(x)\frac{dy}{dx} + p_n(x)y = f(x)$$

where each p is a function of x only. It is thus linear in y and its derivatives y', y'', $\ldots$, $y^{(n)}$.

As examples of such equations we have

$$x\frac{d^3 y}{dx^3} + (x^2 + 3)\frac{d^2 y}{dx^2} + 7y = 4x + 5$$

$$(x^3 + 2)\frac{d^2 y}{dx^2} + (4\sin x)\frac{dy}{dx} = \log_e (x^2 \cos x)$$

It is often convenient to replace dy/dx by Dy, d^2y/dx^2 by D^2y, etc., and if this is done we can write the equation as

$$\{p_0(x)D^n + p_1(x)D^{n-1} + p_2(x)D^{n-2} + \ldots + p_{n-1}(x)D + p_n(x)\}y = f(x)$$

We now consider the solution of equations of this kind. We begin by considering the simplified equation which is obtained if we put $f(x) = 0$, i.e.,

$$\{p_0(x)D^n + p_1(x)D^{n-1} + \ldots + p_{n-1}(x)D + p_n(x)\}y = 0$$

We call this the *reduced equation*.

Now suppose that we can find n *particular solutions of this reduced equation*. Let us denote them by $y_1, y_2, \ldots, y_n$. Then it can be shown that the *general integral of the reduced equation* is given by

$$Y = C_1y_1 + C_2y_2 + \ldots + C_ny_n$$

where the C's are arbitrary constants, *provided* that it is impossible to find a set of C's not all zero which reduce the right-hand side of this solution to zero. (For example, if we found three solutions $y = \sin^2 x$, $y = \cos^2 x$ and $y = \cos 2x$ then we could not combine these to give the general integral of the reduced equation because, if we choose $C_1 = 1$, $C_2 = -1$ and $C_3 = 1$, we would have

$$\sin^2 x - \cos^2 x + \cos 2x$$

which is zero for all values of x.) It will be noticed that the number of particular solutions occurring in this solution is equal to the order of the equation.

Furthermore, it can be shown that the *general integral of the original equation consists of the sum of the general integral of the reduced equation plus any particular integral (not involving an arbitrary constant) of the original equation.*

Let us do an example.

Consider

$$x(x^2 + 1)^2D^2y - (3x^2 - 1)(x^2 + 1)Dy + 4x(x^2 - 1)y = x^4 - 6x^2 + 1$$

The reduced equation is obtained by rewriting the left-hand side equal to zero.

The task now is to find particular solutions. We shall shortly look at methods for doing this. Meanwhile let us accept that two particular solutions (which may be verified by differentiation) are

$$x^2 + 1 \quad \text{and} \quad (x^2 + 1) \log x$$

Now consider $A(x^2 + 1) + B(x^2 + 1) \log x$. There are no values of A and B such that this is zero for all values of x. Consequently, since it contains the same number of arbitrary constants as the order of the equation, we can say that the *general integral of the reduced equation* is

$$y = A(x^2 + 1) + B(x^2 + 1) \log x$$
$$= (x^2 + 1)(A + B \log x)$$

Furthermore, the *original* equation is satisfied by $y = z$, which is therefore a particular integral. And so the general integral of the original equation is given by

$$y = (x^2 + 1)(A + B \log x) + x$$

The general integral of the reduced equation is sometimes called the *complementary function*.

The problem of solving a linear differential equation, therefore consists of

   (i) finding the n particular solutions of the reduced equation,

  (ii) combining them to form the general integral of the reduced equation,

 (iii) finding a particular integral of the original equation.

The solution so obtained has a number of arbitrary constants in it. We shall later see that sometimes these have to be given very definite values.

Since the complete solution of an equation consists of the sum of the complementary function and a particular integral of the original equation, we must consider how to find these. Sometimes this is very difficult. In this chapter we restrict ourselves to indicating one method which is of use in a case of great importance to economists —the case of the linear equation with constant coefficients.

6. The linear equation with constant coefficients

If the functions $p(x)$ in the General Linear Equation are replaced by constants, which we may denote by $a_1, a_2, a_3, \dots$, then the equation becomes

$$\{a_1 D^n + a_2 D^{n-1} + \dots + a_n D + a_{n+1}\} y = f(x)$$

As an example we have

$$\frac{d^2 y}{dx^2} - 3 \frac{dy}{dx} + 2y = \sin x$$

$$(D^2 - 3D + 2)y = \sin x$$

Consider the equation

$$y = e^{px}$$

Note that

$$Dy = pe^{px} = py$$
$$D^2 y = p^2 e^{px} = p^2 y$$
$$\vdots \qquad \vdots \qquad \vdots$$
$$D^n y = p^n e^{px} = p^n y$$

Thus, if y is so defined, then

$$\{a_1 D^n + a_2 D^{n-1} + \ldots + a_{n+1}\} y = (a_1 p^n + a_2 p^{n-2} + \ldots + a_{n+1})y$$

Consequently, our reduced equation is satisfied by

$$y = e^{px}$$

provided that

$$a_1 p^n + a_2 p^{n-2} + \ldots + a_{n+1} = 0$$

We call this the auxiliary equation. In general there will be n values of p which satisfy it. Some of them may be coincident, and some of them complex. Let us denote them by

$$p_1, p_2, \ldots, p_n.$$

If they are all different then these n values of p give us n different particular solutions, and the sum of these, each multiplied by an arbitrary constant, gives us our complementary function, viz.,

$$y = A_1 e^{p_1 x} + A_2 e^{p_2 x} + \ldots + A_n e^{p_n x}$$

Example

If we have

$$\frac{d^2 y}{dx^2} - 3\frac{dy}{dx} + 2y = \sin x$$

then the auxiliary equation is

$$p^2 - 3p + 2 = 0$$

having roots

$$p = 2 \quad \text{and} \quad p = 1$$

The complementary function is therefore

$$y = A e^{2x} + B e^x$$

where A and B are arbitrary constants which may be determined by the initial conditions—if any are given.

If there are *two coincident real roots* (which we may think of as p_1 and p_2 where $p_1 = p_2$) then the complementary function is

$$y = (A_1 + A_2 x)e^{p_1 x} + A_3 e^{p_3 x} + \ldots + A_n e^{p_n x}$$

If there are three real coincident roots the complementary function will contain the term

$$y = (A_1 + A_2 x + A_3 x^2)e^{p_1 x}$$

and so on. These results may be proved along the same lines as those used above.

Example

If the auxiliary equation is

$$p^6 - p^5 - 19p^4 - 15p^3 + 46p^2 + 28p - 40 = 0$$

which may be written

$$(p - 1)^2(p + 2)^3(p - 5) = 0$$

then we have 6 roots given by

$$p_1 = p_2 = 1$$
$$p_3 = p_4 = p_5 = -2$$
$$p_6 = 5$$

and so the complementary function is

$$y = (A_1 + A_2 x)e^x + (A_3 + A_4 x + A_5 x^2)e^{-2x} + A_6 e^{5x}$$

If we have *complex roots* then they will occur in conjugate pairs. If we denote such a pair by

$$p_1 = \alpha + i\omega \quad \text{and} \quad p_2 = \alpha - i\omega$$

then we can easily prove, in a way similar to that used for the case of n real and different roots, that the relevant part of the complementary function becomes

$$A_1 e^{(\alpha + i\omega)x} + A_2 e^{(\alpha - i\omega)x}$$
$$= e^{\alpha x}(A_1 e^{i\omega x} + A_2 e^{-i\omega x})$$
$$= e^{\alpha x}[A_1(\cos \omega x + i \sin \omega x) + A_2(\cos \omega x - i \sin \omega x)]$$
$$= e^{\alpha x}[(A_1 + A_2) \cos \omega x + i(A_1 - A_2) \sin \omega x]$$
$$= e^{\alpha x}[B_1 \cos \omega x + B_2 \sin \omega x]$$

where B_1 and B_2 are new arbitrary constants defined by

$$B_1 = A_1 + A_2 \quad \text{and} \quad B_2 = i(A_1 - A_2)$$

If we wish these to be real, then we have simply to choose our original arbitrary constants in such a way that they are two complex conjugates.

The solution may be put in a somewhat different form by writing

$$B_1 = B \cos \epsilon \quad \text{and} \quad B_2 = B \sin \epsilon$$

when it becomes

$$e^{\alpha x}[B \cos \epsilon \cos \omega x + B \sin \epsilon \sin \omega x]$$
$$= e^{\alpha x}[B \cos (\omega x - \epsilon)]$$
$$= B e^{\alpha x} \cos (\omega x - \epsilon)$$

We shall discuss the interpretation of a result in this form later in the chapter.

EXERCISE 26.2

Find the Complementary Functions in the following cases:

(i) $\dfrac{d^2y}{dx^2} + 4\dfrac{dy}{dx} - 12y = e^x$ (ii) $\dfrac{d^2y}{dx^2} + 4\dfrac{dy}{dx} + 4y = e^x$

(iii) $\dfrac{d^3y}{dx^3} + \dfrac{d^2y}{dx^2} - 8\dfrac{dy}{dx} - 12y = \sin x$ (iv) $\dfrac{d^2y}{dx^2} + 4\dfrac{dy}{dx} + 9y = \cos x$

(b) Finding the particular integral:

Before we can obtain the complete solution we have to find the Particular Integral, which depends on the value of $f(x)$ on the R.H.S. of the original equation. This is often extremely difficult. One way, which is very useful in the case we are considering, is to proceed as follows. For a justification of this procedure the student should consult one of the books mentioned in the bibliography.

(1) Write the equation in the form

$$(a_1 D^n + a_2 D^{n-1} + \ldots + a_n D + a_{n+1})y = f(x)$$

which may be condensed to

$$F(D)y = f(x)$$

where $F(D) \equiv (a_1 D^n + a_2 D^{n-1} + \ldots + a_n D + a_{n+1})$

(2) Rewrite in the form

$$y = \frac{1}{F(D)} f(x)$$

and evaluate the R.H.S. according to the rules below:

If $f(x)$ is an exponential function $e^{\alpha x}$ and $F(\alpha) \neq 0$

then $\dfrac{1}{F(D)} e^{\alpha x} = \dfrac{e^{\alpha x}}{F(\alpha)}$

Example

If $(D^2 + 3D - 1)y = e^{2x}$ then the particular integral is obtained by considering

$$y = \frac{1}{D^2 + 3D - 1} e^{2x}$$

and the above rule shows us that the value of this is given by replacing the operator D by the constant 2, obtaining as a particular integral

$$y = \frac{1}{2^2 + 3(2) - 1} e^{2x} = \frac{e^{2x}}{9}$$

If $f(x)$ is an exponential function $e^{\alpha x}$ and $F(\alpha) = 0$ then it follows that $(D - \alpha)$ is a factor of $F(D)$. We shall let it be a factor r times, so that

$$F(D) = G(D)(D - \alpha)^r \quad \text{where} \quad G(\alpha) = 0$$

Then it can be shown that

$$\frac{1}{F(D)} e^{\alpha v} = \frac{e^{\alpha x} x^r}{r! \, G(\alpha)}$$

Example

If
$$(D^2 + 3D - 10)y = e^{2x}$$

then $\alpha = 2$ and
$$F(\alpha) = 2^2 + 3(2) - 10 = 0$$

But
$$F(D) = (D - 2)(D + 5)$$

Write
$$G(D) = (D + 5)$$

Then
$$F(D) = G(D)(D - 2)$$

and so
$$\frac{1}{F(D)} e^{2x} = \frac{e^{2x} \, x}{1! \, G(\alpha)}$$

$$= \frac{e^{2x} \, x}{(2 + 5)}$$

$$= \frac{e^{2x} \, x}{7}$$

If $f(x)$ is a polynomial in x

then $1/F(D)$ should be expressed in terms of powers of D as in the following examples, and the particular integral then obtained by operating with the various powers of D according to the rules of ordinary differentiation.

Example

$$\tfrac{1}{2} \frac{d^2 y}{dx^2} - 2y = ax^2 + bx + c$$

$$\tfrac{1}{2}(D^2 - 4)y = ax^2 + bx + c$$

P.I.
$$y = \frac{2}{D^2 - 4} (ax^2 + bx + c)$$

$$= \tfrac{1}{2} \left[\frac{1}{D - 2} - \frac{1}{D + 2} \right] (ax^2 + bx + c)$$

$$= \left[-\tfrac{1}{2} \frac{1}{2 - D} - \tfrac{1}{2} \frac{1}{2 + D} \right] ax^2 + bx + c \Big]$$

$$= -\tfrac{1}{4} \left[\frac{1}{1 - \tfrac{1}{2}D} + \frac{1}{1 + \tfrac{1}{2}D} \right] (ax^2 + bx + c)$$

$$= \left[-\tfrac{1}{4}\left(1 - \frac{D}{2}\right)^{-1} - \tfrac{1}{4}\left(1 + \frac{D}{2}\right)^{-1} \right] (ax^2 + bx + c)$$

Use of the Binomial Theorem shows that this is

$$-\tfrac{1}{4}\left[1+\frac{D}{2}+\frac{D^2}{4}+\ldots+1-\frac{D}{2}+\frac{D^2}{4}-\ldots\right](ax^2+bx+c)$$

$$=-\tfrac{1}{2}\left(1+\frac{D^2}{4}\ldots\right)(ax^2+bx+c)$$

where we need go no further than D^2 since subsequent derivatives will be zero. We therefore have

$$y=-\tfrac{1}{2}(ax^2+bx+c)-\frac{D^2}{8}(ax^2+bx+c)$$

$$=-\tfrac{1}{2}(ax^2+bx+c)-\frac{D}{8}(2ax+b)$$

$$=-\tfrac{1}{2}(ax^2+bx+c)-\frac{2a}{8}$$

$$=-\left[\frac{ax^2}{2}+\frac{bx}{2}+\frac{2c+a}{4}\right]$$

It is easily checked that this satisfies the original equation.

Example

If P.I. is $y=\dfrac{D-1}{D(D+1)}(3x+4)$

Then we write $\qquad \dfrac{D-1}{D(D+1)}\equiv\dfrac{A}{D}+\dfrac{B}{D+1}$

where A and B are constants to be determined such that this identity is always true. Multiplying by $D(D+1)$ we have

$$D-1\equiv A(D+1)+BD$$

i.e., $\qquad\qquad D-1\equiv(A+B)D+A$

If this is to be true for all values of D, then

$$A+B=1$$

and $\qquad\qquad\qquad\qquad A=-1$

whence $\qquad\qquad\qquad\qquad B=2$

We may, therefore, write

$$y=\left(-\frac{1}{D}+\frac{2}{D+1}\right)\left(3x+4\right)$$

$$=[-D^{-1}+2(1+D)^{-1}](3x+4)$$
$$=-D^{-1}(3x+4)+2(1+D)^{-1}(3x+4)$$

Remembering that D denotes the process of differentiation, we interpret D^{-1} as the inverse process of integration, thus getting

$$y = -\int(3x+4)\,dx + 2(1+D)^{-1}(3x+4)$$

$$= -\left[\frac{3x^2}{2}+4x+C\right] + 2[1-D+D^2-...](3x+4)$$

$$= -\left[\frac{3x^2}{2}+4x+C\right] + 2[(3x+4)-D(3x+4)+...]$$

$$= -\left[\frac{3x^2}{2}+4x+C\right] + 2[(3x+4)-3]$$

$$= -\frac{3x^2}{2}+2x-C \quad \text{where } C \text{ is a constant}$$

If $f(x)$ is of the form $\sin \omega x$ *or* $\cos \omega x$
then we use the fact that for even powers of D

$$\frac{1}{F(D)}\begin{Bmatrix}\sin\\\cos\end{Bmatrix}\omega x = \frac{1}{F(-\omega^2)}\begin{Bmatrix}\sin\\\cos\end{Bmatrix}\omega x$$

The use of this is illustrated in the following examples. The method cannot be used if $F(-\omega^2)=0$.

Examples

(i) If $(D^2+1)y=\sin 3x$

Then P.I. is
$$y=\frac{1}{D^2+1}\sin 3x$$

$$=\frac{1}{-3^2+1}\sin 3x$$

$$=-\frac{1}{8}\sin 3x$$

(ii) If $(D^2+3D+2)y=\cos 2x$

Then P.I. is
$$y=\frac{1}{D^2+3D+2}\cos 2x$$

We now replace D^2 by $-(2)^2$ wherever this is possible and obtain

$$y=\frac{1}{-2^2+3D+2}\cos 2x$$

$$=\frac{1}{3D-2}\cos 2x$$

It is difficult to handle this as it stands so we rewrite it as

$$y = \frac{3D+2}{(3D-2)(3D+2)} \cos 2x$$

$$= \frac{3D+2}{9D^2-4} \cos 2x$$

We can now remove the D^2 by writing

$$y = \frac{3D+2}{-9(2)^2-4} \cos 2x$$

$$= \frac{3D+2}{-40} \cos 2x$$

and in this form, with no D in the denominator, the solution is straightforward

$$y = -\frac{1}{40}\left[3D \cos 2x + 2 \cos 2x \right]$$

$$= -\frac{1}{40}\left[-6 \sin 2x + 2 \cos 2x \right]$$

$$= -\frac{1}{20}\left[-3 \sin 2x + \cos 2x \right]$$

EXERCISE 26.3

1. Find the Particular Integrals in the following cases:

(i) $(D^2 + 4D - 2)y = = e^{3x}$ (ii) $(D^3 + 3D^2 + D - 3)y = e^{2x}$
(iii) $(D^3 + 3D^2 - 20)y = e^{2x}$ (iv) $(D^3 - 4D^2 + 5D - 2)y = e^x$
(v) $(D^2 - 9)y = 2x^3 + 3x + 1$ (vi) $(D^2 - D - 6)y = 3x^2 + x + 2$
(vii) $(D^2 + 4)y = \sin 5x$ (viii) $(D^2 + 2D + 3)y = \cos 3x$

(c) *A completed example:*

Consider the equation

$$(D^3 - D^2 - D - 2)y = \sin 2x$$

This is a linear equation with constant coefficients, and we may therefore attempt a solution along the lines we have just indicated. We begin by finding the complementary function.

Complementary function

Auxiliary equation is

$$p^3 - p^2 - p - 2 = 0$$

which has one root given by $p = 2$. Division by $p - 2$ shows that the equation is

$$(p-2)(p^2 + p + 1) = 0$$

The second bracket has the complex roots

$$p = \frac{-1 \pm i\sqrt{3}}{2}$$

and so the complementary function is

$$y = A_1 e^{2x} + A_2 e^{[-\frac{1}{2}+i(\sqrt{3}/2)]x} + A_3 e^{[-\frac{1}{2}-i(\sqrt{3}/2)]x}$$

$$y = A_1 e^{2x} + B e^{-x/2} \cos\left(\frac{\sqrt{3}}{2}x - \epsilon\right)$$

Particular integral

The Particular Integral is given by

$$y = \frac{1}{F(D)} \sin 2x$$

$$= \frac{1}{D^3 - D^2 - D - 2} \sin 2x$$

$$= \frac{1}{(D-2)(D^2+D+1)} \sin 2x$$

$$= \frac{1}{(D-2)(-4+D+1)} \sin 2x$$

$$\qquad\qquad \text{(by putting } D^2 = -(2)^2)$$

$$= \frac{1}{(D-2)(D-3)} \sin 2x$$

$$= \frac{1}{D^2 - 5D + 6} \sin 2x$$

$$= \frac{1}{2 - 5D} \sin 2x \quad \text{(by putting } D^2 = -(2)^2)$$

$$= \frac{2 + 5D}{(2-5D)(2+5D)} \sin 2x$$

$$= \frac{2 + 5D}{4 - 25D^2} \sin 2x$$

$$= \frac{2 + 5D}{104} \sin 2x \quad \text{(by putting } D^2 = -(2)^2)$$

$$= \frac{1}{104}\left(2 \sin 2x + 5D \sin 2x\right)$$

$$= \frac{1}{104}\left(2 \sin 2x + 10 \cos 2x\right)$$

$$= \frac{1}{52}\left(\sin 2x + 5 \cos 2x\right)$$

Complete solution

The complete solution is therefore given by the sum of the C.F. and the P.I. which is

$$y = A_1 e^{2x} + B e^{-x/2} \cos\left(\frac{\sqrt{3}}{2}x - \epsilon\right) + \frac{1}{52}\left(\sin 2x + 5\cos 2x\right)$$

It will be noticed that this solution has three arbitrary constants, A_1, B and ϵ. Normally these constants will have to take values which will be dictated by some conditions (usually initial conditions) which are determined by non-mathematical considerations, as explained earlier in the chapter. Let us suppose, for example, that we are told that

$$\text{when } x = 0 \qquad \begin{aligned} y &= 2\tfrac{5}{52} \\ y' &= 1\tfrac{14}{26} \\ y'' &= 3\tfrac{3}{26} \end{aligned}$$

We take the solution involving the arbitrary constants and differentiate it twice to obtain y' and y''. If we put $x = 0$ in these values we obtain the equations below, where we now drop the suffix from A:

$$y = A + B\cos\epsilon + \frac{5}{52}$$

$$y' = 2A - \frac{B}{2}\left(\cos\epsilon - \sqrt{3}\sin\epsilon\right) + \frac{1}{26}$$

$$y'' = 4A - \frac{B}{2}\cos\epsilon - \frac{B}{2}\sqrt{3}\sin\epsilon - \frac{5}{13}$$

But we have already been told that when $x = 0$ the values of y, y' and y'' are to be $2\tfrac{5}{52}$, $1\tfrac{14}{26}$ and $3\tfrac{3}{26}$. Consequently we have that

$$A + B\cos\epsilon \qquad\qquad = 2$$

$$2A - \frac{B}{2}\left(\cos\epsilon - \sqrt{3}\sin\epsilon\right) = 1\tfrac{1}{2}$$

$$4A - \frac{B}{2}\left(\cos\epsilon + \sqrt{3}\sin\epsilon\right) = 3\tfrac{1}{2}$$

These results give us

$$A = 1$$

It follows immediately that $B = 1$ and $\epsilon = 0$.

We thus have the values of the constants that enable our solution to fit the particular case which is characterised by the stated initial conditions. The final solution is, accordingly,

$$y = e^{2x} + e^{-x/2}\cos\frac{\sqrt{3}x}{2} + \frac{1}{52}\left(\sin 2x + 5\cos 2x\right)$$

If we put $(\sin 2x + 5 \cos 2x) \equiv K \sin (2x + \phi)$ we will find that

$$\sin 2x + 5 \cos 2x \equiv K \cos \phi \sin 2x + K \sin \phi \cos 2x$$

which yields $\qquad K \cos \phi = 1 \quad$ and $\quad K \sin \phi = 5$

whence $\qquad\qquad\qquad\qquad K = \sqrt{26}$

and $\qquad\qquad\qquad\qquad \tan \phi = 5$

The solution may therefore be written as

$$y = e^{2x} + e^{-x/2} \cos \frac{\sqrt{3}x}{2} + \frac{\sqrt{26}}{52} \sin \left(2x + \phi\right)$$

where $\phi = \tan^{-1} 5$. We shall later see the value of this form.

EXERCISE 26.4

1. Obtain the complete solutions to the equations of Exercise 26.3.

7. The graphical representation of the solution

It is often useful to be able to visualise the graphical representation of the solution of a differential equation. In economic problems the solution is often given as a function of time, and a graph will show us how production (or whatever other variable may be represented by y) is likely to behave over a long period. The graph can usually be drawn by substitution of numerical values, but this is a slow and tedious process. By using the following notes it will often be possible to determine the principal features of the graph without actually drawing it, or working out any numerical values.

First of all there are two general principles to be noted.

(1) If the solution is of the form

$$y = f_1(x) + f_2(x)$$

then the graph of y may be obtained by adding the ordinates of the separate graphs of $y_1 = f_1(x)$ and $y_2 = f_2(x)$.

(2) If the solution is of the form

$$y = f_1(x) f_2(x)$$

then the graph of y may be obtained by multiplying the ordinates of the separate graphs $y_1 = f_1(x)$ and $y_2 = f_2(x)$.

The truth of these propositions is self-evident.

Secondly we must have some knowledge of the shapes of the graphs of certain commonly occurring functions. Some of these we have already considered elsewhere in this book. The most important at the moment are the graphs of the functions e^{ax}, $\log ax$ and $\frac{\sin}{\cos} (\omega x - \epsilon)$.

If $a>0$ then the value of e^{ax} increases as x increases, and does so increasingly rapidly, since $dy/dx = ae^{ax}$ which is an increasing function of x. If $a<0$ then, conversely, the value of e^{ax} decreases as x increases, and gets closer and closer to zero: but it is always positive. This is shown in Diagram 26.1.

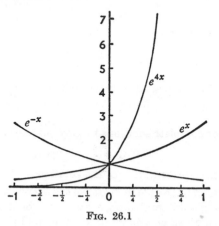

FIG. 26.1

The function $\log ax$ has a real value only if $a>0$. The value increases as x increases, but does so at a decreasing rate (since $dy/dx = a/x$ which decreases as x increases). This is shown in Diagram 26.2.

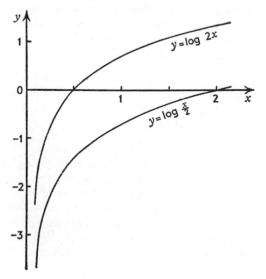

FIG. 26.2

The function $\sin(\omega x - \epsilon)$ has a regular wave-like curve with a period of $2\pi/\omega$. The effect of ϵ is to make the curve lag behind the curve $\sin \omega x$ in such a way that it crosses the horizontal axis at points where $x = \epsilon/\omega$, $(\pi + \epsilon)/\omega$, $(2\pi + \epsilon)/\omega$, ... , instead of at the points 0, π/ω, $2\pi/\omega$, ...

The function $\cos(\omega x - \epsilon)$ has a similar graph, as shown in Diagram 26.3.

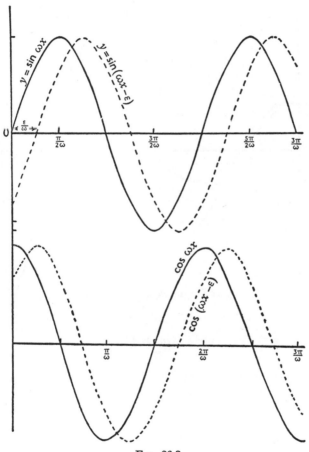

FIG. 26.3

We are now in a position to consider a product which frequently occurs in the solution of a differential equation. It is

$$y = e^{ax} \cos(\omega x - \epsilon)$$

which consists of the product of an exponential term and a trigonometric term. Clearly, for any value of x, the value of y is going to be given by multiplying the appropriate values of e^{ax} and cos $(\omega x - \epsilon)$. Now if $a = 0$, then $e^{ax} = 1$ and in this case the value of the product is simply cos $(\omega x - \epsilon)$ and the graph is as shown in Diagram 26.4, having steady oscillations of period $2\pi/\omega$.

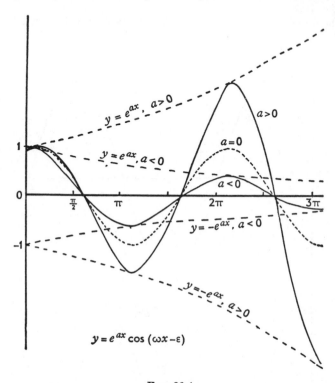

FIG. 26.4

If, however, $a > 0$ then the effect of the exponential term is to magnify the oscillations by a factor which increases at an exponential rate. The oscillations are as shown in Diagram 26.4 and are said to be explosive.

If $a < 0$ then the effect of the exponential term is to diminish the oscillations by a factor which increases at an exponential rate. The oscillations as in Diagram 26.4 are said to be damped.

We may now examine the solution to the example we have just worked, namely

$$y = e^{2x} + e^{-x/2} \cos \frac{\sqrt{3}x}{2} + \frac{\sqrt{26}}{52} \sin (2x + \phi), \quad \phi = \tan^{-1} 5$$

This consists of three parts.

The part corresponding to e^{2x} increases rapidly as x increases.

The part corresponding to $e^{-x/2} \cos (\sqrt{3}x/2)$ has damped oscillations of period $4\pi/\sqrt{3}$. For high values of x the effect of this term is negligible, because of the damping.

The part corresponding to $(\sqrt{26}/52) \sin (2x + \phi)$ consists of a regular oscillation of constant amplitude.

It is clear that when we combine these three parts the result will be an exponential growth, having superimposed upon it a damped oscillation which soon ceases to have any importance, and an oscillation of constant amplitude which (compared with the exponential term) soon becomes relatively unimportant. The result is shown in Diagram 26.5.

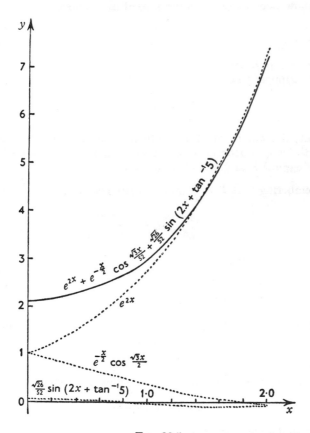

FIG. 26.5

When there is one term which becomes much more important than the other terms for large values of x we describe this term as being *dominant*. In the above example the dominant term is e^{2x}.

Some further light is shed on the graphical interpretation of solutions at the end of the next chapter.

8. An economic example

It is sometimes necessary to enquire into the speed at which one economic variable adapts itself to the changing value of itself, or of another variable. A change in interest rates may have some immediate effect on the demand for mortgages—but it takes time for the full effect to be felt. We now consider an example in which this idea plays a part.

We know that if I denotes the rate of investment

$$I = \frac{dK}{dt}$$

where K is capital stock.

Let us suppose that

$$\frac{dI}{dt} = \lambda^2(\bar{K} - K)$$

represents the way in which the rate of investment grows—the rate of growth being proportional to the difference between some desired stock of capital $\bar{K}$ and an actual stock K.

Remembering that $I = \frac{dK}{dt}$ we can write this as

$$\frac{d^2K}{dt^2} = \lambda^2(\bar{K} - K)$$

This can be written as

$$D^2K + \lambda^2 K = \lambda^2 \bar{K}$$
$$(D^2 + \lambda^2)K = \lambda^2 \bar{K} = \text{constant}$$

To find the complementary function we write the auxiliary equation as

$$p^2 + \lambda^2 = 0$$

which has two imaginary roots $\pm i\lambda$ yielding a solution

$$K = B \cos(\lambda t - \epsilon)$$

where B and ϵ are constants.

For the particular integral we solve

$$K = \frac{1}{D^2 + \lambda^2} \lambda^2 \bar{K}$$

$$= \frac{1}{\lambda^2[1 + (D^2/\lambda^2)]} \lambda^2 \bar{K}$$

$$= [1 - (D^2/\lambda^2) + \ldots] \bar{K}$$

$$= \bar{K}$$

since $D^2\bar{K} = 0$, as $\bar{K} = $ constant.

Thus the Complete Solution is

$$K = B \cos (\lambda t - \epsilon) + \bar{K}$$

This means that if

$$\frac{dI}{dt} - \lambda^2(\bar{K} - K)$$

then the capital stock K fluctuates around the desired level $\bar{K}$, with steady oscillations of period $2\pi/\lambda$. Its phase will depend on ϵ, which will be prescribed by initial conditions, as will B, which will measure the amplitude of the oscillations.

If we know that at time 0, $K = K_0$ then

$$K_0 = B \cos (-\epsilon) + \bar{K}$$
$$= B \cos \epsilon + \bar{K}$$

This is not enough to determine both B and ϵ. We need another initial condition, which is not already embodied in our equations. Suppose that we know something about

$$\frac{d^2I}{dt^2}$$

at time $t = 0$. Let this have a value α. Then as

$$K = B \cos (\lambda t - \epsilon) + \bar{K}$$

$$I = \frac{dK}{dt} = -\lambda B \sin (\lambda t - \epsilon)$$

$$\frac{dI}{dt} = -\lambda^2 B \cos (\lambda t - \epsilon)$$

$$\frac{d^2I}{dt^2} = \lambda^3 B \sin (\lambda t - \epsilon)$$

which, when $t = 0$, gives

$$\frac{d^2I}{dt^2} = -\lambda^3 B \sin \epsilon$$

We know that this is α. We thus have

$$B \cos \epsilon = K_0 - \bar{K}$$

$$B \sin \epsilon = -\frac{\alpha}{\lambda^3}$$

from which to determine B and ϵ. Squaring and adding quickly shows that

$$B^2 = (K_0 - \bar{K})^2 + \frac{\alpha^2}{\lambda^6}$$

while division shows that

$$\tan \epsilon = -\frac{\alpha}{\lambda^3(K_0 - \bar{K})}$$

Our solution

$$K = B \cos (\lambda t - \epsilon) + \bar{K}$$

is thus quite unambiguous.

The rate of investment, $I = \dfrac{dK}{dt}$, is

$$-\lambda B \sin (\lambda t - \epsilon)$$

where B and ϵ are determined as above. This, too, is a steadily oscillating function, which is negative for half of the time—implying disinvestment.

CHAPTER XXVII

DIFFERENCE EQUATIONS

And each the other's difference bears.

ANDREW MARVELL : *Eyes and Tears*

1. Introduction

A general definition of a difference equation was given in the last chapter. We shall now restrict ourselves to a study of certain kinds of difference equations which frequently occur in economics. Because most of the equations we are likely to meet involve time, we shall find it convenient to refer to an independent variable which increases by equal amounts as " time ", but the method of solution is independent of the nature of this variable provided that it does increase by equal amounts. The equations may relate the value of some variable (such as income) at one moment to the value at some other moment. More usually they will relate the value during a period to the value during some other period. The important point is that *either* time is subdivided into a number of *equal* periods which are such that we can meaningfully compare the income (say) of one period with the income of another period *or* a series of *evenly spaced* instants is chosen such that we can meaningfully compare the population (say) at one instant with the population at another. In what follows we shall think of time as being divided into periods, partly for verbal convenience and partly to facilitate reference to economic examples. We may think of each period as being a day, a month, two months, or any other finite period.

Let us suppose that the income of a certain society during the period t (e.g., this week) is given by Y_t and that the income of the same society during the previous period (e.g., last week) is given by Y_{t-1}. Then if these incomes are related by an equation such as

$$Y_t = a Y_{t-1} + A$$

where a and A are constants we may say that the income Y_t satisfies a difference equation of the *first order*. We say that it is of the first order because we can determine the income in any period if we know the income in the previous period.

If the equation is of the form

$$Y_t = a Y_{t-1} + b Y_{t-2} + A$$

then it is of the *second order*, since we need to go back two periods before we have enough information to calculate the income in a given period. The equation

$$Y_t = aY_{t-2} + A$$

is also of the second order, for the same reason. The equation

$$Y_t = \sqrt{aY_{t-1} + bY_{t-3}} + t$$

is of the third order, and the equation

$$Y_t = (aY_{t-1} + bY_{t-2})^2 + cY_{t-3} - Y_{t-5} + t^2$$

is of the fifth order.

More generally, we define a difference equation of the nth order as

$$f(Y_t, Y_{t-1}, Y_{t-2}, \dots, Y_{t-n}, t) = 0$$

2. Alternative forms

Before considering how to solve a difference equation we should mention some alternative forms in which they are sometimes written.

The first of these is simply a device to facilitate printing. Instead of writing Y_t, Y_{t-1}, ... , some authors write $Y(t)$, $Y(t-1)$, ... , with the convention that $Y(t)$ means " the value of Y at time t ". This certainly helps the printer, but the beginner in mathematics is likely to be confused by the fact that the bracket notation is already used to denote multiplication and functional relationships. For this reason we shall not use this notation in this book. It is, however, used by Baumol in his *Economic Dynamics* and in many of the journals.

The second alternative presentation arises out of the fact that just as period $(t - n)$ is n periods behind period t, so period t is n periods behind period $(t + n)$, and so

if $\qquad Y_t = Y_{t-n} + A \quad$ for all values of t

then $\qquad Y_{t+n} = Y_t \quad + A \quad$ for all values of t

and $\qquad Y_{t+n-r} = Y_{t-r} + A \quad$ for all values of t and for any integer r.

All of these equations are identical, as will be seen if, for example, we put $n = 3$ and $A = 6$. The first then becomes

$$Y_t = Y_{t-3} + 6 \qquad \text{for all values of } t$$

and so, if we put $t = 3$

$$Y_3 = Y_0 + 6$$
$$Y_4 = Y_1 + 6$$
$$Y_6 = Y_3 + 6 = Y_0 + 12$$
$$Y_7 = Y_4 + 6 = Y_1 + 12 \quad \text{etc.}$$

The second becomes

$$Y_{t+3} = Y_t + 6$$

and so, putting $t = 0$, we have

$$Y_3 = Y_0 + 6$$

as before.

The third could yield (by putting $r = 1$)

$$Y_{t+3-1} = Y_{t-1} + K$$

i.e., $$Y_{t+2} = Y_{t-1} + K$$

If we put $t = 1$ we have

$$Y_3 = Y_0 + K$$

as before.

More generally this means that

$$f(Y_t, Y_{t-1}, Y_{t-2}, \ldots, Y_{t-n}, t) = 0$$

may be replaced by

$$f(Y_{t+r}, Y_{t+r-1}, Y_{t+r-2}, \ldots, Y_{t+r-n}, t) = 0$$

where r is any positive or negative integer.

The third alternative involves a different conception altogether. Instead of expressing Y_t in terms of $Y_{t-1}, \ldots, Y_{t-n}$, we express it in terms of the differences $\Delta Y_t, \Delta^2 Y_t, \ldots, \Delta^n Y_t$ where these are defined as follows :

We let the difference $Y_{t+1} - Y_t$ be denoted by ΔY_t

$$Y_{t+2} - Y_{t+1} \qquad\qquad \Delta Y_{t+1}$$
$$\vdots \qquad\qquad\qquad \vdots$$
$$Y_{t+n} - Y_{t+n-1} \qquad\qquad \Delta Y_{t+n-1}$$

We call these differences the *first differences*.

Let us now consider the difference between two successive differences.

We let the *second difference* $\Delta Y_{t+1} - \Delta Y_t$ be denoted by $\Delta^2 Y_t$

$$\Delta Y_{t+2} - \Delta Y_{t+1} \qquad\qquad \Delta^2 Y_{t+1}$$
$$\vdots \qquad\qquad\qquad \vdots$$
$$\Delta Y_{t+n} - \Delta Y_{t+n-1} \qquad\qquad \Delta^2 Y_{t+n-1}$$

Similarly we may think of *third differences* where

$$\Delta^2 Y_{t+1} - \Delta^2 Y_t = \Delta^3 Y_t$$

and so on.

Now $\qquad Y_{t+1} = Y_t + \varDelta Y_t$

and $\qquad \begin{aligned} Y_{t+2} &= Y_{t+1} + \varDelta Y_{t+1} \\ &= (Y_t + \varDelta Y_t) + \varDelta Y_{t+1} \end{aligned}$

But $\qquad \varDelta Y_{t+1} = \varDelta Y_t + \varDelta^2 Y_t \qquad$ (from definition of $\varDelta^2 Y_t$)

and so $\qquad Y_{t+2} = (Y_t + \varDelta Y_t) + (\varDelta Y_t + \varDelta^2 Y_t)$

i.e., $\qquad Y_{t+2} = Y_t + 2\varDelta Y_t + \varDelta^2 Y_t$

A similar analysis will show that

$$\begin{aligned} Y_{t+3} &= Y_{(t+1)+2} \\ &= Y_{t+1} + 2\varDelta Y_{t+1} + \varDelta^2 Y_{t+1} \\ &= (Y_t + \varDelta Y_t) + 2(\varDelta Y_t + \varDelta^2 Y_t) + (\varDelta^2 Y_t + \varDelta^3 Y_t) \\ &= Y_t + 3\,\varDelta Y_t + 3\,\varDelta^2 Y_t + \varDelta^3 Y_t \end{aligned}$$

and so on. More generally it may be shown that

$$Y_{t+n} = Y_t + {}_nC_1 \varDelta Y_t + {}_nC_2 \varDelta^2 Y_t + \ldots + {}_nC_{n-1} \varDelta^{n-1} Y_t + \varDelta^n Y_t$$

where the successive coefficients are those of the binomial expansion. It follows that any function

$$f(Y_t, Y_{t+1}, \ldots, Y_{t+n}, t) = 0$$

may be written in terms of Y_t and its first n differences, resulting in some equivalent expression

$$\phi(Y_t, \varDelta Y_t, \ldots, \varDelta^n Y_t, t) = 0$$

For example, the equation

$$Y_t - Y_{t+1} + 2Y_{t+2} = 6t + 4$$

may be put as

$$Y_t - (Y_t + \varDelta Y_t) + 2(Y_t + 2\varDelta Y_t + \varDelta^2 Y_t) = 6t + 4$$

giving

$$2Y_t + 3\,\varDelta Y_t + 2\,\varDelta^2 Y_t = 6t + 4$$

The equation may also, of course, be written as

$$Y_{t-1} - Y_t + 2Y_{t+1} = 6(t-1) + 4 = 6t - 2$$

or as $\qquad Y_{t-2} - Y_{t-1} + 2Y_t = 6(t-2) + 4 = 6t - 8$

EXERCISE 27.1

1. Express the following equations in the alternative form :

$$\phi(Y_t, \varDelta Y_t, \varDelta^2 Y_t, \ldots, \varDelta^n Y_t, t) = 0$$

(i) $Y_t + Y_{t+1} + Y_{t+2} = 6t$ (ii) $2Y_t + Y_{t+1} + Y_{t+2} = 3t - 4$

(iii) $Y_t + Y_{t-1} + 2Y_{t-2} = 3t + 2$ (iv) $2Y_t - Y_{t-1} - Y_{t-2} = 8t + 1$

2. Express the following equations in the alternative form :

$$F(Y_t, Y_{t+1}, Y_{t+2}, ..., Y_{t+n}, t) = 0$$

(i) $Y_t + 2 \Delta Y_t = 3t$ (ii) $Y_t + \Delta^2 Y_t = 3t$

(iii) $Y_t - \Delta Y_t + \Delta^2 Y_t = 4t$ (iv) $2Y_t - 2 \Delta Y_t + \Delta^2 Y_t = 4t$

3. The linear difference equation with constant coefficients

In this chapter we are going to be particularly concerned with difference equations which can be written in the form

$$a_0 Y_t + a_1 Y_{t-1} + a_2 Y_{t-2} + ... + a_n Y_{t-n} = f(t)$$

where the a's are constants. These equations are stated to be linear with constant coefficients. They are linear because they are of the first degree in all Y's. There is no term involving the product of two or Y's, or Y_t^2, or anything else that would mean a departure from linearity. Some of the examples in Section I of this chapter are not linear.

The solution to an equation of this kind falls into two parts, in much the same way as the solution to a differential equation. The complete solution is obtained by adding any *particular integral* of the above equation to the general integral of what we now call the *homogeneous form* obtained by putting $f(t) = 0$—i.e., by adding the particular integral to the *complementary function*.

We must, therefore, consider methods for finding the particular integral and the complementary function. We shall then examine one equation more fully, in order to illustrate the complete method, before passing on to consider the graphical interpretation of solutions. Since the complementary function is often simpler to find than the particular integral we will begin by considering this.

Finding the Complementary Function

We take the homogeneous form (obtained by putting $f(t) = 0$)

$$a_0 Y_t + a_1 Y_{t-1} + a_2 Y_{t-2} + ... + a_n Y_{t-n} = 0$$

and rewrite it in the form of the *auxiliary equation*

$$a_0 \lambda^n + a_1 \lambda^{n-1} + a_2 \lambda^{n-2} + ... + a_n = 0$$

This equation has n roots, which may be real or complex. It can be shown that provided all the roots are different then the complementary function is given by

$$Y_t = A_1 \lambda_1{}^t + A_2 \lambda_2{}^t + ... + A_n \lambda_n{}^t$$

where $\lambda_1, \lambda_2, ... , \lambda_n$ are the roots, and the A's are arbitrary constants, which must later be made to satisfy the initial conditions.

The proof is similar to the proof of the comparable theorem (involving $y = e^{px}$) for differential equations.

If two roots coincide then (as in the case of a differential equation) the solution is

$$Y_t = (A_1 + A_2 t)\lambda_1{}^t + A_3 \lambda_3{}^t + \ldots + A_n \lambda_n{}^t$$

while if three roots coincide the complementary function includes the term

$$(A_1 + A_2 t + A_3 t^2)\lambda_1{}^t$$

and so on.

Example

The equation $\qquad Y_t - 3Y_{t-1} + 4Y_{t-3} = 0$

has an auxiliary equation

$$\lambda^3 - 3\lambda^2 + 4 = 0$$

i.e., $\qquad\qquad (\lambda - 2)^2(\lambda + 1) = 0$

yielding roots $\qquad \lambda = 2$ (twice) and $\lambda = -1$

therefore, C.F. is $\qquad Y_t = (A_1 + A_2 t)2^t + A_3(-1)^t$

If some of the roots are complex it is usually advantageous to obtain the solution in a trigonometric form, as in the case of a differential equation. We shall shortly illustrate this method.

Finding the Particular Integral

(1) When $f(t) = \alpha = $ constant.
We take the original equation

$$a_0 Y_t + a_1 Y_{t-1} + \ldots + a_n Y_{t-n} = \alpha$$

and suppose that a particular integral is given by

$$Y_t = Y$$

for all values of t, so that Y_t is constant.

Substitution yields

$$(a_0 + a_1 + \ldots + a_n) Y = \alpha$$

and so $\qquad\qquad Y = \dfrac{\alpha}{a_0 + a_1 + \ldots + a_n}$

Example

$$3Y_t + 2Y_{t-1} + 5Y_{t-2} = 6$$

P.I. $\qquad\qquad 3Y + 2Y + 5Y = 6$

whence $\qquad\qquad Y = \tfrac{6}{10}$

If, however, $a_0 + a_1 + \ldots + a_n = 0$ then this method leads to a nonsense solution, as will readily be seen if we consider

$$3Y_t + 2Y_{t-1} - 5Y_{t-2} = 6$$

In such a case we try

$$Y_t = Yt$$

Example

P.I.
$$3Y_t + 2Y_{t-1} - 5Y_{t-2} = 6$$
$$3Yt + 2Y(t-1) - 5Y(t-2) = 6$$
$$3Yt + 2Yt - 2Y - 5Yt + 10Y = 6$$
$$8Y = 6$$
$$Y = \frac{6}{8}$$

i.e., P.I. is
$$Y_t = \frac{6t}{8}$$

If this device also fails we try $Y_t = Yt^2$ as in the following example.

Example

$$3Y_t - 5Y_{t-2} + 2Y_{t-5} = 6$$
$$3Yt^2 - 5Y(t-2)^2 + 2Y(t-5)^2 = 6$$

whence
$$30Y = 6$$

$$Y = \frac{1}{5}$$

and P.I. is
$$Y_t = \frac{t^2}{5}$$

If the equation is of order n we may possibly have to persevere until at last we try $Y_t = Yt^n$ which will certainly give us a particular integral if all previous attempts have failed, provided that $f(t) = \alpha$.

(2) When $f(t) = \alpha^t$, α being a constant.
It can be shown that in this case a particular integral is

$$\frac{\alpha^{n+t}}{a_0\alpha^n + a_1\alpha^{n-1} + \ldots + a_n}$$

Example

$$Y_t + 4Y_{t-1} + 3Y_{t-2} = 3^t$$

P.I. is
$$Y_t = \frac{3^{2+t}}{3^2 + 4(3) + 3} = \frac{3^{t+1}}{8}$$

(3) Other cases.

When $f(t)$ takes neither of the above forms it may be possible to find a particular integral by the use of some operational method. There is a useful introduction to these methods in Professor R. G. D. Allen's *Mathematical Economics*, Chapter 6, and Appendix A. For other methods the student may consult the works listed in the bibliography.

EXERCISE 27.2

1. Find the Particular Integrals in the following cases :

    (i) $3Y_t + 3Y_{t-1} + 4Y_{t-2} = 7$

    (ii) $4Y_t - 5Y_{t-1} + Y_{t-2} = 8$

    (iii) $5Y_t - 3Y_{t-1} + Y_{t-2} - 3Y_{t-3} = 4$

    (iv) $4Y_t + 2Y_{t-1} - 5Y_{t-2} = 7$

    (v) $3Y_t + 3Y_{t-1} + 4Y_{t-2} = 7^t$

2. Find the Complementary Functions in the following cases :

    (i) $Y_t - 5Y_{t-1} + 6Y_{t-2} = 7$

    (ii) $Y_t - 6Y_{t-1} + 9Y_{t-2} = 3$

    (iii) $Y_t + 5Y_{t-1} + 6Y_{t-2} = 4^t$

    (iv) $Y_t + 9Y_{t-2} = 5^t$

    (v) $Y_t - Y_{t-1} + 9Y_{t-2} - 9Y_{t-3} = 4^t$

COMPLETE EXAMPLE

We wish to solve the fifth order difference equation

$$Y_t - 10Y_{t-1} + 49Y_{t-2} - 134Y_{t-3} + 188Y_{t-4} - 104Y_{t-5} = \alpha^t$$

We begin by taking the auxiliary equation

$$\lambda^5 - 10\lambda^4 + 49\lambda^3 - 134\lambda^2 + 188\lambda - 104 = 0$$

which can be factorised by the methods of Chapter I to yield

$$(\lambda - 2)^3 (\lambda^2 - 4\lambda + 13) = 0$$

showing that there is the triple factor

$$\lambda = 2$$

and the two complex conjugate factors

$$\lambda = 2 + 3i \quad \text{and} \quad \lambda = 2 - 3i$$

The part of the solution corresponding to the triple root is

$$(A + Bt + Ct^2)2^t$$

while the part corresponding to the complex roots may be written

$$D(2+3i)^t + E(2-3i)^t = (\sqrt{2^2+3^2})^t[e \cos tR + f \sin tR]$$
$$= G(\sqrt{2^2+3^2})^t \cos(tR - \phi)$$

where $\cos R = 2/\sqrt{2^2+3^2}$, and e, f, G and ϕ are related arbitrary constants. It follows that the complementary function is

$$Y_t = (A + Bt + Ct^2)2^t + G(\sqrt{2^2+3^2})^t \cos(tR - \phi)$$

We now have to find the particular integral. It is

$$\frac{\alpha^{t+5}}{\alpha^5 - 10\alpha^4 + 49\alpha^3 - 134\alpha^2 + 188\alpha - 104} = \frac{\alpha^{t+5}}{(\alpha-2)^3(\alpha^2 - 4\alpha + 13)}$$

The complete solution is therefore

$$Y_t = (A + Bt + Ct^2)2^t + G(\sqrt{13})^t \cos(tR - \phi) + \frac{\alpha^{t+5}}{(\alpha-2)^3(\alpha^2 - 4\alpha + 13)}$$

where $\cos R = \dfrac{2}{\sqrt{13}}$.

This solution involves arbitrary constants. Their values will depend on the given initial conditions. Let us suppose that we are told that

$$Y_0 = \frac{\alpha^5}{(\alpha-2)^3(\alpha^2 - 4\alpha + 13)} = \frac{\alpha^5}{[\alpha]}$$

(where we write the denominator as $[\alpha]$ to save trouble)

$$Y_1 = \frac{\alpha^6}{[\alpha]} + 6 + 3\sqrt{3}$$

$$Y_2 = \frac{\alpha^7}{[\alpha]} + 41 + 12\sqrt{3}$$

$$Y_3 = \frac{\alpha^8}{[\alpha]} + 174 + 9\sqrt{3}$$

$$Y_4 = \frac{\alpha^9}{[\alpha]} + 519 - 120\sqrt{3}$$

These five initial conditions should enable us to determine the values of the five constants, A, B, C, G and ϕ.

Since $Y_t = (A + Bt + Ct^2)2^t \quad + G \quad (\sqrt{13})^t \cos(tR - \phi) + \dfrac{\alpha^{t+5}}{[\alpha]}$

$Y_0 = (A \qquad\qquad) \quad + G \qquad\qquad \cos(\quad -\phi) + \dfrac{\alpha^5}{[\alpha]}$

$Y_1 = (A + B + C \quad)2 \quad + G\sqrt{13} \qquad \cos(R - \phi) + \dfrac{\alpha^6}{[\alpha]}$

$Y_2 = (A + 2B + 4C)4 \quad + G \cdot 13 \cdot \qquad \cos(2R - \phi) + \dfrac{\alpha^7}{[\alpha]}$

$Y_3 = (A + 3B + 9C)8 \quad + G \cdot 13\sqrt{13} \ \cos(3R - \phi) + \dfrac{\alpha^8}{[\alpha]}$

$Y_4 = (A + 4B + 16C)16 + G \cdot 169 \qquad \cos(4R - \phi) + \dfrac{\alpha^9}{[\alpha]}$

If we now substitute the given initial conditions, we will find that the terms containing α disappear, and we are left with

$$A \qquad\qquad\qquad + G\cos(-\phi) \qquad\qquad = 0$$
$$2(A + B + C) \qquad + \sqrt{13}G\cos(R - \phi) \qquad = 6 + 3\sqrt{3}$$
$$4(A + 2B + 4C) \quad + 13G\cos(2R - \phi) \qquad = 41 + 12\sqrt{3}$$
$$8(A + 3B + 9C) \quad + 13\sqrt{13}G\cos(3R - \phi) = 174 + 9\sqrt{3}$$
$$16(A + 4B + 16C) + 169G\cos(4R - \phi) \quad = 519 - 120\sqrt{3}$$

It is shown in Appendix 10 that if $\cos R = \dfrac{2}{\sqrt{13}}$ then

$$\cos(R - \phi) \ = \frac{2}{\sqrt{13}}\cos\phi + \frac{3}{\sqrt{13}}\sin\phi$$
$$\cos(2R - \phi) = \frac{12}{13}\sin\phi - \frac{5}{13}\cos\phi$$
$$\cos(3R - \phi) = \frac{9}{13\sqrt{13}}\sin\phi - \frac{46}{13\sqrt{13}}\cos\phi$$
$$\cos(4R - \phi) = -\frac{120}{169}\sin\phi - \frac{119}{169}\cos\phi$$

and, of course,

$$\cos(-\phi) = \cos\phi$$

We, therefore, have

$$A \qquad\qquad\qquad + G\cos\phi \qquad\qquad\qquad\qquad = 0$$
$$2A + 2B + 2C \qquad + G(2\cos\phi + 3\sin\phi) \qquad\qquad = 6 + 3\sqrt{3}$$
$$4A + 8B + 16C \qquad + G(-5\cos\phi + 12\sin\phi) \qquad = 41 + 12\sqrt{3}$$
$$8A + 24B + 72C \qquad + G(-46\cos\phi + 9\sin\phi) \qquad = 174 + 9\sqrt{3}$$
$$16A + 64B + 256C + G(-119\cos\phi - 120\sin\phi) = 519 - 120\sqrt{3}$$

The first of these equations tells us that

$$G \cos \phi = -A$$

We use this to remove $G \cos \phi$ from the remaining four equations, and at the same time write

$$G \sin \phi = K$$

for convenience. We thus obtain

$$
\begin{aligned}
2B + \quad 2C + \quad 3K &= 6 + 3\sqrt{3} \\
+\,9A + \quad 8B + \quad 16C + \quad 12K &= 41 + 12\sqrt{3} \\
54A + 24B + \quad 72C + \quad 9K &= 174 + 9\sqrt{3} \\
135A + 64B + 256C - 120K &= 519 - 120\sqrt{3}
\end{aligned}
$$

These four equations may now be solved for A, B, C and K by the use of determinants, or by using the first of them to eliminate B from the remaining three : and then using one of these to eliminate C from the remaining two—thus obtaining two equations in A and K which may be solved simultaneously. The solution is left as an exercise to the reader, who should obtain

$$A = 1, \; B = 2, \; C = 1, \; K - \sqrt{3}$$

It may be easily verified that these values satisfy the four equations We have put

$$K = G \sin \phi \text{ and have that } A = -G \cos \phi$$

and so $\qquad \sqrt{3} = G \sin \phi$ and $\qquad\qquad 1 = -G \cos \phi$

it follows that

$$
\begin{aligned}
G^2 \sin^2 \phi + G^2 \cos^2 \phi &= (\sqrt{3})^2 + (1)^2 \\
\text{i.e.,} \qquad G^2 (\sin^2 \phi + \cos^2 \phi) &= 4 \\
G^2 &= 4 \\
G &= \pm 2
\end{aligned}
$$

If $G = +2$, then

$$
\begin{aligned}
2 \sin \phi &= \sqrt{3} \\
-2 \cos \phi &= 1
\end{aligned}
$$

whence

$$\phi = \frac{2\pi}{3}$$

If $G = -2$, then

$$
\begin{aligned}
-2 \sin \phi &= \sqrt{3} \\
2 \cos \phi &= 1
\end{aligned}
$$

and so

$$\phi = -\frac{\pi}{3}$$

o

In the former case the trigonometric term of the solution

$$2(\sqrt{13})^t \cos (tR - \phi)$$

becomes
$$2\left(\sqrt{13}\right)^t \cos \left(tR - \frac{2\pi}{3}\right)$$

and in the latter case

$$-2\left(\sqrt{13}\right)^t \cos \left(tR + \frac{\pi}{3}\right)$$

These are identical expressions since $\cos (\theta - \pi) \equiv -\cos \theta$.
We therefore have the complete solution

$$Y_t = (1 + 2t + t^2)2^t + 2(\sqrt{13})^t \cos \left(tR - \frac{2\pi}{3}\right) + \frac{\alpha^{t+5}}{(\alpha - 2)^3 (\alpha^2 - 4\alpha + 13)}$$

EXERCISE 27.3

Find the complete solutions of the following equation

(i) $y_t - 5y_{t-1} + 6y_{t-2} = 7$, given $y_0 = 5$, $y_1 = 9$.

(ii) $y_t - 6y_{t-1} + 9y_{t-2} = 3$, given $y_0 = 1$, $y_1 = 3$.

(iii) $4y_t - 5y_{t-1} + y_{t-2} = 8$, given $y_0 = y_1 = 4$.

(iv) $y_t + 9y_{t-2} = 5_t$, given $y_0 = 59/34$, $y_1 = 23/34$.

4. The graphical representation of solutions

(1) The solution $y_t = A\lambda^t$

The form taken by this solution will depend on the values of A and λ. We must consider each possible case separately.

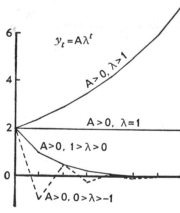

FIG. 27.1

(a) $A > 0$, $\lambda > 1$. Since $\lambda > 1$, λ^t increases indefinitely with t. The effect of A is simply to multiply all values by a constant amount. The graph is as shown in Diagram 27.1 which is of $y_t = 2(1\cdot2)^t$. We call this an *explosive solution* because y_t has no upper limit.

(b) $A > 0$, $\lambda = 1$. In this case $y_t = A$ for all values of t. The solution is shown in Diagram 27.1.

(c) $A > 0$, $1 > \lambda > 0$. Here we have λ positive but less than unity. It follows that λ^t is always positive, less than unity, and approaching zero as t increases indefinitely. The effect of A is as in case (a). The values y_t are always positive, and decrease towards zero as in Diagram 27.1, which shows $y_t = 2(0\cdot5)^t$.

(d) $A > 0$, $\lambda = 0$. This is a trivial solution, y_t being zero for all t.

(e) $A > 0$, $0 > \lambda > -1$. Since λ is negative, odd powers of λ will be negative but even powers positive. It follows that the sign of $y_t = A\lambda^t$ will alternate, being positive for $t = 0, 2, 4, \ldots$, and negative for $t = 1, 3, 5, \ldots$. If we ignore the sign, we will have that the absolute values of y_t decrease towards zero for the reasons given in (c) above: but now there will be a positive y_0, a smaller negative y_1, a smaller positive y_2, and so on, as shown in Diagram 27.1. We refer to such a solution as one exhibiting *damped oscillations*.

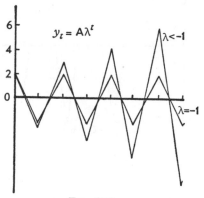

FIG. 27.2

(f) $A > 0$, $\lambda = -1$. Once again there are oscillations, because λ is negative: but now all the values of y_t are *numerically* equal, and the solution shows a constant oscillation A, $-A$, A, $-A$, $\ldots$, as in Diagram 27.2

(g) $A > 0$, $\lambda < -1$. In this case the oscillations are *explosive*. Each value of y_t is numerically larger than the previous one, but the signs alternate. Diagram 27.2 depicts the solution

$$y_t = 2(-1\cdot2)^t$$

(h) $A = 0$ corresponds to the solution $y_t = 0$ and is trivial.

(i)–(o) $A < 0$. Here we have seven further cases, similar to the cases (a) to (g), except that where there was previously a positive y_t there is now a negative y_t, and vice versa.

(2) *The solution $y_t = At\lambda^t$*

This solution is similar to the preceding one but the presence of the factor t results in an element of growth which would not otherwise be present. This can be seen most easily by considering case (b) above. Now the solution will be $y_t = At$, giving $y_0 = 0$, $y_1 = A$, $y_2 = 2A$, $\ldots$, in place of the constant solution $y_t = A$.

The various cases are outlined below :

(a) $A > 0$, $\lambda > 1$. Similar to case 1 (a) but explodes more rapidly, the values of $y_t = 2t(1\cdot2)^t$ being 0, 2·4, 5·76, ... , compared with the values of $y_t = 2(1\cdot2)^t$ which are 2, 2·4, 2·88, ...

(b) $A>0$, $\lambda=1$. Explosive.

(c) $A>0$, $1>\lambda>0$. This case is particularly interesting,

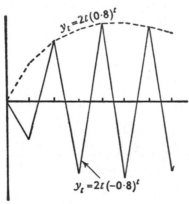

$y_t = 2t(0\cdot8)^t$

$y_t = 2t(-0\cdot8)^t$

Fig. 27.3

because although the powers λ^t become smaller and tend to zero, the value of t increases indefinitely. The reader may care to show (by considering the tests in Chapter II) that provided $0<\lambda<\epsilon<1$, then eventually the values of $t\lambda^t$ tend to zero for larger values of t. Two numerical examples of this are given below, and the solution

$$y_t = 2t(0\cdot8)^t$$

is graphed in Diagram 27.3. The point is that after the first few terms the damping effect of λ^t is more powerful than the explosive effect of t.

t	0	1	2	3	4	5	6	7
$y_t = 2t(0\cdot5)^t$	0	1	1	0·75	0·50	0·31	0·19	0·11
$y_t = 2t(0\cdot8)^t$	0	1·6	2·56	3·07	3·27	3·28	3·15	2·94

(d) $A>0$, $\lambda=0$ is trivial.

(e) $A>0$, $0>\lambda>-1$. Here the remarks of 2 (c) may be applied to case 1 (e). There are oscillations which may appear to be explosive for the first few terms but eventually become damped. Diagram 27.3 shows the solution $y=2t(-0\cdot8)^t$.

(f) $A>0$, $\lambda=-1$. The oscillations are now explosive, the values of y_t being 0, $-A$, $+2A$, $-3A$, $+4A$, ...

(g) $A>0$, $\lambda<-1$. The effect of the factor t is to exaggerate the explosive property of the oscillations.

(h) $A=0$ is trivial.

(i) –(o) $A<0$ may be examined similarly.

(3) *The solution* $y_t = At^p\lambda^t (p>0)$

The general approach to this solution is the same as that to the solution $y_t = At\lambda^t$. Provided p is constant and finite then *eventually* the product $At^p\lambda^t$ will converge towards zero if $|\lambda|<\epsilon<1$. (Cf. cases (c), (e), (k) and (m) of solution (2).) In all other cases the effect of t^p is to introduce or to exaggerate an explosive tendency. Diagram 27.4 shows the course of $y_t = 2t^3(-0.5)^t$, for which the table is

t	0	1	2	3	4	5	6	7	8	9	10
$(-0.5)^t$	1	$-\frac{1}{2}$	$\frac{1}{4}$	$-\frac{1}{8}$	$\frac{1}{16}$	$-\frac{1}{32}$	$\frac{1}{64}$	$-\frac{1}{128}$	$\frac{1}{256}$	$-\frac{1}{512}$	$\frac{1}{1024}$
t^3	0	1	8	27	64	125	216	343	512	729	1000
$y_t = 2t^3(-0.5)^t$	0	-1	4	$-6\frac{3}{4}$	8	$-7\frac{13}{16}$	$6\frac{3}{4}$	$-5\frac{23}{64}$	4	$-2\frac{217}{256}$	$1\frac{61}{64}$

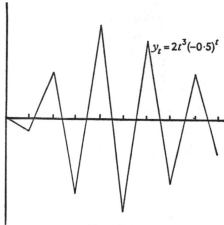

$$y_t = 2t^3(-0.5)^t$$

FIG. 27.4

4) *The solution* $y_t = A \cos(\omega t - \phi)$

If t were allowed to increase continuously the solution would have a smooth curve as shown in Diagram 27.5. This curve would have a

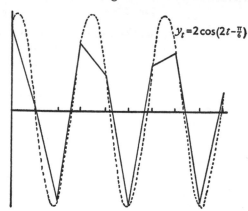

$$y_t = 2\cos\left(2t - \tfrac{\pi}{6}\right)$$

FIG. 27.5

period of $2\pi/\omega$. As we saw in Chapter IV, the effect of ϕ is simply to shift the curve bodily, and we call ϕ the *phase*. Two solutions which are " out of phase " have different ϕ's. If, however, the ϕ's differ by a multiple of $2\pi/\omega$, then the one curve is an exact number

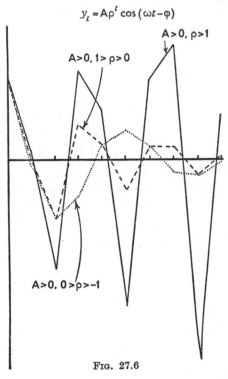

$$y_t = A\rho^t \cos(\omega t - \varphi)$$

A>0, ρ>1

A>0, 1> ρ>0

A>0, 0 >ρ>−1

FIG. 27.6

of cycles ahead of the other, and the two solutions are " in phase ".

In the solution of a difference equation, t normally increases by finite amounts. The values of y_t will give a graph such that all points on it lie on the curve of Diagram 27.5 which illustrates the solution $y_t = 2 \cos[2t - (\pi/6)]$ radians. The important point to notice is that although the values of y_t do not *alternate* in sign they do exhibit fluctuations which are neither explosive nor damped, fitting as they do, into the dotted curve of constant amplitude.

(5) *The solution* $y_t = A\rho^t \cos(\omega t - \phi)$

This is a particularly important solution, and may be recognised as a combination of solutions (1) and (4). We shall consider it in the same way as we considered solution (1).

(a) $A > 0$, $\rho > 1$. The oscillations of case (4) will explode, since as t increases y_t involves a factor ρ^t which increases indefinitely. See Diagram 27.6.

(b) $A > 0$, $\rho = 1$ leads us back to solution (4).

(c) $A > 0$, $1 > \rho > 0$ leads to damped oscillations of the same period and phase as in solution (4). The damping is due to the presence of the factor ρ^t which tends to zero. See Diagram 27.6.

(d) $A > 0$, $\rho = 0$ is trivial.

(e) $A > 0$, $0 > \rho > -1$ leads once again to damped oscillations, because the numerical value of ρ^t decreases towards zero. There are, however, complications. The sign of ρ^t alternates. The sign of the periodic part is sometimes positive and sometimes negative. Sometimes the signs of each part will agree, to give a positive product; sometimes they will differ to give a negative product. Diagram 27.6 shows the solution $y_t = 2(-0.8)^t \cos[2t - (\pi/6)]$.

(f) $A > 0$, $\rho = -1$. There is no damping, but otherwise the remarks of (e) above hold true.

(g) $A > 0$, $\rho < -1$. There will be explosive oscillations. The remarks of (e) above hold true if " damped " is replaced by " explosive. "

(6) The solution $y_t = (A + Bt)\lambda^t$

This is a combination of solutions (1) and (2). When we have a combination of this kind we should always ask two questions. First of all, do the different parts of the solution tend to reinforce each other, each moving y_t in the same way, or to weaken each other? Secondly, is there some part of the solution which dominates the rest of it when t becomes large? It happens that in this solution the two questions are easily answered. If we consider the remarks already made about solutions (1) and (2) we will see that, provided A and B are of the same sign then the two parts of the solution reinforce each other. If they are of opposite sign they weaken each other. We will also see that no matter what the numerical values of A and B may be, eventually, Bt is bound to be very large compared with A, and the part of the solution due to $Bt\lambda^t$ will become much more important than the part due to $A\lambda^t$.

These remarks indicate a general approach, and the detail may now be examined as in solution (2). There is, however, a different way of looking at it.

We can write
$$y_t = (A + Bt)\lambda^t$$
$$= A\left(1 + \frac{B}{A}t\right)\lambda^t$$
$$= A(1 + Ct)\lambda^t \qquad \text{where } C = B/A$$

Now if A and B are the same sign $C>0$ and the term $(1+Ct)$ is bound to be positive. It also increases as t increases. The effect of the bracket $(1+Ct)$ is therefore to act on solution (1) in much the same way as t acted upon it in solution (2).

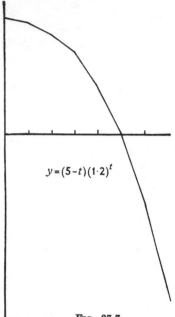

$$y=(5-t)(1\cdot2)^t$$

If, however, A and B are of opposite sign then C is negative. If $|A|>|B|$ then for small values of t the bracket $(1+Ct)$ will be positive but decreasing, and will result in a certain damping of the earlier terms which solution (1) would provide. Eventually, however, (and immediately if $|B|>|A|$) the bracket $(1+Ct)$ will become negative, and increasingly so. The result will then be to change the signs of the terms in the solution (1) and to introduce an explosive element which will be more than balanced by damping if $|\lambda|<1$.

Fig. 27.7

Diagram 27.7 shows the solution $y_t=(5-t)(1\cdot2)^t$.

(7) *The solution* $y_t=(A+Bt+Ct^2)\lambda^t$

This may be approached similarly. Whatever may happen in the earlier terms, the solution will eventually be dominated by $Ct^2\lambda^t$ and become almost indistinguishable from $y_t=C_t^2\lambda^t$.

(8) *The solution* $y_t=A_1\lambda_1{}^t+A_2\lambda_2{}^t$

To consider this fully we would have to consider all possible combinations of the fifteen cases examined in solution (1) in respect of $A_1\lambda_1{}^t$ with fifteen similar cases in respect of $A_2\lambda_2{}^t$. We might have, for example, a case in which $A_1>0, 1>\lambda_1>0$, $A_2<0, \lambda_2<-1$, which would (see solution (1)) result in a damped positive component

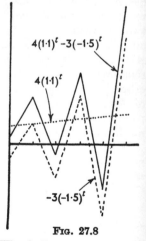

$4(1\cdot1)^t-3(-1\cdot5)^t$

$4(1\cdot1)^t$

$-3(-1\cdot5)^t$

Fig. 27.8

to be added to an explosive oscillation. We shall not examine all possible cases, but it will be useful to note a few points.

If λ_1 and λ_2 are each positive, then the two parts of the solution reinforce each other if A_1 and A_2 are of the same sign. Otherwise they weaken each other.

If λ_1 and λ_2 are each negative, then the two parts of the solution still reinforce each other if A_1 and A_2 are of the same sign, but otherwise they weaken each other.

If λ_1 and λ_2 are of opposite sign then the completely positive (negative) solution arising from the one term may be strong enough to overcome (for a limited period or for ever) the negative (positive) parts of the solution arising from the other term.

The *dominant* term will be the term which has the λ of larger numerical value, and for high values of t the solution will approximate to the solution of this term alone.

The solution $y_t = 4(1 \cdot 1)^t - 3(-1 \cdot 5)^t$ is shown in Diagram 27.8.

(9) *The solution* $y_t = (1 + 2t + t^2)2^t + 2(\sqrt{13})^t \cos \left(tR - \dfrac{2\pi}{3} \right)$

$$+ \frac{\alpha^{t+5}}{(\alpha - 2)^3 (\alpha^2 - 4\alpha + 13)}$$

This solution was obtained on page 404. We now indicate its more important aspects.

There are three parts. The first $(1 + 2t + t^2)2^t$ is clearly positive and explosive. It does not oscillate.

The second, $2(\sqrt{13})^t \cos \left(tR - \dfrac{2\pi}{3} \right)$ oscillates, with change of sign. The factor $(\sqrt{13})^t$ indicates that the oscillations are explosive.

The third depends on the value of α which appears in two guises. The denominator $(\alpha - 2)^3 (\alpha^2 - 4\alpha + 13)$ has a magnitude and sign depending on the value of α. In addition, the numerator indicates oscillations if $\alpha < 0$, explosions if $|\alpha| > 1$, and damping if $|\alpha| < 1$. We now consider the various possibilities.

(a) $\alpha > 2$. Numerator α^{t+5} explodes. Denominator consists of the product of two positive quantities, and is therefore positive. The whole term is therefore positive and explosive.

(b) $\alpha = 2$. Numerator explodes. Denominator is zero. The term is therefore infinitely large.

(c) $2 > \alpha > 1$. Numerator explodes, and is positive. Denominator is negative—since $(\alpha - 2)^3$ will be negative, while $\alpha^2 - 4\alpha + 13 = (\alpha - 2)^2 + 3^2$ will always be positive. The term is therefore negative and explosive.

(d) $\alpha = 1$. Numerator is unity. Denominator is -10, and so this part of the solution yields $-\frac{1}{10}$.

(e) $1 > \alpha > 0$. Numerator is positive and damped. Denominator negative. Whole term negative and damped.

(f) $\alpha = 0$. Term is zero.

(g) $0 > \alpha > -1$. Numerator gives damped oscillations of alternating sign. Denominator is negative and therefore reverses the sign of these.

(h) $\alpha = -1$. As (g) but of steady amplitude, being alternatively

$$\pm \frac{1}{27 \times 18}$$

(i) $\alpha < -1$. As (g) but explosive.

We must now consider how to add these parts of the solution.

For low values of t it is difficult to say what will happen without working out each part of the solution separately. Clearly the result will depend on the value of α, but whether, for example, the positive explosion from the first part of the solution will (in the early periods) more than cancel the negative explosion of case (c) above cannot really be ascertained without calculation; and even then we must add the element due to the trigonometric term.

For higher values of t, however, the term involving $(\sqrt{13})^t$ dominates the term involving 2^t. It will also dominate the term involving α^{t+5} unless $|\alpha| \geqslant \sqrt{13}$, in which case the third part of the solution will dominate. We therefore have that for large values of t the solution will eventually approximate to the explosive oscillations prescribed by the trigonometric term provided $|\alpha| < \sqrt{13}$. If, however, $\alpha \geqslant \sqrt{13}$ then the solution will eventually approximate to a positive explosion of type (a). If instead $\alpha \leqslant -\sqrt{13}$ then the solution will eventually approximate to an oscillating explosion of type (i).

SECTION VI

CHIEFLY LINEAR ALGEBRA

This section can be read at any time after Sections I and II. There is hardly any calculus in it. The first chapter looks at a few ideas of " modern algebra ". It is very simple, provided that one is not frightened by unusual symbols. It is not used much in the rest of the book, but a reading of it may help to put the student into the right frame of mind for what follows. The next four chapters are chiefly concerned with the problem of solving simultaneous linear equations: but they do so by developing a very powerful form of algebra. They include a discussion of linear programming. The last chapter takes this discussion further and ends with a few more algebraic ideas of use to the economist.

CHAPTER XXVIII

SETS, NUMBERS AND BINARY OPERATIONS

A hen is only an egg's way of making another egg.
SAMUEL BUTLER

1. Introduction

We have come across the following types of real number:

> Integers
> Rational Fractions
> Irrational Numbers
> Surds

In all cases it has been possible to associate a corresponding negative number with every positive number, and to perform the operations of addition, subtraction, multiplication and division.

We have also discussed imaginary numbers, and by operating on any real number with i, as in Chapter XXV, we can produce an associated imaginary number.

In this chapter we take a more general look at the concepts of " number ", " addition " and so on. From the point of view of a purist, these are matters which should probably have been discussed in our first chapter. The argument for dealing with them now is that so far we have been able to get along without them, but in the next few chapters we shall progress more easily if we have some familiarity with the kind of thinking that this chapter involves. It is possible to understand the rest of the book with knowledge of only a small part of this chapter. But the ideas in later chapters will be better appreciated if the reader invests some time in this one.

2. Sets

We define a set as follows:

Definition:

A set is any collection of entities.

As a few examples we have:
(1) Real numbers
(2) General de Gaulle, President Johnson and Ringo
(3) All people

413

(4) All men

(5) All men with black hair

(6) Postage stamps

(7) Used French postage stamps

(8) Imaginary numbers

(9) All mathematical ideas

(10) The mathematical ideas understood now by you.

In order to have exact knowledge of the set we must be able to say whether a specified entity is an _element_ of it—i.e., whether it belongs to the set. This means that the set must be well defined, either through listing the elements or through some definition, which may be a general description. Thus the first set mentioned above is defined through the definition of its elements, while the second is defined by listing its elements.

Symbolically we write that the set may be denoted by A where

$$A = \{\text{list of elements}\}$$

and/or

$$A = \{a \mid a \text{ is } Q\}$$

where, in the second case, the right hand side should be read " the set of elements a, where a has the property specified by the description Q."

Notice that a set may have other sets as its elements. We have above a set of ten sets. Sometimes it is convenient to refer to a set which is an element of another set as a _sub-set_. This, in turn, may have several elements, and some or all of these may be sets of other elements.

The important point is that every set should have its elements well defined.

If a is an element of the set A we write

$$a \in A$$

If B is a sub-set of A then we write

$$B \subseteq A$$

For this to be so

$$b \in B \quad \text{implies} \quad b \in A$$

which is sometimes written

$$b \in B \rightarrow b \in A$$

If there are some elements of A which are not elements of B then

$$a \in A \nrightarrow a \in B$$

where the sign means " does not imply ".

If $$b \in B \to b \in A$$

and $$a \in A \to a \in B$$

then any entity which belongs to either set must also belong to the other. We say that the sets are *equal* and write

$$A = B$$

In this case

$$B \subseteq A$$

and $$A \subseteq B$$

If, however,

$$B \subseteq A$$

but

$$B \ne A$$

then we write

$$B \subset A$$

and call B a *proper sub-set* of A, in that A contains B and more than B.

Thus, if we use numbers to refer to the ten sets listed above, then if

$$A = \{a \mid a \in (4)\}$$

and $$B = \{b \mid b \in (5)\}$$

then $$B \subset A$$

If $$P = \{p \mid p \in (9)\}$$

and $$Q = \{q \mid q \in (10)\}$$

then, quite certainly

$$Q \subseteq P$$

possibly $$Q = P$$

probably $$Q \subset P$$

and it is impossible for

$$P \subset Q$$

There are two sets which are of important analytical use:

(1) The *empty set*, having no elements, which we may denote by 0.
(2) The *universal set*, which includes all of the elements relative to our problem, which we denote by U.

It follows from the above definitions that for any set A

$$0 \subseteq A \subseteq U$$

We have so far been concerned with defining and illustrating the idea of a set. We may notice that all of the kinds of numbers which we listed at the beginning of this chapter are sets; and that one can form a set of these sets. When considering numbers we found it necessary to define certain operations on them—such as addition. We now consider operations on sets. First we shall do so in a rather restricted way, and then we shall look at a more general approach.

There are two questions in which we may often be interested.

(1) What elements are common to both sets? e.g. Who is both a Trotskyist and an Economist?

(2) What elements are in at least one of the two sets? e.g. Who qualifies for free medicine on grounds of either age or severity of sickness (or both)?

To help us in the analysis of these questions we introduce two definitions.

Definitions:

(1) The *intersection* of two sets A and B to be the set C where C is such that its elements c are elements of A and of B.

$$C = \{c \mid c \in A \quad \text{and} \quad c \in B\}$$

(2) The *union* of two sets A and B to be the set D where D is such that its elements d are elements of A or of B or of both

$$D = \{d \mid d \in A \quad \text{or} \quad d \in B\}$$

where " or " includes " and " in the sense just described.

Symbolically we write

$$C = A \cap B \qquad \text{(intersection)}$$
$$D = A \cup B \qquad \text{(union)}$$

For example, if

$$A = \{a \mid a \text{ is a man}\}$$
$$B = \{b \mid b \text{ is a civil servant}\}$$

then

$$C = A \cap B$$
$$= \{c \mid c \text{ is a male civil servant}\}$$

and

$$D = A \cup B$$
$$= \{d \mid d \text{ is a man or a civil servant or both}\}$$

A third idea which eases analysis is the complement of a set.

Definition:

(3) The *complement* of the set A is the set A' containing all of those elements not in A, and of none other. Clearly

$$A \cap A' = 0$$
$$A \cup A' = U$$

In terms of the last example, if U is the universal set of all people, then A' is the set of women, B' the set of non-civil servants, C' the set of all women plus male non-civil servants, and D' the set of all female non-civil servants.

Notice that

$$A' \cap B' = D'$$

(compared with $A \cup B = D$)

and that $A' \cup B' = C'$

(compared with $A \cap B = C$)

These relationships can be illustrated graphically by means of " Venn diagrams ". In each case the points in the square represent the elements of the universal set.

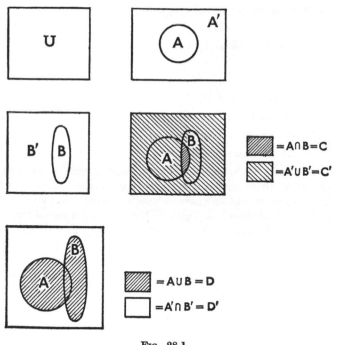

FIG. 28.1

The fact that if

$$C = A \cap B$$

then $C' = A' \cup B'$

prompts us to enquire into the rules that the operations of union and intersection must obey *as a consequence of our definitions of sets and these operations*. Before pursuing the rules which these diagrams may suggest, let us consider the operations of addition, subtraction, multiplication and division of real numbers.

We know that for real numbers, our definitions of these operations are such that the following laws hold.

Laws of Association for Addition

$$a + (b + c) = (a + b) + c$$

i.e.   if c is added to b, to form $(b + c)$ and this is then added to a, the result is the same as if b is added to a, to form $(a + b)$ and c is added to this.

Law of Commutation for Addition

$$a + b = b + a$$

Law of Association for Multiplication

$$a \times (b \times c) = (a \times b) \times c$$

i.e.   if c is multiplied by b, to form $(b \times c)$, and this product is multiplied by a, the result is the same as if b were multiplied by a, to form $(a \times b)$ and then c were multiplied by this product.

Law of Commutation for Multiplication

$$a \times b = b \times a$$

Law of Distribution

$$a \times (b + c) = (a \times b) + (a \times c)$$

All of these laws are so familiar to most students that even the statement of them may seem to be superfluous. But before this book is over we shall come across occasions where we shall be adding or multiplying entities in a way which will violate some of these laws.

Let us now consider a generalisation of these operations. Multiplication, addition, division and subtraction all involve two (or more) numbers. When more than two are involved we can begin with two, to get a new number, and then combine this with the third of the original numbers, and so on. We can, for example, look upon the operation

$$a + b + c + d$$

as forming

$$x = c + d$$

then

$$y = b + x$$

then

$$z = a + y$$
$$= a + (b + x)$$
$$= a + (b + [c + d])$$

Without defining it in any more detail, let us suppose that there is some way of *operating* on an *element a* with an *element b*. If we seek a concrete example we may, for convenience, turn to elements which are real numbers. If we put $a = 3$ and $b = 5$ then examples of the kind of operation we have in mind are

$$\begin{array}{lll} \text{Addition} & 3 + 5 = 8 \\ \text{Multiplication} & 3 \times 5 = 15 \\ \text{Raising to a Power} & 3^5 = 243 \end{array}$$

and so on. If we seek non-numerical examples we may define a and b as two people, and operations may be such as

> b sees a
> b employs a
> b marries a
> b divorces a
> b kills a

These operations, all involving two elements, are called *binary operations*. In some way or the other they combine one element with another. The result may be another element of the same set (as when $3 + 5 = 8$) or it may not (as when $3 \div 5 = \frac{3}{5}$ combines two integers to form a non-integer). It is important to notice the direction of the operation. Sometimes it may be reversible, as in addition, when

$$a + b = b + a,$$

but at other times it is not, as when

$$b \div a \neq a \div b$$

To take another example

$$(b \text{ marries } a) \rightarrow (a \text{ marries } b)$$

but

$$(b \text{ sees } a) \nrightarrow (a \text{ sees } b)$$

Let us be more formal.

Definition:

We define a *binary operation* to be a rule for combining two elements of a set. We shall denote a binary operation in which b operates on a by

$$a^0b$$

The notation a^0b is a general way of representing a whole set of binary operations which will include $a + b$, a^b, $a \times b$, but not (in general) $b+a$, b^a, $b \times a$, (which would be represented by b^0a) unless it happens that the particular operation concerned is defined so that

$$a^0b = b^0a$$

which is true if $\qquad a^0b$ means $a+b$

but untrue if $\qquad a^0b$ means a^b

Binary operations need not obey any rules other than those implied by the definition, which is very general and implies very little. But if we interest ourselves in only those binary operations which do obey certain rules then some useful theorems emerge. The following rules are useful ones, and we are often concerned with operations which obey all or most of them. We shall list them and then discuss them.

(1) *Closure*

$$\text{if } a \in S \quad \text{and} \quad b \in S \text{ then}$$
$$c = a^0b \in S$$

i.e. The result of the operation is an element which belongs to the same set as a and b.

(2) *Commutative*

$$a^0b = b^0a$$

(3) *Associative*

$$a^0(b^0c) = (a^0b)^0c$$

(4) *Identity*

There is a unique element e of the set S, such that for every element a of the set S

$$a^0e = a = e^0a$$

This element is called the *identity* or *unit*,

(5) *Inverse*

For every element a of the set S there is an associated element (which we shall denote by a^{-1}), such that

$$a^0a^{-1} = (a^{-1})^0a = e$$

where e is the identity or unit.

(6) *Cancellation*

$$\text{if } a^0b = a^0c \quad \text{then} \quad b = c$$

(7) *Distributive*

If we denote one binary operation by ^{0} and another by $*$ then if ^{0} *is distributive over* $*$

$$a^0(b*c) = (a^0b)*(a^0c) \quad \text{and} \quad (a*b)^0c = (a^0c)*(b^0c)$$

if $*$ *is distributive over* ^{0} then

$$a*(b^0c) = (a*b)^0(a*c) \quad \text{and} \quad (a^0b)*c = (a*c)^0(b*c)$$

We now consider these rules very briefly. The first is useful if we are essentially concerned with the study of elements of a set. It means that binary operations which result in elements outside the set are of no interest to us. If, for example, we are interested only in the set of integers, the operation of ordinary division, which may result in a fraction (which is not in the set of integers) will not be of any interest to us.

The second rule is often not obeyed. We have already seen an example of this ($3^5 \neq 5^3$). Another example is multiplication in matrix algebra, as we shall shortly see.

The third rule says that if you first operate on b with c (to produce b^0c); and then operate on a with the result of this operation (to produce $a^0(b^0c)$), then the answer is the same as if you first operate on a with b (to get a^0b) and then operate on this with c (to get $(a^0b)^0c$). An example of an operation which obeys this is ordinary addition of real numbers

$$a + (b + c) = (a + b) + c$$

An example of an operation in which it does not work is adding the square of b to a, so that

$$a^0b = a + b^2$$

Consider
$$a = 1, b = 2, c = 3.$$
$$(a^0b)^0c = (1 + 2^2) + 3^2$$
$$= 5 + 3^2$$
$$= 14$$

while
$$a^0(b^0c) = 1 + (2 + 3^2)^2$$
$$= 1 + 11^2$$
$$= 122$$

The next two rules are useful in that they enable us to account for " no change " and " reversibility " as will later become apparent.

The identity rule is a statement of " no change ". It asserts that we are interested in cases where, if we operate on a with e to form a^0e then there is some e which will produce an a^0e identical to a. In ordinary multiplication of real numbers, $e = 1$. In the addition of real numbers $e = 0$.

i.e.
$$a \times 1 = a$$
$$a + 0 = a$$

This does not hold if the operation is simply one of replacement, so that

$$a^0b = b$$

for then there would be no unique e such that for all a

$$a^0e = a$$

The next rule specifies that every element has its inverse, such that if we operate on the one with the other we produce the unit element. Clearly, if there is no unit element then this rule cannot hold.

The rule of cancellation states that if one gets the same result by operating on any (and every) element a with the element b as one does with the element c then the two elements b and c are equal.

If the operation is marriage and the result of Miss Edmunds marrying Mr. White is the same as the result of Mrs. Evans marrying Mr. White then as far as we are concerned (and presumably as far as Mr. White is concerned) Miss Edmunds equals Mrs. Evans.

The distributive rule differs from the others in that it concerns two operations. As an example, consider the two operations of $+$ and $\times$. We have

$$a \times (b + c) = (a \times b) + (a \times c)$$

but note that

$$a + (b \times c) \neq (a + b) \times (a + c)$$

Thus $+$ is distributive over $\times$, but
$\times$ is not distributive over $+$.

The reader may now consider whether all of the above rules are valid for addition, subtraction, multiplication and division of (i) real numbers (ii) imaginary numbers.

We now consider whether they apply to the union and intersection of sets. We consider the rules in turn.

If the sets A and B are sub-sets (or elements) of U, then so are the sub-sets $A \cup B$ and $A \cap B$. The rule of closure holds.

Clearly, from our definitions of $\cup$ and $\cap$

$$A \cup B = B \cup A$$
$$A \cap B = B \cap A$$

and so the operations are commutative.

The law of association requires that

and $$A \cup (B \cup C) = (A \cup B) \cup C$$
$$A \cap (B \cap C) = (A \cap B) \cap C$$

The validity of these statements can best be demonstrated with Venn diagrams.

In Diagram 28.2 the area $B \cup C$ is shaded horizontally, and the area $A \cup (B \cup C)$ is shaded vertically. We would have had the same area of vertical shading if we had begun by shading $(A \cup B)$ and then $(A \cup B) \cup C$.

In Diagram 28.3 the area $B \cap C$ is shaded horizontally, and the area $A \cap (B \cap C)$ is shaded vertically.

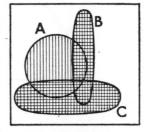

FIG. 28.2

The student may satisfy himself that first shading $(A \cap B)$ and then $(A \cap B) \cap C$ will produce the same area of vertical shading.

Is there an identity element such that

$$A \cup e = A?$$

Obviously if e is the empty set 0 then this is so.

For the intersection the identity element is the universal set U, for

$$A \cap U = A$$

In using the union we now have to see if there is an inverse to A such that

$$A \cup A^{-1} = 0$$

Clearly this cannot be so.

For the intersection we need an inverse such that

$$A \cap A^{-1} = U$$

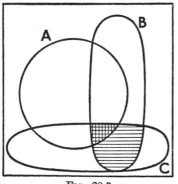

FIG. 28.3

This cannot be, for, by definition, it is impossible for an intersection to contain elements not in each of the sets. Yet U contains elements not in A. Therefore U cannot result from the intersection of A and another set.

The law of cancellation will hold if the following are true:

(i) if $A \cup B = A \cup C$ then $B = C$
(ii) if $A \cap B = A \cap C$ then $B = C$.

Clearly, as is shown in Diagrams 28.4 and 28.5 these need not be true, and the law of cancellation does not hold.

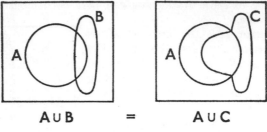

$$A \cup B \quad = \quad A \cup C$$

Fig. 28.4

On the other hand the student may show by using Venn diagrams that the law of distribution does hold, for

$$A \cap (B \cup C) = (A \cap B) \cup (A \cap C)$$
and
$$A \cup (B \cap C) = (A \cup B) \cap (A \cup C)$$

We shall not be using these ideas very much, but it is useful for the student to appreciate how many seemingly different ideas may be special cases of more general ideas. This will become apparent in the next few chapters. Meanwhile we may note that if, for example, we can evolve theorems about sets of elements, with an associated binary operation, which are such that (say) the laws of closure, association, identity and inverse hold, then these theorems will be applicable whatever the precise natures of the elements and operations may be.

A set of the kind just described, with these four laws holding, is called a *group*. It is possible to develop the remaining chapters of this book entirely in terms of sets, and to identify various groups, but we shall not do so.

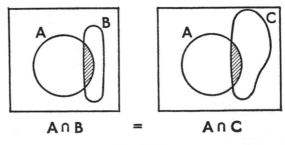

$$A \cap B \quad = \quad A \cap C$$

Fig. 28.5

CHAPTER XXIX

VECTORS, MATRICES AND DETERMINANTS

Here is my space.

W. SHAKESPEARE

1. Vectors and spaces

Let us now consider a problem which has an obvious economic content but which will also allow us to extend these ideas. We shall begin with a very simple version of the problem.

Suppose that there are only two goods—ale and brandy. Let their prices be p_a and p_b per bottle. In a given market at a given time these prices will have some definite values, and we may describe this local and momentary price situation at time t by the expression $(p_a, p_b)_t$. In this notation it is easy to see that the information about the prices could be equally well represented by a point on a graph, along whose axes we measure the values of p_a and p_b.

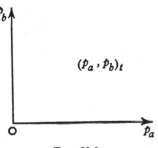

FIG. 29.1

Clearly we are concerned only with positive prices. All possible combinations of them could be represented by different points on this graph. Every single point would represent a different price combination. The whole set of prices, of which $(p_a, p_b)_t$ would be a typical element, would correspond to the whole set of points which form the " space " to the right of Op_b and above Op_a.

To economise in writing we could denote a typical element by

$$\mathbf{p}_t = (p_a, p_b)_t$$

where $\mathbf{p}_t$ would be used to denote any pair of prices, it being understood that the precise values of p_a and p_b would be provided when necessary.

Now let us introduce a third good—cheese—whose price may be denoted by p_c. A typical price situation would be given by $(p_a, p_b, p_c)_t$ and, provided we know that we are now talking of three prices rather than of two we could, once again, denote this by $\mathbf{p}_t$.

425

We could also represent this typical element by a point in a diagram. But here we have to be careful. We have to find a diagram which will enable us to distinguish between (p_a, p_b, p_c) and (p_a, p_b, p_c'). A change in just one price will call for a different point in the diagram. Now in Diagram 29.1 every point is precisely and uniquely defined by the values of p_a and p_b. The value of p_c is not needed, and plays no part. Furthermore, there is no way of introducing it by considering some third axis in the plane of the paper. Suppose, for example, that we tried to measure p_a and p_b as before, and p_c along the axis Op_c in Diagram 29.2, where all the lines are in the same plane. How would one represent the set of prices (4, 3, 7)? A similar difficulty arises if Op_c has any other direction on the plane of the paper: because the points on Op_c—which are supposed to represent values of p_c—all represent unique pairs of values p_a and p_b. Only if every pair of values (p_a, p_b)

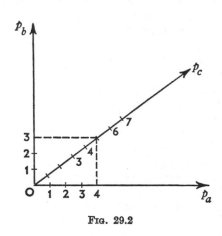

FIG. 29.2

uniquely defines the price p_c can we make sense of this diagram: and in that case—when a knowledge of p_a and p_b tells us the value of p_c—there is no reason for using p_c.

We need a diagram in which the axis Op_c is such that (apart from the origin) no point on it can represent values of p_a and p_b. It has to be a three-dimensional diagram. It happens that we can *represent* a three-dimensional problem in two dimensions, and Diagram 29.3 shows how the prices (4, 3, 7) may be associated uniquely with a point in a three-dimensional space. Every point in this three-dimensional space would represent a unique triad of prices

(p_a, p_b, p_c). Each triad would be an element in the set of prices which would correspond to the total space.

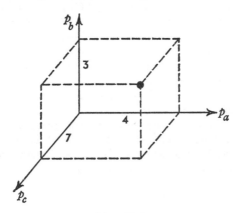

FIG. 29.3

Suppose now that we have a fourth commodity, whose price is p_d. How can we represent the prices (p_a, p_b, p_c, p_d) in a diagram? We cannot. But it is always useful if we can extend the meanings of our words in a way which makes sense and allows us to tackle a wider range of problems through the use of the same language and ideas. Just as the pair of prices (p_a, p_b) can be represented by a point in a two-dimensional space, and the triad (p_a, p_b, p_c) by a point in a three-dimensional space, so we shall introduce the concept of a four-dimensional space (which we cannot draw or envisage) consisting of all possible four-dimensional points which can be obtained by giving all possible values to the four prices (p_a, p_b, p_c, p_d). The usefulness of this extension of our language will soon become apparent. We need not, of course, stop at four dimensions.

Let us now consider some other quantities in which we are interested. So far we have spoken only of prices, and our spaces have been sets of points corresponding to prices. They have been price-spaces. Suppose that we had, instead, measured the quantities purchased. In this case each basket of goods, defined by the quantities (q_a, q_b, q_c, q_d)—if there are four goods—would be represented by a point in the appropriately dimensioned quantity-space.

It is easy to see that, in the two-dimensional and three-dimensional spaces, there corresponds to each point an appropriate *vector*, which is the straight line joining that point to the origin, and which is unambiguously defined by the co-ordinates of the point concerned. In two dimensions a vector can always be represented by a complex number. We shall let the co-ordinates of points in four and more dimensions also define vectors in these spaces. Thus the quantities $(4, 2, 1, 5)$ define a single vector in the four-dimensional quantity-space. A vector clearly has both length and direction.

But if we are going to extend the meaning of the word " space " to cover a concept which lies beyond our powers of physical representation it is important for us to say what properties we are going to give this space which we have created.

Let us notice a property of spaces of two and three dimensions, as we ordinarily consider them to be.

If we have two quantity vectors q_1 and q_2 then it is meaningful to introduce a third quantity vector q_3 and *to define the operation of addition of vectors* in an appropriate way, which makes q_3 the sum of q_1 and q_2. The definition that we adopt is

if
$$q_1 = (q_{a1}, q_{b1})$$
$$q_2 = (q_{a2}, q_{b2})$$

then
$$q_3 = q_1 + q_2 = (q_{a1} + q_{a2}, q_{b1} + q_{b2})$$

Diagrammatically the definition is illustrated in Diagram 29.4. It is easy to see that the vector representing the sum of the other two is

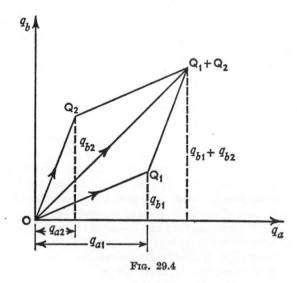

FIG. 29.4

the diagonal of the completed parallelogram. Study of the definition, or of the diagram, will show that the following rules hold

$$q_1 + q_2 = q_2 + q_1$$
$$(q_1 + q_2) + q_3 = q_1 + (q_2 + q_3)$$

We can also define a null vector

$$0 = (0, 0)$$

such that $\qquad\qquad q + 0 = q$

and a negative $-q$ such that

$$q + (-q) = 0$$

Clearly if $\qquad\qquad q = (q_a, q_b)$

then $\qquad\qquad -q = (-q_a, -q_b)$

where (in this case) the value $-q_a$ would denote a net sale of (rather than purchase of) the goods a.

The reader may care to relate these ideas to those of the addition of complex numbers, and to consider the binary operation of vector addition in terms of the last chapter.

We define our spaces of three and more dimensions to have the properties which we have just listed. For example, defining a four-dimensional vector to be an ordered set of four numbers (q_a, q_b, q_c, q_d) we define a four-dimensional space to be a set of such vectors which are subject to a law of addition such that the laws of association and commutation hold, such that a unique null vector exists, and such that to every vector there is a unique negative. With this definition before us, we need not try to visualise such a space. We think of it as being simply a set of vectors, where each vector is an ordered set of numbers.

Let us now consider the relationship between price, quantity and expenditure. We know that for a single good

$$m_a = p_a q_a$$

We can obtain the expenditures m_b, m_c and m_d in a similar way. Each time, we obtain one element of the expenditure vector m by multiplying an element of p by the appropriate element of q.

$$m = (m_a, m_b, m_c, m_d)$$
$$= (p_a q_a, p_b q_b, p_c q_c, p_d q_d)$$

Now it is possible that we will be interested in the expenditure-space—in knowing how much is spent on each good. But we may also be interested in knowing the *total expenditure*. We want to know the sum

$$p_a q_a + p_b q_b + p_c q_c + p_d q_d$$

rather than the values of the individual elements that are in it. This kind of question arises so often that it is convenient to have a notation for it. Remembering that we obtain the total expenditure by multiplying each element of **p** by the corresponding element of **q** and then adding,

we *define* the *inner product* of **p** and **q**
to be $p_a q_a + p_b q_b + p_c q_c + \dots$.
and denote it by **p . q**.

We note that the inner product of two vectors is a *scalar*. It is a pure number and does not represent a point in space.

We may also note that we can form the inner product of two vectors only if they are of the same dimensions. If we have four goods and only three prices we cannot evaluate the total expenditure; while four prices and three goods means that we have superfluous data, and the price vector can be replaced by one of fewer dimensions.

It is useful if we look more carefully at this idea, beginning with two vectors in two dimensions. If we draw the two axes at right angles then the lengths p_a, p_b, q_a, q_b are as shown.

Let θ be the angle between the vectors.

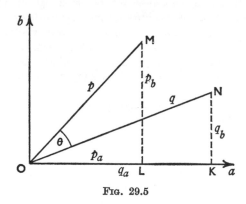

FIG. 29.5

Then
$$\cos \theta = \cos (\angle MOL - \angle NOK)$$
$$= \cos \angle MOL \cos \angle NOK + \sin \angle MOL \sin \angle NOK$$
$$= \frac{OL}{OM} \frac{OK}{ON} + \frac{ML}{OM} \frac{NK}{ON}$$

whence

$$OM.ON \cos \theta = OL.OK + ML.NK$$
$$= p_a q_a + p_b q_b = \mathbf{p} \cdot \mathbf{q}$$

But OM is the *magnitude* of $\mathbf{p}$. It is given by $\sqrt{p_a{}^2 + p_b{}^2}$ and may be denoted by $|\mathbf{p}|$. Similarly ON is the magnitude of $\mathbf{q}$, denoted by $|\mathbf{q}|$.
Thus

$$\cos \theta = \frac{\mathbf{p} \cdot \mathbf{q}}{|\mathbf{p}||\mathbf{q}|}$$

This is a result which does not depend for its validity on our initial assumption that the two axes are right angles—even though this particular proof does.

We can, in fact, *define* the *angle* between any two vectors $\mathbf{p}$ and $\mathbf{q}$ to be θ such that

$$\cos \theta = \frac{\mathbf{p} \cdot \mathbf{q}}{|\mathbf{p}||\mathbf{q}|}$$

where

$$|\mathbf{p}| = \sqrt{\sum_{i=1}^{n} p_i{}^2}$$

and

$$|\mathbf{q}| = \sqrt{\sum_{i=1}^{n} q_i{}^2}$$

For example the angle between

$$\mathbf{p} = (1, 3, 7, 4)$$

and

$$\mathbf{q} = (2, 0, 8, 3)$$

is defined to be θ, such that

$$\cos \theta = \frac{(1 \times 2) + (3 \times 0) + (7 \times 8) + (4 \times 3)}{\sqrt{1^2 + 3^2 + 7^2 + 4^2} \cdot \sqrt{2^2 + 0^2 + 8^2 + 3^2}}$$

$$= \frac{2 + 0 + 56 + 12}{\sqrt{75 \times 77}}$$

$$\doteqdot \frac{70}{76}$$

Here we may note two consequences of this definition.

(1) If $\mathbf{pq} = 0$ the two vectors are at right angles, or, as we more often say, they are *orthogonal*.

(2) If we have two sets of elements, $(p_1 ... p_n)$ and $(q_1 ... q_n)$ then the correlation between p and q is given by

$$r = \frac{\sum pq}{\sqrt{\sum p^2 \cdot \sum q^2}}$$

if $\sum p = \sum q = 0$. Thus, if we have two vectors measured from an origin such that $\sum p = \sum q = 0$, the correlation coefficient measures the cosine of the angle between them.

2. Linear programming

Let us now consider the following problem. A man spends his money on various foods, buying quantities q_i of good i at unit price p_i. Each good has a different calorific content c_i, a different protein content b_i (for body-building), and a different vitamin content v_i. His aim is to provide himself with a diet containing at least C calories, at least B units of protein, and at least V units of vitamin, at the lowest possible cost $\sum p_i q_i$. Given the price vector $\mathbf{p}$, what should be his vector $\mathbf{q}$?

Let the calorific content vector be $\mathbf{c}$
„ „ protein „ „ „ $\mathbf{b}$
„ „ vitamin „ „ „ $\mathbf{v}$

Then his aim is to minimise $\mathbf{p} \cdot \mathbf{q} = \sum p_i q_i$

subject to
$$\mathbf{q} \cdot \mathbf{c} \geqslant C$$
$$\mathbf{q} \cdot \mathbf{b} \geqslant B$$
$$\mathbf{q} \cdot \mathbf{v} \geqslant V$$

In the simple case of only two goods, denoted by subscripts 1 and 2, his aim is to minimise

$$m = p_1 q_1 + p_2 q_2$$

subject to
$$q_1 c_1 + q_2 c_2 \geqslant C$$
$$q_1 b_1 + q_2 b_2 \geqslant B$$
$$q_1 v_1 + q_2 v_2 \geqslant V$$

Let us consider this simple case geometrically. In Diagram 29.6 we represent q_1 and q_2 along the axes. The condition

$$q_1 c_1 + q_2 c_2 \geqslant C$$

means that we need consider only those points (q_1, q_2) which lie above (or on) the line

$$q_1c_1 + q_2c_2 = C \quad \text{(i.e., } \mathbf{q} \cdot \mathbf{c} = C)$$

Any point beneath this line will contain too few calories.

The next condition requires us to choose a point above (or on) the line

$$q_1b_1 + q_2b_2 = B \quad (\mathbf{q} \cdot \mathbf{b} = B)$$

which we shall suppose to cross the other line, as shown. Whether it does depends on the values of b_1, b_2, B, c_1, c_2 and C.

There is also the constraint that the selected pattern of purchases (q_1, q_2) should lie above or on the line

$$\mathbf{q} \cdot \mathbf{v} = V$$

This means that we need consider only those points which lie above, or on, the heavy line.

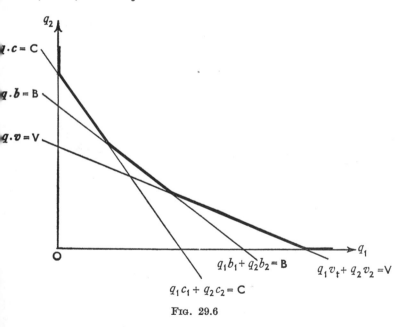

FIG. 29.6

Before we go further with this two-dimensional problem let us extend the idea into three and more dimensions.

In three dimensions the equation

$$q_1c_1 + q_2c_2 + q_3c_3 = C$$

P

defines, not a line, but a plane. The condition

$$\mathbf{q} \cdot \mathbf{c} \geqslant C$$

means that the point q must lie above or on this plane.

We say that in more than three dimensions the equation

$$\mathbf{q} \cdot \mathbf{c} = C$$

defines a hyperplane, and the condition

$$\mathbf{q} \cdot \mathbf{c} \geqslant C$$

means that the point q must lie on or " above " this hyperplane.

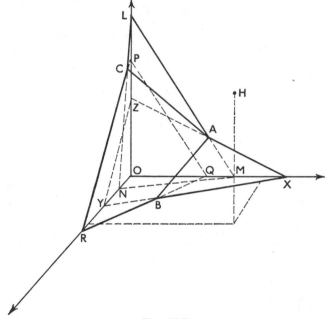

Fig. 29.7

Thus in three dimensions the existence of several constraints of the kind

$$\mathbf{q} \cdot \mathbf{x} \geqslant X$$

where X and $\mathbf{x}$ may be $C, B...\mathbf{c}, \mathbf{b}...$etc. means that instead of considering only the area on or above the heavy line, we should consider only that on or above a surface consisting of several planes, one defined by each constraint. Thus if the boundary planes are LMN, XYZ, and PQR, as shown in Diagram 29.7, then we have to choose a point lying "above" the surface whose perimeter is described by $LAXBRCL$. Such a point might be the point H.

Our job now is to find the position of H which will minimise expenditure.

In more than three dimensions the constraint

$$\mathbf{q} \cdot \mathbf{x} \geqslant X$$

means that $\mathbf{q}$ has to lie on or above the *hyperplane*

$$\mathbf{q} \cdot \mathbf{x} = X$$

The expenditure incurred through purchasing quantities $\mathbf{q}$ at prices $\mathbf{p}$ will be the scalar

$$\mathbf{p} \cdot \mathbf{q}$$

Consider any specified set of prices $\mathbf{p}$ and let us choose those quantities of goods which will make the total expenditure just equal to M, so that

$$\mathbf{p} \cdot \mathbf{q} = M$$

If there are only two goods the equation becomes that of a straight line

$$p_a q_a + p_b q_b = M$$

For three goods the equation is that of a plane

$$p_a q_a + p_b q_b + p_c q_c = M$$

while for four or more goods it is a hyperplane:

$$p_a q_a + p_b q_b + p_c q_c + p_d q_d \ldots = M$$

For any of these equations, provided we keep to our given set of prices, there will be an infinite set of values for $\mathbf{q}$, such that if our purchases are represented by an element of this set then our expenditure will just equal M.

In the case of two goods, these elements are all of those points lying on the line whose equation is given. For three goods they are all of the points lying on the plane; and so on.

If we increase the value of M, but keep our prices constant, we obtain a different line, plane or hyperplane. It will be parallel to the previous one—and will be further out from the origin.

We are now able to solve our problem. Diagram 29.8 illustrates the solution in the case of two goods

Our constraints are satisfied if we purchase quantities q_a and q_b corresponding to a point on or above the line $KLMN$. We call the area on or above this line the "feasible zone"; such purchases

corresponding to points in it are " feasible " in the sense that they
satisfy our constraints.

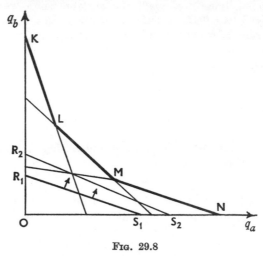

FIG. 29.8

Let the equation

$$p_a q_a + p_b q_b = M$$

produce a line such as $R_1 S_1$ for a given value of M (say M_1). This
shows all values of q_a and q_b which can be bought (at current prices)
for a specified sum of money M_1. Clearly there is no set (q_a, q_b)
which is obtainable for M_1 and which also satisfies our constraints.

Let us now shift the line RS parallel to itself, and away from O
until it touches the line $KLMN$. As we do so we are increasing the
amount of money we can spend. In the position $R_2 S_2$ we have the
purchases which are possible if we spend an amount M_2. As soon as
this line just touches the feasible zone then we will, for the first
time, be able to purchase quantities which will bring us enough
vitamins, etc. Any " lower " position of RS will not allow this.
Any " higher " position will mean spending more.

It follows that the point where the outward shifting line RS just
touches the feasible zone corresponds to those purchases (q_a, q_b)
which satisfy our constraints yet minimise the expenditure. We
cannot get what we want for less.

The point at which the line RS touches the feasible zone will
depend on the positions and slopes of the various constraint lines.
It is worth looking at this a little further.

We may note, first, that if we are concerned only with points
which lie on or above a set of straight lines then the " lower bound "

of the feasible zone (e.g. the line *KLMN*) will have no re-entrant angles. One could not have a feasible zone such as that bounded by *EFGH*.

It is easy to prove this. We do so by showing that if we take any two points in the feasible zone and join them with a straight line then all points on that line are also in the zone. Once this is proved then it follows that *EFGH* cannot be the lower bound of a feasible zone, for clearly points on the straight line joining *E* to *G* would be outside the zone.

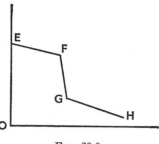

Fig. 29.9

The proof, which we give for the perfectly general case of *n* dimensions, is simple. It rests on an extension into *n* dimensions of our idea of a straight line. Let us first look at the case of two dimensions. Consider two points q_1 and q_2. Take any value of λ and consider the point λq_1. It must lie along the vector q_1, and be a fraction λ along it. Mark this point (*C*), and through it draw a line *CD* parallel to q_2 and of the same length as q_2. Join q_1 and q_2 with a straight line, to cut *CD* at *E*.

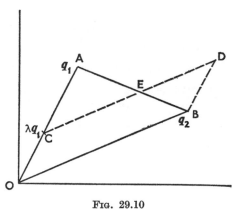

Fig. 29.10

Clearly triangles *ACE*, *AOB* and *EDB* are similar. Also

$$\frac{OC}{OA} = \frac{\lambda}{1}$$

and

$$\frac{OC}{CA} = \frac{\lambda}{1-\lambda} \quad \text{and} \quad \frac{CA}{OA} = \frac{1-\lambda}{1}$$

By similar triangles we have that

$$\frac{CE}{OB} = \frac{CA}{OA} = \frac{1-\lambda}{1}$$
$$\therefore CE = (1-\lambda)OB$$

Now we can define the vector OE to be the vector sum of OC and CE. If we denote it by $\mathbf{q}$ we have that

$$\mathbf{q} = \lambda\mathbf{q}_1 + (1-\lambda)\mathbf{q}_2$$

As we alter λ so the line CD shifts parallel to itself, and the point E moves along the line CD. If λ exceeds unity the point E lies on AB (or BA) produced.

An identical result can be proved for three dimensions in exactly the same way.

In more than three dimensions we *define* the straight line joining $\mathbf{q}_1$ to $\mathbf{q}_2$ to be the set of points

$$\mathbf{q} = \lambda\mathbf{q}_1 + (1-\lambda)\mathbf{q}_2$$

and for $0 < \lambda < 1$ these points lie between $\mathbf{q}_1$ and $\mathbf{q}_2$.

Now consider two constraints such as

$$\mathbf{q}\cdot\mathbf{v} \geqslant V \quad \text{and} \quad \mathbf{q}\cdot\mathbf{b} \geqslant B$$

Let us choose a point $\mathbf{q}_1$ such that

$$\mathbf{q}_1\cdot\mathbf{v} = V \quad \text{and} \quad \mathbf{q}_1\cdot\mathbf{b} > B$$

and another point $\mathbf{q}_2$ such that

$$\mathbf{q}_2\cdot\mathbf{v} > V \quad \text{and} \quad \mathbf{q}_2\cdot\mathbf{b} = B$$

The significance of these requirements is illustrated in Diagram 29.11. Consider $\mathbf{q}$, a point on the straight line joining $\mathbf{q}_1$ and $\mathbf{q}_2$.

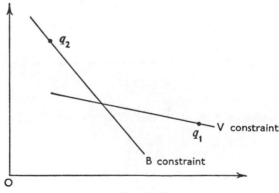

Fɪɢ. 29.11

It will be such that

$$q = \lambda q_1 + (1 - \lambda) q_2$$

We must now see if this point is bound to lie in the feasible zone. First we consider whether q satisfies the V constraint. We have

$$q = \lambda q_1 + (1 - \lambda) q_2$$

$\therefore$ 　　　　$q \cdot v = \lambda q_1 \cdot v + (1 - \lambda) q_2 \cdot v$

But 　　　　$q_1 \cdot v = V$

　　　　　　$1 - \lambda > 0$

and 　　　　$q_2 \cdot v > V$

It follows that

$$q \cdot v > V$$

and so q satisfies the V constraint. It is similarly proved that q satisfies the B constraint. Consequently any point on the line joining q_1 to q_2 lies in the feasible zone. In this proof we have taken extreme cases, in which the points q_1 and q_2 are *just* within the zone. The more general proof for q_1 and q_2 anywhere within the zone follows identical lines, with $q_1 \cdot v \geqslant V$ instead of $q_1 \cdot v = V$.

This allows us to assert the non-existence of re-entrant angles in a feasible zone.

It follows that in Diagram 29.12 the shifting line RS must touch the feasible zone in one of two ways. Either it touches it at a corner, such as M, or it may be parallel to an edge (such as LM) in which case it will touch it along this edge, including the corners (such as L and M) which define its ends.

In either case, RS touches the feasible zone at at least one corner. This means that the expenditure is minimised subject to the constraints being satisfied by a pattern of expenditures corresponding to some corner of the feasible zone, *i.e.* to some point where two constraints just hold simultaneously.

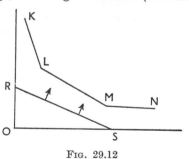

Fig. 29.12

If one considers the three dimensional case for a few minutes one will easily see that the solution to our problem must lie at some corner of the three dimensional feasible zone, at which the upward shifting plane

$$p \cdot q = M$$

(as M increases) just touches the feasible zone (which has no re-entrant corners).

It can be shown that, mathematically, the same idea is true in more than three dimensions.

Our problem is thus reduced to a simpler one. We know that the quantities which minimise our total expenditure subject to our constraints must correspond to one of the corners of our feasible zone. If we can identify these corners (of which there will be a finite number) then we can work out the expenditure at each corner and choose the one which has the lowest expenditure.

But an even greater simplification is possible. Suppose that we locate one corner and evaluate the expenditure there. We do not know (because we have not worked out) the locations of the other corners, but we do know which constraints are defining the corner we have located. Let us move a little from this corner along one of the edges which terminate at it. Does the expenditure increase or decrease? If it increases then a further movement along this edge will increase it even further (as should be obvious if one considers first the two-dimensional and then the three-dimensional case, remembering that as the line RS moves outwards from O the expenditure increases). Similarly if movement along an edge reduces expenditure then the further one goes the greater the reduction.

Our method of solution is now clear.

1. Locate one corner of the feasible zone.
2. Locate an edge from that corner.
3. See if a small movement along it increases or decreases the expenditure.
4. If it increases it or leaves it unaltered choose another edge from that corner and repeat step 3.
5. If it decreases it then locate the corner at the other end of that edge.
6. Locate another edge from this second corner.
7. Repeat steps 3 onwards.
8. Continue in this way until one reaches a corner such that no movement away from it will reduce expenditure.

This problem has been discussed chiefly in order to familiarise the student with certain ideas. It is a problem in Linear Programming, which is concerned with the maximising or minimising of some linear function of certain variables subject to a set of linear constraints. Further treatments of it may be found in books listed in the bibliography. For our present purposes we may say that often there are substantial computational problems, but in principle one can attain a solution along the lines we have indicated and standard computational techniques have been devised for this purpose.

The feasible zone is an example of a *convex set*, which is defined to be a set such that if q_1 and q_2 are any two points within it then

$$q = q_1\lambda + (1-\lambda)q_2 \quad (1 > \lambda > 0)$$

also lies within it. A great deal of advanced mathematical economic theory is developed in terms of convex sets and the hyperplanes which touch them.

We may note that each constraint defines a set of vectors q such that

$$qx \geqslant X$$

where x and X depend on the constraint concerned. The feasible zone is simply the intersection of these sets.

The expenditure plane defines three sets of vectors—those " above " it, those " beneath " it and those " on " it.

Our problem is that of locating this plane such that the intersection of our feasible zone and the set of points beneath it is empty, but such that the intersection of the feasible zone and the plane itself is non-empty. We then have to identify the element or elements which lie in this non-empty intersection.

3. Matrices

We now go back to the idea of a vector as an ordered set of elements, and to the concept of the inner product, defined as

$$p \cdot q = p_1 q_1 + p_2 q_2 + \dots.$$

Let us now consider several different price vectors, each being a set of different values of the prices of m goods. One can think of there being a different vector for each of several days. Let the price vector for day i be p_i.

Let there also be several different quantity vectors denoting the quantities of m goods bought by z different people. The typical vector will be q_i.

We are interested in the following questions.

(i) How much is spent each day?
(ii) What is the total expenditure of each person?

Let us write out the price vectors for n days thus:

$$\mathbf{p}_1 = p_{11}, \; p_{12}, \; p_{13}, \; p_{14} \cdots p_{1m}$$
$$\mathbf{p}_2 = p_{21}, \; p_{22}, \; p_{23}, \; p_{24} \cdots p_{2m}$$
$$\mathbf{p}_3 = p_{31}, \; p_{32}, \; p_{33}, \; p_{34} \cdots p_{3m}$$
$$\vdots$$
$$\mathbf{p}_n = p_{n1}, \; p_{n2}, \; p_{n3}, \; p_{n4} \cdots p_{nm}$$

where p_{ij} is the price on day i of good j. All of these prices are, of course, elements in a set of prices. If we write them out as

$$\begin{bmatrix} p_{11} & p_{12} & p_{13} & p_{14} \cdots p_{1m} \\ p_{21} & p_{22} & p_{23} & p_{24} \cdots p_{2m} \\ \vdots & \vdots & \vdots & \vdots \quad \vdots \\ p_{n1} & p_{n2} & p_{n3} & p_{n4} \cdots p_{nm} \end{bmatrix}$$

then we can think of this array of $n \times m$ elements as a table, or *matrix*, which spells out the elements of the n vectors $\mathbf{p}_i$ which constitute its rows. But it is quite meaningful to consider

$$\mathbf{p}_j = \begin{pmatrix} p_{1j} \\ p_{2j} \\ p_{3j} \\ \vdots \\ p_{nj} \end{pmatrix}$$

as a vector. Its elements, which all come from one column of the matrix, tell us the prices of good j on successive days.

For a reason which will soon appear let us set out our price and quantity data thus:

$$\mathbf{P} = \begin{bmatrix} p_{11} & p_{12} & \cdots & p_{1m} \\ p_{21} & p_{22} & \cdots & p_{2m} \\ \vdots & \vdots & & \vdots \\ p_{i1} & p_{i2} & \cdots p_{ij} \cdots & p_{im} \\ \vdots & \vdots & & \vdots \\ p_{n1} & p_{n2} & \cdots & p_{nm} \end{bmatrix} \quad *$$

where $p_{ij} = $ price on day i of good j.
n rows, one for each day
m columns, one for each good.

$$*$$

$$Q = \begin{bmatrix} q_{11} & q_{12} \cdots q_{1k} \cdots q_{1z} \\ q_{21} & q_{22} \cdots q_{2k} \cdots q_{2z} \\ \vdots & \vdots \quad\quad q_{jk} \quad \vdots \\ \vdots & \vdots \quad\quad\quad\quad \vdots \\ q_{m1} & q_{m2} \cdots q_{mk} \cdots q_{mz} \end{bmatrix}$$

where $q_{jk} =$ quantity of good j bought by person k.   There are m rows, one for each good and z columns, one for each person.   Q is said to be of *order* $(m \times z)$.

Consider now the problem of finding how much money is spent on each day by each person.   There are n days and z persons, so we need an answer consisting of $(n \times z)$ elements.   Each element will denote the total expenditure of a specified person on a specified day. For person k on day i the expenditure may be represented by e_{ik} and will be equal to

(price on day i of good 1) $\times$ (quantity of good 1 bought by k)
$+ (\quad ,, \quad ,, \quad ,, \quad i ,, \quad ,, \quad 2) \times (\quad ,, \quad\quad ,, \quad ,, \quad 2 \quad ,, \quad\quad ,, \quad k)$
$+ (\quad ,, \quad ,, \quad ,, \quad i ,, \quad ,, \quad 3) \times (\quad ,, \quad\quad ,, \quad ,, \quad 3 \quad ,, \quad\quad ,, \quad k)$
etc.
$= p_{i1}q_{1k} + p_{i2}q_{2k} + p_{i3}q_{3k} + \ldots + p_{im}q_{mk}$

which is the inner product

$$\mathbf{p}_i\mathbf{q}_k$$

where $\mathbf{p}_i$ is the row vector marked by the asterisk
and $\mathbf{q}_k$ is the column vector marked by the asterisk.

Thus our full table of results will appear as

$$E = \begin{bmatrix} e_{11} & e_{12} \cdots e_{1k} \cdots e_{1z} \\ e_{21} & e_{22} \cdots e_{2k} \cdots e_{2z} \\ e_{i1} & e_{i2} \cdots e_{ik} \cdots e_{iz} \\ e_{n1} & e_{n2} \cdots e_{nk} \cdots e_{nz} \end{bmatrix}$$

where $e_{ik} = \mathbf{p}_i\mathbf{q}_k$
It is worth while to do a simple numerical example.   Suppose that

$$\xleftarrow{\quad} \text{goods} \xrightarrow{\quad}$$

$$\mathbf{P} = \begin{bmatrix} 1 & 2 & 3 & 7 \\ 4 & 6 & 5 & 8 \\ 9 & 10 & 11 & 12 \end{bmatrix} \text{days}$$

<←persons→>

$$Q = \begin{bmatrix} 20 & 35 \\ 30 & 45 \\ 40 & 55 \\ 50 & 65 \end{bmatrix} \text{goods}$$

Then, by application of the above procedure, we have, for example, that the first person's expenditure on the second day will be

$$e_{12} = (4, 6, 5, 8) \begin{pmatrix} 20 \\ 30 \\ 40 \\ 50 \end{pmatrix}$$

$$= (4 \times 20) + (6 \times 30) + (5 \times 40) + (8 \times 50)$$
$$= \quad 80 \quad + \quad 180 \quad + \quad 200 \quad + \quad 400$$
$$= \quad 860$$

and, more fully, the various daily expenditures by the two persons will be given by

<←—persons—→>

$$E = \begin{bmatrix} 550 & 745 \\ 860 & 1205 \\ 1520 & 2150 \end{bmatrix} \text{days}$$

This process of combining two matrices to form a third by the formation of various inner products is used very frequently. We call the procedure *matrix multiplication*, which we now define, and then examine in some detail.

Definition:

> *We define* the *post-multiplication* of matrix **A** by matrix **B**, and the *pre-multiplication* of matrix **B** by matrix **A**, to be the formation of the matrix **C** where the (i, k)th element of **C** is the inner product of the ith row of **A** and the kth column of **B**
>
> $$i.e. \quad c_{ik} = \sum_j a_{ij} b_{jk}$$

and we write

$$C = AB$$

We must now consider some consequences of this definition.

First, it is possible to form **AB** only if the number of columns in **A** is equal to the number of rows in **B**. One cannot form **AB** if

$$A = \begin{bmatrix} 2 & 1 & 2 & 3 \\ 1 & 0 & 2 & 4 \end{bmatrix} \quad \text{and} \quad B = \begin{bmatrix} 1 & 2 & 3 & 1 \\ 3 & 2 & 1 & 1 \\ 1 & 2 & 4 & 2 \end{bmatrix}$$

If A is an $(m \times n)$ matrix then B must be $(n \times p)$ if AB is to be possible.

Let us now discover whether the operation of matrix multiplication obeys some of the rules which we have previously considered.

First let us ask if AB = BA. Obviously the answer cannot be an unqualified yes, for it may be possible to form only one of these products, as in the case where

$$A = \begin{bmatrix} 2 & 1 & 3 \\ 1 & 2 & 4 \end{bmatrix} \quad B = \begin{bmatrix} 3 & 1 & 1 & 1 \\ 1 & 2 & 3 & 1 \\ 1 & 1 & 1 & 4 \end{bmatrix}$$

AB can be formed, but BA cannot be formed. For AB to be possible the matrices must be of order $(m \times n)$ and $(n \times p)$ respectively. For BA also to be possible they must be of order $(m \times n)$ and $(n \times m)$, for then B has as many rows as A has columns, and A as many rows as B has columns.

But consider

$$A = \begin{bmatrix} 2 & 3 & 1 \\ 1 & 2 & 4 \end{bmatrix}$$

$$B = \begin{bmatrix} 1 & 2 \\ 2 & 2 \\ 3 & 1 \end{bmatrix}$$

$$AB = \begin{bmatrix} (2 \times 1) + (3 \times 2) + (1 \times 3) & (2 \times 2) + (3 \times 2) + (1 \times 1) \\ (1 \times 1) + (2 \times 2) + (3 \times 4) & (1 \times 2) + (2 \times 2) + (4 \times 1) \end{bmatrix}$$

$$= \begin{bmatrix} 11 & 11 \\ 17 & 10 \end{bmatrix}$$

but

$$BA = \begin{bmatrix} 1 & 2 \\ 2 & 2 \\ 3 & 1 \end{bmatrix} \begin{bmatrix} 2 & 3 & 1 \\ 1 & 2 & 4 \end{bmatrix}$$

$$= \begin{bmatrix} (1 \times 2) + (2 \times 1) & (1 \times 3) + (2 \times 2) & (1 \times 1) + (2 \times 4) \\ (2 \times 2) + (2 \times 1) & (2 \times 3) + (2 \times 2) & (2 \times 1) + (2 \times 4) \\ (3 \times 2) + (1 \times 1) & (3 \times 3) + (1 \times 2) & (3 \times 1) + (1 \times 4) \end{bmatrix}$$

$$= \begin{bmatrix} 4 & 7 & 9 \\ 6 & 10 & 10 \\ 7 & 11 & 7 \end{bmatrix}$$

which clearly is not the same as **AB**.

It is worth looking at this lack of symmetry a little further. Suppose **A** tells us how many times various bishops go to see certain films, and that **B** tells us how many times various actresses appear on the screen alone in each film.*

	Sorry Sight	Naughty Night	Pretty Plight
Bishop of Exe	1	3	6
Bishop of Wye	2	4	7
Bishop of Zed	1	5	8

	Alice	Bernice	Clarice
Sorry Sight	2	4	1
Naughty Night	0	1	85
Pretty Plight	3	2	6

The product **AB** tells us how many times each bishop sees each actress alone. For example, Wye sees Clarice on

$$(2 \times 1) + (4 \times 85) + (7 \times 6) \quad \text{occasions}$$

But **BA** tells us something quite different. The inner product of the first row of **B** and the first column of **A** tells us something about Sorry Sight, namely the number of times that Exe sees Alice, plus the number of times that Wye sees Bernice, plus the number of times that Zed sees Clarice, all in this film. The first row of **B** and the second column of **A** yield an inner product which makes no sense at all.

Before leaving this point we should emphasise that even where multiplication is *possible* it may not yield a *meaningful* result.

Before considering other properties of multiplication we define, and consider, the addition of matrices. We do so only for two matrices of the same *order* (i.e. having the same number of rows and the same number of columns).

We *define* the *sum* of the $(m \times n)$ matrix **A** and the $(m \times n)$ matrix **B** to be an $(m \times n)$ matrix **C** such that

$$c_{ij} = a_{ij} + b_{ij}$$

and we call the process of forming **C** the *addition* of **B** to **A**.

* This idea is developed in my forthcoming edition of *Algebra for Actresses* by Arch Bishop.

Suppose that **A** denotes the quantities bought in each of three consecutive weeks of four different goods by a certain man. Let **B** denote the quantities of the same goods bought in these weeks by his wife. Then **C** would denote their total purchases.

$$A = \begin{bmatrix} 1 & 2 & 3 \\ 3 & 1 & 4 \\ 2 & 1 & 0 \\ 0 & 0 & 1 \end{bmatrix} \qquad B = \begin{bmatrix} 2 & 1 & 1 \\ 3 & 4 & 0 \\ 2 & 1 & 0 \\ 1 & 1 & 1 \end{bmatrix}$$

$$C = A + B = \begin{bmatrix} 3 & 3 & 4 \\ 6 & 5 & 4 \\ 4 & 2 & 0 \\ 1 & 1 & 2 \end{bmatrix}$$

It is left to the reader to prove that, when matrix addition is defined as above, then

$$A + B = B + A$$
$$(A + B) + C = A + (B + C)$$

We also need two other definitions.

Definitions:

(1) We *define* a *null* matrix **0** of order $(m \times n)$ such that

$$A + 0 = 0 + A = A$$

(2) We *define* the *negative* of **A** to be a matrix $(-A)$ such that

$$A + (-A) = 0$$

By considering the (i, j)th element the reader may easily show that

(i) every element of **0** is zero
(ii) the (ij)th element of $(-A)$ is $-a_{ij}$ where a_{ij} is the (ij)th element of **A**.
(iii) $A0 = 0$

We now consider whether matrix multiplication and addition together obey the rule analogous to the rule for scalar multiplication and addition, namely

$$a(b + c) = ab + ac$$

Let **B** and **C** be two matrices of order $(n \times p)$. They can be added to form $B + C$, which we may denote by **D**. This will also be of order

$(n \times p)$. Let $\mathbf{A}$ be of order $(m \times n)$. Then we can form $\mathbf{AD}$, of order $(m \times p)$—call it $\mathbf{E}$. We can also form $\mathbf{AB}$ and $\mathbf{AC}$. Each of these products will also be of order $(m \times p)$ and they can be added to form a matrix $\mathbf{F}$ which will be of order $(m \times p)$, and so of the same order as $\mathbf{AD}$. Thus, to sum up, we *can* form $\mathbf{A(B+C)}$ and we *can* form $\mathbf{AB+AC}$. Furthermore, in each case we get a matrix of order $(m \times p)$. Thus it is possible, at least, for $\mathbf{A(B+C)}$ and $\mathbf{AB+AC}$ to be equal. Each expression has m rows and p columns. We have now to consider whether the corresponding elements are equal.

Consider the typical elements e_{ij} (of $\mathbf{E=A(B+C)}$) and f_{ij} (of $\mathbf{F=AB+AC}$)). We have

$$\mathbf{E=A(B+C)=AD}$$

where $\mathbf{D=(B+C)}$

The (ij)th element of $\mathbf{E}$ is the inner product of the ith row of $\mathbf{A}$ and the jth column of $\mathbf{D}$.

i.e.

$$e_{ij}=a_{i1}d_{1j}+a_{i2}d_{2j}+\ldots+a_{in}d_{nj}$$
$$=a_{i1}(b_{1j}+c_{1j})+a_{i2}(b_{2j}+c_{2j})+\ldots+a_{in}(b_{nj}+c_{nj})$$

But
$$\mathbf{F=AB+AC}$$

The (ij)th element of $\mathbf{F}$ is the sum of the (ij)th element of $\mathbf{AB}$ and the (ij)th element of $\mathbf{AC}$.

The (ij)th element of $\mathbf{AB}$ is the inner product of the ith row of $\mathbf{A}$ and the jth column of $\mathbf{B}$, *i.e.* it is

$$a_{i1}b_{1j}+a_{i2}b_{2j}+\ldots+a_{in}b_{nj}$$

The (ij)th element of $\mathbf{AC}$ is, similarly,

$$a_{i1}c_{1j}+a_{i2}c_{2j}+\ldots+a_{in}c_{nj}$$

Therefore, by summation of these

$$f_{ij}=a_{i1}b_{1j}+a_{i2}b_{2j}+\ldots+a_{in}b_{nj}$$
$$+a_{i1}c_{1j}+a_{i2}c_{2j}+\ldots+a_{in}c_{nj}$$
$$=(a_{i1}b_{1j}+a_{i1}c_{1j})+\ldots+(a_{in}b_{nj}+a_{in}c_{nj})$$
$$=e_{ij}$$

Thus $\mathbf{E=A(B+C)}$ is of the same order as $\mathbf{F=AB+AC}$, and every element e_{ij} of $\mathbf{E}$ is the same as the corresponding element f_{ij} of $\mathbf{F}$.

Thus
$$\mathbf{A(B+C)=AB+AC}$$

The student should show for himself, as an exercise, that

$$(A \cdot B) \cdot C = A \cdot (B \cdot C)$$

where

A is of order $(m \times n)$
B „ „ „ $(n \times p)$
C „ „ „ $(p \times q)$

Usually we write $(A \cdot B) \cdot C$ simply as **ABC**, since the result just quoted shows that whether we form $D = AB$, and then obtain DC, or first form $E = BC$ and then form **AE**, the result is the same.

Let us now see to what extent the results to date enable us to draw parallels between the addition and multiplication of matrices and of scalar numbers.

We have that for scalar quantities

(i) $a + b = b + a$
(ii) $(a + b) + c = a + (b + c)$
(iii) $ab = ba$
(iv) $(ab)c = a(bc)$
(v) $a(b + c) = ab + ac$
(vi)   there is a unique zero element such that
       $a + 0 = a$ and $a0 = 0$
(vii) there is a negative $(-a)$ such that
       $a + (-a) = 0$
(viii) there is a unique identity 1 such that
       $1 \times a = a \times 1 = a$

We have shown that if a, b and c are replaced by the matrices **A**, **B** and **C** then results (i), (ii), (iv), (v), (vi) and (vii) are valid. Use of (vii) allows us to re-write the other rules with minus signs. We have also shown that result (iii) does not hold for matrices.

i.e. $$AB \neq BA$$

4. The identity matrix

We still have to enquire into (viii). Is there a unique identity matrix **I** such that

$$AI = IA = A?$$

Consider a matrix **A** of order $(m \times n)$ post-multiplied by some other matrix **I**.

Clearly **I** must be of order $(n \times p)$.

The product **AI** is of order $(m \times p)$

If this product **AI** is to be equal to **A** then it must be of order $(m \times n)$, and so $n = p$.

Therefore **I** must be of order $(n \times n)$—*i.e.* it is *square*.

Consider the following matrix

$$I = \begin{bmatrix} 1 & 0 & 0 \\ 0 & 1 & 0 \\ 0 & 0 & 1 \end{bmatrix}$$

Use it to post-multiply the matrix

$$A = \begin{bmatrix} a_{11} & a_{12} & a_{13} \\ a_{21} & a_{22} & a_{23} \\ a_{31} & a_{32} & a_{33} \\ a_{41} & a_{42} & a_{43} \end{bmatrix}$$

The reader may easily show that

$$AI = A$$

Let us now suppose that there is some other matrix K such that

$$AK = A$$

Then consider

$$AI - AK = A - A = 0$$

Use of rules (v) and (vii) shows us that

$$AI - AK = A(I - K)$$

Thus

$$A(I - K) = 0$$

If one follows the rules of scalar algebra one might be tempted to think that this implies that either $A = 0$ or $I - K = 0$, but here we have to be careful.

Consider, for example,

$$P = \begin{bmatrix} 4 & 2 \\ 6 & 3 \end{bmatrix} , \quad Q = \begin{bmatrix} -1 & -2 \\ 2 & 4 \end{bmatrix}$$

The reader may verify that

$$PQ = \begin{bmatrix} 0 & 0 \\ 0 & 0 \end{bmatrix} = 0$$

It follows that

$$A(I - K) = 0$$

does *not* necessarily imply that

either $A = 0$ *or* $(I - K) = 0$

and so we have not established that if $\mathbf{AI} = \mathbf{AK} = \mathbf{A}$, then $\mathbf{K}$ must equal $\mathbf{I}$.

It may also be verified by expansion that the following results are true

$$\begin{bmatrix} 4 & 2 & 5 \\ 1 & 3 & 1 \\ 6 & 8 & 7 \end{bmatrix} \begin{bmatrix} 1 & 0 & 0 \\ 0 & 1 & 0 \\ 0 & 0 & 1 \end{bmatrix} = \begin{bmatrix} 4 & 2 & 5 \\ 1 & 3 & 1 \\ 6 & 8 & 7 \end{bmatrix}$$

$$\begin{bmatrix} 4 & 2 & 5 \\ 1 & 3 & 1 \\ 6 & 8 & 7 \end{bmatrix} \begin{bmatrix} -12 & 0 & -26 \\ 1 & 1 & 2 \\ 10 & 0 & 21 \end{bmatrix} = \begin{bmatrix} 4 & 2 & 5 \\ 1 & 3 & 1 \\ 6 & 8 & 7 \end{bmatrix}$$

This shows quite conclusively that if

$$AB = A$$

where both matrices are of order (3×3) then it is not necessary for $\mathbf{B}$ to equal $\mathbf{I}$.

To examine this further we consider two matrices, each of order (2×2) whose product is zero. Let us take

$$\begin{bmatrix} a_{11} & a_{12} \\ a_{21} & a_{22} \end{bmatrix} \begin{bmatrix} b_{11} & b_{12} \\ b_{21} & b_{22} \end{bmatrix} = 0$$

Expansion and comparison of the elements on each side yields

$$a_{11}b_{11} + a_{12}b_{21} = 0 \quad \text{(i)}$$
$$a_{11}b_{12} + a_{12}b_{22} = 0 \quad \text{(ii)}$$
$$a_{21}b_{11} + a_{22}b_{21} = 0 \quad \text{(iii)}$$
$$a_{21}b_{12} + a_{22}b_{22} = 0 \quad \text{(iv)}$$

The first and third of these equations may be solved for b_{11} and b_{21}. Assume $a_{21} \neq 0$, $a_{11} \neq 0$. Then

$$a_{21}a_{11}b_{11} + a_{21}a_{12}b_{21} = 0 \quad \text{from (i)}$$
$$a_{21}a_{11}b_{11} + a_{22}a_{11}b_{21} = 0 \quad \text{from (iii)}$$

Subtraction yields

$$(a_{21}a_{12} - a_{22}a_{11})b_{21} = 0$$

whence   *either* $b_{21} = 0$ *or* $a_{21}a_{12} - a_{22}a_{11} = 0$

The other two equations yield

either $b_{12} = 0$ *or* $a_{21}a_{12} - a_{22}a_{11} = 0$

If $a_{21}a_{12} - a_{22}a_{11} \neq 0$ then substitution of $b_{21} = b_{12} = 0$ in the other equations yields

$$b_{11} = b_{22} = 0$$

Thus

either $a_{21}a_{12} - a_{22}a_{11} = 0$

or $B = \begin{bmatrix} 0 & 0 \\ 0 & 0 \end{bmatrix}$

In other words, if $AB = 0$ then either a certain relationship exists between the elements of A, or $B = 0$. We shall denote the left-hand side of this relationship by $|A|$, so that we can say

either $|A| = 0$ *or* $B = 0$.

It follows that if

$$B = I - K$$

and $AB = 0$ then either $|A| = 0$ or $I - K = 0$; and this means that either $|A| = 0$ or $I = K$. Thus, if $AI = AK = A$ then either $|A| = 0$ or $K = I$. Unless $|A| = 0$, $AK = A$ means that $K = I$.

We have shown this for A of order (2×2). A similar exercise would show that if $AB = A$ and A is a square matrix of order (3×3) then

$$B = \begin{bmatrix} 1 & 0 & 0 \\ 0 & 1 & 0 \\ 0 & 0 & 1 \end{bmatrix}$$

unless $a_{11}(a_{22}a_{33} - a_{32}a_{23}) - a_{12}(a_{21}a_{33} - a_{3j}a_{23}) + a_{13}(a_{21}a_{32} - a_{31}a_{22}) = 0$. We can also write this as $|A| = 0$, it being understood that $|A|$ denotes this special arrangement of the elements of A. Later we shall consider this notation further.

It can be shown that a similar result holds for any square matrix A, of any order. If $AB = A$ then either some special arrangement of the elements of A produces zero, or B is a square matrix with units along the principal diagonal and all other elements zero.

If A is not square and $AB = A$ then a similar result is true, but now the arrangement of the elements of A has to be approached somewhat differently, as we shall see later.

We conclude for an $(m \times n)$ matrix A there is a square matrix of order $(n \times n)$, which we shall henceforward denote by I_n, with unit elements along the principle diagonal and all other elements zero, such that

$$AI_n = A$$

But if $AB = A$ it does not necessarily follow that $B = I_n$.

The reader may also prove that there is a square matrix of order $(m \times m)$, which is identical to I_n except that it is of a different order, such that

$$I_m A = A$$

but if $BA = A$ then B need not be equal to I_m.

Thus for example,

$$\begin{bmatrix} 1 & 3 \\ 2 & 5 \\ 4 & 7 \end{bmatrix} \begin{bmatrix} 1 & 0 \\ 0 & 1 \end{bmatrix} = \begin{bmatrix} 1 & 0 & 0 \\ 0 & 1 & 0 \\ 0 & 0 & 1 \end{bmatrix} \begin{bmatrix} 1 & 3 \\ 2 & 5 \\ 4 & 7 \end{bmatrix} = \begin{bmatrix} 1 & 3 \\ 2 & 5 \\ 4 & 7 \end{bmatrix}$$

Usually we write both I_m and I_n as I. The order of I is the order that is necessary if multiplication is to be possible.

Using the result that

$$(AB)C = A(BC)$$

we have such results as

$$AB = ABI = AIB = IAB = IAIB = IABI = IAIBI$$

For example

$$\begin{aligned} IAIBI &= AIBI & \text{(since } IA = A) \\ &= ABI & \text{(since } AI = A) \\ &= AB & \text{(since } BI = B) \end{aligned}$$

We may also note what happens if we pre- or post-multiply a matrix by a square matrix derived from I by interchanging some of its rows or columns. We give a few numerical results and leave it to the reader to derive the general result, that if we use $E_{(rs)}$ to denote the identity of matrix I with the (rs) columns interchanged then $E_{(rs)}A$ results in a matrix A with its r and s rows interchanged, while $AE_{(rs)}$ produces an interchange of the r and s columns of A.

e.g.

$$\begin{bmatrix} 1 & 0 & 0 \\ 0 & 0 & 1 \\ 0 & 1 & 0 \end{bmatrix} \begin{bmatrix} a_{11} & a_{12} & a_{13} & a_{14} \\ a_{21} & a_{22} & a_{23} & a_{24} \\ a_{31} & a_{32} & a_{33} & a_{34} \end{bmatrix} = \begin{bmatrix} a_{11} & a_{12} & a_{13} & a_{14} \\ a_{31} & a_{32} & a_{33} & a_{34} \\ a_{21} & a_{22} & a_{23} & a_{24} \end{bmatrix}$$

$$\begin{bmatrix} 1 & 0 & 0 & 0 \\ 0 & 0 & 1 & 0 \\ 0 & 1 & 0 & 0 \\ 0 & 0 & 0 & 1 \end{bmatrix} \begin{bmatrix} a_{11} & a_{12} \\ a_{21} & a_{22} \\ a_{31} & a_{32} \\ a_{41} & a_{42} \end{bmatrix} = \begin{bmatrix} a_{11} & a_{12} \\ a_{31} & a_{32} \\ a_{21} & a_{22} \\ a_{41} & a_{42} \end{bmatrix}$$

$$\begin{bmatrix} a_{11} & a_{12} & a_{13} & a_{14} \\ a_{21} & a_{22} & a_{23} & a_{24} \\ a_{31} & a_{32} & a_{33} & a_{34} \end{bmatrix} \begin{bmatrix} 1 & 0 & 0 & 0 \\ 0 & 0 & 1 & 0 \\ 0 & 1 & 0 & 0 \\ 0 & 0 & 0 & 1 \end{bmatrix} = \begin{bmatrix} a_{11} & a_{13} & a_{12} & a_{14} \\ a_{21} & a_{23} & a_{22} & a_{24} \\ a_{31} & a_{33} & a_{32} & a_{34} \end{bmatrix}$$

Before proceeding to the problem of how to define matrix division it will be useful for us to look at an example of matrix multiplication which brings out a geometrical meaning.

5. Matrices and geometry

Consider the matrix P (order 4×3) denoting the prices of 3 different goods in each of 4 consecutive weeks. Let matrix Q (order 3×2) denote the quantities of these 3 different goods bought by 2 people, who are assumed to buy the same quanties each week.

$$P = \begin{bmatrix} p_{11} & p_{12} & p_{13} \\ p_{21} & p_{22} & p_{23} \\ p_{31} & p_{32} & p_{33} \\ p_{41} & p_{42} & p_{43} \end{bmatrix} \qquad Q = \begin{bmatrix} q_{11} & q_{12} \\ q_{21} & q_{22} \\ q_{31} & q_{32} \end{bmatrix}$$

The product PQ will have as its (ij)th element

$$p_{i1}q_{1j} + p_{i2}q_{2j} + \dots$$

which is the total expenditure in week i of person j. The product matrix is of order (4×2).

Now P can be looked upon as either a set of 3 column vectors or a set of 4 row vectors. In forming PQ we take the inner products of rows of P with columns of Q. Let us, therefore, think of P as a set of 4 row vectors, but of Q as a set of two column vectors.

The 4 row vectors in P are easily interpreted in the space of 3 dimensions. The vector (p_{11}, p_{12}, p_{13}) is the price vector for week 1. There is a point in 3-dimensional space uniquely denoting the state of prices in that week. There is another point for the second week, and so on. The matrix **P** is a set of four price vectors. Similarly the two columns of **Q** denote two quantity vectors, one for each man. The vector (q_{11}, q_{21}, q_{31}) denotes a point in three-dimensional space which corresponds to the first man's weekly purchases—in terms of physical quantities.

The product **PQ**, of order (4×2), can be regarded in two ways. It can be looked upon as 4 column vectors, each of 2 dimensions, or as a pair of 4-dimensional row vectors. In the former case there is one vector for each week. Each of them denotes the total expenditures of the two men. The set of four such vectors will be four points in two-dimensional space, as shown in Diagram 29.13. If, however, we look at **PQ** as two vectors in four dimensions, one vector will show how the first man varies his weekly expenditure from week to week. For every possible set of four expenditures in the four different weeks there is a unique point in space. The one vector, corresponding to a particular point, defines the weekly expenditures of the first man. The second vector defines the second man's expenditures. As they are in four dimensions we cannot draw these vectors.

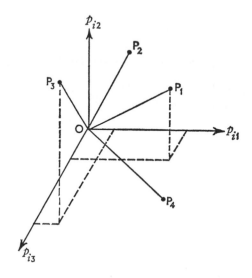

FIG. 29.13
P as a set of 4 price vectors in 3 dimensions

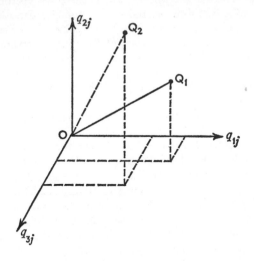

FIG. 29.14
Q as a set of 2 quantity vectors in 3 dimensions

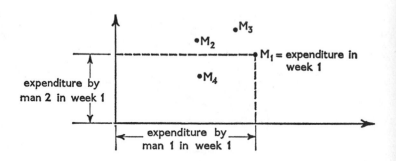

FIG. 29.15
PQ as a set of 4 expenditure vectors in 2 dimensions
PQ is also a set of 2 vectors in 4 dimensions

6. Inverse matrices and determinants

We now turn to a question which implies a need for division. Suppose that we know the quantity matrix **Q** and the expenditure matrix **PQ**. How can we find the price matrix **P**?

Let us begin by asking whether it is possible to find a matrix, which we will denote by Q^{-1}, such that

$$QQ^{-1} = I$$

If this is possible, then post-multiplication of PQ by Q^{-1} will yield

$$(PQ)Q^{-1} = P(QQ^{-1})$$
$$= PI$$
$$= P$$

In other words, if we know Q and PQ we can find P if we can obtain Q^{-1} such that

$$QQ^{-1} = I$$

It is to this problem, of finding the *inverse* Q^{-1} of Q to which we now turn.

First let us consider a simple example. Let us define A and B such that $AB = I$.

$$\begin{bmatrix} a_{11} & a_{12} \\ a_{21} & a_{22} \end{bmatrix} \begin{bmatrix} b_{11} & b_{12} \\ b_{21} & b_{22} \end{bmatrix} = \begin{bmatrix} 1 & 0 \\ 0 & 1 \end{bmatrix}$$

Expansion shows that

$$a_{11}b_{11} + a_{12}b_{21} = 1 \qquad \text{(i)}$$
$$a_{11}b_{12} + a_{12}b_{22} = 0 \qquad \text{(ii)}$$
$$a_{21}b_{11} + a_{22}b_{21} = 0 \qquad \text{(iii)}$$
$$a_{21}b_{12} + a_{22}b_{22} = 1 \qquad \text{(iv)}$$

Solution of these equations for the b's in terms of the a's will depend on whether certain combinations of the a's are zero. Assuming, for the moment, that these combinations are not zero it is easy to obtain the results

$$b_{11} = \frac{a_{22}}{a_{11}a_{22} - a_{12}a_{21}}$$

$$b_{21} = \frac{-a_{21}}{a_{11}a_{22} - a_{12}a_{21}}$$

$$b_{22} = \frac{a_{11}}{a_{11}a_{22} - a_{12}a_{21}}$$

$$b_{12} = \frac{-a_{12}}{a_{11}a_{22} - a_{12}a_{21}}$$

Denoting the denominator of these expressions by $|A|$ we have that

$$B = \begin{bmatrix} \dfrac{a_{22}}{|A|} & \dfrac{-a_{12}}{|A|} \\[2ex] \dfrac{-a_{21}}{|A|} & \dfrac{a_{11}}{|A|} \end{bmatrix}$$

It is easy to check that

$$AB = \begin{bmatrix} \dfrac{a_{11}a_{22}}{|A|} - \dfrac{a_{12}a_{21}}{|A|} & - \dfrac{a_{11}a_{12}}{|A|} + \dfrac{a_{12}a_{11}}{|A|} \\[2ex] \dfrac{a_{21}a_{22}}{|A|} - \dfrac{a_{22}a_{21}}{|A|} & - \dfrac{a_{21}a_{12}}{|A|} + \dfrac{a_{22}a_{11}}{|A|} \end{bmatrix}$$

$$= \begin{bmatrix} 1 & 0 \\ 0 & 1 \end{bmatrix} \quad \text{from the definition of } |A|.$$

We may note that if $|A| = 0$ then the method breaks down. This may remind us of the case in which $A(I - K) = 0$ does not necessarily mean that either $A = 0$ or $I - K = 0$. It looks as if it may be worth while to consider $|A|$ a little further

First let us repeat that $|A|$ is a special scalar quantity associated with the vector A. If A is of order 2×2 then we define

$$|A| = a_{11}a_{22} - a_{12}a_{21} = \begin{vmatrix} a_{11} & a_{12} \\ a_{21} & a_{22} \end{vmatrix}$$

If A is of order 3×3 then we define

$$|A| = a_{11}(a_{22}a_{33} - a_{32}a_{23}) - a_{12}(a_{21}a_{33} - a_{31}a_{23}) + a_{13}(a_{21}a_{32} - a_{31}a_{22})$$

It may be recalled that this expression is the one whose value has to be considered if we are enquiring whether $AB = A$ implies that $B = I$ when A is of order (3×3).

Let us now consider how these expressions may be derived from the matrices.

In the case of a (2×2) matrix the value of $|A|$ may be obtained by finding the product of the terms in the principle diagonal $(a_{11}a_{22})$ and subtracting the product of the other two terms $(a_{12}a_{21})$. We call it a second-order *determinant*.

$$\begin{vmatrix} a_{11} & a_{12} \\ a_{21} & a_{22} \end{vmatrix}$$

Now consider the value of $|A|$ for a 3×3 matrix. It can be written as

$$a_{11} \begin{vmatrix} a_{22} & a_{23} \\ a_{32} & a_{33} \end{vmatrix} - a_{12} \begin{vmatrix} a_{21} & a_{23} \\ a_{31} & a_{33} \end{vmatrix} + a_{13} \begin{vmatrix} a_{21} & a_{22} \\ a_{31} & a_{32} \end{vmatrix}$$

since $\quad \begin{vmatrix} a_{21} & a_{23} \\ a_{31} & a_{33} \end{vmatrix} = a_{21}a_{33} - a_{31}a_{23}$, etc.

The determinant which is the coefficient of a_{11} may be obtained from **A** by deleting the row and column containing a_{11} and forming the determinant of the elements that are left. We may denote $\quad \begin{vmatrix} a_{11} & a_{12} & a_{13} \\ a_{21} & a_{22} & a_{23} \\ a_{31} & a_{32} & a_{33} \end{vmatrix}$
this determinant by $|A_{11}|$. It is called the *minor* of **A** corresponding to a_{11}.

The coefficient of a_{12} is obtained from **A** by deleting the first row and second column and forming the determinant of the remaining elements. We denote it by $|A_{12}|$.

With this convention we have that

$$|A| = a_{11}|A_{11}| - a_{12}|A_{12}| + a_{13}|A_{13}|$$

With this value of a third-order determinant before us, let us proceed to a more general definition.

Let there be associated with any square matrix **A** a scalar value derived in a certain way from its elements. Let us denote this by $|A|$, or, if we wish to spell out the elements, by

$$\begin{vmatrix} a_{11} & a_{12} & \cdots & a_{1n} \\ a_{21} & a_{22} & \cdots & a_{2n} \\ \vdots & \vdots & & \vdots \\ a_{n1} & a_{n2} & \cdots & a_{nn} \end{vmatrix}$$

Let us denote by $|A_{ij}|$ the scalar value associated with the matrix derived from **A** if we delete the row and column containing a_{ij}. We call this a *minor* of $|A|$.

For a matrix of order $(n \times n)$ let us define the associated scalar value to be

$$|A| = a_{11}|A_{11}| - a_{12}|A_{12}| + a_{13}|A_{13}| \ldots (-)a_{1n}{}^{n-1}|A_{1n}|$$

with the understanding that if **A** consists of only one element then the associated scalar value is the scalar value of that element. We call $|A|$ the *determinant* of **A**.

Thus

(1) if $\quad \mathbf{A} = [a_{11}]$ then $|A| = a_{11}$

(2) if $\mathbf{A} = \begin{bmatrix} a_{11} & a_{12} \\ a_{21} & a_{22} \end{bmatrix}$ then $|A| = a_{11} |A_{11}| - a_{12} |A_{12}|$

and as $|A_{11}| = a_{22}$ and $|A_{12}| = a_{21}$, $|A| = a_{11}a_{22} - a_{12}a_{21}$

(3) if $\mathbf{A} = \begin{bmatrix} a_{11} & a_{12} & a_{13} \\ a_{21} & a_{22} & a_{23} \\ a_{31} & a_{32} & a_{33} \end{bmatrix}$ then $|A| = a_{11}|A_{11}| - a_{12}|A_{12}| + a_{13}|A_{13}|$

$$= a_{11} \begin{vmatrix} a_{22} & a_{23} \\ a_{32} & a_{33} \end{vmatrix} - a_{12} \begin{vmatrix} a_{21} & a_{23} \\ a_{31} & a_{33} \end{vmatrix} + a_{13} \begin{vmatrix} a_{21} & a_{22} \\ a_{31} & a_{32} \end{vmatrix}$$

(4) if $\mathbf{A} = \begin{bmatrix} a_{11} & a_{12} & a_{13} & a_{14} \\ a_{21} & a_{22} & a_{23} & a_{24} \\ a_{31} & a_{32} & a_{33} & a_{34} \\ a_{41} & a_{42} & a_{43} & a_{44} \end{bmatrix}$

then $|A| = a_{11} \begin{vmatrix} a_{22} & a_{23} & a_{24} \\ a_{32} & a_{33} & a_{34} \\ a_{42} & a_{43} & a_{44} \end{vmatrix} - a_{12} \begin{vmatrix} a_{21} & a_{23} & a_{24} \\ a_{31} & a_{33} & a_{34} \\ a_{41} & a_{43} & a_{44} \end{vmatrix}$

$$+ a_{13} \begin{vmatrix} a_{21} & a_{22} & a_{24} \\ a_{31} & a_{32} & a_{34} \\ a_{41} & a_{42} & a_{44} \end{vmatrix} - a_{14} \begin{vmatrix} a_{21} & a_{22} & a_{23} \\ a_{31} & a_{32} & a_{33} \\ a_{41} & a_{42} & a_{43} \end{vmatrix}$$

In short, the value of an nth-order determinant $|A|$ can be obtained as a sum of n determinants of order $(n-1)$, each with an appropriate coefficient and sign.

Before going further we may consider a few numerical examples.

(i) if $\mathbf{A} = \begin{bmatrix} 3 & 1 & -2 \\ 0 & 4 & 1 \\ 8 & 2 & 3 \end{bmatrix}$

$|A| = \begin{vmatrix} 3 & 1 & -2 \\ 0 & 4 & 1 \\ 8 & 2 & 3 \end{vmatrix} = 3 \begin{vmatrix} 4 & 1 \\ 2 & 3 \end{vmatrix} - 1 \begin{vmatrix} 0 & 1 \\ 8 & 3 \end{vmatrix} + (-2) \begin{vmatrix} 0 & 4 \\ 8 & 2 \end{vmatrix}$

$= 3(12-2) - 1(0-8) - 2(0-32)$
$= (3 \times 10) + 8 \qquad + 64$
$= 102$

(ii) if $\mathbf{A} = \begin{bmatrix} 1 & 2 & 3 & 4 \\ 3 & 5 & -1 & 2 \\ 0 & 4 & -4 & 1 \\ 4 & 7 & 2 & 6 \end{bmatrix}$ then

$$|A| = 1 \begin{vmatrix} 5 & -1 & 2 \\ 4 & -4 & 1 \\ 7 & 2 & 6 \end{vmatrix} -2 \begin{vmatrix} 3 & -1 & 2 \\ 0 & -4 & 1 \\ 4 & 2 & 6 \end{vmatrix} +3 \begin{vmatrix} 3 & 5 & 2 \\ 0 & 4 & 1 \\ 4 & 7 & 6 \end{vmatrix} -4 \begin{vmatrix} 3 & 5 & -1 \\ 0 & 4 & -4 \\ 4 & 7 & 2 \end{vmatrix}$$

which the reader may evaluate. He should obtain the value $|A| = 0$.
We now have to consider some properties of these determinants.
If we write out the third-order determinant we have

$$|A| = a_{11} \begin{vmatrix} a_{22} & a_{23} \\ a_{32} & a_{33} \end{vmatrix} - a_{12} \begin{vmatrix} a_{21} & a_{23} \\ a_{31} & a_{33} \end{vmatrix} + a_{13} \begin{vmatrix} a_{21} & a_{22} \\ a_{31} & a_{32} \end{vmatrix}$$

$$= a_{11}(a_{22}a_{33} - a_{23}a_{32}) - a_{12}(a_{21}a_{33} - a_{23}a_{31}) + a_{13}(a_{21}a_{32} - a_{22}a_{31})$$

$$= a_{11}a_{22}a_{33} - a_{11}a_{23}a_{32} - a_{12}a_{21}a_{33} + a_{12}a_{23}a_{31} + a_{13}a_{21}a_{32} - a_{13}a_{22}a_{31}$$

Notice that this has six terms, all different, half with positive signs and half with negative signs.

Each term is the product of three elements.

In each term there is one and only one element from each row, and one and only one element from each column.

Now consider the fourth-order determinant whose value is

$$a_{11} |A_{11}| - a_{12} |A_{12}| + a_{13} |A_{13}| - a_{14} |A_{14}|$$

All of the minors are third-order determinants and therefore their expansions have the properties just listed.

The six terms in the expansion of $|A_{11}|$ will now all be multiplied by a_{11}. They will clearly be different from the six terms arising from $|A_{12}|$, all multiplied by $-a_{12}$. And so on. In short, there will be $4 \times 6 = 24 = 4!$ terms. Each will be the product of four elements.

The terms coming from $|A_{11}|$ will be half positive and half negative. Multiplication by a_{11} will not change their signs. Those from $|A_{12}|$ will have their signs changed, but half will still be positive, and half will be negative. In short, the fourth-order determinant will have 12 positive and 12 negative terms.

The terms in $|A_{11}|$ are drawn from all rows and columns of **A** except the first row and first column. But they are multiplied by a_{11} and so each of the first six terms of $|A|$ will contain a_{11} and one and only one element from every other row and column. Similarly, the next six will contain a_{12} and one and only one element from every other row and column. Thus, every term will consist of the product of four elements, one from each of the four rows, chosen so that no column is duplicated.

It is easy to see that a fifth-order determinant will have $5!$ terms, half positive and half negative, all different, each consisting of five elements chosen so that rows and columns are not duplicated.

In general, the nth-order determinant will have $n!$ such terms. Each will have n elements. One (and only one) of these will be from row 1. Suppose it is from column i. Another will be from row 2. Let it also be in column j. A third comes from row 3 and (say) column k. And so on, until all the rows are used up. By then, the rule of "no duplication" will have used all of the columns. Thus, the typical term can be written as

$$a_{1i}a_{2j}a_{3k}\ldots a_{nc}$$

where a_{nc} comes from row n and (say) column c. In this expression no (row) number is duplicated. The (column) letters are in any order, but without duplication.

Now we can write out n letters in $n!$ different orders. If we do so, and insert them in the value of the typical term, we obtain the $n!$ different terms in the value of $|A|$.

Thus $|A|$ consists of the $n!$ different possible combinations of n elements drawn one from each row and one from each column.

We know that one term in the expansion of (say) a fifth-order determinant is

$$a_{13}a_{21}a_{34}a_{42}a_{55}$$

How can we tell its sign? Provided that the first subscripts (denoting the rows) are written in natural order, the sign depends on the order of the second subscripts.

When any two numbers are out of their natural order we say that an *inversion of order* has occurred.

In the expression 3 1 4 2 there are three inversions, namely 3 before 1, 3 before 2, 4 before 2.

The sign of a term in the expansion of a determinant is *positive* if the number of inversions in the order of the second subscript is even, and *negative* if it is odd, provided that the first subscripts are in natural order.

Thus

$$a_{13}a_{21}a_{34}a_{42}a_{55}$$

has its first subscripts in natural order. Its second subscripts are in the order

$$3\ 1\ 4\ 2\ 5$$

which has

3 before 1
3 ,,   2
4 ,,   2

and so has three inversions. This term, therefore, has a negative sign.

It is possible to justify this rule by a tedious explanation in which one first considers the signs of terms in the expansion of a second-order matrix, and then how these lead to the signs in a third-order matrix, and so on. The reader who experiments with a few low-order matrices will see the validity of the rule in less time than he would take to read the explanation.

Essentially the point is that if one starts at the top row and works downwards to the bottom of the matrix then every movement to the left introduces not only one inversion of the column subscript but also one negative sign into the value of that term.

7. Properties of determinants

We are now able, quite quickly, to prove some important results which will help us to evaluate determinants much more speedily, and also allow us to divide matrices.

1. *If a determinant is transposed so that its rows become columns and its columns become rows then its value is unchanged.*

Consider any term of $|A|$, say

$$a_{14}a_{21}a_{33}a_{55}$$

In the transposed determinant this term will still appear, and have the same numerical value. Will its sign be different? Consider the tabular presentation below.

$$|A| = \begin{vmatrix} a_{11} & a_{12} & a_{13} & \boxed{a_{14}} \\ \boxed{a_{21}} & a_{22} & a_{23} & a_{24} \\ a_{31} & a_{32} & \boxed{a_{33}} & a_{34} \\ a_{41} & \boxed{a_{42}} & a_{43} & a_{44} \end{vmatrix} \qquad |A'| = \begin{vmatrix} a_{11} & \boxed{a_{21}} & a_{31} & a_{41} \\ a_{12} & a_{22} & a_{32} & \boxed{a_{42}} \\ a_{13} & a_{23} & \boxed{a_{33}} & a_{43} \\ \boxed{a_{14}} & a_{24} & a_{34} & a_{44} \end{vmatrix}$$

Let us rewrite this term so that its row subscripts (which are now the second subscripts) are in natural order. It becomes

$$a_{21}a_{42}a_{33}a_{14}$$

In the former case we have to consider the inversions in

$$4 \quad 1 \quad 3 \quad 2$$

and in the latter case the number of inversions in

$$2 \quad 4 \quad 3 \quad 1$$

Starting with

$$4 \quad 1 \quad 3 \quad 2$$

move 1 two places to the right, getting

$$4 \quad 3 \quad 2 \quad 1$$

which has added two inversions. Now move 2 two places to the left, getting

$$2 \quad 4 \quad 3 \quad 1$$

which has removed two inversions. The total number of inversions is thus unchanged, and so is the sign of the term.

The result will, in fact, apply to any term in the determinant. If the first subscripts are in order and there are p inversions in the second subscript, then rearrangement so that the second subscripts are in order will introduce p inversions of the first subscript.

Thus both the sign and the numerical value of each term in the expansion of $|\ A'\ |$ is the same as in $|\ A\ |$.

2. *Any rule about the rows of a determinant will also hold for its columns*, since one has only to transpose the determinant in order to make it applicable.

3. *Shifting a row (or column) p places will result in no change of sign if p is even, but a change if p is odd.*

This is easily seen. Suppose we shift a row one place downwards. Then row i and row $i+1$ change places. Thus in the term

$$a_{1c}a_{2d}a_{3e}\ldots a_{if}a_{i+1g}\ldots$$

the only change is that it becomes

$$a_{1c}a_{2d}a_{3e}\ldots a_{i+1g}a_{if}\ldots$$

Whatever inversions existed before exist now—except that if $f<g$ then there was previously no inversion out of a_{ii} and a_{i+1g}, whereas there is now; and if $f \geqslant g$ then the inversion that used to exist no longer exists. Thus, shifting row i one place downwards changes the number of inversions (and therefore the sign) associated with every term. It follows that the determinant changes sign. If we shift it another place, the determinant changes sign again— and so retains its original sign.

In general, a shift (upwards or downwards) of p places will result in p changes of sign. If p is even there is no net change. If p is odd there is a net change of sign.

4. *If two rows (or columns) are interchanged there is a change of sign.*

Suppose we interchange row i and row $i+m$. Effectively, we move row i to position $i+m+1$ (passing over m rows to do so) and then we move row $i+m$ into position i (passing over $m-1$ rows). This means a total of $m+m-1$ changes of sign: and this is an odd number, indicating a reversal of sign.

5. *If two rows (or columns) are identical the value of the determinant is zero.*

If the original value is $|\,A\,|$ and we interchange two rows we obtain $-|\,A\,|$. But if the rows are equal there is no change and so

$$|\,A\,| = -|\,A\,|$$

which can be so only if $|\,A\,| = 0$.

6. *If every element of a row (or column) is multiplied by a scalar λ then the determinant is multiplied by λ.*

This is easily proved. Suppose that in the determinant $|\,A\,|$ we multiply every element of row i by λ, and call this new determinant $|\,B\,|$. Shift row i to the top position. This gives a determinant $|\,B\,|^*$ which may differ from $|\,B\,|$ in sign but not otherwise. Also shift the ith row of $|\,A\,|$ to the top position. This new matrix will possibly differ from $|\,A\,|$ in sign, but not otherwise. Call it $|\,A\,|^*$.

Compare $|\,A\,|^*$ and $|\,B\,|^*$. They are identical except for their top rows. In A^* the top row will be

$$a_{i1} a_{i2} a_{i3} \ldots$$

and in B^* it will be

$$\lambda a_{i1} \lambda a_{i2} \lambda a_{i3} \ldots$$

Consider the expansions of these determinants. For $|\,A\,|^*$ we have

$$|\,A\,|^* = a_{i1}\,|\,A_{i1}\,| \; - a_{i2}\,|\,A_{i2}\,| \; + \ldots \text{ etc.}$$

for

$$|\,B\,|^* = \lambda a_{i1}\,|\,A_{i1}\,| \; - \lambda a_{i2}\,|\,A_{i2}\,| \; + \ldots \text{ etc.}$$

which shows that

$$|\,B\,|^* = \lambda\,|\,A\,|^*$$

It follows that

$$|\,B\,| = \lambda\,|\,A\,|$$

7. *If one row (or column) is multiplied by λ, and a second row (or column) by μ then the new determinant has the value*

$$\lambda\mu\,|\,A\,|$$

Q

8. *A determinant can be expanded in terms of the elements of any row (or column) and their associated minors, with appropriate change of sign.*

This follows directly from the method of proof of result 6. For example

$$\begin{vmatrix} 1 & 2 & 3 & 5 \\ 4 & 6 & 7 & 8 \\ 2 & 0 & 1 & 9 \\ 8 & 2 & 1 & 3 \end{vmatrix} = -4 \begin{vmatrix} 2 & 3 & 5 \\ 0 & 1 & 9 \\ 2 & 1 & 3 \end{vmatrix} + 6 \begin{vmatrix} 1 & 3 & 5 \\ 2 & 1 & 9 \\ 8 & 1 & 3 \end{vmatrix}$$

$$-7 \begin{vmatrix} 1 & 2 & 5 \\ 2 & 0 & 9 \\ 8 & 2 & 3 \end{vmatrix} + 8 \begin{vmatrix} 4 & 6 & 7 \\ 2 & 0 & 1 \\ 8 & 2 & 1 \end{vmatrix}$$

where the signs have been reversed because it would take one change of sign to get the second row into the top position.

9. *The addition of a constant multiple of one row (or column) to another row (or column) leaves the determinant unchanged.*

Suppose we have a third-order determinant and form a new one by adding to the top row λ times the second row, getting

$$\begin{vmatrix} a_{11} + \lambda a_{21} & a_{12} + \lambda a_{22} & a_{13} + \lambda a_{23} \\ a_{21} & a_{22} & a_{23} \\ a_{31} & a_{32} & a_{33} \end{vmatrix}$$

The expansion of this is

$$(a_{11} + \lambda a_{21}) \mid A_{11} \mid + (a_{12} + \lambda a_{22}) \mid A_{12} \mid + (a_{13} + \lambda a_{23}) \mid A_{13} \mid$$
$$= a_{11} \mid A_{11} \mid + a_{12} \mid A_{12} \mid + a_{13} \mid A_{13} \mid$$
$$+ \lambda(a_{21} \mid A_{11} \mid + a_{22} \mid A_{12} \mid + a_{23} \mid A_{13} \mid)$$
$$= \mid A \mid + 0$$

since the bracketed expression is the expansion of

$$\begin{vmatrix} a_{21} & a_{22} & a_{23} \\ a_{21} & a_{22} & a_{23} \\ a_{31} & a_{32} & a_{33} \end{vmatrix}$$

which has two identical rows.

10. *If a determinant is expanded in terms of the minors of a wrong row (or column) it vanishes.*

Let us first put this result in more acceptable language. The element a_{ij} has associated with it a *minor* $\mid A_{ij} \mid$. We call the signed minor $(-1)^{i+j} \mid A_{ij} \mid$ the co-factor. It is easy to see that

$$\mid A \mid = a_{11} \mid A_{11} \mid - a_{12} \mid A_{12} \mid + a_{13} \mid A_{13} \mid - \ldots$$
$$= a_{11} \mid B_{11} \mid + a_{12} \mid B_{12} \mid + a_{13} \mid B_{13} \mid + \ldots$$

where $\mid B_{1j} \mid = (-1)^{1+j} \mid A_{1j} \mid$

Consideration of result (8) now allows us to write

$$|A| = \sum a_{ij}(-1)^{i+j} |A_{ij}|$$

The result that we are about to prove is that

$$\sum a_{kj}(-1)^{k+j} |A_{ij}| = 0$$

e.g. that for

$$\begin{vmatrix} a_{11} & a_{12} & a_{13} \\ a_{21} & a_{22} & a_{23} \\ a_{31} & a_{32} & a_{33} \end{vmatrix}$$

$$a_{21}|A_{11}| - a_{22}|A_{12}| + a_{23}|A_{13}|$$
$$= a_{21}(a_{22}a_{33} - a_{23}a_{32}) - a_{22}(a_{21}a_{33} - a_{23}a_{31}) + a_{23}(a_{21}a_{32} - a_{31}a_{22}) = 0$$

The proof is identical to the second part of the proof of result 9. The procedure is usually described as " expansion in terms of alien co-factors ".

8. The Evaluation of Determinants

We now consider one or two tricks to reduce the amount of work involved in the evaluation of determinants: for whenever we perform matrix division we shall need to evaluate one. To begin, we consider a few tricks which will help us to see some of the uses of the rules we have just proven. Then we demonstrate a more systematic procedure.

(1) Consider $\quad |A| = \begin{vmatrix} 1 & 3 & 7 & 4 \\ 2 & 1 & 1 & 1 \\ 1 & 2 & 3 & 1 \\ 5 & 5 & 9 & 6 \end{vmatrix}$

Form a new Row 1 by adding to the original Row 1 twice the original Row 2. We denote this procedure by

$$R1^* = R1 + 2R2$$

We get

$$\begin{vmatrix} 5 & 5 & 9 & 6 \\ 2 & 1 & 1 & 1 \\ 1 & 2 & 3 & 1 \\ 5 & 5 & 9 & 6 \end{vmatrix}$$

which is zero—as two rows are the same.

$$(2) \quad |A| = \begin{vmatrix} 2 & 4 & 6 & 8 \\ 1 & 5 & 1 & 4 \\ 3 & 7 & 3 & 0 \\ 1 & 3 & 2 & 1 \end{vmatrix} = 2 \begin{vmatrix} 1 & 2 & 3 & 4 \\ 1 & 5 & 1 & 4 \\ 3 & 7 & 3 & 0 \\ 1 & 3 & 2 & 1 \end{vmatrix} \quad \text{(rule 7)}$$

$$= 2 \begin{vmatrix} 3 & 7 & 3 & 0 \\ 1 & 2 & 3 & 4 \\ 1 & 5 & 1 & 4 \\ 1 & 3 & 2 & 1 \end{vmatrix} \quad \text{(by shifting row 3 upwards)}$$

$$= -2 \begin{vmatrix} 0 & 3 & 7 & 3 \\ 4 & 1 & 2 & 3 \\ 4 & 1 & 5 & 1 \\ 1 & 1 & 3 & 2 \end{vmatrix} \quad \text{(by shifting column 4 frontwards)}$$

$$= -2 \begin{vmatrix} 0 & 0 & 7 & 3 \\ 4 & -2 & 2 & 3 \\ 4 & 0 & 5 & 1 \\ 1 & -1 & 3 & 2 \end{vmatrix} \quad (C2^* = C2 - C4)$$

$$= -2 \times 7 \begin{vmatrix} 4 & -2 & 3 \\ 4 & 0 & 1 \\ 1 & -1 & 2 \end{vmatrix} \quad -2 \times (-3) \begin{vmatrix} 4 & -2 & 2 \\ 4 & 0 & 5 \\ 1 & -1 & 3 \end{vmatrix}$$

$$= -14 \begin{vmatrix} 0 & -2 & 2 \\ 4 & 0 & 1 \\ 1 & -1 & 2 \end{vmatrix} \quad +6 \begin{vmatrix} 0 & -2 & -3 \\ 4 & 0 & 5 \\ 1 & -1 & 3 \end{vmatrix} \quad \begin{matrix} (R1^* = R1 - R2 \\ \text{in each case)} \end{matrix}$$

$$= -14 \begin{vmatrix} 0 & 0 & 2 \\ 4 & 1 & 1 \\ 1 & 1 & 2 \end{vmatrix} \quad +6 \begin{vmatrix} 0 & -2 & -3 \\ 0 & 4 & -7 \\ 1 & -1 & 3 \end{vmatrix}$$

$$= -14 \times 2 \begin{vmatrix} 4 & 1 \\ 1 & 1 \end{vmatrix} \quad +6 \begin{vmatrix} 0 & 0 & 1 \\ -2 & 4 & -1 \\ -3 & -7 & 3 \end{vmatrix} \quad \text{(transposing)}$$

$$= -28(4 - 1) \quad\quad +6 \begin{vmatrix} -2 & 4 \\ -3 & -7 \end{vmatrix}$$

$$= -84 \quad\quad\quad +6(14 + 12)$$
$$= -84 \quad\quad\quad +156$$
$$= 72$$

(3) We may obtain this value more systematically by a procedure which is useful if a calculating machine is being used. We reduce the order of the determinant by obtaining a top row which has all

but one of its elements as zero. We then act similarly on this smaller determinant. Consider, once again,

$$|A| = \begin{vmatrix} 2 & 4 & 6 & 8 \\ 1 & 5 & 1 & 4 \\ 3 & 7 & 3 & 0 \\ 1 & 3 & 2 & 1 \end{vmatrix}$$

Form a new determinant (of the same value) by subtracting those multiples of column 1 from columns 2, 3 and 4 which will produce a top row with zeros everywhere except in the first position.

$$|A| = \begin{vmatrix} 2 & 0 & 0 & 0 \\ 1 & 3 & -2 & 0 \\ 3 & 1 & -6 & -12 \\ 1 & 1 & -1 & -3 \end{vmatrix} = 2 \begin{vmatrix} 3 & -2 & 0 \\ 1 & -6 & -12 \\ 1 & -1 & -3 \end{vmatrix}$$

$$= \tfrac{2}{3} \begin{vmatrix} 3 & -6 & 0 \\ 1 & -18 & -12 \\ 1 & -3 & -3 \end{vmatrix}$$

where we have altered column (2) to make its top element a multiple of the new top left-hand element.

Now add twice the first column to the second, to get

$$\tfrac{2}{3} \begin{vmatrix} 3 & 0 & 0 \\ 1 & -16 & -12 \\ 1 & -1 & -3 \end{vmatrix} = \tfrac{2}{3} \times 3 \begin{vmatrix} -16 & -12 \\ -1 & -3 \end{vmatrix}$$

$$= 2(-1)^2 \begin{vmatrix} 16 & 12 \\ 1 & 3 \end{vmatrix} = 72$$

EXERCISE 29.1

By using at least two different methods, evaluate the determinants below and check that the different methods give the same results.

(i) $\begin{vmatrix} 1 & 3 & 7 \\ 2 & 1 & 1 \\ 1 & 2 & 3 \end{vmatrix}$ (ii) $\begin{vmatrix} 1 & 3 & 8 \\ 4 & 1 & 2 \\ 8 & 1 & 3 \end{vmatrix}$ (iii) $\begin{vmatrix} 2 & 4 & 6 \\ 8 & 10 & 12 \\ 10 & 14 & 18 \end{vmatrix}$

(iv) $\begin{vmatrix} 2 & 4 & 8 & 16 \\ 32 & 8 & 4 & 8 \\ 2 & 16 & 8 & 4 \\ 4 & 2 & 16 & 8 \end{vmatrix}$ (v) $\begin{vmatrix} 27 & 81 & 27 & 81 \\ 81 & 27 & 27 & 81 \\ 27 & 81 & 81 & 27 \\ 81 & 81 & 27 & 27 \end{vmatrix}$

9. Division of matrices

We can now return to the division of matrices. We have shown that the problem is that of finding A^{-1} such that

$$AA^{-1} = I$$

Consider the square matrix A whose typical element is a_{ij}.

Form a new matrix by replacing each element by its co-factor—which we denote by c_{ij}.

For example, if

$$A = \begin{bmatrix} 1 & 2 & 3 \\ 4 & 0 & 5 \\ 6 & 7 & 8 \end{bmatrix} \qquad \text{then} \quad \begin{aligned} c_{11} &= (0 \times 8 - 5 \times 7) = -35 \\ c_{12} &= -(4 \times 8 - 5 \times 6) = -2 \\ c_{13} &= (4 \times 7 - 0 \times 6) = 28 \\ &\text{etc.} \end{aligned}$$

and our new matrix becomes

$$\begin{bmatrix} -35 & -2 & 28 \\ 5 & -10 & 5 \\ 10 & 7 & -8 \end{bmatrix}$$

Now transpose this matrix, so that the (ij)th element becomes c_{ji}.

$$\begin{bmatrix} c_{11} & c_{21} \dots c_{n1} \\ c_{12} & c_{22} \dots c_{n2} \\ \vdots & \vdots \\ c_{1n} & c_{2n} \dots c_{nn} \end{bmatrix} \qquad \text{e.g.} \quad \begin{bmatrix} -35 & 5 & 10 \\ -2 & -10 & 7 \\ 28 & 5 & -8 \end{bmatrix}$$

We call this new matrix the *adjugate* of A and write it as adj A

Now consider the matrix product

$$A \,(\text{adj } A) = \begin{bmatrix} a_{11} & a_{12} & \dots & a_{1n} \\ a_{21} & a_{22} & \dots & a_{2n} \\ \vdots & \vdots & & \vdots \\ a_{n1} & a_{n2} & \dots & a_{nn} \end{bmatrix} \begin{bmatrix} c_{11} & c_{21} & \dots & c_{n1} \\ c_{12} & c_{22} & \dots & c_{n2} \\ \vdots & \vdots & & \vdots \\ c_{1n} & c_{2n} & \dots & c_{nn} \end{bmatrix}$$

In the expansion of this, the first row into the first column gives

$$a_{11}c_{11} + a_{12}c_{12} + \dots + a_{1n}c_{1n}$$

which, remembering that $c_{11} = |\,A_{11}\,|$, is simply $|\,A\,|$.

But the first row into the second column gives

$$a_{11}c_{21} + a_{12}c_{22} + \dots + a_{1n}c_{2n}$$

which is an expansion in terms of alien co-factors, and is therefore zero. Proceeding in this way we find that all but the diagonal elements are zero. We obtain

$$A \,(\text{adj } A) = \begin{bmatrix} |\,A\,| & 0 & 0 & \dots & 0 \\ 0 & |\,A\,| & 0 & \dots & 0 \\ 0 & 0 & |\,A\,| & \dots & 0 \\ \vdots & \vdots & \vdots & & \vdots \\ 0 & 0 & 0 & \dots & |\,A\,| \end{bmatrix}$$

$$= |A| \begin{bmatrix} 1 & 0 & 0 & ... & 0 \\ 0 & 1 & 0 & ... & 0 \\ 0 & 0 & 1 & ... & 0 \\ \vdots & \vdots & \vdots & & \vdots \\ 0 & 0 & 0 & ... & 1 \end{bmatrix} = |A| \mathbf{I}$$

For example, if **A** has the value just considered,

$$A \, (\mathrm{adj} \, A) = \begin{bmatrix} 1 & 2 & 3 \\ 4 & 0 & 5 \\ 6 & 7 & 8 \end{bmatrix} \begin{bmatrix} -35 & 5 & 10 \\ -2 & -10 & 7 \\ 28 & 5 & -8 \end{bmatrix}$$

$$= \begin{bmatrix} -35-4+84 & 5-20+15 & 10+14-24 \\ -140+0+140 & 20+0+25 & 40+0-40 \\ -210-14+224 & 30-70+40 & 60+49-64 \end{bmatrix}$$

$$= \begin{bmatrix} 45 & 0 & 0 \\ 0 & 45 & 0 \\ 0 & 0 & 45 \end{bmatrix} = 45\mathbf{I}$$

where, it may be noted,

$$|A| = \begin{vmatrix} 1 & 2 & 3 \\ 4 & 0 & 5 \\ 6 & 7 & 8 \end{vmatrix}$$

$$= \begin{vmatrix} 1 & 0 & 0 \\ 4 & -8 & -7 \\ 6 & -5 & -10 \end{vmatrix} = 80-35 = 45$$

The reader may care to show that

$$(\mathrm{adj} \, A)A = |A| \mathbf{I}$$

Thus, if we define A^{-1} as $\dfrac{1}{|A|} \mathrm{adj} \, \mathbf{A}$

$$= \begin{bmatrix} \dfrac{c_{11}}{|A|} & \dfrac{c_{21}}{|A|} & ... & \dfrac{c_{n1}}{|A|} \\ \dfrac{c_{12}}{|A|} & \dfrac{c_{22}}{|A|} & ... & \dfrac{c_{n2}}{|A|} \\ \vdots & \vdots & & \vdots \end{bmatrix}$$

then

$$\mathbf{AA^{-1} = A^{-1}A = I}$$

In terms of our example

$$A^{-1} = \begin{bmatrix} \dfrac{-35}{45} & \dfrac{5}{45} & \dfrac{45}{10} \\ \dfrac{-2}{45} & \dfrac{-10}{45} & \dfrac{7}{45} \\ \dfrac{28}{45} & \dfrac{5}{45} & \dfrac{-8}{45} \end{bmatrix}$$

Note that A^{-1} has been defined only for a square matrix, since $A \mid$ has been defined only for a square determinant.

We call A^{-1} the *inverse* or *reciprocal* of A.

Note, also, that A^{-1} is unique, for suppose that there is a matrix B such that

$$AB = I$$

We have that

$$A^{-1} = A^{-1}I = A^{-1}AB$$
$$= \quad IB$$
$$= \quad B$$

Similarly if there is a matrix C such that

$$CA = I$$

then

$$C = A^{-1}$$

But we must note one other point, which is absolutely crucial. A^{-1} can be defined as $\dfrac{1}{\mid A \mid}$ adj A only if $\mid A \mid$ is not zero. It is this which has already been illustrated on several occasions—as when we found that

$$AB = 0$$

does not necessarily mean that A or B is 0. We can now see what happens here. Suppose

$$AB = 0$$

Let A be such that $\mid A \mid \neq 0$. Then A^{-1} exists and $A^{-1}A = I$. Therefore, if

$$AB = 0$$
$$A^{-1}AB = A^{-1}0 = 0$$
$$IB = \quad 0$$
$$\therefore \quad B = \quad 0$$

Alternatively, let $\mid B \mid = 0$. It follows that

$$A = 0$$

But another alternative is $|A| = 0$. In this case A^{-1} does not exist, and we can conclude no general result about A or B.

Thus if $AB = 0$ then either $A = 0$, or $B = 0$ or $|A| = 0$ or $|B| = 0$, or more than one of these results holds.

Consider now the problem we have already cited. Given Q and the value of $M = PQ$, what is P? For example, the quantity matrix is

$$Q = \begin{bmatrix} 5 & 8 \\ 4 & 2 \\ 6 & 3 \end{bmatrix} \quad \text{(2 men, 3 goods)}$$

and the expenditure matrix is

$$M = \begin{bmatrix} 31 & 21 \\ 21 & 16 \\ 20 & 21 \\ 29 & 20 \end{bmatrix} \quad \text{(2 men, 4 weeks)}$$

What is the price matrix, showing the prices of the three different goods in each of the four weeks?

The answer comes thus:

$$PQ = M$$

if Q^{-1} exists then

$$PQQ^{-1} = MQ^{-1}$$

i.e.
$$PI = MQ^{-1}$$
$$P = MQ^{-1}$$

and so the evaluation of P involves finding Q^{-1} and pre-multiplying this by M.

But in the example we have given Q is not square, and we have defined the inverse of a matrix only for square matrices. Is a simple solution possible?

Let the matrix we are seeking be

$$P = \begin{bmatrix} p_{11} & p_{12} & p_{13} \\ p_{21} & & \vdots \\ p_{31} & & \vdots \\ p_{41} & \cdots & p_{43} \end{bmatrix}$$

Then
$$PQ = \begin{bmatrix} p_{11} & p_{12} & p_{13} \\ p_{21} & & \vdots \\ p_{31} & & \vdots \\ p_{41} & \cdots & p_{43} \end{bmatrix} \begin{bmatrix} 5 & 8 \\ 4 & 2 \\ 6 & 3 \end{bmatrix}$$

$$= \begin{bmatrix} 5p_{11} + 4p_{12} + 6p_{13} & 8p_{11} + 2p_{12} + 3p_{13} \\ & \text{etc.} \end{bmatrix}$$

Comparison of this with the known matrix M shows that

$$5p_{11} + 4p_{12} + 6p_{13} = 31$$
$$8p_{11} + 2p_{12} + 3p_{13} = 21$$

and there are no more equations involving these three p's. If we subtract the first equation from double the second equation we get

$$11p_{11} = 11 \quad \text{whence} \quad p_{11} = 1$$

but we can go little further, for both equations now tell us the same thing, namely

$$2p_{12} + 3p_{13} = 13$$

There is an infinite set of values of p_{12} and p_{13} which satisfy this. In other words, there is an infinite set of answers to the question,

given $\mathbf{Q} = \begin{bmatrix} 58 \\ 42 \\ 63 \end{bmatrix}$ and $\mathbf{M} = \begin{bmatrix} 31 & 21 \\ 21 & 16 \\ 20 & 21 \\ 29 & 20 \end{bmatrix}$, what is $\mathbf{P}$?

If, on the other hand, we have information about one more man, there is hope—provided that the new, square, matrix $\mathbf{Q}$ does not have a zero $|Q|$—i.e. provided it is non-singular. If, for example, the third man results in a new $\mathbf{Q}$ and a new $\mathbf{M}$ such that

$$\mathbf{Q} = \begin{bmatrix} 5 & 8 & 13 \\ 4 & 2 & 5 \\ 6 & 3 & 8 \end{bmatrix}$$

and
$$\mathbf{M} = \begin{bmatrix} 31 & 21 & 47 \\ 21 & 16 & 34 \\ 20 & 21 & 39 \\ 29 & 20 & 44 \end{bmatrix}$$

then $\mathbf{Q}^{-1}$ exists, and

$$\mathbf{P} = \mathbf{MQ}^{-1}$$

which the student may show to have the value $\begin{bmatrix} 1 & 2 & 3 \\ 1 & 1 & 2 \\ 2 & 1 & 1 \\ 1 & 3 & 2 \end{bmatrix}$

But if our third man buys slightly different quantities, yielding a matrix

$$Q = \begin{bmatrix} 5 & 8 & 13 \\ 4 & 2 & 6 \\ 6 & 3 & 9 \end{bmatrix}$$

or even

$$Q = \begin{bmatrix} 5 & 8 & 18 \\ 4 & 2 & 10 \\ 6 & 3 & 15 \end{bmatrix}$$

then we are no further forwards. In each case Q is *singular* since $|Q| = 0$. In the first case his spending is simply the sum of the spending of the other two—on each item. The information we derive from a knowledge of the purchases and expenditures of the first two men tells us everything that a knowledge of the third man produces. In the second case, $C_3 = 2C_1 + C_2$, and once again we learn nothing new.

But we may have a different variation of this problem. Suppose that we have

$$Q = \begin{bmatrix} 5 & 1 & 2 \\ 2 & 3 & 1 \end{bmatrix} \qquad \text{3 men, 2 goods}$$

$$M = \begin{bmatrix} 9 & 7 & 4 \\ 12 & 5 & 5 \\ 17 & 6 & 7 \\ 24 & 10 & 10 \end{bmatrix} \qquad \text{4 weeks, 3 men}$$

Once again, Q is not square and we have not defined Q^{-1}.

$$\text{Let } P = \begin{bmatrix} p_{11} & p_{12} \\ p_{21} & p_{22} \\ p_{31} & p_{32} \\ p_{41} & p_{42} \end{bmatrix}$$

then we can evaluate PQ and compare terms with M to obtain

$$5p_{11} + 2p_{12} = 9$$
$$p_{11} + 3p_{12} = 7$$
$$2p_{11} + p_{12} = 4$$
$$\text{etc.}$$

The first two of these equations yield

$$p_{11} = 1, \quad p_{12} = 2$$

and these values satisfy the third equation, which gives us no new information. Here the third man is redundant. The quantity vector $\begin{bmatrix} 5 & 1 \\ 2 & 3 \end{bmatrix}$ tells us all that we want to know—and it is non-singular.

But suppose that our expenditure vector had had a value of 6 for m_{13}, instead of 4. Our equations would then have been

$$5p_{11} + 2p_{12} = 9$$
$$p_{11} + 3p_{12} = 7$$
$$2p_{11} + \ p_{12} = 6$$

The first two yield $p_{11} = 1$, $p_{12} = 2$, but these values do not satisfy the third equation. Our data are not consistent with each other—and no solution is possible.

We shall shortly consider the tests which have to be applied to a matrix equation of the form $\mathbf{PQ} = \mathbf{M}$, to see whether (a) it is consistent, (b) it contains enough information for solution to be possible, and (c) it contains any superfluous information.

CHAPTER XXX

LINEAR DEPENDENCE

Nothing in the world is single;
All things, by a law divine,
In one another's being mingle—
P. B. SHELLEY

1. Introduction

While it is possible to illustrate some ideas by the methods of Chapter XXIX, there is a great deal to be said for presenting them more formally and proving them; for, apart from anything else, in doing so we come across several concepts which are frequently used in the linear algebraic development of economics.

We must first define the concept of linear dependence.

Definitions:

(1) We define the set of elements $(x_1, x_2,...,x_n)$ to be *linearly dependent* if at least one element in the set—say x_i—can be written as

$$x_i = \sum_r \lambda_r x_r$$

where x_r are other members of the set and λ_r are constant scalar quantities.

(2) We define a set of vectors to be linearly dependent if we can write one of them as a linear combination of the others, thus

$$\mathbf{x}_1 = \sum_r \lambda_r \mathbf{x}_r \qquad (r \neq 1)$$

where the λ's are not all zero. Another statement of this definition is that if there exists a set of λ's not all zero, such that

$$\sum \lambda_r \mathbf{x}_r = 0$$

then the set of vectors is linearly dependent. If there is no such set of λ's then the vectors are *linearly independent*.

477

For example, if

$$x_1 = (3, 4)$$

and $\qquad\qquad\qquad x_2 = (5, 2)$

and $\qquad\qquad\qquad x_3 = (9, 11)$

then consider whether there exist λ_1 and λ_2 such that

$$x_3 = \lambda_1 x_1 + \lambda_2 x_2$$

We have $\qquad\qquad 9 = 3\lambda_1 + 5\lambda_2$

$$11 = 4\lambda_1 + 2\lambda_2$$

which are satisfied by $\lambda_1 = \dfrac{37}{14}$, $\lambda_2 = \dfrac{3}{14}$ which shows that the vectors

are linearly dependent. We could write

$$x_3 = \frac{37}{14} x_1 + \frac{3}{14} x_2$$

whence $\qquad\qquad 14x_3 = 37x_1 + 3x_2$

whence $\qquad\qquad 37x_1 + 3x_2 - 14x_3 = 0$

which is in the form of the alternative statement of the definition of linear dependence.

We may illustrate the idea of the linear dependence of vectors by considering any three vectors drawn in a two-dimensional space. It is easy to see that, provided none of these vectors is simply a scalar multiple of one of the others, then each vector is a linear combination of the other two. By going a certain distance from the origin in the direction of one vector, and then the correct distance parallel to the second vector, one can reach the point denoting the end of the third vector. If P_1 denotes quantities of two goods bought by Man 1, and P_2 those bought by Man 2, then the purchases of Man 3 can be thought of as a combination of a certain multiple

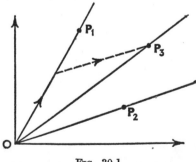

FIG. 30.1

of those of Man 1 and a multiple of those of Man 2. Another way of looking at it is to say that if all people were of Type 1, 2 or 3, then if a certain number of those of Type 1 and some number of those of Type 2 were to pool their purchases, it would enable them exactly to satisfy the needs of a specified number of men of Type 3.

We shall shortly need a general statement of the result just illustrated—that there can be m, and no more than m, linearly independent vectors in a space of m dimensions. For example, if there are four different goods, and various people buy them in different quantities, we may denote the basket of goods bought by the ith man by

$$\mathbf{x}_i = (\mathbf{x}_{i1}, \mathbf{x}_{i2}, \mathbf{x}_{i3}, \mathbf{x}_{i4})$$

where $\mathbf{x}_{ij}(j=1,...,4)$ is the quantity of good j bought by man i. Our theorem states that (a) we can find four men such that the basket of goods bought by one of them cannot be made up by taking multiples (or fractions) of the other baskets, and (b) the basket of goods purchased by any fifth man can be made up from the baskets of these four.

For example, if the purchases are indicated by

$$\mathbf{x}_1 = (1, 3, 4, 2)$$
$$\mathbf{x}_2 = (0, 0, 5, 3)$$
$$\mathbf{x}_3 = (4, 1, 2, 0)$$
$$\mathbf{x}_4 = (1, 1, 1, 2)$$

then if there is a set of λ's such that $\Sigma \lambda_i \mathbf{x}_i = 0$, we have

$$\lambda_1 + 0\lambda_2 + 4\lambda_3 + \lambda_4 = 0$$
$$3\lambda_1 + 0\lambda_2 + \lambda_3 + \lambda_4 = 0$$
$$4\lambda_1 + 5\lambda_2 + 2\lambda_3 + \lambda_4 = 0$$
$$2\lambda_1 + 3\lambda_2 + 0\lambda_3 + 2\lambda_4 = 0$$

The reader may solve these by any method he knows, or wait until later in the book, and discover that there is no set of λ's not all zero satisfying these equations. It follows that the vectors are independent. It is impossible to construct (a multiple of) the purchases of Man 4 by taking multiples of the purchases of other men.

But now let a fifth man make purchases

$$\mathbf{x}_5 = (8, 15, 1, 15)$$

The test for dependence yields

$$\lambda_1 + 0\lambda_2 + 4\lambda_3 + \lambda_4 + 8\lambda_5 = 0$$
$$3\lambda_1 + 0\lambda_2 + \lambda_3 + \lambda_4 + 15\lambda_5 = 0$$
$$4\lambda_1 + 5\lambda_2 + 2\lambda_3 + \lambda_4 + \lambda_5 = 0$$
$$2\lambda_1 + 3\lambda_2 + 0\lambda_3 + 2\lambda_4 + 15\lambda_5 = 0$$

and these equations certainly have at least one solution (other than all zero), as may be checked by writing it $\lambda_1 = 2$, $\lambda_2 = -3$, $\lambda_3 = -1$, $\lambda_4 = 10$, $\lambda_5 = -1$. In other words

$$2x_1 - 3x_2 - x_3 + 10x_4 - x_5 = 0$$

whence $$x_5 = 2x_1 - 3x_2 - x_3 + 10x_4.$$

The purchases of Man 5 can be obtained exactly by adding twice the purchases of Man 1 to ten times the purchases of Man 4, and then subtracting the purchases of Man 3 and three times the purchases of Man 2. As customers, men of Type 5 can be synthesised out of men of the other Types. They bring no new dimension into the market. The vectors are not independent.

2. The basis for a space

We shall now prove this important theorem. We do so for a space of three dimensions, but the proof for any other number of dimensions follows exactly the same lines.

Consider the vectors

$$e_1 = (1, 0, 0)$$
$$e_2 = (0, 1, 0)$$
$$e_3 = (0, 0, 1)$$

We call these *unit vectors*. It is obvious, and may easily be checked from the definition, that they are independent. It is equally obvious that any vector in the space of three dimensions may be expressed as a weighted sum of these unit vectors, for

$$x = (x_1, x_2, x_3)$$
$$= x_1 e_1 + x_2 e_2 + x_3 e_3$$

Because of these properties we say that the e's form a *basis* for the space of four dimensions.

Definition:

> If a set of linearly independent vectors, each of n dimensions, is such that no other vector in the same space can be independent of them then that set constitutes a *basis* for that space. We say that the vectors in this basis *span* the space.

It is easily shown that any given vector can be written as a linear combination of the basis vectors in only one way. Let us have three linearly independent vectors a_1, a_2 and a_3 forming a basis for a space.

Suppose that another vector a_4, in this space can be expressed in terms of the basis in more than one way. This means that in addition to

$$a_4 = \lambda_1 a_1 + \lambda_2 a_2 + \lambda_3 a_3$$

we have at least one expression of the form

$$a_4 = \mu_1 a_1 + \mu_2 a_2 + \mu_3 a_3$$

Subtraction yields

$$0 = (\lambda_1 - \mu_1)a_1 + (\lambda_2 - \mu_2)a_2 + (\lambda_3 - \mu_3)a_3$$

But a_1, a_2 and a_3 are linearly independent and so, by the definition of this property,

$$(\lambda_1 - \mu_1) = (\lambda_2 - \mu_2) = (\lambda_3 - \mu_3) = 0$$

yielding

$$\lambda_1 = \mu_1, \quad \lambda_2 = \mu_2 \quad \text{and} \quad \lambda_3 = \mu_3.$$

Thus the representation of the vector a_4 in terms of the basis vectors is unique.

What we have now to consider is how to obtain a basis for a given space. One basis is the set of n unit vectors; but it is sometimes convenient to use some other basis. There is a simple procedure.

Let a space of n dimensions be spanned by a set of m vectors

$$a_1, a_2, \ldots, a_m.$$

Choose some other vector in this space. Denote it by b. It can be written

$$b = \lambda_1 a_1 + \lambda_2 a_2 + \ldots + \lambda_m a_m$$

Choose one of the basis vectors (say a_2) for which the associated λ is not zero.
Then the m vectors

$$a_1, b, a_3, \ldots, a_m$$

also form a basis.

It is easy to prove this. We do so for $m = 3$.

Any vector x in the space can be written as

$$x = x_1 a_1 + x_2 a_2 + x_3 a_3$$

We have to show that it can also be written as

$$x = p_1 a_1 + p_2 b + p_3 a_3$$

where

$$b = \lambda_1 a_1 + \lambda_2 a_2 + \lambda_3 a_3 \qquad (\lambda_2 \neq 0).$$

Our expression for b enables us to write

$$a_2 = \frac{b - \lambda_1 a_1 - \lambda_3 a_3}{\lambda_2}$$

and so

$$x = x_1 a_1 + \frac{x_2}{\lambda_2}(b - \lambda_1 a_1 - \lambda_3 a_3) + x_3 a_3$$

$$= \left(x_1 - x_2 \frac{\lambda_1}{\lambda_2}\right) a_1 + \frac{x_2}{\lambda_2} b + \left(x_3 - x_2 \frac{\lambda_3}{\lambda_2}\right) a_3$$

which is of the required form, proving the correctness of our procedure for forming a new basis, provided that we can establish that x_1, b and x_3 are, like x_1, x_2 and x_3, linearly independent.

We know that there is no non-zero solution to

$$\theta_1 a_1 + \theta_2 a_2 + \theta_3 a_3 = 0$$

(where we use θ instead of the conventional λ to avoid confusion with the λ we have already used in our definition of b).

We have to show that the same is true of

$$\phi_1 a_1 + \phi_2 b + \phi_3 a_3 = 0$$

This can be written as

$$\phi_1 a_1 + \phi_2 \lambda_1 a_1 + \phi_2 \lambda_2 a_2 + \phi_2 \lambda_3 a_3 + \phi_3 a_3 = 0$$
i.e.
$$(\phi_1 + \phi_2 \lambda_1)a_1 + \phi_2 \lambda_2 a_2 + (\phi_2 \lambda_3 + \phi_3)a_3 = 0$$

We know that a_1, a_2 and a_3 are independent. Therefore

$$(\phi_1 + \phi_2 \lambda_1) = 0$$
$$\phi_2 \lambda_2 = 0$$
$$(\phi_2 \lambda_3 + \phi_3) = 0$$

But we have stipulated that $\lambda_2 \neq 0$. Therefore

$$\phi_2 = 0$$

Substitution in the two remaining equations gives $\phi_1 = \phi_3 = 0$ and so

$$a_1, \text{ b and } a_3$$

are linearly independent. They thus form a basis.

Our next step is to prove that a basis for a space of n dimensions consists of exactly n linearly independent vectors. We know that the n unit vectors form a basis. Our proof consists of showing that any two bases for a space of n dimensions must have the same number of vectors, and that therefore, as the n unit vectors form a basis, every basis must have n vectors.

Suppose that we have two bases

$$a_1,...,a_p \quad \text{of } p \text{ vectors}$$

and $\qquad\qquad b_1,...,b_q \quad\text{,, } q \text{ ,,}$

Since the a's form a basis, every b can be expressed in terms of the p a's. Let us suppose, in particular, that b_q has been so expressed, as

$$b_q = \sum \lambda_i a_i$$

and that

$$\lambda_p \neq 0$$

Then we can replace a_p by b_q to obtain a new basis

$$a_1,...,a_{p-1}, b_q$$

Proceeding in this way we can successively replace the elements until we get

$$a_1,...,a_{p-x}, b_1...b_v$$

or $\qquad\qquad b_1,...,b_q$

as a new basis. If the former arises then it implies that there are more a's than b's. In this case we could have begun at the other end, replacing b_q by a_p in our basis, and ending (by the same procedure) with

$$a_{p-q},...,a_p$$

as a basis. This would contradict our assertion that

$$a_1,...,a_p$$

is a basis, for it would mean that $a_1,...,a_{p-q-1}$ could be expressed in terms of the remaining a's and so would not be independent of them. Thus we cannot have $p > q$. Similarly, arguing the other way, we cannot have $q > p$.

It follows that $p = q$, and so any two bases have the same number of vectors. Since the n unit vectors form a base, all bases of the space of n dimensions have n vectors.

We have thus shown that any vector in the space of n dimensions may be uniquely represented as a weighted sum of n linearly independent vectors spanning that space. We can go a little further than this.

Suppose we have any set of n vectors $x_1,...,x_n$ which are linearly independent. If we begin with the basis of unit vectors then we can replace them one at a time by the procedure above until this set of x's forms a new basis. This means that any set of n linearly independent vectors in the space of n dimensions forms a basis.

It follows at once that one can have n and no more than n linearly independent vectors in the space of n dimensions.

Before proceeding to use these results let us note an implication of our ideas. If we have a set of n linearly independent vectors in m dimensions then the property of linear independence says that

$$\sum \lambda_i \mathbf{x}_i = 0$$

has only zero solutions for λ.

This, written more fully, is

$$\lambda_1(x_{11}\mathbf{e}_1 + x_{12}\mathbf{e}_2 + \ldots + x_{1m}\mathbf{e}_m)$$
$$+ \lambda_2(x_{21}\mathbf{e}_1 + x_{22}\mathbf{e}_2 + \ldots + x_{2m}\mathbf{e}_m)$$
$$+ \ldots\ldots\ldots\ldots\ldots\ldots\ldots\ldots\ldots$$
$$+ \lambda_n(x_{n1}\mathbf{e}_1 + x_{n2}\mathbf{e}_2 + \ldots + x_{nm}\mathbf{e}_m) = 0$$

has only zero solutions. This, in turn, means that the equations

$$\lambda_1 x_{11} + \lambda_2 x_{21} + \ldots + \lambda_n x_{n1} = 0$$
$$\lambda_1 x_{12} + \lambda_2 x_{22} + \ldots + \lambda_n x_{n2} = 0$$
$$\ldots\ldots\ldots\ldots\ldots\ldots\ldots\ldots\ldots\ldots$$
$$\lambda_1 x_{1m} + \lambda_2 x_{2m} + \ldots + \lambda_n x_{nm} = 0$$

have only zero solutions if the n vectors whose elements form the columns of x's are independent. We also know that there cannot be more than m such vectors. Thus if either $n > m$ or the equations have solutions for λ other than zero the vectors formed in turn from the columns of x's are not independent.

3. The independence and consistency of equations

Let us now consider a matrix $\mathbf{A}$ of order $(m \times n)$. We can regard it as a set of m row vectors of order $(1 \times n)$ or as a set of n column vectors of order $(m \times 1)$.

If $m > n$ then the number of row vectors exceeds the number of dimensions in each vector. From the argument just developed this means that only, at the most, n of the m row vectors can be independent. At least $(m - n)$ of the rows in the matrix are dependent on the others, and consequently are weighted sums of them.

If $n > m$ a similar statement holds for the columns.

Now suppose that we have a set of linear equations written as

$$\mathbf{Ax} = \mathbf{b}$$

We consider two cases. In one of them $\mathbf{b} = 0$, while in the other case $\mathbf{b} \neq 0$.

In the former case our system of equations is one such as

$$a_{11}x_1 + a_{12}x_2 + \ldots + a_{1n}x_n = 0$$
$$a_{21}x_1 + a_{22}x_2 + \ldots + a_{2n}x_n = 0$$
$$\vdots \qquad\qquad \vdots$$
$$a_{m1}x_1 + a_{m2}x_2 + \ldots + a_{mn}x_n = 0$$

which is the system of equations which we have just examined in the last section.

Suppose that $m > n$. Then at least $(m-n)$ of the rows of **A** are dependent on the other rows. This means that at least $(m-n)$ of the above equations can be built up from the other equations. They tell us nothing new, and can be discarded.

If exactly $(m-n)$ equations are dependent on the others then we are left with n independent equations, each involving n unknowns. Let the matrix of coefficients in these equations be $\mathbf{A_1}$. Then, after elimination of the superfluous equations, our problem is to solve

$$\mathbf{A_1 x = 0}$$

where $\mathbf{A}_i$ is square. One solution is the trivial solution $\mathbf{x} = 0$. Is there another?

We may note, first, that if there is a non-zero solution of the form

$$\mathbf{x = p}$$

where $\mathbf{p}$ is a column of constants, then there will be an infinite set of solutions, given by $\mu\mathbf{p}$ where μ can take any value. The point is that we will not be able to find a unique value for any x_i, but we may be able to find a unique ratio between any pair of x_i. As an illustration consider

$$4x_1 + 3x_2 = 0$$
$$12x_1 + 9x_2 = 0$$

We may note that here the second equation is a restatement of the first. They are not independent. Each is satisfied by $x_1 = x_2 = 0$. But it is also true that each is satisfied by

$$x_2 = -\tfrac{3}{4}x_1$$

where x_1 has any value. Thus $x_1 = 3$, $x_2 = -4$ is a solution. So are $x_1 = 6$, $x_2 = -8$, and $x_1 = 7$, $x_2 = -\dfrac{28}{3}$.

Yet these two equations do not quite fit the bill—because they do not illustrate an $\mathbf{A_1}$ of order (2×2) with independent rows. Consider, then,

$$4x_1 + 3x_2 = 0$$
$$12x_1 + 8x_2 = 0$$

for which the equations are independent.   One solution is $x_1 = x_2 = 0$.
To find any other solution one may multiply the first equation by 3
and subtract the second getting

$$12x_1 + 9x_2 = 0$$
$$12x_1 + 8x_2 = 0$$
whence $\qquad\qquad\qquad\qquad x_2 = 0$

and so, by substitution in either equation, $x_1 = 0$.   There is no non-
zero solution.

Thus, in the case we have considered, we have failed to find a
non-zero solution for

$$A_1 x = 0$$

where A_1 is a square matrix of order $(n \times n)$ with n independent
rows.

We can shed some light on this if we now consider the case in
which $b \neq 0$, so that our equations

$$Ax = b$$

look like

$$a_{11}x_1 + a_{12}x_2 + \ldots + a_{1n}x_n = b_1$$
$$a_{21}x_1 + a_{22}x_2 + \ldots + a_{2n}x_n = b_2$$
$$\vdots \qquad\qquad \vdots$$
$$a_{m1}x_1 + a_{m2}x_2 + \ldots + a_{mn}x_n = b_m$$

where at least one b is non-zero.

Here, if one row of A can be constructed out of other rows, it
does not follow that one equation can be constructed out of other
equations, for we have also to consider the values of b.   For example,
in

$$3x_1 + 2x_2 + 3x_3 = 7$$
$$2x_1 + x_2 + 3x_3 = 6$$
$$3x_1 + x_2 + 6x_3 = 12$$

the third equation is such that its left-hand side may be formed by
subtracting the L.H.S. of the first equation from three times that
of the second.   (e.g. $3x_1 = 3(2x_1) - 3x_1$.)   But the third equation
cannot be built up out of the other two, for if we operate on the
right-hand side in the same way we would get

$$(3 \times 6) - 7 = 11$$
instead of 12.

A moment's thought shows that these equations are not consistent.
If the first two are correct then it follows, from the above procedure,
that

$$3x_1 + x_2 + 6x_3 = 11$$

If this is so then

$$3x_1 + x_2 + 6x_3 = 12$$

cannot at the same time be true.

This demonstrates that for the set of equations

$$\mathbf{Ax = b}$$

to be consistent it is necessary for the following condition to hold:

Write out the matrix **A** and add an additional column, consisting of b, increasing the order of the matrix. Call this *augmented matrix* **B**. Then if the equations are to be consistent, any linear dependence between rows of **A** must also exist for the same rows of **B**.

Let us now suppose that we have a set of equations

$$\mathbf{Ax = b}$$

which are consistent. **A** is of order $(m \times n)$ and we suppose that $m > n$. As before, there are at most n linearly independent rows in **A**, and as the equations are consistent this means at most n linearly independent rows in **B**. There are also, at most, n linearly independent equations.

Let us suppose that there are precisely n linearly independent equations, each of n unknowns, and that **A₁** is the matrix of coefficients, while $\mathbf{b_1}$ is the vector consisting of the elements of b which appear in the independent equations. All other equations are superfluous and we need to solve

$$\mathbf{A_1 x = b_1}$$

where **A₁** is square and therefore has an inverse $\mathbf{A_1^{-1}}$ unless **A₁** is singular. If it is not singular there are solutions

$$\mathbf{x = A^{-1} b}$$

But we have still to consider what happens if there are fewer than n independent equations.

We may now, in this first view of the situation, go back to

$$\mathbf{Ax = 0}$$

Let us choose a particular element of x, say x_k, and rewrite our equations as

$$a_{11}x_1 + a_{12}x_2 + \ldots + a_{1n}x_n = -a_{1k}x_k$$
$$a_{21}x_1 + a_{22}x_2 + \ldots + a_{2n}x_n = -a_{2k}x_k$$
$$\ldots$$
$$a_{m1}x_1 + a_{m2}x_2 + \ldots + a_{mn}x_n = -a_{mk}x_k$$

where the L.H.S. has no x_k and consists of a set of sums, each being of $n-1$ elements.

Let us give x_k any constant value.

Then the equations become of the form

$$\mathbf{A}\mathbf{x} = \mathbf{b}$$

but now $\mathbf{A}$ has m rows and only $n - 1$ columns. At least one row must be dependent on the others, and so the system reduces to

$$\mathbf{A_1}\mathbf{x} = \mathbf{b_1}$$

where $\mathbf{A_1}$ is either a square matrix of order $(n-1) \times (n-1)$ or a matrix of $n - 1$ columns and fewer rows.

Clearly if we can find rules for establishing the existence of solutions to

$$\mathbf{A_1}\mathbf{x} = \mathbf{b} \qquad (\mathbf{b} \neq 0)$$

then the above procedure allows us to evolve rules for considering

$$\mathbf{A}\mathbf{x} = 0$$

Finally let us look at the kind of problem that arises if $m > n$. As an example we have

$$a_{11}x_1 + a_{12}x_2 + a_{13}x_3 = b_1$$
$$a_{21}x_1 + a_{22}x_2 + a_{23}x_3 = b_2$$

The matrix $\mathbf{A}$ has fewer rows than columns. Look upon it as a set of n column vectors each of m dimensions, $n > m$. There are at most m independent columns. This means that (say) the values of a_{13}, a_{23} are linear sums of a_{11}, a_{12} and a_{21}, a_{22}. What does this mean?

Suppose that the a's represent quantities of goods bought at prices given by x, and that there is one equation for each person. If we have 3 goods and 2 persons then the dependence of a_{i3} on a_{i1} and a_{i2} means that the information we have about spending is no better than we would get if the good 3 were never sold separately, but always appeared in some package deal with the other goods. In this case it would have no individual price, since this would figure in the price of the package. Let us consider an example.

Suppose that when I buy toothpaste at an apparent price of x_1 per tube I am " given " a tablet of soap; and that if I buy 3 bottles of disinfectant at an apparent price of x_2 per bottle I am " given " a tablet of soap.

Let me buy 2 tubes of toothpaste and 3 bottles of disinfectant. My total expenditure is (say) 140 p. Then

$$2x_1 + 3x_2 = 140$$

Another person buys 5 tubes of toothpaste and 6 bottles of disinfectant. Let his bill be $320p$. , Then

$$5x_1 + 6x_2 = 320$$

We can solve these equations for x_1 and x_2 getting

$$x_1 = 40 \quad x_2 = 20$$

but really these are not the prices of the toothpaste and the disinfectant, since they include something for the soap. For the amount x_1 one gets a tube of paste plus a bar of soap; for x_2 one gets a bottle of disinfectant and one-third of a bar of soap. Denoting the true prices of paste, disinfectant and soap by p_1, p_2 and p_3 we have

$$p_1 + p_3 = x_1 = 40$$
$$p_2 + \tfrac{1}{3}p_3 = x_2 = 20$$

i.e.
$$p_1 + p_3 = 40$$
$$3p_2 + p_3 = 60$$

and we cannot solve these for p_1, p_2 or p_3. If we decide to choose $p_1 = 10$ (say), then the first equation gives us $p_3 = 30$, and so, from the second equation, we find that $p_2 = 10$. But this is one of many possible solutions.

Alternatively we could have begun by stipulating that each of three goods has a price, and so obtained equations as follows:

Buying 2 tubes of paste and 3 bottles of disinfectant involves me in the involuntary purchase of 3 bars of soap. My total expenditure is 140 *p*. and so

$$2p_1 + 3p_2 + 3p_3 = 140$$

For the other person

$$5p_1 + 6p_2 + 7p_3 = 320$$

Here are two equations in three unknowns—and the coefficients of p_3 are determined by those of p_1 and p_2 and the rules of the package deal. If we subtract double the first equation from the second we get

$$p_1 + p_3 = 40$$

while 5 times the first minus twice the second gives

$$3p_1 + p_3 = 60$$

These are the equations obtained a moment ago.

It is fairly obvious from this that if one has m equations in n unknowns, and $n > m$, then there is no unique solution; but if one specifies values for $n - m$ of the unknowns then, usually, the remaining m unknowns become determined.

We have said enough to present a feel for the problem, and to indicate that the answer has something to do with the numbers of independent rows and columns in the matrix A and the augmented matrix B (made up out of A and the column vector b). It is now time to consider this more formally. We shall begin by stating the rule for determining the solubility of a set of equations. Then we shall produce a proof which involves some useful ideas that will have to be introduced en route. They have, as one of their uses, the advantage of helping us to apply the rule more easily than might otherwise be possible.

CHAPTER XXXI

EQUIVALENCE, PARTITIONING AND RANK

To make the third she join'd the former two.

J. DRYDEN

1. Rank and the solution of equations

First we state (and shall later prove) that in any matrix of any order, square or otherwise, the maximum number of linearly independent rows that can be found is also the maximum number of linearly independent columns. If, for example, in a 7×5 matrix one can find 3, but no more than 3, linearly independent rows then one will be able to find 3, but no more than 3, linearly independent columns.

We call this maximum number of linearly independent rows (or columns) the *rank* of the matrix. We could define rank in this way, but we prefer to use a different starting point and our definition (to appear shortly) will enable us to prove that the rank of a matrix is the number we have just mentioned.

We shall show that if A is of order $(m \times n)$ and rank r

(i) the equations $Ax = b$ will be inconsistent if the rank of B, the augmented matrix, exceeds r.

(ii) if the rank of B is r, the same as that of A, then precisely r of the x's can be expressed in terms of the remaining $n - r$ variables; so if $r = n$ then all of the x's can be found uniquely.

(iii) the equations $Ax = 0$ will have a non-zero solution only if $r > n$.

(iv) if $m < n$ then $Ax = 0$ can always be solved, although the solution will not necessarily be unique and may simply specify some of the x's in terms of the others.

(v) if $m = n$ then $Ax = 0$ has a non-zero solution if A is singular.

(vi) if $m > n$ then $Ax = 0$ has a solution provided $r < n$.

The reader may benefit by going over our earlier discussion in Section VI with these results in mind. We now set about proving them.

From any matrix of order $(m \times n)$ one can obtain *sub-matrices* by deleting some of the rows and|or columns. By deleting appropriate numbers of rows and columns one can obtain *square* sub-matrices of order $(r \times r)$. For example, if we have

$$A = \begin{bmatrix} 1 & 2 & 3 & 4 \\ 8 & 7 & 6 & 5 \\ 9 & 0 & 1 & 2 \end{bmatrix}$$

then by omitting one column we can get square sub-matrices of order (3×3). There are four possible ones, these being

$$\begin{bmatrix} 1 & 2 & 3 \\ 8 & 7 & 6 \\ 9 & 0 & 1 \end{bmatrix}, \quad \begin{bmatrix} 1 & 2 & 4 \\ 8 & 7 & 5 \\ 9 & 0 & 2 \end{bmatrix}, \quad \begin{bmatrix} 1 & 3 & 4 \\ 8 & 6 & 5 \\ 9 & 1 & 2 \end{bmatrix} \quad \text{and} \quad \begin{bmatrix} 2 & 3 & 4 \\ 7 & 6 & 5 \\ 0 & 1 & 2 \end{bmatrix}$$

Alternatively, we could have omitted one row and two columns. If we omit the first row and the first two columns we get the square sub-matrix of order (2×2),

$$\begin{bmatrix} 6 & 5 \\ 1 & 2 \end{bmatrix}$$

while a different choice of the one row and two columns for deletion would lead to a different sub-matrix. (The student may prove for himself that there can be 18 different sub-matrices of order (2×2) arising out of A). Finally, if we omit two rows and three columns (which can be done in 12 ways) we get a square sub-matrix of order (1×1), which will obviously be simply a single element of A.

Thus, from a matrix A one can obtain several different square sub-matrices, of order ranging from (1×1) up to the smaller of $(m \times m)$ and $(n \times n)$.

Each of these square sub-matrices of A has an associated determinant. Some of these determinants may have zero value; some may not—i.e. some are singular and some are not.

Definition:

We define the *rank* of a matrix A to be r if, and only if, there is at least one square sub-matrix of order $(r \times r)$ which is non-singular, and no square sub-matrix of higher order which is non-singular.

For example, the matrix

$$A = \begin{bmatrix} 1 & 2 & 3 & 4 \\ 5 & 7 & 6 & 8 \\ 6 & 9 & 9 & 12 \end{bmatrix}$$

has four sub-matrices of order (3×3). The reader may check that they all have zero determinants. Consequently the rank of A cannot be as high as 3. It also has eighteen sub-matrices of order (2×2). One of these is obviously

$$\begin{bmatrix} 1 & 2 \\ 5 & 7 \end{bmatrix}$$

which has a non-zero determinant. There is no need to go further. By our definition, the rank of A is 2.

Obtaining the rank in this way is often a very tedious process. Can we find a shorter method? In any case, how does this definition tie up with the number of independent rows or columns?

2. Partitioning

To answer these questions we first consider the partitioning of matrices. Any rectangular matrix may be *partitioned*, by means of horizontal and vertical lines, into a set of smaller matrices, provided that the lines go completely across the matrix, from side to side or from top to bottom.

We may divide

$$\left[\begin{array}{ccc:c} a_{11} & a_{12} & a_{13} & a_{14} \\ a_{21} & a_{22} & a_{23} & a_{24} \\ \hdashline a_{31} & a_{32} & a_{33} & a_{34} \\ \hdashline a_{41} & a_{42} & a_{43} & a_{44} \\ a_{51} & a_{52} & a_{53} & a_{54} \end{array} \right]$$

into (for example) six matrices

$$A_{11} = \begin{bmatrix} a_{11} & a_{12} & a_{13} \\ a_{21} & a_{22} & a_{23} \end{bmatrix} \quad A_{12} = \begin{bmatrix} a_{14} \\ a_{24} \end{bmatrix}$$

$$A_{21} = \begin{bmatrix} a_{31} & a_{32} & a_{33} \end{bmatrix} \quad A_{22} = \begin{bmatrix} a_{34} \end{bmatrix}$$

$$A_{31} = \begin{bmatrix} a_{41} & a_{42} & a_{43} \\ a_{51} & a_{52} & a_{53} \end{bmatrix} \quad A_{32} = \begin{bmatrix} a_{44} \\ a_{54} \end{bmatrix}$$

and write

$$A = \begin{bmatrix} A_{11} & A_{12} \\ A_{21} & A_{22} \\ A_{31} & A_{32} \end{bmatrix}$$

Now suppose that we have the matrix A, just defined, and a matrix B

$$\begin{bmatrix} b_{11} & b_{12} & b_{13} \\ b_{21} & b_{22} & b_{23} \\ b_{31} & b_{32} & b_{33} \\ b_{41} & b_{42} & b_{43} \end{bmatrix}$$

Let us partition B as shown, and write it as

$$\begin{bmatrix} B_1 \\ B_2 \end{bmatrix}$$

Consider whether A and B can be multiplied

(i) when written out fully

(ii) when partitioned.

Clearly they can be multiplied when written out in full, since A is of order 5×4 and B is of order 4×3. The product AB will be of order 5×3.

When they are partitioned in the way just described, we have A as a matrix of (3×2) elements (where each element is itself a matrix, such as A_{11}) and B is a matrix of (2×1) elements. Thus it is possible to form AB according to the usual rules of matrix multiplication, yielding

$$AB = \begin{bmatrix} A_{11}B_1 + A_{12}B_2 \\ A_{21}B_1 + A_{22}B_2 \\ A_{31}B_1 + A_{32}B_2 \end{bmatrix}$$

provided that it is possible to form the six individual multiplications such as $A_{11}B_1$.

This happens to be possible in the case we are considering, for A_{11} is of order (2×3) and B_1 is of order (3×3). For similar reasons the other products can also be formed.

But suppose that we had partitioned B into two matrices each of 2 rows. It would still be written as

$$\begin{bmatrix} B_1 \\ B_2 \end{bmatrix}$$

but now $A_{11}B_1$ could not be formed, as A_{11} would be of order (2×3) while B_1 would be of an order—(2×3)—which would not allow multiplication.

Thus, matrices can meaningfully be multiplied in their partitioned forms only if they are *conformably* partitioned, i.e. only if they are partitioned so that

(i) the number of rows and columns in the partitioned forms allow multiplication.

and (ii) each matrix product arising out of this multiplication is possible.

The reader may convince himself by a few experiments that these requirements will be satisfied provided that

(i) the original matrices A and B can be multiplied

and (ii) the partitioning lines separating columns of A correspond to those separating rows of B—i.e. if there is a partition between row n and $n+1$ of A there must be one between column n and $n+1$ of B, and vice versa. The partitions between rows of A and columns of B may be anywhere.

He may also convince himself that, when the product of the partitioned matrices is written out in full, the result is the same as the product of the original matrices.

Example

$$A = \begin{bmatrix} 1 & 2 & 4 & 3 \\ 1 & 0 & 1 & 2 \\ 2 & 3 & 1 & 0 \end{bmatrix} \quad B = \begin{bmatrix} 1 & 2 \\ 4 & 1 \\ 7 & 9 \\ 8 & 1 \end{bmatrix}$$

Ignoring the partitions we have

$$AB = \begin{bmatrix} 1+ 8+28+24 & 2+2+36+3 \\ 1+ 0+ 7+16 & 2+0+ 9+2 \\ 2+12+ 7+ 0 & 4+3+ 9+0 \end{bmatrix}$$

If we partition we get

$$A = \begin{bmatrix} A_{11} & A_{12} & A_{13} \\ A_{21} & A_{22} & A_{23} \end{bmatrix} \quad B = \begin{bmatrix} B_1 \\ B_2 \\ B_3 \end{bmatrix}$$

and so

$$AB = \begin{bmatrix} A_{11}B_1 + A_{12}B_2 + A_{13}B_3 \\ A_{21}B_1 + A_{22}B_2 + A_{23}B_3 \end{bmatrix}$$

where

$$A_{11} = \begin{bmatrix} 1 \\ 1 \end{bmatrix} \qquad A_{12} = \begin{bmatrix} 2 & 4 \\ 0 & 1 \end{bmatrix} \qquad A_{13} = \begin{bmatrix} 3 \\ 2 \end{bmatrix}$$

$$B_1 = \begin{bmatrix} 1 & 2 \end{bmatrix} \qquad B_2 = \begin{bmatrix} 4 & 1 \\ 7 & 9 \end{bmatrix} \qquad B_3 = \begin{bmatrix} 8 & 1 \end{bmatrix}$$

and so

$$\begin{bmatrix} A_{11}B_1 \end{bmatrix} = \begin{bmatrix} 1 \\ 1 \end{bmatrix} \begin{bmatrix} 1 & 2 \end{bmatrix} = \begin{bmatrix} 1 & 2 \\ 1 & 2 \end{bmatrix}$$

$$\begin{bmatrix} A_{12}B_2 \end{bmatrix} = \begin{bmatrix} 2 & 4 \\ 0 & 1 \end{bmatrix} \begin{bmatrix} 4 & 1 \\ 7 & 9 \end{bmatrix} = \begin{bmatrix} 8+28 & 2+36 \\ 0+7 & 0+9 \end{bmatrix}$$

$$\begin{bmatrix} A_{13}B_3 \end{bmatrix} = \begin{bmatrix} 3 \\ 2 \end{bmatrix} \begin{bmatrix} 8 & 1 \end{bmatrix} = \begin{bmatrix} 24 & 3 \\ 16 & 2 \end{bmatrix}$$

and so the sum of these three products, element by corresponding element, is

$$\begin{bmatrix} 1+8+28+24 & 2+2+36+3 \\ 1+0+7+16 & 2+0+9+2 \end{bmatrix}$$

which corresponds to the first two rows of **AB** obtained without partitioning. The bottom row follows similarly.

We shall shortly see how this idea of partitioning can help us in our analysis. Meanwhile, the reader is advised to invent a few examples for himself, by taking a matrix of order (say) 3×4 and another of order 4×5, and multiplying them (*a*) without partitioning and (*b*) with partitioning. Note that a given pair of matrices may be capable of being conformably partitioned in several different ways.

3. Elementary operations and equivalence

There are a few more ideas which we need for our proof, and which also have much wider application. When we introduced determinants we found that

(1) interchange of two rows (or columns) reverses the sign of the determinant.

(2) the addition of h times one row (or column) to another leaves the value of the determinant unaltered.

(3) the multiplication of a row (or column) by a constant k increases the value of the determinant by k as a factor.

Clearly if the initial determinant is singular (with zero value), the application of these operations will leave it singular. If the initial determinant is non-singular then (unless $k = 0$) none of these operations can turn it into a singular determinant.

We call these three operations, *elementary operations*. They can be applied to matrices just as easily as to determinants, although we have yet to examine the consequences of applying them. To do so we begin by applying them to the unit matrix I. It will help us if we consider I to be of order (4×4), so that

$$I = \begin{bmatrix} 1 & 0 & 0 & 0 \\ 0 & 1 & 0 & 0 \\ 0 & 0 & 1 & 0 \\ 0 & 0 & 0 & 1 \end{bmatrix}$$

Let $E_{(rs)}$ denote the matrix obtained from I by the interchange of rows r and s, so that, for example

$$E_{(13)} = \begin{bmatrix} 0 & 0 & 1 & 0 \\ 0 & 1 & 0 & 0 \\ 1 & 0 & 0 & 0 \\ 0 & 0 & 0 & 1 \end{bmatrix}$$

Also let $H_{(rs)}$ be derived from I by the second elementary operation so that, for example

$$H_{13} = \begin{bmatrix} 1 & 0 & 0 & 0 \\ 0 & 1 & 0 & 0 \\ h & 0 & 1 & 0 \\ 0 & 0 & 0 & 1 \end{bmatrix}$$

and let $K_{(r)}$ stem from the third elementary operation applied to I so that

$$K_{(3)} = \begin{bmatrix} 1 & 0 & 0 & 0 \\ 0 & 1 & 0 & 0 \\ 0 & 0 & k & 0 \\ 0 & 0 & 0 & 1 \end{bmatrix}$$

We call E, H and K *elementary matrices*. Some authors use E, F, G. The reader should explore the consequences of pre- or post-multiplying a matrix A by these elementary matrices, of appropriate order, to obtain a matrix B. We say that then B is *equivalent* to A, for a reason which will shortly be clear.

R

For example,

$$\mathbf{B} = \begin{bmatrix} 1 & 0 & 0 \\ 0 & 0 & 1 \\ 0 & 1 & 0 \end{bmatrix} \begin{bmatrix} 1 & 0 & 0 \\ 0 & 1 & 1 \\ 0 & 0 & 1 \end{bmatrix} \begin{bmatrix} 4 & 2 & 6 \\ 3 & 1 & 1 \\ 2 & 4 & 3 \end{bmatrix} \begin{bmatrix} 4 & 0 & 0 \\ 0 & 1 & 0 \\ 0 & 0 & 1 \end{bmatrix} = \begin{bmatrix} 16 & 2 & 6 \\ 8 & 4 & 3 \\ 20 & 5 & 4 \end{bmatrix}$$

is equivalent to $\mathbf{A} = \begin{bmatrix} 4 & 2 & 6 \\ 3 & 1 & 1 \\ 2 & 4 & 3 \end{bmatrix}$ since the three other matrices are,

respectively,

$$\mathbf{E}_{(23)}, \mathbf{H}_{(32)}, \mathbf{K}_{(1)}$$

We can now show, with the help of partitioning, how equivalence will enable us quickly to find the rank of a matrix. Then we shall see how this enables us to consider the possibility of solutions to a set of linear equations.

We begin with an important theorem. It is that any matrix $\mathbf{A}$ of order $(m \times n)$ can be reduced (by a series of elementary operations) to an equivalent matrix

$$\mathbf{P} = \begin{bmatrix} 1 & 0 & 0 & ... & 0 & 0 & 0 ... \\ 0 & 1 & 0 & ... & 0 & 0 & 0 \\ 0 & 0 & 1 & & 0 & 0 & 0 \\ \vdots & & & & \vdots & \vdots & \\ 0 & 0 & 0 & ... & 1 & 0 & 0 \\ 0 & 0 & 0 & ... & 0 & 0 & 0 \\ 0 & 0 & 0 & ... & 0 & 0 & 0 \\ \vdots & & & & & & \end{bmatrix} = \begin{bmatrix} \mathbf{I}_r & 0 \\ 0 & 0 \end{bmatrix}$$

where $\mathbf{I}_r$ is a unit matrix of order $(r \times r)$, with

$$r \leqslant \min(m, n)$$

where $\min(m, n)$ is the lower of the two values m and n. We shall show that r is the rank of the matrix $\mathbf{A}$.

As an example, the theorem states that

$$\begin{bmatrix} 3 & 4 & 2 & 1 \\ 1 & 4 & 5 & 6 \\ 6 & 5 & 3 & 2 \end{bmatrix}$$ is equivalent to

$$\begin{bmatrix} 0 & 0 & 0 & 0 \\ 0 & 0 & 0 & 0 \\ 0 & 0 & 0 & 0 \end{bmatrix} \text{ or to } \begin{bmatrix} 1 & 0 & 0 & 0 \\ 0 & 0 & 0 & 0 \\ 0 & 0 & 0 & 0 \end{bmatrix} \text{ or to } \begin{bmatrix} 1 & 0 & 0 & 0 \\ 0 & 1 & 0 & 0 \\ 0 & 0 & 0 & 0 \end{bmatrix} \text{ or to } \begin{bmatrix} 1 & 0 & 0 & 0 \\ 0 & 1 & 0 & 0 \\ 0 & 0 & 1 & 0 \end{bmatrix}$$

where r as taken the values 0, 1, 2 and 3.

The proof of this theorem is simple but instructive. Remembering the definition of equivalence, and its dependence on the elementary operations, we consider any matrix **A**. Unless it has only zero elements (in which case the result is obvious), we can find at least one non-zero element. Let us perform the elementary operations of interchanging rows and columns until we get this non-zero element into the leading position (the top left-hand corner). Now multiply the top row by the reciprocal of this element, so that the new top row begins with 1.

So far we have performed only elementary operations on **A**, and so the new matrix is equivalent to **A**. The only other thing that we know about it is that its leading element is unity.

Let the second element in the top row of this new determinant be (say) x. Then if we subtract x times the first column from the second column we will obtain a new second column whose top term is zero. This is an elementary operation and so this gives us a matrix equivalent to **A**, and such that we know its first two elements to be as below

$$\begin{bmatrix} 1 & 0 & ? & ? & ? & \cdots \\ ? & ? & ? & ? & ? & \cdots \\ ? & ? & ? & ? & ? & \cdots \\ \vdots & \vdots & \vdots & \vdots & \vdots & \end{bmatrix}$$

Repeat the procedure to obtain an equivalent matrix with the third element of row 1 equal to zero. Then again to get another zero. And so on. We eventually reach

$$\begin{bmatrix} 1 & 0 & 0 & 0 & 0 & \cdots \\ ? & ? & ? & ? & ? & \cdots \\ ? & ? & ? & ? & ? & \cdots \\ \vdots & \vdots & \vdots & \vdots & \vdots & \end{bmatrix}$$

Now turn to the first column. Elementary operations on the rows will enable us to produce a set of zero elements, so that we obtain an equivalent matrix of the form

$$\begin{bmatrix} 1 & 0 & 0 & 0 & 0 & \cdots \\ 0 & ? & ? & ? & ? & \cdots \\ 0 & ? & ? & ? & ? & \cdots \\ \vdots & \vdots & \vdots & \vdots & \vdots & \end{bmatrix}$$

which can be written more compactly as

$$\begin{bmatrix} 1 & 0 \\ \hline 0 & Q \end{bmatrix}$$

where $\mathbf{Q}$ is some unknown matrix. Either $\mathbf{Q}$ is a null matrix, in which case the theorem is proved, or it is not. If it is not then, by the same process, it can be written as

$$\mathbf{Q} = \begin{bmatrix} 1 & 0 \\ \hline 0 & \mathbf{R} \end{bmatrix}$$

and so $\mathbf{A}$ is equivalent to a matrix which may be written as

$$\begin{bmatrix} 1 & 0 & \\ & 1 & 0 \\ \hline 0 & 0 & \mathbf{R} \end{bmatrix} = \begin{bmatrix} 1 & 0 & 0 & 0 & \ldots & 0 \\ 0 & 1 & 0 & 0 & \ldots & 0 \\ 0 & 0 & ? & ? & \ldots & ? \\ 0 & 0 & ? & ? & \ldots & ? \\ \vdots & \vdots & \vdots & \vdots & & \vdots \\ 0 & 0 & ? & ? & \ldots & ? \end{bmatrix}$$

which can clearly be rewritten

$$\begin{bmatrix} 1 & 0 & 0 \\ 0 & 1 & \\ \hline & 0 & \mathbf{R} \end{bmatrix}$$

where either $\mathbf{R} = 0$ or $\mathbf{R}$ can be put equivalent to

$$\begin{bmatrix} 1 & 0 \\ \hline 0 & \mathbf{S} \end{bmatrix}$$

And so on. The theorem is proved.

Example

(i) $\begin{bmatrix} 9 & 2 & 3 \\ 6 & 5 & 4 \\ 8 & 7 & 1 \end{bmatrix}$ is equivalent to $\begin{bmatrix} 3 & 2 & 9 \\ 4 & 5 & 6 \\ 1 & 7 & 8 \end{bmatrix}$

and so to $\begin{bmatrix} 1 & 7 & 8 \\ 4 & 5 & 6 \\ 3 & 2 & 9 \end{bmatrix}$ and to $\begin{bmatrix} 1 & 7-7 & 8 \\ 4 & 5-28 & 6 \\ 3 & 2-21 & 9 \end{bmatrix}$

and to $\begin{bmatrix} 1 & 0 & 8 \\ 4 & -23 & 6 \\ 3 & -19 & 9 \end{bmatrix}$ and to $\begin{bmatrix} 1 & 0 & 0 \\ 4 & -23 & -26 \\ 3 & -19 & -15 \end{bmatrix}$

and to $\begin{bmatrix} 1 & 0 & 0 \\ 0 & -23 & -26 \\ 3 & -19 & -15 \end{bmatrix}$ and to $\begin{bmatrix} 1 & 0 & 0 \\ 0 & -23 & -26 \\ 0 & -19 & -15 \end{bmatrix}$

and to $\qquad$
$$\begin{array}{ccc} 1 & 0 & 0 \\ 0 & 1 & \dfrac{26}{23} \\ 0 & 19 & 15 \end{array}$$

and to $\qquad$
$$\begin{array}{ccc} 1 & 0 & 0 \\ 0 & 1 & 0 \\ 0 & 19 & k \end{array}$$

and to $\qquad$
$$\begin{array}{ccc} 1 & 0 & 0 \\ 0 & 1 & 0 \\ 0 & 0 & k \end{array}$$

where $k = 15 - (26 \times 19)/23$

and so to $\qquad$
$$\begin{array}{ccc} 1 & 0 & 0 \\ 0 & 1 & 0 \\ 0 & 0 & 1 \end{array}$$

(ii) $\quad$
$$\begin{array}{cccc} 1 & 4 & 6 & 8 \\ 5 & 2 & 3 & 4 \\ 6 & 6 & 9 & 12 \end{array}$$
is equivalent to
$$\begin{array}{cccc} 1 & 0 & 0 & 0 \\ 5 & -18 & -27 & -36 \\ 6 & -18 & -27 & -36 \end{array}$$

and to $\qquad$
$$\begin{array}{cccc} 1 & 0 & 0 & 0 \\ 0 & -18 & -27 & -36 \\ 0 & -18 & -27 & -36 \end{array}$$
and to $\qquad$
$$\begin{array}{cccc} 1 & 0 & 0 & 0 \\ 0 & -18 & -27 & -36 \\ 0 & 0 & 0 & 0 \end{array}$$

and to $\qquad$
$$\begin{array}{cccc} 1 & 0 & 0 & 0 \\ 0 & -2 & -3 & -4 \\ 0 & 0 & 0 & 0 \end{array}$$
and to $\qquad$
$$\begin{array}{cccc} 1 & 0 & 0 & 0 \\ 0 & 1 & \frac{3}{2} & 2 \\ 0 & 0 & 0 & 0 \end{array}$$

and to $\qquad$
$$\begin{array}{cccc} 1 & 0 & 0 & 0 \\ 0 & 1 & 0 & 0 \\ 0 & 0 & 0 & 0 \end{array}$$
by a succession of elementary operations.

We now come to the use of equivalence in determining the rank of a matrix, and so, as we shall shortly see, in helping to examine the solution of simultaneous equations. The reader is advised to go back to the definition of rank and to re-read that section.

Suppose that **A** and **B** are two equivalent matrices. This means that **B** can be derived from **A** by a series of elementary operations. But these are all such that, when applied to a determinant, they leave a singular determinant singular and never turn a non-singular determinant into a singular one. This means that whether a sub-matrix of **B** has a singular or non-singular determinant will depend entirely on whether the corresponding sub-matrix of **A** is, or is not, singular. It follows that A and B must have the same rank. In particular, if any matrix A has been reduced to its equivalent matrix

$$P = \left[\begin{array}{c|c} I & 0 \\ \hline 0 & 0 \end{array} \right].$$

then the rank of A is the rank of P which is clearly the number of rows or columns of I.

For example, the reader can show that

$$A = \begin{bmatrix} 1 & 2 & 3 & 4 \\ 5 & 7 & 6 & 8 \\ 6 & 9 & 9 & 12 \end{bmatrix}$$

is equivalent to

$$P = \begin{bmatrix} 1 & 0 & 0 & 0 \\ 0 & 1 & 0 & 0 \\ 0 & 0 & 0 & 0 \end{bmatrix}$$

This matrix has one square sub-matrix of order (2×2) which has a non-zero determinant, namely

$$\begin{bmatrix} 1 & 0 \\ 0 & 1 \end{bmatrix}$$

All its sub-matrices of order (3×3) yield zero. It follows that the rank of P, and therefore of A, is 2.

The problem of finding the rank of a matrix is thus simplified. We systematically reduce the matrix to the form

$$P_r = \left[\begin{array}{c|c} I_r & 0 \\ \hline 0 & 0 \end{array} \right].$$

where I_r is a unit matrix of order $(r \times r)$. The rank of the matrix is given immediately by r.

It follows that if any two matrices are of the same rank then they are each equivalent to the same reduced form, and therefore to each other.

Let us now show that the rank, defined as the order of the highest order non-vanishing determinant, is the number of independent rows, which is also the number of independent columns.

Consider any matrix A. Let it have p independent rows. Use of elementary operations on A will not change the number of independent rows. Therefore any matrix equivalent to A has the same number of independent rows as A does. But A can always be reduced to the equivalent form

$$P = \left[\begin{array}{c|c} I_r & 0 \\ \hline 0 & 0 \end{array}\right]$$

where $\quad I_r = \begin{bmatrix} 1 & 0 & 0 & \dots & 0 \\ 0 & 1 & 0 & \dots & \vdots \\ 0 & 0 & 1 & \dots & \vdots \\ \vdots & \vdots & \vdots & \vdots & \vdots \\ & & & \dots\dots 1 & 0 \\ & & & \vdots & \vdots \\ 0 & & \dots\dots 0 & 1 \end{bmatrix} \quad$ of order $(r \times r)$.

By our definition, the rank of A is r. But I_r has r independent rows. So, therefore, does P. Therefore A does.

But the same argument applies to columns, and so A has r independent columns and r independent rows, where r is the rank of A—the order of the highest order non-vanishing determinant.

CHAPTER XXXII

THE SOLUBILITY OF LINEAR EQUATIONS

my offence is rank,
W. SHAKESPEARE

1. The solubility of Ax = b

We can now use our knowledge of rank and linear independence to determine rules for the solubility of a set of linear equations.

Consider any matrix A of order $m \times n$. If $m > n$ then the order of the highest order determinant (zero or otherwise) which can be constructed out of the rows and columns of A is obviously n (for a determinant must be square, if there are only n columns). Similarly, if $m < n$ then the order is m. Thus it is impossible for the rank of A to exceed the lower of m and n. Consequently the maximum possible number of linearly independent rows (and columns) is the lower of m and n.

If we have a system of equations

$$Ax = b$$

where $b \neq 0$, and A is of order $(m \times n)$ and rank r then precisely r rows of A are independent. This means that the remaining $(m - r)$ rows can be constructed from them. If the equations are written out in full, in the usual way, then this result means that the left hand sides of $m - r$ of these equations are linear combinations of the left hand sides of other r equations. Unless the corresponding right hand sides can be obtained in the same way the equations will be inconsistent. We saw an example of this in section 3 of Chapter XXX.

This means that the number of independent rows in the augmented matrix $B = [A \vdots b]$ must be the same as the number in A.

The rank of B must therefore equal the rank of A if the equations are to be consistent. The reader should be able to prove for himself that this condition is not only necessary (as we have shown) but also sufficient.

Suppose then that we have a set of consistent equations, and that the ranks of A and of B are r.

If $r = n$, where n is the number of columns, and consequently the number of variables, then we have n independent equations from which to find n unknowns. There may also be other equations, but these will be dependent on the n we have mentioned and may be disregarded.

From our n equations in n unknowns we can form a square matrix of L.H.S. coefficients, denoted by A_1. It will be of order $(n \times n)$ and rank n. From the definition of rank, $| A_1 | \neq 0$ and so the inverse of A_1 exists. The solution to

$$Ax = b$$

is therefore the solution of

$$A_1x = b$$

which is

$$x = A_1^{-1}b$$

All of the unknowns can be found, as elements of x.

Now consider the case in which $r < n$. Now any matrix of order $(n \times n)$ will be singular and so its inverse will not exist. But there will be at least one matrix of order $(r \times r)$ which will not be singular. The r columns of this matrix will define r of our variables, and the r rows will define r equations containing them. These rows and columns will be independent and (since our equations are consistent) there will thus be r equations in r unknowns. Denoting the relevant sub-matrix of order $(r \times r)$ by A_2 we have that

$$A_2x_2 = c$$

where x_2 is a column vector of the x's associated with the independent columns of A, and c consists of elements which are the remaining parts of the appropriate equations. For example, if we begin with

$$a_{11}x_1 + a_{12}x_2 + a_{13}x_3 + a_{14}x_4 = b_1$$
$$a_{21}x_1 + a_{22}x_2 + a_{23}x_3 + a_{24}x_4 = b_2$$
$$a_{31}x_1 + a_{32}x_2 + a_{33}x_3 + a_{34}x_4 = b_3$$
$$a_{41}x_1 + a_{42}x_2 + a_{43}x_3 + a_{44}x_4 = b_4$$

and the ranks of A and B are 2, with

$$\begin{vmatrix} a_{11} & a_{13} \\ a_{21} & a_{23} \end{vmatrix} \neq 0$$

then we pick out the first and second rows and rewrite them as

$$a_{11}x_1 + a_{13}x_3 = b_1 - a_{12}x_2 - a_{14}x_4$$
$$a_{21}x_1 + a_{23}x_3 = b_2 - a_{22}x_2 - a_{24}x_4$$

The remaining rows are dependent on these; and these equations have the independent columns of a on their left. We can now say that, for any given values of x_2 and x_4 the right-hand sides may be written as constants c_1 and c_2 so that

$$a_{11}x_1 + a_{13}x_3 = c_1$$
$$a_{21}x_1 + a_{23}x_3 = c_2$$

being an example of

$$A_2 x_2 = c$$

We know that A_2 is non-singular, and so we obtain the solution

$$x_2 = A_2^{-1} c$$

which expresses precisely r of the x's in terms of the a's and the c's —and the values of c depend on the values of the other x's.

Thus if A and B are each of rank r then precisely r of the x's can be expressed in terms of the remaining $n-r$ variables.

2. The solubility of $Ax = 0$

We now consider what happens if $b = 0$ so that

$$Ax = 0.$$

We have already explained that if there is a non-zero set of x's satisfying these equations then we can multiply all of the x's by an arbitrary constant to obtain a new solution. We have to consider the circumstances in which a non-zero set may be found.

Clearly, if we choose any variable x_j which has at least one non-zero coefficient x_{ij} then we can give x_j an arbitrary value and then transfer all the x_j column to the R.H.S. to get

$$A_j x = b$$

where A_j is a matrix of order $m \times (n-1)$ derived from A by omitting the jth column, and

$$b_i = -a_{ij}x_j$$

But the system

$$A_j x = b$$

can be solved provided that the rank of A_j is equal to the rank of the augmented matrix (which is A) for consistency. The maximum possible rank for A_j is $(n-1)$—if m is large enough. Certainly the rank of A_j cannot exceed its number of columns, which is $n-1$. Thus, for the equations

$$Ax = 0$$

to be soluble (in non-trivial terms) the rank of A must be less than n, the number of columns.

We now consider the three possibilities given by $m = n$, $m > n$ and $m < n$.

If $m = n$ then $\mathbf{A}$ is square and $\mathbf{A}_j$ is of order $n \times (n-1)$. If the rank of $\mathbf{A}$ is less than n then the equations are consistent. This will be so if $|\,\mathbf{A}\,|$ is singular, from the definition of rank. There will be r independent equations which may be solved to give values of r of the unknowns in terms of the remaining unknowns by writing the equations in the form

$$\mathbf{A}_j\mathbf{x} = \mathbf{b}$$

and then as

$$\mathbf{A}_{j2}\mathbf{x}_2 = \mathbf{c}$$

where $\mathbf{A}_{j2}$ is related to $\mathbf{A}_j$ in the same way as $\mathbf{A}_2$ used above was related to $\mathbf{A}$.

If $m > n$ then if the equations are consistent at least $(m - n)$ equations are dependent on the others (because the coefficients of $\mathbf{x}$ form m vectors in n dimensions, and only n of them can be independent). We thus have at most n independent equations. If these are considered as a set they have a matrix $\mathbf{A}$ of order $(m \times n)$ where either $m = n$ or $m < n$. In the former case there is a solution, provided $r < n$, as we have just shown.

If, however, the number of independent equations $m < n$ then the rank of the associated matrix can be no more than m. We can thus pick out r $(\leqslant m)$ unknowns and express them in terms of the remaining unknowns, as indicated above.

The reader should now go back to the rules printed on page 491 and satisfy himself that we have given a proof of them. To summarise, we have

(1) If $\mathbf{Ax} = \mathbf{b}$ then for consistency $r(\mathbf{A} : \mathbf{b}) = r(\mathbf{A})$ where $r(\mathbf{A})$ is the rank of $\mathbf{A}$.

(2) If $r(\mathbf{A}) = r(\mathbf{B}) = r$, then r of the $\mathbf{x}$'s can be expressed in terms of the remaining $n - r$ $\mathbf{x}$'s.

(3) If $\mathbf{Ax} = 0$ then there is a non-zero solution provided $r < n$.

CHAPTER XXXIII

MORE LINEAR PROGRAMMING AND ALGEBRA

Who brought me hither
Will bring me hence; no other guide I seek.

J. MILTON

1. Introduction

We have considered only a few of the properties of sets, vectors and matrices. In doing so we have had two limited but important objectives. One of these has been to give the reader a feel of things which enables him to think of a matrix as a set of vectors, and of a vector as a set of elements; to appreciate that a problem in n dimensions may be viewed as one in geometry or as one in algebra; and to be able to think of a problem in whatever terms may be most appropriate. The other objective has been that of developing a few results, and introducing a few ideas, which are fundamental in that so much else is built upon them. If the reader, as a consequence, really understands linear independence, equivalence and rank he will be able to understand other theory which is developed in these terms.

In order to underline our first objective we shall now consider in more detail some of the parallels between set theory, matrix algebra and geometry. Then, we shall conclude with a few more results which the student may find useful.

2. Linear programming again

Let us return to the problem of Linear Programming considered in Chapter XXIX. The problem was that of minimising total expenditure

$$\Sigma p_i q_i$$

subject to certain constraints of the form

$$\Sigma p_i a_i \geqslant K \quad (a_i, K = \text{const.})$$

We discussed it in geometrical terms, and it was clear that out of the set of all possible combinations of quantities q, some were ruled out because they also belonged to the set of quantities which would not

508

satisfy one of the constraints. Eventually we found a set of q's which would satisfy the constraints, and we discovered a method for choosing, out of this set, that combination of quantities which would minimise expenditure. The reader is advised to go over Chapter XXIX, sections 1 and 2 again, before proceeding to our next paragraph.

Now consider several goods, purchased in quantities q_i at prices p_j. If we form a price vector

$$\mathbf{p} = (p_1, p_2, ..., p_n)$$

and a quantity vector

$$\mathbf{q} = \begin{pmatrix} q_1 \\ q_2 \\ \vdots \\ q_n \end{pmatrix}$$

then

$$\mathbf{pq} = p_1 q_1 + p_2 q_2 + \dots + p_n q_n = \Sigma p_i q_i$$

Thus the aim is to minimise

$$\mathbf{pq}$$

subject to certain constraints. Let us write these our in the form

$$a_{11}q_1 + a_{12}q_2 + \dots + a_{1n}q_n \geqslant K_1$$
$$a_{21}q_1 + a_{22}q_2 + \dots + a_{2n}q_n \geqslant K_2$$
$$\vdots \qquad \vdots \qquad \qquad \vdots \qquad \vdots$$
$$a_{m1}q_1 + a_{m2}q_2 + \dots + a_{mn}q_n \geqslant K_m$$

which is a generalised expression of constraints of the kind

$$q_1 c_1 + q_2 c_2 \geqslant C$$
$$q_1 b_1 + q_2 b_2 \geqslant B$$
$$q_1 v_1 + q_2 v_2 \geqslant V$$

Then our constraints can be summarised as

$$\mathbf{Aq} \geqslant \mathbf{K}$$

and so our linear programming problem is to minimise $\mathbf{pq}$ subject to $\mathbf{Aq} \geqslant \mathbf{K}$. (We should add that some Linear Programming problems involve maximising a function—say $\mathbf{xy}$—subject to constraints of the form $\mathbf{Ay} \leqslant \mathbf{K}$—i.e. to the $\mathbf{y}$'s being below certain constraints, rather than above them.)

Thus our L.P. problem, conceived as an attempt to locate the best mix of purchases out of that set which satisfied the basic requirements, and developed as a problem in geometry, can be stated as a problem in matrix algebra.

Let us look at this algebraic version, and compare it with the geometrical version.

Suppose that we know our prices p_i and choose quantities q_i so that the total expenditure is some known amount E. Then

$$pq = E$$

is a matrix equation of which the right hand side is a single scalar element. We would like to find the quantities q_i which satisfy it. Our rules for the solution of matrix equations tell us that the solubility will depend on the rank of p, and as this is a single row vector with at least one non-zero element its rank is quite simply one. Thus, we can determine one of the q's in terms of E and all other q's. This is obvious from the form

$$p_1 q_1 + p_2 q_2 + \ldots + p_n q_n = E$$

whence, for example,

$$p_1 = \frac{E - (p_2 q_2 + \ldots + p_n q_n)}{q_1}$$

But in geometrical terms we have that, given the value of E, there is a hyperplane such that at all points on it the quantities purchased yield a total expenditure of E; and these points are denoted by their coordinates, such as $(q_1, q_2, \ldots, q_n)$. Thus, any set of q's emerging from

$$pq = E$$

will denote a particular point on the hyperplane which corresponds to an expenditure E. In short, $pq = E$ is the equation of that hyperplane.

We have, similarly, that if we take the limiting cases of the constraints, replacing the inequality by equality, then the set of equation

$$Aq = K$$

denotes a set of hyperplanes, one corresponding to each constraint. Given A and K one can write

$$q = A^{-1} K$$

and the solubility of the equations will depend on the rank of A. Clearly this cannot exceed the number of columns—of which there is one for each q which appears in a constraint. If there is a good which does not appear in any of the constraints then the purchase of it is irrelevant to the problem, and so we can suppose that if there are n goods then all of them appear somewhere in the constraints. The number of linearly independent equations in the system

$$Aq = K$$

cannot exceed the rank of $\mathbf{A}$ which cannot exceed n, the number of goods, or m the number of constraints. If the rank is r then one can express r of the q's in terms of the other q's—provided the equations are consistent.

In geometrical terms this means that if there are n goods, each defining a dimension in which the quantity purchased is measured, then if $r = n$, so that the number of independent constraints is equal to the number of goods, then all of the q's can be found. There will be just one solution, giving the coordinates of the one point at which all constraints are just satisfied.

But if the number of constraints is fewer than the number of goods then the rank r is less than n. Now there will be no unique solution but an infinite set of them, with r of the quantities expressed as a function of whatever values we care to give the remaining $(n - r)$ quantities. Instead of there being a single point at which all of the constraints are just satisfied there will be some line or plane (or hyperplane) at which they are all just satisfied.

Consider, for example, the case of two constraints and two goods (or two dimensions) illustrated by

$$2q_1 + 3q_2 = 7$$
$$4q_1 + q_2 = 9$$

We can manipulate these equations in the following way.

Solution through the elimination of q_1 yields

$$6q_2 - q_2 = 14 - 9$$

yielding $\qquad\qquad 5q_2 = 5$

whence $\qquad\qquad q_2 = 1$

and so, by substitution, $q_1 = 2$. At this combination of purchases each constraint is just satisfied, and the combination corresponds to the point where the two lines

$$2q_1 + 3q_2 = 7 \quad \text{and} \quad 4q_1 + q_2 = 9$$

cross.

Now consider the case of two constraints and three goods (or dimensions) illustrated by

$$2q_1 + 3q_2 + 4q_3 = 13$$
$$4q_1 + q_2 + 2q_3 = 14$$

Here we have two planes forming lower constraints. The first has " corners " on the three axes at $(\frac{13}{2}, 0, 0)$, $(0, \frac{13}{3}, 0)$ and $(0, 0, \frac{13}{4})$, while the second has " corners " at $(\frac{14}{4}, 0, 0)$, $(0, 14, 0)$ and $(0, 0, 7)$, as in Diagram 33.1.

These two planes intersect along the line LM and at all combinations of the q's giving rise to points on this line the two constraints are simultaneously just satisfied.

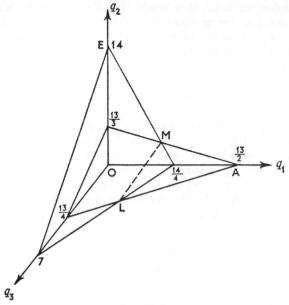

FIG. 33.1

If we consider the two equations above from a matrix point of view, we find that the rank of the matrix **A** is 2, which is also the rank of the augmented matrix **B**. We can express two of the q's in terms of the third q. To do so we can write the equations as

$$2q_1 + 3q_2 = 13 - 4q_3$$
$$4q_1 + q_2 = 14 - 2q_3$$

Multiplication of the second by 3, and subtraction from the first, yields

$$-10q_1 = (13 - 4q_3) - 3(14 - 2q_3)$$
$$= -29 + 2q_3$$

whence

$$q_1 = \frac{29 - 2q_3}{10}$$

Similarly,

$$q_2 = \frac{12 - 6q_3}{5}$$

Alternatively the equations may be solved by writing

$$c_1 = 13 - 4q_3$$
$$c_2 = 14 - 2q_3$$

and then

$$Aq = c$$

yielding

$$q = A^{-1}c$$

where, as $A = \begin{bmatrix} 2 & 3 \\ 4 & 1 \end{bmatrix}$

$$A^{-1} = \frac{\begin{bmatrix} 1 & -3 \\ -4 & 2 \end{bmatrix}}{\begin{vmatrix} 2 & 3 \\ 4 & 1 \end{vmatrix}} = -\frac{1}{10} \begin{bmatrix} 1 & -3 \\ -4 & 2 \end{bmatrix}$$

Thus

$$q = -\frac{1}{10} \begin{bmatrix} 1 & -3 \\ -4 & 2 \end{bmatrix} \begin{bmatrix} c_1 \\ c_2 \end{bmatrix}$$

$$= -\frac{1}{10} \begin{bmatrix} c_1 & -3c_2 \\ -4c_1 & +2c_2 \end{bmatrix}$$

and so

$$q_1 = -\frac{1}{10}\{(13 - 4q_3) - 3(14 - 2q_3)\} = -\frac{1}{10}\{2q_3 - 29\}$$

i.e. $\qquad q_1 = \dfrac{29 - 2q_3}{10} \qquad$ as before.

Similarly,

$$q_2 = \frac{12 - 6q_3}{5}$$

These results mean that if we choose any value for q_3 then the constraints will be just satisfied if both q_1 and q_2 are chosen according to these equations. What is their graphical interpretation?

If we consider the equation

$$q_1 = \frac{29 - 2q_3}{10}$$

we see that it contains no q_2. It represents a plane, parallel to the q_2-axis, such that there is this simple relationship between q_1 and q_3, independently of the value of q_2. It is the plane $URST$ in Diagram 33.2, having the points $S = (2\cdot9, 0, 0)$ and $R = (0, 0, 14\cdot5)$ at its base.

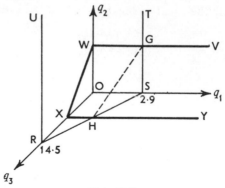

FIG. 33.2

Similarly, the equation

$$q_2 = \frac{12 - 6q_3}{5}$$

represents a plane parallel to q_3, as shown by $VWXY$. The point W has co-ordinates $(0, 2\cdot4, 0)$ while X is $(0, 0, 2)$.

These two planes are such that, for any q_3, the two constraints will be simultaneously satisfied only if q_1 lies on $URST$ and q_2 on $VWXY$. This can be so only along the line GH where these two planes intersect.

The line GH has one " end " at the point where ST cuts WV. It therefore has co-ordinates $(2\cdot9, 2\cdot4, 0)$. The other end has co-ordinates $(0, 2\cdot5, 2)$. The reader may check that these are also the co-ordinates of the " ends " of the line ML in the previous diagram.

We can now compare our two diagrams. The first is the " geometrical " approach. We have two planes—one for each of the imposed constraints—and we locate their line of intersection along which both constraints are satisfied.

In the second diagram we also have two planes: but they do not correspond directly to the imposed constraints. They arise out of the following argument.

We have two linearly independent constraints in three unknowns. They can be simultaneously satisfied by an infinite number of points. To locate these we can give one of the unknowns (q_3) any value, and then specify the relationships which the other unknowns must bear to q_3. Each of these is such that the other unknown is not involved in it; but they must each be satisfied.

One of these relationships, relating q_2 to q_3, is the equation of the plane $VWXY$. For both constraints to be satisfied the quantities q_2 and q_3 must be such that the point (q_1, q_2, q_3) lies somewhere on this plane. But, also, the quantities q_1 and q_3 must be such that the point (q_1, q_2, q_3) lies somewhere on the plane $URST$, which relates q_1 to q_3 independently of q_2.

In a more general case each hyperplane would specify a relationship between one of the r independent unknowns and the $n-r$ dependent ones.

It may be useful to think of the geometrical approach as " thinking along the constraints ", and of the matrix approach as " thinking parallel to the axes ".

But this is not the end of the story, for in the L.P. programme the constraints do not necessarily have to be just satisfied. The line LM (or GH) is simply an " edge " of the feasible zone. We do not need to satisfy the equations

$$\mathbf{Aq} = \mathbf{K}$$

but we do need to satisfy the inequalities

$$\mathbf{Aq} \geqslant \mathbf{K}$$

To embark seriously on the algebra that arises out of this inequality would take us too far afield, but we may make one point about it.

The trick is to turn the inequalities into equations. In terms of our example, if we wish to ensure that the total supply of vitamins

$$q_1 v_1 + q_2 v_2$$

is at least equal to V, we could do so by specifying that we always gave away a certain non-negative quantity of vitamins, for no reward, and always made our purchases so that we were left with exactly V. Thus if we give away a quantity λ_v, we have

$$q_1 v_1 + q_2 v_2 - \lambda_v = V$$

The other inequalities can be manipulated similarly into

$$q_1 c_1 + q_2 c_2 - \lambda_c = C$$
$$q_1 b_1 + q_2 b_2 - \lambda_b = B$$

But λ_b, λ_c and λ_v measure different things and so are in different dimensions. We have added three dimensions to the inequalities in order to turn them into equations. We can now write

$$\mathbf{Aq} = \mathbf{K}$$

where $\mathbf{A}$ is derived from the L.H.S. of

$$
\begin{aligned}
q_1 v_1 + q_2 v_2 - \lambda_v \quad\quad\quad &= V \\
q_1 c_1 + q_2 c_2 \quad\quad - \lambda_c \quad\quad &= C \\
q_1 b_1 + q_2 b_2 \quad\quad\quad\quad - \lambda_b &= B
\end{aligned}
$$

and is

$$\begin{bmatrix} v_1 & v_2 & -1 & 0 & 0 \\ c_1 & c_2 & 0 & -1 & 0 \\ b_1 & b_2 & 0 & 0 & -1 \end{bmatrix}$$

while q is now

$$\begin{bmatrix} q_1 \\ q_2 \\ \lambda_v \\ \lambda_c \\ \lambda_b \end{bmatrix}$$

and we are interested only in those solutions with λ_v, λ_c, $\lambda_b \geqslant 0$. We now seek to minimise

$$pq = E$$

subject to $Aq = K$ with non-negative λ's and zero prices for the λ's.

In geometric terms the λ's indicate the distance which we are above the relevant constraint. We minimise E subject to all of these distances being zero or positive.

We call our λ's slack variables. For a fuller discussion of the problem one may consult some of the works cited in the Bibliography (pages 561 f.)

3. Theorems about inverse matrices

Finally we present a few miscellaneous properties of matrices. We begin with some properties of inverse matrices.

1. $$(AB)^{-1} = B^{-1}A^{-1}$$

This is easily shown, by using the laws which we have found for matrices. Thus,

$$\begin{aligned} (AB)(B^{-1}A^{-1}) &= A(BB^{-1})A^{-1} \\ &= AIA^{-1} \\ &= AA^{-1} \\ &= I \end{aligned}$$

Whence, as
$$\begin{aligned} (AB)(AB)^{-1} &= I \\ (AB)^{-1} &= B^{-1}A^{-1} \end{aligned}$$

For example, if
$$C = \begin{bmatrix} 1 & 2 & 3 \\ 2 & 3 & 1 \\ 4 & 1 & 0 \end{bmatrix} \begin{bmatrix} 1 & 1 & 4 \\ 2 & 3 & 1 \\ 1 & 1 & 2 \end{bmatrix}$$

then
$$C^{-1} = B^{-1}A^{-1}$$

where
$$B^{-1} = \begin{bmatrix} 1 & 1 & 4 \\ 2 & 3 & 1 \\ 1 & 1 & 2 \end{bmatrix}^{-1}$$

and
$$A^{-1} = \begin{bmatrix} 1 & 2 & 3 \\ 2 & 3 & 1 \\ 4 & 1 & 0 \end{bmatrix}^{-1}$$

We may recall that to find B^{-1} we form a new matrix by replacing each element by its co-factor, and then transposing, before dividing each element by $|B|$.

It follows from this result that
$$(ABC)^{-1} = C^{-1}B^{-1}A^{-1}$$
Another result is
$$(A')^{-1} = (A^{-1})'$$
i.e. The inverse of the transpose is the transpose of the inverse. This is so because, if we transpose both sides of
$$AA^{-1} = I$$
we get
$$(A^{-1})'A' = I' = I$$
If we post-multiply by $(A')^{-1}$ we get
$$(A^{-1})'A'(A')^{-1} = I(A')^{-1}$$
whence
$$(A^{-1})'I = I(A')^{-1}$$
$$\therefore (A^{-1})' = (A')^{-1}$$

4. The quadratic form and vector differentiation

The student of economics may often come across a matrix expression of the form
$$x'Ax$$
where x is a column vector, x' its transpose, and A is a (necessarily) square matrix. We may consider, as an example,

$$\begin{bmatrix} x_1, & x_2, & x_3 \end{bmatrix} \begin{bmatrix} a_{11} & a_{12} & a_{13} \\ a_{21} & a_{22} & a_{23} \\ a_{31} & a_{32} & a_{33} \end{bmatrix} \begin{bmatrix} x_1 \\ x_2 \\ x_3 \end{bmatrix}$$

Multiplication of this will yield its value as

$$\sum_j \sum_i a_{ij} x_i x_i$$

i.e. $a_{11} x_1^2 + a_{12} x_1 x_2 + a_{13} x_1 x_3 + a_{21} x_2 x_1 + a_{22} x_2^2 + \ldots + a_{33} x_3^2$

It is the sum of all possible combinations of the **x**'s taken two at a time (including squares) with appropriate coefficients.

It is called a quadratic form.

We also come across vector differentiation. Essentially this is differentiating each component of the vector with respect to the appropriate variable. For example, let **a**′ be the transpose of a column vector **a** such that

$$\mathbf{a}'\mathbf{x} = \begin{bmatrix} a_1 & a_2 & a_3 \end{bmatrix} \begin{bmatrix} x_1 \\ x_2 \\ x_3 \end{bmatrix}$$

$$= a_1 x_1 + a_2 x_2 + a_3 x_3$$

which is a scalar function of x_1, x_2 and x_3 and may be differentiated.

Form the partial derivatives

$$\frac{\partial(\mathbf{ax})}{\partial x_1} = a_1$$

$$\frac{\partial(\mathbf{ax})}{\partial x_2} = a_2$$

$$\frac{\partial(\mathbf{ax})}{\partial x_3} = a_3$$

Let us now *define* the process of vector differentiation to be the formation of the vector whose elements are these partial derivatives (a_1, a_2, a_3)—which is clearly—in this case—the vector **a**.

Symbolically we have

$$\frac{\partial(\mathbf{a}'\mathbf{x})}{\partial \mathbf{x}} = \mathbf{a}$$

Usually, for reasons which a more advanced use of this idea would make clear, we write the derivative as a column vector.

If we now turn to matrices we can extend the idea. For example if we write out

$$\mathbf{x}'\mathbf{A}\mathbf{x}$$

in full, form the partial derivatives with respect to the various x's, and then assemble these as the elements of a vector we obtain

$$\frac{\partial}{\partial \mathbf{x}}(\mathbf{x}'\mathbf{A}\mathbf{x}) = 2\mathbf{A}\mathbf{x} = 2\mathbf{x}'\mathbf{A}$$

which may be compared with $\dfrac{d}{dx}(Ax^2) = 2Ax$.

For the sign of a quadratic form see Appendix 7.

5. Characteristic roots and vectors

Another concept which frequently arises is this.   Given a square matrix A, are there vectors x and scalars λ such that

$$Ax = \lambda x?$$

We call such vectors and scalars " characteristic vectors " and " characteristic roots ".   Some authors use " latent " or " eigen " in place of " characteristic ".

Note that if
$$Ax = \lambda x$$
$$= \lambda I x$$

then
$$(A - \lambda I)x = 0$$

Provided that $A - \lambda I$ is non-singular we can form its inverse, but in that case

$$x = (A - \lambda I)^{-1}0$$
$$= 0$$

Thus if x is to be non-null, the matrix $A - \lambda I$ must be singular, which means that the determinant

$$|A - \lambda I| = 0$$

This determinant, on expansion, yields a polynomial in λ, and the various roots become the characteristic roots.   The associated x's are the characteristic vectors.

For example, suppose we begin with

$$A = \begin{bmatrix} 2 & 4 \\ 1 & 5 \end{bmatrix}$$

Consider
$$A - \lambda I = \begin{bmatrix} 2 & 4 \\ 1 & 5 \end{bmatrix} - \begin{bmatrix} \lambda & 0 \\ 0 & \lambda \end{bmatrix}$$

$$= \begin{bmatrix} 2-\lambda & 4 \\ 1 & 5-\lambda \end{bmatrix}$$

Then
$$|A - \lambda I| = (2-\lambda)(5-\lambda) - 4$$
$$= \lambda^2 - 7\lambda + 10 - 4$$
$$= (\lambda - 6)(\lambda - 1)$$

Thus the roots of

$$(A - \lambda I) = 0$$

are $\lambda_1 = 6$, $\lambda_2 = 1$ and these are the characteristic roots.

We now find a characteristic vector for each λ. Letting that associated with $\lambda_1 = 6$ be x_1, with elements x_{11} and x_{12}, we have

$$A \quad x = \lambda x$$

whence

$$\begin{bmatrix} 2 & 4 \\ 1 & 5 \end{bmatrix} \begin{bmatrix} x_{11} \\ x_{12} \end{bmatrix} = 6 \begin{bmatrix} x_{11} \\ x_{12} \end{bmatrix}$$

and so

$$2x_{11} + 4x_{12} = 6x_{11}$$
$$x_{11} + 5x_{12} = 6x_{12}$$

whence

$$x_{11} = x_{12}$$

(This comes from either equation. Both are bound to tell the same story as λ is chosen to produce a singular matrix, and so the equations must be dependent.) Thus the characteristic vector associated with $\lambda_1 = 6$ is

$$x_1 = (x_{11}, x_{11})$$

where x_{11} can have any value. It is usual to " standardise " it by choosing x_{11} so that, in this case,

$$x_{11}^2 + x_{11}^2 = 1$$

which gives

$$2x_{11}^2 = 1$$

$$x_{11} = \pm \sqrt{\frac{1}{2}}$$

and so the characteristic vector is

$$\left(\pm \sqrt{\frac{1}{2}}, \ \pm \sqrt{\frac{1}{2}} \right)$$

The student should check whether all the combinations of signs are possible. There is also another characteristic vector associated with $\lambda_2 = 1$. It is

$$\left(\pm \sqrt{\frac{16}{17}}, \ \mp \sqrt{\frac{1}{17}} \right)$$

An important result is that if A is a symmetric matrix (which means that $a_{ij} = a_{ji}$) then the product of the characteristic vectors is zero. A geometrical interpretation of this is that they are ortho-gonal (or perpendicular).

The Factors of $Ax^2 + Bx + C$

If the expression

$$Ax^2 + Bx + C$$

has real factors, they may often be found quite easily by using a simple device which forms the subject of this Appendix.

We know that

$$(ax + \alpha)(bx + \beta) \equiv abx^2 + (a\beta + b\alpha)x + \alpha\beta$$

If we compare the right-hand side of this identity with $Ax^2 + Bx + C$ we have

$$A = ab$$
$$B = a\beta + b\alpha$$
$$C = \alpha\beta$$

We note that if A is positive, then a and b will be of the same sign. If A is negative, a and b have different signs.

If C is positive, α and β have the same sign. If C is negative, α and β have different signs.

If B is negative, then at least one of the products $a\beta$ and $b\alpha$ must be negative. This cannot be unless either a and β have opposite signs, or b and α have opposite signs. A positive B implies only that at least one of the products is positive, and that therefore either a and β, or b and α have the same signs.

All of this information is readily used by writing the result in the form

$$abx^2 + (a\beta + b\alpha)x + \alpha\beta$$

$$
\begin{array}{cc}
+ ax & + \alpha \\
+ bx & + \beta \\
\end{array}
$$

$$(ax + \alpha)(bx + \beta)$$

The term abx^2 is the product of the terms on the left extremities of the cross.

The term $\alpha\beta$ is the product of the terms on the right extremities of the cross.

The term $(a\beta + b\alpha)x$ is obtained by multiplying together the terms at the ends of the one diagonal, and adding to their product the product of the terms at the ends of the other diagonal, i.e.,

$$(ax \times \beta) + (bx \times \alpha)$$

The rest of the method is best explained by taking some examples.

(i) $6x^2 + 19x + 10$

In this all the terms are positive. This means that a and b are of the same sign and that α and β are of the same sign.

Now since a has the same sign as b, and β the same sign as α, then $a\beta$ has the same sign as $b\alpha$.

But the middle term is positive, which indicates that $a\beta + b\alpha$ is positive. It follows that both $a\beta$ and $b\alpha$ are positive, and that therefore a and β have the same signs. We thus have that a, b, α and β are all of one sign. *This is always the case when A, B and C in the expression $Ax^2 + Bx + C$ are of the same sign.*

Knowing that they are all of one sign, we shall assume a, b, α and β to be positive. We shall later justify this.

Then we commence our solution by writing

$$6x^2 + 19x + 10$$

$$+ x \qquad +$$

$$+ x \qquad +$$

$$(x + \quad)(x + \quad)$$

Now the possible factors of 6 are 6 and 1, and 2 and 3. The possible factors of 10 are 10 and 1 and 5 and 2. We now consider the possible combination of these factors.

Try $Ax^2 + Bx + C$

$$+ 6x \qquad + 10$$

$$+ x \qquad + 1$$

This gives us

$$Ax^2 = abx = 6x^2$$
$$Bx = (a\beta + b\alpha)x = (6x \times 1) + (x \times 10) = 16x$$
$$C = \alpha\beta = 10$$

It follows that

$$(6x + 10)(x + 1) = 6x^2 + 16x + 10$$

which is not what we want.

Therefore, try

$$\begin{array}{ccc} +6x & \diagdown\diagup & +1 \\ +x & \diagup\diagdown & +10 \end{array}$$

This gives us

$$6x^2 + [(6x \times 10) + (x \times 1)] + 10 = 6x^2 + 61x + 10$$

which is not the result we require.
Now try

$$\begin{array}{ccc} +3x & \diagdown\diagup & +10 \\ +2x & \diagup\diagdown & +1 \end{array}$$

which gives $6x^2 + 23x + 10$

The scheme

$$\begin{array}{ccc} +3x & \diagdown\diagup & +1 \\ +2x & \diagup\diagdown & +10 \end{array}$$

is clearly not what we want, because this yields

$$(3x + 1)(2x + 10)$$

which is divisible by 2, while the expression $6x^2 + 19x + 10$ is not.
Therefore, try

$$\begin{array}{ccc} +3x & \diagdown\diagup & +2 \\ +2x & \diagup\diagdown & +5 \end{array}$$

which yields

$$6x^2 + [(3x \times 5) + (2x \times 2)] + 10$$

which is $6x^2 + 19x + 10$, which is the result we require. It follows that

$$6x^2 + 19x + 10 = (3x + 2)(2x + 5)$$

It is also true that

$$\begin{aligned} 6x^2 + 19x + 10 &= (-3x - 2)(-2x - 5) \\ &= [-1(3x + 2)][-1(2x + 5)] \\ &= (-1)^2(3x + 2)(2x + 5) \\ &= (3x + 2)(2x + 5) \end{aligned}$$

The assumption that, when a, b, α and β are all of the same sign then that sign is positive is therefore quite valid.

(ii) $6x^2 - 19x + 10$

Here we have that a and b are of the same sign, and that α and β are of the same sign. But the negative B shows that either $a\beta$ or $b\alpha$ is negative. A little thought along the lines of the previous example will show us that in such a case it is perfectly general to begin

$$6x^2 - 19x + 10$$

$$\begin{matrix} +\,x & \diagdown & - \\ & \times & \\ +\,x & \diagup & - \end{matrix}$$

If we try the possible pairs of factors we have

$$\begin{matrix} +3x & \diagdown & -5 \\ & \times & \\ +2x & \diagup & -2 \end{matrix}$$

which gives $6x^2 + [(3x \times -2) + (2x \times -5)] + 10$

$$= 6x^2 - 16x + 10$$

which is not what we want—as we might have foreseen because of the factor 2.

Now try

$$\begin{matrix} +3x & \diagdown & -2 \\ & \times & \\ +2x & \diagup & -5 \end{matrix}$$

which gives $6x^2 + [(3x \times -5) + (2x \times -2)] + 10$

$$= 6x^2 - 19x + 10$$

It follows that $6x^2 - 19x + 10 = (3x - 2)(2x - 5)$

$$= (2 - 3x)(5 - 2x)$$

(iii) $6x^2 + 11x - 10$

Here we begin, for reasons similar to those above, with

$$\begin{matrix} +\,x & \diagdown & - \\ & \times & \\ +\,x & \diagup & + \end{matrix}$$

Trying

$$\begin{matrix} +3x & \diagdown & -10 \\ & \times & \\ +2x & \diagup & +1 \end{matrix}$$

we obtain $6x^2 + [(3x \times 1) + (2x \times -10)] - 10$

$$= 6x^2 - 17x - 10$$

which is not what we want.

We next try

$$+3x \diagdown -5$$
$$+2x \diagup +2$$

and obtain

$$6x^2 - 4x - 10$$

which is not what we want. We could have foretold this.
We now try

$$+3x \diagdown -2$$
$$+2x \diagup +5$$

which yields the result $6x^2 + 11x - 10$. The factors are therefore $(3x - 2)$ and $(2x + 5)$.

(iv) $6x^2 - 59x - 10$

Once again we begin

$$+ x \diagdown -$$
$$+ x \diagup +$$

and obtain

$$+x \diagdown -10$$
$$+6x \diagup +1$$

which yields

$$(x - 10)(6x + 1)$$

Exercise A1.1

Factorise :

  (i) $2x^2 + 5x + 3$ (ii) $2x^2 + 7x + 3$

 (iii) $2x^2 + 13x + 15$ (iv) $4x^2 + 16x + 15$

  (v) $4x^2 + 23x + 15$ (vi) $8x^2 + 21x + 10$

 (vii) $8x^2 + 18x + 10$ (viii) $8x^2 + 42x + 10$

  (ix) $8x^2 + 10x - 3$ (x) $8x^2 - 2x - 3$

  (xi) $8x^2 + 5x - 3$ (xii) $20x^2 - 31x + 12$

(xiii) $20x^2 - 34x + 12$ (xiv) $20x^2 + 14x - 12$

On the Roots of Equations of Degree n

If an equation $f(x) = 0$ has x_1 as a root, then clearly

$$(x - x_1)$$

is a factor of $f(x)$, which can therefore be written as

$$f(x) \equiv (x - x_1)g(x) = 0$$

where
$$g(x) \equiv \frac{f(x)}{x - x_1}$$

If it also has x_2 as a root then the equation can be written as

$$f(x) \equiv (x - x_1)(x - x_2)h(x) = 0$$

where
$$h(x) \equiv \frac{f(x)}{(x - x_1)(x - x_2)}$$

If the n roots are $x_1, x_2, \ldots, x_n$ then we may write

$$f(x) \equiv (x - x_1)(x - x_2) \ldots (x - x_n)\lambda = 0$$

where
$$\lambda = \frac{f(x)}{(x - x_1)(x - x_2) \ldots (x - x_n)} \qquad \text{and is a constant.}$$

Let us now consider

$$f(x) \equiv (x - x_1)(x - x_2) \ldots (x - x_n)\lambda = 0$$

The expansion of this may be written as

$$\lambda\{x^n - x^{n-1}[x_1 + x_2 + x_3 + \ldots + x_n]$$
$$+ x^{n-2}[x_1x_2 + x_1x_3 + \ldots + x_1x_n$$
$$+ x_2x_3 + \ldots + x_2x_n$$
$$+ \ldots$$
$$+ x_{n-1}x_n]$$

$- x^{n-3}$[sum of all possible combinations of the roots taken 3 at a time]

$+ x^{n-4}$[sum of all possible combinations of the roots taken 4 at a time]

$- \ldots$

$$+ (-1)^{n-1}x[x_2x_3x_4 \ldots x_n + x_1x_3x_4 \ldots x_n$$
$$+ x_1x_2x_4 \ldots x_n + \ldots + \ldots + x_1x_2x_3 \ldots x_{n-1}]$$
$$+ (-1)^n[x_1x_2x_3 \ldots x_n]\}$$

This may be compared with the standard form

$$f(x) \equiv a_0 x^n + a_1 x^{n-1} + a_2 x^{n-2} + \ldots + a_n = 0$$

Since the two expressions are identically equal it follows that

$$a_0 = \lambda$$

$$\frac{a_1}{\lambda} = -[x_1 + x_2 + x_3 + \ldots + x_n]$$

$$\frac{a_2}{\lambda} = [(x_1 x_2 + x_1 x_3 + \ldots + x_1 x_n) + (x_2 x_3 + \ldots + x_2 x_n) + \ldots + x_{n-1} x_n]$$

etc.

We may illustrate the importance of these results by considering the equation

$$4x^2 - 16x + 15 = 0$$

If this has roots x_1 and x_2 then

$$-\frac{16}{4} = -[x_1 + x_2]$$

and

$$\frac{15}{4} = x_1 x_2$$

i.e.,

$$x_1 + x_2 = 4$$

$$x_1 x_2 = \frac{15}{4}$$

It is clear that $x_1 = \frac{3}{2}$ and $x_2 = \frac{5}{2}$ will satisfy these requirements, and so we have these as our solutions.

Example

Suppose that the equation

$$24x^3 - 14x^2 - 63x + 45 = 0$$

is known to have one root double another. Let the roots be x_1, $2x_1$ and x_3. Then we have

$$-(x_1 + 2x_1 + x_3) = -\tfrac{14}{24}$$

$$(2x_1^2 + x_1 x_3 + 2x_1 x_3) = -\tfrac{63}{24}$$

$$-2x_1^2 x_3 = \tfrac{45}{24}$$

from which we obtain

$$3x_1 + x_3 = \tfrac{7}{12}, \quad 2x_1^2 + 3x_1 x_3 = -\tfrac{21}{8}, \quad 2x_1^2 x_3 = -\tfrac{15}{8}$$

The first two of these yield

$$8x_1^2 - 2x_1 - 3 = 0$$

whence $x_1 = \frac{3}{4}$ or $-\frac{1}{2}$, and $x_3 = -\frac{5}{3}$ or $\frac{25}{12}$.

Substitution in the third equation $2x_1^2 x_3 = -\frac{15}{8}$ will show that the values $x_1 = -\frac{1}{2}$, $x_3 = \frac{25}{12}$ do not satisfy it, and so we have to accept as the only roots

$$x_1 = \tfrac{3}{4}, \quad x_2 = 2x_1 = \tfrac{3}{2}, \quad x_3 = -\tfrac{5}{3}$$

APPENDIX 3

The Equation $x^n = A$

The solution of $x^n = \pm a$

It can be shown, by using de Moivre's Theorem, that if n is any integer and a is positive and real then the equation

$$x^n + a = 0$$

has the n roots

$$x_1 = \sqrt[n]{a}\left(\cos\frac{\pi}{n} + i\sin\frac{\pi}{n}\right)$$

$$x_2 = \sqrt[n]{a}\left(\cos\frac{\pi + 2\pi}{n} + i\sin\frac{\pi + 2\pi}{n}\right)$$

$$\vdots$$

$$x_n = \sqrt[n]{a}\left(\cos\frac{(2n-1)\pi}{n} + i\sin\frac{(2n-1)\pi}{n}\right)$$

It can be shown that the equation

$$x^n - a = 0$$

has the n roots

$$x_1 = \sqrt[n]{a}$$

$$x_2 = \sqrt[n]{a}\left(\cos\frac{2\pi}{n} + i\sin\frac{2\pi}{n}\right)$$

$$\vdots$$

$$x_n = \sqrt[n]{a}\left(\cos\frac{(2n-2)\pi}{n} + i\sin\frac{(2n-2)\pi}{n}\right)$$

Examples

(i) $$x^3 + 27 = 0$$

has the three roots

$$x_1 = \sqrt[3]{27}\left(\cos\frac{\pi}{3} + i\sin\frac{\pi}{3}\right)$$

$$= 3\left(\tfrac{1}{2} + i\frac{\sqrt{3}}{2}\right)$$

$$x_2 = \sqrt[3]{27}\left(\cos\frac{3\pi}{3} + i\sin\frac{3\pi}{3}\right)$$

529

s

$$= 3(-1+0) = -3$$

$$x_3 = \sqrt[3]{27}\left(\cos\frac{5\pi}{3} + i\sin\frac{5\pi}{3}\right)$$

$$= 3\left(\tfrac{1}{2} - i\frac{\sqrt{3}}{2}\right)$$

(ii) The equation $x^3 - 27 = 0$

has the three roots

$$x_1 = \sqrt[3]{27} = 3$$

$$x_2 = \sqrt[3]{27}\left(\cos\frac{2\pi}{3} + i\sin\frac{2\pi}{3}\right)$$

$$= 3\left(-\tfrac{1}{2} + i\frac{\sqrt{3}}{2}\right)$$

$$x_3 = \sqrt[3]{27}\left(\cos\frac{4\pi}{3} + i\sin\frac{4\pi}{3}\right)$$

$$= 3\left(-\tfrac{1}{2} - i\frac{\sqrt{3}}{2}\right)$$

The cube roots of unity

The solutions of the equation

$$x^3 - 1 = 0$$

are of especial importance. Application of the above results will show them to be

$$x_1 = 1, \quad x_2 = -\tfrac{1}{2} + i\frac{\sqrt{3}}{2}, \quad x_3 = -\tfrac{1}{2} - i\frac{\sqrt{3}}{2}$$

A little algebra will show that in this case

$$x_3 = x_2{}^2$$

It is customary to denote x_2—in this particular case—by ω and to write the result in the form

$$x_1 = 1, \quad x_2 = \omega, \quad x_3 = \omega^2$$

where

$$\omega = -\tfrac{1}{2} + i\frac{\sqrt{3}}{2}$$

$$\omega^2 = -\tfrac{1}{2} - i\frac{\sqrt{3}}{2}$$

and, of course, $\omega^3 = 1$

It will be noticed that $1 + \omega + \omega^2 = 0$.

The Cubic Equation

1. The cubic equation

The cubic equation may be written in its most general form as

$$x^3 + a_1 x^2 + a_2 x + a_3 = 0$$

where a_1, a_2 and a_3 are constants.

If we now write

$$y = x + \frac{a_1}{3}$$

so that

$$x = y - \frac{a_1}{3}$$

then the equation may be reduced to

$$y^3 + \left(a_2 - \frac{a_1^2}{3} \right) y + \left(\frac{2a_1^3}{27} - \frac{a_1 a_2}{3} + a_3 \right) = 0$$

This equation may be written

$$y^3 + Ay + B = 0$$

where A and B are constants defined by

$$A = a_2 - \frac{a_1^2}{3}$$

and

$$B = \frac{2a_1^3}{27} - \frac{a_1 a_2}{3} + a_3$$

It is possible to perform this reduction in every case, and we can therefore consider the *reduced form*

$$y^3 + Ay + B = 0$$

to be the perfectly general cubic equation.

It can be shown that the three solutions of this reduced cubic equation are given by

(i) $y_1 = z + v$

(ii) $y_2 = \omega z + \omega^2 v$

(iii) $y_3 = \omega^2 z + \omega v$

where 1, ω and ω^2 are the roots of $x^3 - 1 = 0$, as indicated earlier in Appendix 3, and

$$z = \sqrt[3]{(-B/2) + \sqrt{[(B^2/4) + (A^3/27)]}}$$

and
$$v = \sqrt[3]{(-B/2) - \sqrt{[(B^2/4) + (A^3/27)]}}$$

Example

$$4x^3 + 12x^2 - 60x = 208$$

may be written

$$x^3 + 3x^2 - 15x - 52 = 0$$

Put
$$x = y - \frac{3}{3} = y - 1$$

We have

$$(y-1)^3 + 3(y-1)^2 - 15(y-1) - 52 = 0$$
$$(y^3 - 3y^2 + 3y - 1) + (3y^2 - 6y + 3) - 15y + 15 - 52 = 0$$

whence
$$y^3 - 18y - 35 = 0$$

We now take the equation in this reduced form. Its three solutions include z and v where

$$\begin{aligned}
z &= \sqrt[3]{-(-35/2) + \sqrt{[(-35^2/4) + (-18^3/27)]}} \\
&= \sqrt[3]{(35/2) + \sqrt{[(1225/4) - (5832/27)]}} \\
&= \sqrt[3]{(35/2) + \sqrt{[(1225/4) - (216)]}} \\
&= \sqrt[3]{(35/2) + \sqrt{[(361/4)]}} \\
&= \sqrt[3]{(35/2) + (19/2)} \\
&= \sqrt[3]{54/2} \\
&= \sqrt[3]{27} \\
&= 3
\end{aligned}$$

and
$$\begin{aligned}
v &= \sqrt[3]{-(-35/2) - \sqrt{[(-35^2/4) + (-18^3/27)]}} \\
&= \sqrt[3]{(35/2) - (19/2)} \\
&= \sqrt[3]{16/2} \\
&= \sqrt[3]{8} \\
&= 2
\end{aligned}$$

The solutions are

$$\begin{aligned}
y_1 &= z + v = 3 + 2 = 5 \\
y_2 &= \omega z + \omega^2 v = 3\omega + 2\omega^2 = -\frac{5}{2} + i\frac{\sqrt{3}}{2} \\
y_3 &= \omega^2 z + \omega v = 3\omega^2 + 2\omega = -\frac{5}{2} - i\frac{\sqrt{3}}{2}
\end{aligned}$$

It will be noticed that the sum of the three roots is zero, which is always true of the roots of the reduced cubic.

It follows that the solutions to the original equation are

$$x_1 = y_1 - 1 = 4$$
$$x_2 = y_2 - 1 = -\frac{7}{2} + i\frac{\sqrt{3}}{2}$$
$$x_3 = y_3 - 1 = -\frac{7}{2} - i\frac{\sqrt{3}}{2}$$

It can be shown that if A and B are real then the reduced cubic equation

$$y^3 + Ay + B = 0$$

has

(i) three real roots if $\dfrac{B^2}{4} + \dfrac{A^3}{27} < 0$

(ii) two equal roots if $\dfrac{B^2}{4} + \dfrac{A^3}{27} = 0$

(iii) one real and two complex conjugate roots if $\dfrac{B^2}{4} + \dfrac{A^3}{27} > 0$.

The method of solution which we have given will work in the three cases, but if case (i) above is true the solutions are more easily obtained from

$$y_1 = 2 \sqrt[6]{(-A^3/27)} \cos \theta$$
$$y_2 = 2 \sqrt[6]{(-A^3/27)} \cos (\theta + \tfrac{2}{3}\pi)$$
$$y_3 = 2 \sqrt[6]{(-A^3/27)} \cos (\theta + \tfrac{4}{3}\pi)$$

where
$$\cos \theta = -\frac{B}{2\sqrt{(-A^3/27)}}$$

For a fuller treatment of this subject students may consult Professor H. W. Turnbull's *Theory of Equations*, on whose Chapter IX this section is based.

Newton's Method for Numerical Solution

If $x = a$ is an approximate solution to

$$f(x) = 0$$

then a closer approximation is given by

$$b = a - \frac{f(a)}{f'(a)}$$

where $f(a)$ is the value of $f(x)$ when $x = a$
and $f'(a)$ is the value of the derivative $f'(x)$ when $x = a$.

Successive application of this result will frequently enable a root to be approximated to quite closely.

Example

Consider $x^3 - 2x - 6 = 0$

In this case
$$f(2) = 8 - 4 - 6 = -2$$
$$f(3) = 27 - 6 - 6 = 15$$

There is therefore at least one root between 2 and 3.

Let $a = 2$
$$f(x) = x^3 - 2x - 6$$
$$f'(x) = 3x^2 - 2$$

when $a = 2$
$$f(a) = 8 - 4 - 6 = -2$$
$$f'(a) = 12 - 2 = 10$$
$$b = a - \frac{f(a)}{f'(a)} = 2 - \frac{-2}{10} = 2 \cdot 2$$

We now take 2·2 as a closer approximation
$$f(2 \cdot 2) = (2 \cdot 2)^3 - 2(2 \cdot 2) - 6$$
$$= 10 \cdot 648 - 4 \cdot 4 - 6 = 0 \cdot 248$$
$$f'(2 \cdot 2) = 3(2 \cdot 2)^2 - 2$$
$$= 14 \cdot 52 - 2 = 12 \cdot 52$$

Therefore, a closer approximation is
$$c = b - \frac{f(b)}{f'(b)} = 2 \cdot 2 - \frac{0 \cdot 248}{12 \cdot 52}$$
$$= 2 \cdot 18019169$$

We now take this and obtain

$$f(2 \cdot 18019169) = 0 \cdot 002\ 581\ 86$$
$$f'(2 \cdot 18019169) = 12 \cdot 259\ 707\ 45$$
$$d = 2 \cdot 180\ 191\ 69 - \frac{0 \cdot 002\ 581\ 86}{12 \cdot 259\ 707\ 45}$$
$$= 2 \cdot 179\ 981\ 09$$

If we take this as a further approximation we will obtain

$$f(2 \cdot 179\ 981\ 09) = 0 \cdot 000\ 000\ 219$$
$$f'(2 \cdot 179\ 981\ 09) = 12 \cdot 256\ 952\ 659$$

and

$$e = d - \frac{f(d)}{f'(d)}$$

$$= 2 \cdot 179\ 981\ 07$$

as a still closer approximation.

Determinants and the Solution of Equations

1. Introduction

There are two problems which, in some form or the other, have frequently occurred in this book. One is the solution of simultaneous equations (often involving several variables); the other is the problem of deciding whether a given expression is positive or negative. Both of these problems are soluble most easily when determinants are used. Although the determinant notation looks a trifle formidable, there is, in fact, very little difficulty in its use, and the solution of the simplest simultaneous equations is simplified still further by the introduction of this notation. There are various ways of introducing these ideas. The student who is seeking a thorough understanding of the subject should read Chapter XXIX, but he who is prepared to take the theory completely on trust and to be content with an ability to use determinants, even though he does not know why they work as they do, may find it easier to read this Appendix and selected parts of Chapter XXIX which will be mentioned later.

Suppose that we have two simultaneous equations

$$a_1 x + b_1 y = h_1$$
$$a_2 x + b_2 y = h_2$$

where $a_1, a_2, b_1, b_2, h_1, h_2$ are constants, which are written in this form for reasons that will become apparent later on. If we solve these by the method given in Chapter I we will obtain the solution

$$x = \frac{h_1 b_2 - h_2 b_1}{a_1 b_2 - a_2 b_1} \quad \text{and} \quad y = \frac{h_2 a_1 - h_1 a_2}{a_1 b_2 - a_2 b_1}$$

This solution, of course, depends only on the coefficients a and b and the constants h. Let us consider the denominators of the two expressions, which are the same. If we compare them with the equations themselves, as set out above, we will see that the denominator is obtained from the coefficients of x and y by taking the product indicated by the downward arrow $(a_1 b_2)$ and then subtracting the product indicated by the upward arrow $(a_2 b_1)$

$$a_1 x + b_1 y = h_1$$
$$a_2 x + b_2 y = h_2$$

Having noted this we could agree to write the denominator as

$$\begin{vmatrix} a_1 & b_1 \\ a_2 & b_2 \end{vmatrix}$$

with the understanding that this is just a convenient way of writing the expression formed by taking the product of the terms a_1 and b_2, and then subtracting the product of the terms a_2 and b_1.

Inspection of the numerators of these expressions will show that, with the same notation, we can put the solutions as

$$x = \frac{\begin{vmatrix} h_1 & b_1 \\ h_2 & b_2 \end{vmatrix}}{\begin{vmatrix} a_1 & b_1 \\ a_2 & b_2 \end{vmatrix}} \quad \text{and} \quad y = \frac{\begin{vmatrix} a_1 & h_1 \\ a_2 & h_2 \end{vmatrix}}{\begin{vmatrix} a_1 & b_1 \\ a_2 & b_2 \end{vmatrix}}$$

and this immediately shows us the worth of this notation. For example, to solve

$$\begin{array}{c} 3x + 4y = 8 \\ \text{and} \quad 2x + 3y = 5 \end{array} \quad \begin{array}{c} a_1 x + b_1 y = h_1 \\ a_2 x + b_2 y = h_2 \end{array}$$

we write the solutions *immediately* as

$$x = \frac{\begin{vmatrix} h_1 & b_1 \\ h_2 & b_2 \end{vmatrix}}{\begin{vmatrix} a_1 & b_1 \\ a_2 & b_2 \end{vmatrix}} = \frac{\begin{vmatrix} 8 & 4 \\ 5 & 3 \end{vmatrix}}{\begin{vmatrix} 3 & 4 \\ 2 & 3 \end{vmatrix}} \quad \text{and} \quad y = \frac{\begin{vmatrix} a_1 & h_1 \\ a_2 & h_2 \end{vmatrix}}{\begin{vmatrix} a_1 & b_1 \\ a_2 & b_2 \end{vmatrix}} = \frac{\begin{vmatrix} 3 & 8 \\ 2 & 5 \end{vmatrix}}{\begin{vmatrix} 3 & 4 \\ 2 & 3 \end{vmatrix}}$$

which give at once

$$x = \frac{8 \times 3 - 4 \times 5}{3 \times 3 - 4 \times 2} = \frac{24 - 20}{9 - 8} = 4 \quad \text{and} \quad y = \frac{3 \times 5 - 8 \times 2}{3 \times 3 - 4 \times 2} = \frac{15 - 16}{9 - 8} = -1$$

Because the coefficients written out in this manner help us to determine the solution, we refer to arrays of this kind as *determinants*. The determinants we have just used are determinants of the *second order*, since they consist of two rows and two columns. All determinants have the same number of rows as of columns.

EXERCISE A6.1

1. Evaluate the determinants :

(i) $\begin{vmatrix} 4 & 2 \\ 3 & 4 \end{vmatrix}$ (ii) $\begin{vmatrix} 4 & 7 \\ 5 & 4 \end{vmatrix}$ (iii) $\begin{vmatrix} 4 & 8 \\ 2 & 4 \end{vmatrix}$ (iv) $\begin{vmatrix} -3 & 1 \\ 2 & 4 \end{vmatrix}$ (v) $\begin{vmatrix} -3 & -1 \\ 2 & -4 \end{vmatrix}$

(vi) $\begin{vmatrix} -3 & -1 \\ -2 & -4 \end{vmatrix}$ (vii) $\begin{vmatrix} 3 & 1 \\ 2 & 4 \end{vmatrix}$ (viii) $\begin{vmatrix} 3 & 5 \\ 1 & 0 \end{vmatrix}$ (ix) $\begin{vmatrix} x & 1 \\ 1 & x \end{vmatrix}$ (x) $\begin{vmatrix} -x & 1 \\ 1 & x \end{vmatrix}$

2. Solve the equations :

(i) $3x + 2y = 7$
    $x + 3y = 8$
(iv) $5x - 3y = 0$
     $2x + 6y = 7$

(ii) $2x - 5y = 4$
     $3x + 8y = 5$
(v) $3x - 2y = 0$
    $4x + y = 1$

(iii) $x - 3y + 7 = 0$
      $2x - 5y + 8 = 0$
(vi) $7x = 18$
     $5x + y = 14$

Let us now consider a determinant of the third order, which may be written as

$$\begin{vmatrix} a_1 & b_1 & c_1 \\ a_2 & b_2 & c_2 \\ a_3 & b_3 & c_3 \end{vmatrix} \quad \text{or sometimes as} \quad \begin{vmatrix} a_{11} & a_{12} & a_{13} \\ a_{21} & a_{22} & a_{23} \\ a_{31} & a_{32} & a_{33} \end{vmatrix}$$

It will be shown later that determinants of this kind arise when we solve three simultaneous equations. But first of all let us see how to interpret this array. We *want* the determinants of the third order to give us the solutions to the problem of solving three simultaneous equations ; and so we define the expansion of such a determinant in such a way that they do give us the solutions. We define the expansion of a third order determinant to be

$$\begin{vmatrix} a_1 & b_1 & c_1 \\ a_2 & b_2 & c_2 \\ a_3 & b_3 & c_3 \end{vmatrix} = a_1 \begin{vmatrix} b_2 & c_2 \\ b_3 & c_3 \end{vmatrix} - b_1 \begin{vmatrix} a_2 & c_2 \\ a_3 & c_3 \end{vmatrix} + c_1 \begin{vmatrix} a_2 & b_2 \\ a_3 & b_3 \end{vmatrix}$$

$$= a_1 (b_2 c_3 - b_3 c_2) - b_1 (a_2 c_3 - a_3 c_2) + c_1 (a_2 b_3 - a_3 b_2)$$

It will be noticed that the expansion is obtained by taking each term in the top row in turn and multiplying it by the *second order determinant that is left when the row and column containing that term is crossed out* ; and then prefacing the second product with a minus sign before adding them. For example, the second order determinant corresponding to the term b_1 in the top row is given by the terms that are not crossed out in

$$\begin{vmatrix} \cancel{a_1} & \cancel{b_1} & \cancel{c_1} \\ a_2 & b_2 & c_2 \\ a_3 & b_3 & c_3 \end{vmatrix}$$

A numerical example may be useful.

$$\begin{vmatrix} 3 & 4 & 8 \\ 2 & 1 & 3 \\ 7 & -2 & 0 \end{vmatrix} = 3 \begin{vmatrix} 1 & 3 \\ -2 & 0 \end{vmatrix} - 4 \begin{vmatrix} 2 & 3 \\ 7 & 0 \end{vmatrix} + 8 \begin{vmatrix} 2 & 1 \\ 7 & -2 \end{vmatrix}.$$

$$= 3[1 \times 0 - 3(-2)] - 4[2 \times 0 - 3 \times 7] + 8[2 \times (-2) - 1 \times 7]$$

$$= 3(0 + 6) - 4(0 - 21) + 8(-4 - 7)$$

$$= 18 + 84 - 88$$

$$= 14$$

We call the second order determinants arising in this expansion the *minors* of the third order determinant. When prefaced by the appropriate sign they are called the *co-factors*.

A determinant of the fourth order is expanded in terms of its third order minors, each minor being multiplied by the appropriate term in the first row, the signs alternating, thus :

$$\begin{vmatrix} a_1 & b_1 & c_1 & d_1 \\ a_2 & b_2 & c_2 & d_2 \\ a_3 & b_3 & c_3 & d_3 \\ a_4 & b_4 & c_4 & d_4 \end{vmatrix} = a_1 \begin{vmatrix} b_2 & c_2 & d_2 \\ b_3 & c_3 & d_3 \\ b_4 & c_4 & d_4 \end{vmatrix} - b_1 \begin{vmatrix} a_2 & c_2 & d_2 \\ a_3 & c_3 & d_3 \\ a_4 & c_4 & d_4 \end{vmatrix} + c_1 \begin{vmatrix} a_2 & b_2 & d_2 \\ a_3 & b_3 & d_3 \\ a_4 & b_4 & d_4 \end{vmatrix}$$

$$- d_1 \begin{vmatrix} a_2 & b_2 & c_2 \\ a_3 & b_3 & c_3 \\ a_4 & b_4 & c_4 \end{vmatrix}$$

More generally we may expand a determinant of order n, consisting of n rows and n columns, by multiplying the elements of the top row by the corresponding minors, alternating the signs as above. Actually there is no real need to take the elements of the top row : but it avoids confusion to do so. We shall now see how determinants may often be expanded by shorter methods.

EXERCISE A6.2

1. Evaluate :

(i) $\begin{vmatrix} 3 & 4 & 7 \\ 2 & 1 & 3 \\ 7 & 2 & 1 \end{vmatrix}$ (ii) $\begin{vmatrix} 3 & 4 & 7 \\ 2 & 1 & 3 \\ -5 & -1 & 2 \end{vmatrix}$ (iii) $\begin{vmatrix} 2 & 1 & 3 \\ 4 & 2 & 6 \\ 1 & 2 & 3 \end{vmatrix}$

(iv) $\begin{vmatrix} 1 & 2 & 3 \\ 0 & 1 & 2 \\ 2 & 1 & 0 \end{vmatrix}$ (v) $\begin{vmatrix} x & 1 & 2 \\ 2 & x & 2 \\ 3 & 1 & x \end{vmatrix}$ (vi) $\begin{vmatrix} x & 2x & x \\ 3 & 1 & 1 \\ 4 & 5 & 1 \end{vmatrix}$

(vii) $\begin{vmatrix} 9 & -9 & 3 \\ 1 & 1 & 1 \\ 4 & 4 & 5 \end{vmatrix}$ (viii) $\begin{vmatrix} 2 & 3 & 1 \\ 4 & 6 & 1 \\ 6 & 9 & 1 \end{vmatrix}$ (ix) $\begin{vmatrix} a & h & g \\ h & b & f \\ g & f & c \end{vmatrix}$

(x) $\begin{vmatrix} 1 & 4 & 5 \\ 4 & -2 & 3 \\ 5 & 3 & 1 \end{vmatrix}$ (xi) $\begin{vmatrix} 3 & 1 & -1 \\ 2 & 1 & -2 \\ 1 & 1 & -1 \end{vmatrix}$ (xii) $\begin{vmatrix} 7 & 5 & 0 \\ 2 & 1 & -2 \\ 3 & 3 & 4 \end{vmatrix}$

2. Elementary properties of determinants

We here list, without proof, some of the more important properties of determinants.

(1) If the rows and columns are interchanged, the value of the determinant is unchanged,

e.g.
$$\begin{vmatrix} 1 & 4 & 7 \\ 3 & 2 & 1 \\ 4 & 5 & 9 \end{vmatrix} = \begin{vmatrix} 1 & 3 & 4 \\ 4 & 2 & 5 \\ 7 & 1 & 9 \end{vmatrix}$$
the first row becoming the first column, etc.

(2) If any column (or row) is shifted over p adjacent columns (or rows) to occupy a new position then the numerical value of the determinant is unchanged : but if p is odd the sign of the determinant changes ; if p is even there is no change at all,

e.g.
$$\begin{vmatrix} 1 & 3 & 4 & 8 \\ 1 & 2 & 3 & 5 \\ 2 & 3 & 5 & 6 \\ 4 & 5 & 3 & 3 \end{vmatrix} = \begin{vmatrix} 3 & 4 & 1 & 8 \\ 2 & 3 & 1 & 5 \\ 3 & 5 & 2 & 6 \\ 5 & 3 & 4 & 2 \end{vmatrix} = - \begin{vmatrix} 4 & 1 & 8 & 3 \\ 3 & 1 & 5 & 2 \\ 5 & 2 & 6 & 3 \\ 3 & 4 & 2 & 5 \end{vmatrix} = - \begin{vmatrix} 3 & 1 & 5 & 2 \\ 5 & 2 & 6 & 3 \\ 4 & 1 & 8 & 3 \\ 3 & 4 & 2 & 5 \end{vmatrix}$$

where the arrows indicate the successive changes of position.

(3) If any two rows (or columns) are interchanged then the determinant changes sign,

e.g.
$$\begin{vmatrix} 3 & 2 & -3 & 6 \\ 3 & 3 & 2 & 1 \\ 2 & 3 & 5 & 2 \\ 2 & 3 & 4 & 6 \end{vmatrix} = - \begin{vmatrix} 3 & 6 & -3 & 2 \\ 3 & 1 & 2 & 3 \\ 2 & 2 & 5 & 3 \\ 2 & 6 & 4 & 3 \end{vmatrix}$$

(4) If any two rows or columns are identical, the value of the determinant is zero.

$$* \begin{vmatrix} 1 & 3 & 7 & 8 \\ 4 & 5 & 6 & 7 \\ 3 & 4 & 5 & 6 \\ 4 & 5 & 6 & 7 \end{vmatrix} = 0$$

(5) If a new row (or column) is formed by taking an existing row and adding to it (or subtracting from it) a constant multiple of any other row (or column) the value of the determinant is unchanged,

e.g.
$$\begin{vmatrix} 1 & 3 & 4 & 6 \\ 1 & 2 & 2 & 1 \\ 2 & 3 & 3 & 2 \\ 1 & 1 & 1 & 3 \end{vmatrix} = \begin{vmatrix} 2 & 5 & 6 & 7 \\ 1 & 2 & 2 & 1 \\ 2 & 3 & 3 & 2 \\ 5 & 7 & 7 & 7 \end{vmatrix}$$

where *the new row 1 consists of the old row 1 + old row 2* and the *new row 4 consists of the old row 4 + twice old row 3*.

(6) If any row (or column) has a factor p common to all its

elements then this factor may be removed, the determinant being p times the new determinant,

e.g., $\begin{vmatrix} 3 & 6 & 9 & 15 \\ 2 & 4 & 2 & 6 \\ 1 & 4 & 5 & 1 \\ 1 & 2 & 3 & 4 \end{vmatrix} = 3 \begin{vmatrix} 1 & 2 & 5 & 5 \\ 2 & 4 & 2 & 6 \\ 1 & 4 & 5 & 1 \\ 1 & 2 & 3 & 4 \end{vmatrix} = 6 \begin{vmatrix} 1 & 1 & 3 & 5 \\ 2 & 2 & 2 & 6 \\ 1 & 2 & 5 & 1 \\ 1 & 1 & 3 & 4 \end{vmatrix} = 12 \begin{vmatrix} 1 & 1 & 3 & 5 \\ 1 & 1 & 1 & 3 \\ 1 & 2 & 5 & 1 \\ 1 & 1 & 3 & 4 \end{vmatrix}$

EXERCISE A6.3

1. Show that the following statements are true, *without* evaluating the determinants:

(i) $\begin{vmatrix} 3 & 4 & 7 \\ 2 & 1 & 3 \\ 7 & 2 & 1 \end{vmatrix} = \begin{vmatrix} 6 & 4 & 2 \\ 2 & 1 & 3 \\ 9 & 3 & 4 \end{vmatrix} = \begin{vmatrix} 3 & 2 & 1 \\ 4 & 2 & 6 \\ 9 & 3 & 4 \end{vmatrix} = \begin{vmatrix} 1 & 2 & 1 \\ -8 & 2 & 6 \\ 1 & 3 & 4 \end{vmatrix}$

(ii) $\begin{vmatrix} 0 & -2 & 1 \\ -2 & -3 & 1 \\ 2 & -1 & -2 \end{vmatrix} = \begin{vmatrix} 1 & -2 & 1 \\ 1 & 4 & 1 \\ 0 & 1 & 2 \end{vmatrix} = 2\begin{vmatrix} 1 & -2 & 1 \\ 1 & 4 & 1 \\ 1 & 1 & 2 \end{vmatrix} = 2\begin{vmatrix} 3 & 3 & 4 \\ 1 & 4 & 1 \\ 1 & 1 & 2 \end{vmatrix}$

3. The expansion of determinants

The student should now read Section 8 of Chapter XXIX and perform the exercise which follows it. He can obtain further practice by expanding the determinants which appear above.

A small amount of practice in these methods can lead to a considerable degree of skill. There are other methods, which we need not consider here; but we should notice that if we have a determinant involving x and by putting $x = a$ we can make two rows the same, then clearly putting $x = a$ makes the determinant zero, and so $(x - a)$ is a factor of the expansion. For example, we can see that the determinant

$$\begin{vmatrix} 1 & 1 & 1 \\ a & b & c \\ a^2 & b^2 & c^2 \end{vmatrix}$$

has two identical columns if we put $a = b$, $b = c$ or $c = a$. This means that $(a - b)$, $(b - c)$ and $(c - a)$ must be factors of the expansion. Now clearly the expansion of this determinant cannot lead to any terms of degree higher than three. But the multiplication of these factors will lead to third degree terms. Consequently the determinant must equal the product of these factors multiplied by some term that contains no a, b or c which would raise the power of the expansion beyond three, i.e., by some purely numerical term. We may therefore write the determinant as equal to

$$k(a - b)(b - c)(c - a)$$

Now, as will be obvious if a few examples are tried, the *principal* diagonal of a determinant (i.e., the diagonal from the top left to the bottom right) has the property that the product of its elements appears unchanged in the expansion of the determinant. This means that this determinant must have the term $+bc^2$ in its expansion. But expansion of the above expression gives the term $+kbc^2$; and these two can be compatible only if $k=1$. This means that the expansion of the determinant is

$$(a-b)(b-c)(c-a)$$

EXERCISE A6.4

1. Use the rules of the last two sections to evaluate the determinants of Exercises A6.2 and A6.3.

2. Evaluate:

(i) $\begin{vmatrix} 1 & 2 & 3 & 4 \\ 4 & 3 & 2 & 1 \\ 6 & 4 & 4 & 6 \\ 1 & 1 & 1 & 1 \end{vmatrix}$ (ii) $\begin{vmatrix} 2 & 3 & 6 & 1 \\ 3 & -1 & 3 & 2 \\ 4 & -2 & 3 & 1 \\ 5 & -3 & 3 & 1 \end{vmatrix}$ (iii) $\begin{vmatrix} 1 & -1 & 2 & 1 \\ 4 & -2 & 12 & 2 \\ 3 & 1 & 10 & 3 \\ 1 & 0 & 4 & 4 \end{vmatrix}$

(iv) $\begin{vmatrix} 1 & 1 & 1 & 1 \\ a & b & c & d \\ a^2 & b^2 & c^2 & d^2 \\ a^3 & b^3 & c^3 & d^3 \end{vmatrix}$ (v) $\begin{vmatrix} 1 & 2 & 3 & 4 \\ x & 2 & 3 & 4 \\ 1 & x & 3 & 4 \\ 1 & 2 & x & 4 \end{vmatrix}$ (vi) $\begin{vmatrix} a & b & c & d \\ b & c & d & a \\ c & d & a & b \\ d & a & b & c \end{vmatrix}$

4. The solution of simultaneous equations

We can now return to the solution of simultaneous equations. Just as we found a simple rule allowing the solution of simultaneous equations in two unknowns by the use of determinants, so there is a simple rule allowing the solution of simultaneous equations in n unknowns by the use of determinants. It is that the solution of the set of n equations

$$a_{11}x_1 + a_{12}x_2 + a_{13}x_3 + \ldots + a_{1n}x_n = h_1$$
$$a_{21}x_1 + a_{22}x_2 + a_{23}x_3 + \ldots + a_{2n}x_n = h_2$$
$$\ldots \qquad \ldots \qquad \ldots \qquad \ldots$$
$$a_{n1}x_1 + a_{n2}x_2 + a_{n3}x_3 + \ldots + a_{nn}x_n = h_n$$

is given by

$$x_j = \frac{|A_j|}{|A|}$$

where $|A|$ is the determinant formed by the coefficients of the x's arranged as in the equations themselves when set out as shown, and $|A_j|$ is the same determinant with the h's substituted for the coefficients of x_j.

For example the solution to

$$3x_1 + 4x_2 + 3x_3 + 2x_4 = 7$$
$$3x_1 + 3x_2 + 2x_3 + 3x_4 = -3$$
$$2x_1 - 3x_2 - 4x_3 + x_4 = 6$$
$$3x_1 + x_2 - x_3 = 0$$

is given by the following, where it is borne in mind that a complete statement of the fourth equation would involve an additional term $0x_4$, and this zero must figure in the determinants.

$$x_1 = \frac{\begin{vmatrix} 7 & 4 & 3 & 2 \\ -3 & 3 & 2 & 3 \\ 6 & -3 & -4 & 1 \\ 0 & 1 & -1 & 0 \end{vmatrix}}{|A|} \qquad x_2 = \frac{\begin{vmatrix} 3 & 7 & 3 & 2 \\ 3 & -3 & 2 & 3 \\ 2 & 6 & -4 & 1 \\ 3 & 0 & -1 & 0 \end{vmatrix}}{|A|}$$

$$x_3 = \frac{\begin{vmatrix} 3 & 4 & 7 & 2 \\ 3 & 3 & -3 & 3 \\ 2 & -3 & 6 & 1 \\ 3 & 1 & 0 & 0 \end{vmatrix}}{|A|} \qquad x_4 = \frac{\begin{vmatrix} 3 & 4 & 3 & 7 \\ 3 & 3 & 2 & -3 \\ 2 & -3 & -4 & 6 \\ 3 & 1 & -1 & 0 \end{vmatrix}}{|A|}$$

where

$$|A| = \begin{vmatrix} 3 & 4 & 3 & 2 \\ 3 & 3 & 2 & 3 \\ 2 & -3 & -4 & 1 \\ 3 & 1 & -1 & 0 \end{vmatrix}$$

It should be noted that if the determinant A is zero, then either one of the equations is in some way derivable from one or more of the others (so that really there are $n-1$ instead of n equations) or one of the equations is incompatible with another (such as $x+y+z=2$ and $2x+2y+2z=5$). In such cases the equations are insoluble.

EXERCISE A6.5

1. Solve:

(i)
$$\left.\begin{array}{l} x+2y+3z=1 \\ 2x+3y+4z=2 \\ x-y-z=0 \end{array}\right\}$$

(ii)
$$\left.\begin{array}{l} x+y+3z=0 \\ x+2y=6 \\ x+3y+z=8 \end{array}\right\}$$

(iii)
$$\left.\begin{array}{l} w+x+y+z=4 \\ w+2x-2y+2z=5 \\ 2w+x+3y-z=4 \\ w+x+2y+8z=0 \end{array}\right\}$$

(iv)
$$\left.\begin{array}{l} a+2b+c+3d=1 \\ a-2b-c-3d=2 \\ 2a+b+c+d=4 \\ 2a-b+c+d=0 \end{array}\right\}$$

(v)
$$\left.\begin{array}{l} a+2b+c+3d=1 \\ a-2b-c-3d=2 \\ 2a+b+c+d=4 \\ 2a-b-c-d=6 \end{array}\right\}$$

(vi)
$$\left.\begin{array}{l} a+c+d=3 \\ b+2c-d=4 \\ c+d-a=5 \\ a+2b+3c=0 \end{array}\right\}$$

5. Two special determinants

There are two determinants which are of especial interest to us. The first of these occurs in Chapter XXI, and is

$$\Delta = \begin{vmatrix} A & 0 & 0 & 0 & 0 & a_1 \\ 0 & B & 0 & 0 & 0 & b_1 \\ 0 & 0 & C & 0 & 0 & c_1 \\ 0 & 0 & 0 & D & 0 & d_1 \\ 0 & 0 & 0 & 0 & E & e_1 \\ a & b & c & d & e & 0 \end{vmatrix}$$

which we proceed to expand in stages, thus:

$$\Delta = A \begin{vmatrix} B & 0 & 0 & 0 & b_1 \\ 0 & C & 0 & 0 & c_1 \\ 0 & 0 & D & 0 & d_1 \\ 0 & 0 & 0 & E & e_1 \\ b & c & d & e & 0 \end{vmatrix} - a_1 \begin{vmatrix} 0 & B & 0 & 0 & 0 \\ 0 & 0 & C & 0 & 0 \\ 0 & 0 & 0 & D & 0 \\ 0 & 0 & 0 & 0 & E \\ a & b & c & d & e \end{vmatrix}$$

and if we shift the first column of the second of these determinants to occupy the last position we see that this second determinant equals $BCDEa_1$. We therefore have

$$\Delta = A \left\{ B \begin{vmatrix} C & 0 & 0 & c_1 \\ 0 & D & 0 & d_1 \\ 0 & 0 & E & e_1 \\ c & d & e & 0 \end{vmatrix} + b_1 \begin{vmatrix} 0 & C & 0 & 0 \\ 0 & 0 & D & 0 \\ 0 & 0 & 0 & E \\ b & c & d & e \end{vmatrix} \right\} - a_1 a BCDE$$

and if we treat the second determinant as before we find that it is $-CDEb_1$.

$$\Delta = AB \left\{ C \begin{vmatrix} D & 0 & d_1 \\ 0 & E & e_1 \\ d & e & 0 \end{vmatrix} - c_1 \begin{vmatrix} 0 & D & 0 \\ 0 & 0 & E \\ c & d & e \end{vmatrix} \right\} - b_1 b ACDE \\ - a_1 a BCDE$$

$$= ABC \left\{ D \begin{vmatrix} E & e_1 \\ e & 0 \end{vmatrix} + d_1 \begin{vmatrix} 0 & E \\ d & e \end{vmatrix} \right\} - c_1 c ABDE - b_1 b ACDE \\ - a_1 a BCDE$$

$$= - e_1 e ABCD - d_1 d ABCE - c_1 c ABDE - b_1 b ACDE - a_1 a BCDE$$

It is clear that if we were to consider the minor obtained by deleting the first row and column, then the value of this would be

$$- e_1 e BCD - d_1 d BCE - c_1 c BDE - b_1 b CDE$$

The next minor would have the value $- e_1 e CD - d_1 d CE - c_1 c DE$ and so on.

In particular, if the last row and column are identical, so that $a_1 = a$, etc., then the above expansions will contain terms like $e^2 ABCD$.

The second kind of determinant which calls for our special attention is

$$\begin{vmatrix} A & 0 & 0 & 0 & 0 & a & \alpha \\ 0 & B & 0 & 0 & 0 & b & \beta \\ 0 & 0 & C & 0 & 0 & c & \gamma \\ 0 & 0 & 0 & D & 0 & d & \delta \\ 0 & 0 & 0 & 0 & E & e & \epsilon \\ a & b & c & d & e & 0 & 0 \\ \alpha & \beta & \gamma & \delta & \epsilon & 0 & 0 \end{vmatrix}$$

which we now proceed to expand by a method similar to the one we have just employed, but which is made shorter than it would otherwise be if we use the result we have just found,

$$= \begin{vmatrix} 0 & 0 & a & b & c & d & e \\ 0 & 0 & \alpha & \beta & \gamma & \delta & \epsilon \\ a & \alpha & A & 0 & 0 & 0 & 0 \\ b & \beta & 0 & B & 0 & 0 & 0 \\ c & \gamma & 0 & 0 & C & 0 & 0 \\ d & \delta & 0 & 0 & 0 & D & 0 \\ e & \epsilon & 0 & 0 & 0 & 0 & E \end{vmatrix} = a\begin{vmatrix} 0 & 0 & \beta & \gamma & \delta & \epsilon \\ a & \alpha & 0 & 0 & 0 & 0 \\ b & \beta & B & 0 & 0 & 0 \\ c & \gamma & 0 & C & 0 & 0 \\ d & \delta & 0 & 0 & D & 0 \\ e & \epsilon & 0 & 0 & 0 & E \end{vmatrix} - b\begin{vmatrix} 0 & 0 & \alpha & \gamma & \delta & \epsilon \\ a & \alpha & A & 0 & 0 & 0 \\ b & \beta & 0 & 0 & 0 & 0 \\ c & \gamma & 0 & C & 0 & 0 \\ d & \delta & 0 & 0 & D & 0 \\ e & \epsilon & 0 & 0 & 0 & E \end{vmatrix} + \text{ etc.}$$

$$= -a\begin{vmatrix} a & \alpha & 0 & 0 & 0 & 0 \\ 0 & 0 & \beta & \gamma & \delta & \epsilon \\ b & \beta & B & 0 & 0 & 0 \\ c & \gamma & 0 & C & 0 & 0 \\ d & \delta & 0 & 0 & D & 0 \\ e & \epsilon & 0 & 0 & 0 & E \end{vmatrix} - b\begin{vmatrix} b & \beta & 0 & 0 & 0 & 0 \\ 0 & 0 & \alpha & \gamma & \delta & \epsilon \\ a & \alpha & A & 0 & 0 & 0 \\ c & \gamma & 0 & C & 0 & 0 \\ d & \delta & 0 & 0 & D & 0 \\ e & \epsilon & 0 & 0 & 0 & E \end{vmatrix} + \text{ etc.}$$

The first of these determinants gives

$$-a^2\begin{vmatrix} 0 & \beta & \gamma & \delta & \epsilon \\ \beta & B & 0 & 0 & 0 \\ \gamma & 0 & C & 0 & 0 \\ \delta & 0 & 0 & D & 0 \\ \epsilon & 0 & 0 & 0 & E \end{vmatrix} + a\alpha\begin{vmatrix} 0 & \beta & \gamma & \delta & \epsilon \\ b & B & 0 & 0 & 0 \\ c & 0 & C & 0 & 0 \\ d & 0 & 0 & D & 0 \\ e & 0 & 0 & 0 & E \end{vmatrix}$$

and the previous result enables us to write this as

$$-a^2(-\beta^2 CDE - \gamma^2 BDE - \delta^2 BCE - \epsilon^2 BCD)$$
$$+ a\alpha(-b\beta CDE - c\gamma BDE - d\delta BCE - e\epsilon BCD)$$

giving
$$(a^2\beta^2 - ab\alpha\beta)CDE + (a^2\gamma^2 - ac\alpha\gamma)BDE$$
$$+ (a^2\delta^2 - ad\alpha\delta)BCE + (a^2\epsilon^2 - ae\alpha\epsilon)BCD$$
with similar expressions arising from the four remaining minors.

The Sign of a Quadratic Expression

Often we shall wish to know whether a quadratic expression of a certain kind is positive, zero or negative. There is a simple rule for deciding this which we will now state, first of all in the case of a quadratic equation in two variables, then in three and then in n, illustrating the statement with examples.

The quadratic equation in two variables is of the form

$$ax^2 + by^2 + 2hxy + 2gx + 2fy + c = 0$$

It can be shown that the substitutions $X = \alpha x + \beta$, $Y = \gamma y + \delta$ will reduce this to the form

$$AX^2 + 2HXY + BY^2 = C$$

if
$$\frac{\beta}{\alpha} = \frac{bg - hf}{ab - h^2} \quad \text{and} \quad \frac{\delta}{\gamma} = \frac{af - hg}{ab - h^2}$$

We can therefore consider this reduced form to be quite general. In order to determine the sign of the left-hand side, we may note that with a slight change of notation,

$$ax^2 + by^2 + 2hxy$$

may be written as

$$a\left(x + \frac{hy}{a}\right)^2 + \left(\frac{ab - h^2}{a}\right)y^2$$

and this is definitely positive for all values of x and y if, (and only if) both a and $ab - h^2$ are positive. If both a and $ab - h^2$ are negative, then the sign of the left-hand side will depend on the values of x and y. But if a is negative and $ab - h^2$ is positive then the left-hand side is negative for all values of x and y (except, of course, zero).

Noting that $ab - h^2$ may be written as $\begin{vmatrix} a & h \\ h & b \end{vmatrix}$ we may state that

$$ax^2 + 2hxy + by^2$$

is *positive definite* (i.e. positive for all values of x and y) if

$$a > 0 \quad \text{and} \quad \begin{vmatrix} a & h \\ h & b \end{vmatrix} > 0$$

but *negative definite* if

$$a < 0 \quad \text{and} \quad \begin{vmatrix} a & h \\ h & b \end{vmatrix} > 0$$

If we now take the quadratic form involving three variables, we will find that it can always be written as

$$ax^2 + by^2 + cz^2 + 2fyz + 2gxz + 2hxy$$

It can be shown that this is *positive definite* if, and only if,

$$a > 0, \quad \begin{vmatrix} a & h \\ h & b \end{vmatrix} > 0 \quad \text{and} \quad \begin{vmatrix} a & h & g \\ h & b & f \\ g & f & c \end{vmatrix} > 0$$

It is *negative definite* if, and only if, these three expressions are negative, positive and negative respectively.

Finally, in the case of n variables, the quadratic form may be written as

$$a_{11}x_1{}^2 + a_{22}x_2{}^2 + a_{33}x_3{}^2 + \ldots + a_{nn}x_n{}^2$$
$$+ 2(a_{12}x_1x_2 + a_{13}x_1x_3 + a_{14}x_1x_4 + \ldots + a_{1n}x_1x_n)$$
$$+ 2(a_{23}x_2x_3 + a_{24}x_2x_4 + a_{22}x_2x_2 + \ldots + a_{2n}x_2x_n)$$
$$+ \ldots$$
$$+ 2a_{n-1\,n}x_{n-1}x_n$$

which is more symmetrically put as

$$a_{11}\ x_1{}^2 + a_{12}x_1x_2 + \ldots + a_{1n}x_1x_n$$
$$+ a_{21}x_2x_1 + a_{22}\ x_2{}^2 + \ldots + a_{2n}x_2x_n$$
$$+ \ldots$$
$$+ a_{n1}x_nx_1 + a_{n2}x_nx_2 + \ldots + a_{nn}\ x_n{}^2$$

where, in this case $a_{21} = a_{12}$, $a_{n1} = a_{1n}$ and, more generally $a_{rs} = a_{sr}$.

It can be shown that the quadratic form above is *positive definite* if and only if

$$a_{11} > 0, \quad \begin{vmatrix} a_{11} & a_{12} \\ a_{12} & a_{22} \end{vmatrix} > 0, \quad \begin{vmatrix} a_{11} & a_{12} & a_{13} \\ a_{12} & a_{22} & a_{23} \\ a_{13} & a_{23} & a_{33} \end{vmatrix} > 0, \ldots, \quad \begin{vmatrix} a_{11} & a_{12} \ldots a_{1n} \\ a_{12} & a_{22} \ldots a_{2n} \\ \cdots\cdots\cdots\cdots \\ a_{1n} & a_{2n} \ldots a_{nn} \end{vmatrix} > 0$$

where there are, in fact, n such expressions, each being a determinant obtained by adding one more row and one more column to the preceding one, the last being the determinant of the coefficients as set out above. These determinants are called the *principal minors* of the large determinant, which is known as the *discriminant* of the form.

The quadratic form is *negative definite* if, and only if, the above minors are alternatively negative and positive. For example, the expression

$$p^2 + 2q^2 + 3r^2 + s^2 + 2pq + 2qr + 4rs + 4sq + 4pr + 2ps$$

yields the conditions for positive definiteness as

$$1 > 0, \quad \begin{vmatrix} 1 & 1 \\ 1 & 2 \end{vmatrix} > 0, \quad \begin{vmatrix} 1 & 1 & 2 \\ 1 & 2 & 1 \\ 2 & 1 & 3 \end{vmatrix} > 0 \text{ and } \begin{vmatrix} 1 & 1 & 2 & 1 \\ 1 & 2 & 1 & 2 \\ 2 & 1 & 3 & 2 \\ 1 & 2 & 2 & 1 \end{vmatrix} > 0$$

It happens that the third of these determinants is negative, while the preceding two are positive. We need look no further. Neither the condition for a positive definite form nor the condition for a negative definite form holds. The sign of the expression will depend on the values of p, q, r and s.

EXERCISE A7.1

1. Consider whether the following expressions are of definite sign:

(i) $2x^2 + 6xy + 5y^2$ (ii) $2x^2 - 6xy + 5y^2$

(iii) $2x^2 + 6xy + 4y^2$ (iv) $-2x^2 + 6xy + 5y^2$

(v) $2x^2 - 6xy - 8y^2$ (vi) $2x^2 - 6xy - y^2$

(vii) $2x^2 + 5y^2 + z^2 + 3yz + 3xz + 6xy$

(viii) $2x^2 - 5y^2 + z^2 - 2yz - 4xz - 6xy$

(ix) $3x^2 + 4y^2 + 6z^2 + 2yz + 4xz + 2xy$

Summation of Series

1. The arithmetic progression

The sum of n terms is

$$S_n = a + (a + d) + (a + 2d) + \ldots + \{(a + (n - 1)d\}$$

This can also be written backwards as

$$S_n = \{a + (n - 1)d\} + \{a + (n - 2)d\} + \ldots + (a + 2d) + (a + d) + a$$

Addition, term by term yields

$$2S_n = \{2a + (n - 1)d\} + \{2a + (n - 1)d\} + \ldots$$
$$+ \{2a + (n - 1)d\} + \{2a + (n - 1)d\}$$
$$= n\{2a + (n - 1)d\}$$

whence
$$S_n = \frac{n}{2}\{2a + (n - 1)d\}$$

$$= \frac{n}{2}\{a + a + (n - 1)d\}$$

$$= \frac{n}{2}(a + l)$$

2. The geometric progression

The sum of n terms is

$$S_n = a + ar + ar^2 + \ldots + ar^{n-1}$$

whence
$$rS_n = \quad ar + ar^2 + \ldots + ar^{n-1} + ar^n$$

Subtraction yields

$$S_n(1 - r) = a - ar^n$$

whence
$$S_n = \frac{a(1 - r^n)}{1 - r}$$

3. The sum of squares of the natural numbers

Consider $(x+1)^3 - (x)^3 \equiv 3x^2 + 3x + 1$

Giving x the values $0, 1, 2, 3 \ldots n$ we have

$$1^3 - 0^3 = \qquad\qquad\qquad 1$$
$$2^3 - 1^3 = 3(1)^2 + 3(1) + 1$$
$$3^3 - 2^3 = 3(2)^2 + 3(2) + 1$$
$$4^3 - 3^3 = 3(3)^2 + 3(3) + 1$$
$$\cdots\cdots\cdots$$
$$(n+1)^3 - (n)^3 = 3(n)^2 + 3(n) + 1$$

whence, by addition, and noting that a great deal cancels out on the left, we have

$$(n+1)^3 - 0^3 = 3 \sum_{p=1}^{n} p^2 + 3 \sum_{p=1}^{u} p + (n+1)$$

whence

$$3 \sum_{p=1}^{n} p^2 = (n+1)^3 - 3 \sum_{p=1}^{n} p - (n+1)$$

$$= (n+1)^3 - \frac{3n}{2}(n+1) - (n+1)$$

$$= (n+1)\left[(n+1)^2 - \frac{3n}{2} - 1\right]$$

$$= (n+1)\left[n^2 + \frac{n}{2}\right]$$

$$= (n+1)\left(\frac{2n^2+n}{2}\right)$$

$$= \frac{n(n+1)(2n+1)}{2}$$

whence

$$\sum_{p=1}^{n} p^2 = \frac{n(n+1)(2n+1)}{6}$$

Appendix 9

Proofs of Formulae for Differentiation

1. The derivative of $y = ax^n$

In the text we have used the Binomial Theorem to show that

$$\frac{dy}{dx} = nax^{n-1}$$

for n a positive integer.

There are other ways of deriving this result. For example, we know that if

$$y = uv$$

then

$$\frac{dy}{dx} = u\frac{dv}{dx} + v\frac{du}{dx}$$

Put

$$u = v = x$$

then

$$\frac{du}{dx} = \frac{dv}{dx} = 1$$

and now

$$y = x^2$$

and so

$$\frac{dy}{dx} = u\frac{dv}{dx} + v\frac{du}{dx}$$

$$= u \quad + v$$

$$= x \quad + x$$

$$= 2x$$

Now put

$$u = x^2, \quad v = x$$

From the result just obtained

$$\frac{du}{dx} = 2x \quad \text{and} \quad \frac{dv}{dx} = 1$$

Consider

$$y = x^3$$

$$= x^2 \times x$$

$$= uv$$

$$\frac{dy}{dx} = u\frac{dv}{dx} + v\frac{du}{dx}$$

$$= x^2 \times 1 + x \times 2x$$

$$= 3x^2$$

551

One can now consider

$$y = x^4 = x^3 \times x$$

in the same way.

Alternatively one can prove the result by induction. Suppose that for some integer n, then if $y = x^n$,

$$\frac{dy}{dx} = nx^{n-1}$$

Consider
$$y = x^{n+1}$$
$$= x^n \times x$$

Then if the above assumption is correct we have

$$\frac{dy}{dx} = x^n \frac{dx}{dx} + x \cdot \frac{d(x^n)}{dx}$$
$$= x^n \quad + x \cdot nx^{n-1}$$
$$= x^n \quad + nx^n$$
$$= (n+1)x^n$$

Thus, if
$$\frac{d}{dx} x^n = nx^{n-1} \quad \text{for a given } n$$

then
$$\frac{d}{dx} x^{n+1} = (n+1)x^n$$

Having shown that if the result is true for $x = n$ then it must be true for $x = n+1$, we go back to the case of $n = 2$ and see, as above, that it is true. It must therefore be true for $n+1 = 3$. And so, as it is true for 3, it must be true for 4. And so on.

Another way of proceeding is to consider

$$y = x^m$$
$$y + \Delta y = (x + \Delta x)^m$$

whence

$$(y + \Delta y) - y = (x + \Delta x)^m - x^m$$

Now define

$$\frac{dy}{dx} = \underset{\Delta x \to 0}{\text{Limit}} \frac{\Delta y}{\Delta x} = \underset{\Delta x \to 0}{\text{Limit}} \frac{(y + \Delta y) - y}{(x + \Delta x) - x}$$
$$= \underset{\Delta x \to 0}{\text{Limit}} \frac{(x + \Delta x)^m - x^m}{(x + \Delta x) \quad - x}$$
$$= \underset{b \to x}{\text{Limit}} \frac{b^m - x^m}{b - x}$$

where we write

$$b = x + \Delta x$$

This is really what we did in the text. If we perform the division we get

$$\frac{dy}{dx} = \underset{b \to x}{\text{Limit}} \quad (x^{m-1} + bx^{m-2} + \ldots + b^{m-2}x + b^{m-1})$$

where, it may be noted, the bracketed expression is a G.P with common ratio b/x.

This expression has m terms, and as b tends to x each term separately tends to the value x^{m-1}

Thus

$$\frac{dy}{dx} = mx^{m-1}$$

Now consider m as a fraction

$$m = \frac{p}{q}$$

where p and q are positive integers. In the expression

$$L = \frac{b^m - x^m}{b - x}$$

write

$$x^m = x^{p/q} = z^p$$

where

$$x \quad = z^q$$

and

$$b^m = b^{p/q} = c^p$$

where

$$b \quad = c^q$$

Then

$$L = \frac{c^p - z^p}{c^q - z^q}$$

$$= \frac{c^p - z^p}{c - z} \bigg/ \frac{c^q - z^q}{c - z}$$

In this expression, as $c \to z$, the numerator tends to the limit pz^{p-1} while the denominator tends to the limit qz^{q-1}.

If we assume that the limit of the quotient is the quotient of the limits then

$$L \to \frac{pz^{p-1}}{qz^{q-1}}$$

$$= \frac{p}{q} z^{p-q}$$

$$= \frac{p}{q} x^{(p-q)/q}$$

$$= \frac{p}{q} x^{\frac{p}{q}-1} = mx^{m-1}$$

This assumption about the limit of the quotient is one which needs a fuller study of limits if it is to be justified. It is the kind of assumption that is easy to make but is not always correct. The serious student of mathematics should refer to a standard work on limits and convergence if he finds it necessary to make an assumption of this kind.

If m is negative then write $m = -n$ and we have

$$\frac{b^m - x^m}{b - x} = \frac{b^{-n} - x^{-n}}{b - x}$$

$$= -\frac{1}{b^n x^n} \frac{b^n - x^n}{b - x}$$

Here we have a product of two terms. As $b \to x$

$$\frac{1}{b^n x^n} \to \frac{1}{x^{2n}}$$

while

$$\frac{b^n - x^n}{b - x} \to n x^{n-1}$$

If we assume that the limit of the product is the product of the limits (to which the above caution applies) then we obtain

$$-\frac{1}{x^{2n}} n x^{n-1}$$

$$= -n x^{-(n+1)}$$

$$= m x^{m-1}$$

Thus we have shown that for m a positive or negative integer, or a positive fraction

$$\frac{dy}{dx} = m x^{m-1}$$

The proof for a negative fraction follows the same lines.

The reader may care to evolve some proofs by induction for negative m and for fractions. He may also proceed by satisfying himself that the result is true when $m = 0$, and then using the formula for the derivative of a quotient, derived below.

2. The derivative of a quotient

Let

$$y = \frac{u}{v}$$

where $u = u(x)$, $v = v(x)$.

Let x increase by Δx, resulting in increments Δu, Δv and Δy in u, v and y. Then

$$y + \Delta y = \frac{u + \Delta u}{v + \Delta v}$$

whence

$$\Delta y = \frac{u + \Delta u}{v + \Delta v} - \frac{u}{v}$$

$$= \frac{uv + v\,\Delta u - uv - u\,\Delta v}{v(v + \Delta v)}$$

$$= \frac{v\,\Delta u - u\,\Delta v}{v(v + \Delta v)}$$

Thus

$$\frac{\Delta y}{\Delta x} = \frac{v(\Delta u / \Delta x) - u(\Delta v / \Delta x)}{v(v + \Delta v)}$$

$$= \frac{v}{v(v + \Delta v)} \frac{\Delta u}{\Delta x} - \frac{u}{v(v + \Delta v)} \frac{\Delta v}{\Delta x}$$

as $\Delta x \to 0$,

$$\frac{\Delta u}{\Delta x} \to \frac{du}{dx}, \quad \frac{\Delta v}{\Delta x} \to \frac{dv}{dx}$$

and

$$v + \Delta v \to v$$

If we make the same kind of assumption as in the note above, this leads to

$$\frac{dy}{dx} = \frac{1}{v} \frac{du}{dx} - \frac{u}{v^2} \frac{dv}{dx}$$

$$= \frac{v(du/dx) - u(dv/dx)}{v^2}$$

Alternatively we could write

$$\frac{1}{v + \Delta v} \frac{\Delta u}{\Delta x} = \frac{1}{v} \frac{\Delta u}{\Delta x} \left(1 + \frac{\Delta v}{v}\right)^{-1}$$

$$= \frac{1}{v} \frac{\Delta u}{\Delta x} \left(1 - \frac{\Delta v}{v} + \left(\frac{\Delta v}{v}\right)^2 - \ldots\right)$$

and argue that as $\Delta x \to 0$ the bracket $\to 1$, yielding the same result with the same kind of assumption.

Another procedure is to consider

$$y = \frac{u}{v} = uv^{-1}$$

as a product. Then

$$\frac{dy}{dx} = v^{-1} \frac{du}{dx} + u \frac{d(v^{-1})}{dx}$$

$$= \frac{1}{v} \frac{du}{dx} + u \cdot (-v^{-2}) \frac{dv}{dx}$$

from which the result follows.

APPENDIX 10

Some Trigonometric Results

1. Two basic identities

This appendix derives some trigonometric results which are required in Chapter XXVII. They are all derived from the two identities

$$\sin (A \pm B) \equiv \sin A \cos B \pm \cos A \sin B$$

and

$$\cos (A \pm B) \equiv \cos A \cos B \mp \sin A \sin B$$

These identities may be proved by geometrical methods, as is done in most books on elementary trigonometry. Alternatively one may proceed as follows:

Since

$$e^{i\theta} \equiv \cos \theta + i \sin \theta \quad \text{for all } \theta$$

we have, by putting $\theta = A$, B and $(A + B)$

$$e^{iA} \equiv \cos A + i \sin A$$
$$e^{iB} \equiv \cos B + i \sin B$$
$$e^{i(A+B)} \equiv \cos (A + B) + i \sin (A + B)$$

But

$$e^{i(A+B)} \equiv e^{iA} \cdot e^{iB}$$
$$\equiv (\cos A + i \sin A)(\cos B + i \sin B)$$
$$\equiv (\cos A \cos B - \sin A \sin B) + i(\sin A \cos B + \cos A \sin B)$$

Equating the real parts of these two complex expressions for $e^{i(A+B)}$ we obtain

$$\cos (A + B) \equiv \cos A \cos B - \sin A \sin B$$

while the imaginary parts yield

$$\sin (A + B) \equiv \sin A \cos B + \cos A \sin B.$$

The results for $(A - B)$ can be proved similarly.

2. Some other results

It follows from these identities, by putting $A = B$, that

$$\sin 2A \equiv 2 \sin A \cos A$$

and

$$\cos 2A \equiv \cos^2 A - \sin 2A$$

Furthermore, $\sin 3A \equiv \sin (2A + A)$

$$\equiv \sin 2A \cos A + \cos 2A \sin A$$
$$\equiv 2 \sin A \cos^2 A + (\cos^2 A - \sin^2 A) \sin A$$
$$\equiv 3 \sin A \cos^2 A - \sin^3 A$$

and $\cos 3A \equiv \cos (2A + A)$

$$\equiv \cos 2A \cos A - \sin 2A \sin A$$
$$\equiv (\cos^2 A - \sin^2 A) \cos A - (2 \sin A \cos A) \sin A$$
$$\equiv \cos^3 A - 3 \cos A \sin^2 A$$

We also have

$$\sin 4A \equiv \sin (2A + 2A)$$
$$\equiv 2 \sin 2A \cos 2A$$
$$\equiv 4 \sin A \cos A (\cos^2 A - \sin^2 A)$$

and $\cos 4A \equiv \cos (2A + 2A)$

$$\equiv \cos^2 2A - \sin^2 2A$$
$$\equiv 1 - 2 \sin^2 2A \quad (\text{since } \cos^2 2A + \sin^2 2A \equiv 1)$$
$$\equiv 1 - 2(2 \sin A \cos A)^2$$
$$\equiv 1 - 8 \sin^2 A \cos^2 A$$

We now use these results to obtain the values of $\cos (nR - \phi)$ where n takes the values 1, 2, 3 and 4 for use in Chapter XXVII.

To begin with we have that

$$\cos (R - \phi) \equiv \cos R \cos \phi + \sin R \sin \phi$$
$$\cos (2R - \phi) \equiv \cos 2R \cos\phi + \sin 2R \sin \phi$$
$$\cos (3R - \phi) \equiv \cos 3R \cos \phi + \sin 3R \sin \phi$$
$$\cos (4R - \phi) \equiv \cos 4R \cos \phi + \sin 4R \sin \phi$$

Now in Chapter XXVII, $\cos R = \dfrac{2}{\sqrt{13}}$ and $\sin R = \dfrac{3}{\sqrt{13}}$. It follows that

$$\sin 2R = 2 \sin R \cos R$$

$$= 2 \cdot \frac{3}{\sqrt{13}} \cdot \frac{2}{\sqrt{13}} = \frac{12}{13}$$

$$\cos 2R = \cos^2 R - \sin^2 R$$

$$= \frac{4}{13} - \frac{9}{13} = -\frac{5}{13}$$

$$\sin 3R = 3 \sin R \cos^2 R - \sin^3 R$$

$$= 3 \cdot \frac{3}{\sqrt{13}} \frac{4}{13} - \frac{27}{13\sqrt{13}} = \frac{9}{13\sqrt{13}}$$

$$\cos 3R = \cos^3 R - 3 \cos R \sin^2 R$$

$$= \frac{8}{13\sqrt{13}} - 3 \cdot \frac{2}{\sqrt{13}} \frac{9}{13} = -\frac{46}{13\sqrt{13}}$$

$$\sin 4R = 4 \sin R \cos R (\cos^2 R - \sin^2 R)$$

$$= 4 \cdot \frac{3}{\sqrt{13}} \frac{2}{\sqrt{13}} \left(-\frac{5}{13} \right) = -\frac{120}{169}$$

$$\cos 4R = 1 - 8 \sin^2 R \cos^2 R$$

$$= 1 - 8 \cdot \frac{9}{13} \cdot \frac{4}{13} = -\frac{119}{169}$$

Using these results we obtain that when $\cos R = \dfrac{2}{\sqrt{13}}$ and $\sin R = \dfrac{3}{\sqrt{13}}$

then
$$\cos (R - \phi) = \frac{2}{\sqrt{13}} \cos \phi + \frac{3}{\sqrt{13}} \sin \phi$$

$$\cos (2R - \phi) = -\frac{5}{13} \cos \phi + \frac{12}{13} \sin \phi$$

$$\cos (3R - \phi) = -\frac{46}{13\sqrt{13}} \cos \phi + \frac{9}{13\sqrt{13}} \sin \phi$$

$$\cos (4R - \phi) = -\frac{119}{169} \cos \phi - \frac{120}{169} \sin \phi$$

Beta and Gamma Functions

There are two important integrals which often occur in advanced work. Here we simply define them and indicate some of their properties and a few results derived from these.

Provided that n is a positive quantity the integral

$$\int_0^\infty e^{-x} x^{n-1} \, dx$$

converges. It is a function of n, and is called the *Gamma Function*. Usually it is denoted by

$$\Gamma(n)$$

The reader should be able to show for himself that

$$\Gamma(1) = 1 \tag{1}$$
$$\Gamma(n) = (n-1)\Gamma(n-1) \quad \text{provided } n-1 > 0 \tag{2}$$
$$\Gamma(n) = (n-1)! \quad \text{if } n \text{ is a positive integer} \tag{3}$$

The second important integral is

$$\int_0^\infty x^{m-1}(1-x)^{n-1} \, dx$$

which converges provided both m and n are positive. It is a function of m and n, and is called the *Beta Function*, denoted by

$$B(m, n)$$

It is obvious from the definition that

$$B(m, n) = B(n, m) \tag{4}$$

Substitution of $x = \sin^2 \theta$ enables us to see that

$$B(m, n) = 2 \int_0^{\pi/2} \sin^{2m-1} \theta \, \cos^{2n-1} \theta \, d\theta \tag{5}$$

and use of reduction formulae techniques enable us to derive from this the formula

$$B(m, n) = \frac{(m-1)(n-1)}{(m+n-1)(m+n-2)} B(m-1, n-1) \tag{6}$$

It follows that if m and n are positive integers then

$$B(m, n) = \frac{(m-1)! \, (n-1)!}{(m+n-1)!} \tag{7}$$

which immediately demonstrates for positive integers a result which is, in fact, true for all m and n for which Beta and Gamma functions are definable—namely that

$$B(m, n) = \frac{\Gamma(m)\Gamma(n)}{\Gamma(m+n)} \tag{8}$$

We may also see from (7) that

$$B(\tfrac{1}{2}, \tfrac{1}{2}) = \pi \tag{9}$$

and now result (8) helps us to establish that

$$\Gamma(\tfrac{1}{2}) = \sqrt{\pi} \tag{10}$$

Another result of importance is that

$$\Gamma(p)\Gamma(1-p) = \frac{\pi}{\sin p\pi} \tag{11}$$

provided $0 < p < 1$.

BIBLIOGRAPHY

There is a variety of useful introductions to mathematics amongst which some of the more easily obtained, and profitably read, are L. Hogben's *Mathematics for the Million* (Allen & Unwin) and G. L. S. Shackle's *Mathematics at the Fireside* (C.U.P.). In the Pelican series are two very good books by W. W. Sawyer, *Mathematician's Delight* and *Prelude to Mathematics*. Two other books which shed most interesting light on the subject are W. W. Rouse Ball's *A Short History of Mathematics* (Macmillan) and E. T. Bell's *Men of Mathematics* which is available as a Pelican.

The place of mathematics in economics is discussed in many places, including an article by W. W. Leontief called Mathematics in Economics, appearing in the *Bulletin of the American Mathematical Society*, volume 60, and another of the same title by P. A. Samuelson in the *Review of Economics and Statistics*, volume 36. Marshall, Wicksteed, Pareto and many other have made both their comments and their contributions. The important point is that the student should be able to do likewise.

Inevitably this is a selective book, and it is likely that for further clarity, or more detailed exposition, the student will look elsewhere. He will obtain a more rigorous insight into elementary calculus from R. G. D. Allen's *Mathematical Analysis for Economists* (Macmillan) but on the whole if he wishes to go further in calculus he would do well to look not at books which have been written for economists, or engineers, or any other specialist group, but to books on mathematics. At the more elementary level there are many good school textbooks, amongst which we must list the *Teach Yourself* volumes. At the more advanced level we have a very useful series of University Mathematical Texts, published by Oliver & Boyd, amongst which we might mention R. P. Gillespie's *Integration*, E. L. Ince's *Integration of Ordinary Differential Equations*, J. M. Hyslop's *Infinite Series*, H. W. Turnbull's *Theory of Equations*, R. P. Gillespie's *Partial Differentiation* and E. G. Phillips's *Functions of a Complex Variable*, which is most useful but a trifle difficult at first.

For a somewhat easier introduction to complex numbers the reader might look at one of the more general introductions mentioned at the beginning of this bibliography, or to *Advanced Algebra* by C. V. Durell and A. Robson (Bell). A good introduction is also given in R. G. D. Allen's *Mathematical Economics* (Macmillan). A standard work is T. M. MacRobert's *Functions of a Complex Variable* (Macmillan), but beginners will not need books of this level.

Differential equations are discussed in R. G. D. Allen's books, and also in W. J. Baumol's most useful *Economic Dynamics* (Macmillan), which is highly readable. A standard work is H. T. H. Piaggio's *An*

561

T

Elementary Treatise on Differential Equations (Bell, 1920), which has a somewhat modest title.

Difference equations are discussed in Allen's *Mathematical Economics*, in Durell and Robson (*op. cit.*), in a most readable introduction to the subject in Baumol (*op. cit.*) and in appendices to J. R. Hicks's *A Contribution to the Theory of the Trade Cycle* and P. A. Samuelson's *Foundations of Economic Analysis* (Harvard, 1947). A useful book is H. Levy and F. Lessman, *Finite Difference Equations* (Pitman).

On Set Theory and Linear Algebra the economics student should consult Allen's *Mathematical Economics* and his more recent *Basic Mathematics* (Macmillan). He will also find G. Hadley's *Linear Algebra* (Addison–Wesley) very useful. In the Oliver & Boyd series there is A. C. Aitken's *Determinants and Matrices*, which is a trifle compressed. The same publishers have a new series of Mathematical Economics and Econometrics Texts which will include a book on Linear Algebra written partly by me. For the uses of matrices in econometric theory one should consult J. Johnston's *Econometric Methods* (McGraw-Hill).

A number of other good books go unmentioned here, partly because we must keep in mind the object of this volume which is to introduce students of economics to an auxiliary subject. They would be well advised to read this book carefully, doing all of the exercises, and working from one line of text to the next, with paper and pencil if necessary. The person who does this conscientiously should then be able to benefit greatly by reading Professor Allen's books in the order in which they were written. If he perseveres to the second he will find that he is reading modern economic theory of a very advanced level, and should have little difficulty in reading any paper in the learned journals. Alternatively he might prefer to dip into the journals before arming himself in this way; if he does he will soon find both pleasure at what he can understand and disappointment at what is still beyond him. However gentle the lower slopes of a mountain may be made, at the summit breathing must still be difficult. It is the conquest, and the views, which compensate for the effort.

ANSWERS TO EXERCISES

1.4 (1) $5\frac{2}{3}$;   (2) $-5\frac{2}{3}$;   (3) 1;   (4) $1\frac{1}{2}$;   (5) 3;   (6) 1;
     (7) 1;   (8) -12;   (9) -8;   (10) -16;   (11) $-1\frac{1}{3}$;   (12) -13.

1.6 (1) -2, $-\frac{1}{3}$, $(x+2)(x+\frac{1}{3})=0$;   (2) 2, $-\frac{1}{3}$, $(x-2)(x+\frac{1}{3})=0$;
     (3) -2, $\frac{1}{3}$, $(x+2)(x-\frac{1}{3})=0$;   (4) $1\frac{1}{2}$, $-\frac{1}{2}$, $(x-\frac{3}{2})(x+\frac{1}{2})=0$;
     (5) $-\frac{1}{2}$, $-1\frac{1}{2}$;   (6) $-1\frac{1}{2}$, $\frac{1}{2}$;   (7) $1\frac{1}{2}$, $\frac{1}{2}$;   (8) $\frac{1}{4}$, 3;
     (9) $-\frac{1}{4}$, -3;   (10) $\frac{3}{4}$, 1;   (11) $-\frac{3}{4}$, -1;   (12) $-\frac{3}{4}$, 1;
     (13) $1\frac{1}{2}$, 1;   (14) $\frac{1}{2}$, 3;   (15) $\frac{3}{4}$, 2;   (16) $\frac{3}{4}$, -2;
     (17) $1\frac{1}{2}$;   (18) 9, $\frac{1}{4}$.

1.8 (2) $-2\cdot12$, $0\cdot79$; $-1\cdot46$, $5\cdot46$; $-1\cdot27$, $2\cdot77$; $0\cdot42$, $3\cdot58$.

1.9 (1) -2, 2, 4;   (2) -5, 3, 7;   (3) -3, 1, 2, 3.

1.10 (1) $x=2\frac{7}{17}$, $y=-\frac{15}{17}$;   (2) $x=y=2\frac{4}{5}$;   (3) $x=56\frac{2}{3}$, $y=50$;
      (4) $x=9$, $y=0$;   (5) $x=2$, $y=3$.

1.11 (1) $p_1=£0\cdot25$, $p_2=£1$;   (2) $n=40$, $p=1\cdot05$.

1.12 (1) $x=3$, $y=4$; $x=\frac{3}{14}$, $y=-4\frac{5}{14}$;
      (2) $x=3$, $y=4$; $x=-2\frac{13}{31}$, $y=-4\frac{4}{31}$;
      (3) $x=3$, $y=30$; $x=15$, $y=6$;
      (4) $x=2$, $y=-5$; $x=\frac{4}{11}$, $y=-11\frac{6}{11}$;
      (5) $x=1$, $y=-4$; $x=2\frac{23}{81}$, $y=-3\frac{1}{27}$;
      (6) $x=5$, $y=3$; $x=-20\frac{1}{5}$, $y=19\frac{4}{5}$.

1.13 (1) $x=13$;   (2) $x=1$;   (3) $x=5$;   (4) $x=-5$;   (5) $x=6$;
      (6) $x=-6$.

2.1 (1) 1470 ;   (2) 490 ;   (3) n^2+n ;   (4) 42, -60 ;   (5) 6440, 0.

2.2 (1) 1365 ;   (2) -819, 3277, -13107 ;   (3) 126, 127, $127\frac{1}{2}$;
      (4) 42, 43, $42\frac{1}{2}$. The third series cannot reach 128. The fourth series
cannot be as low as $42\cdot6$ or as high as $42\cdot7$ for 50 terms or more.

2.3 (1) 240 ;   (2) 372 ;   (3) 3316.

2.4 Series (ii), (iv), (v) and (vi) converge.

3.1 (1) 12;   (2) 96;   (3) 59,049;   (4) 3380;   (5) 24;   (6) 81 weeks.

3.2 (1) 720;   (2) 56:
      (3) 2520 ; 840 ; 60 ; $13!/3!(2!)^4=64,864,800$; 908,107,200 ; 83,160.

3.3 (1) 126 ;   (2) 847,660,528.

3.4 (1) $x^7+7x^6+21x^5+35x^4+35x^3+21x^2+7x+1$;
      (2) $10^7+7\cdot10^6+21\cdot10^5+\ldots=19,487,171$;
      (3) $1-2y+3y^2-4y^3+5y^4-\ldots$ for $|y|<1$.

4.1 (2) (i) $BC=3\cdot58''$, $AC=4\cdot67''$;   (ii) $BC=2\cdot30''$, $AB=1\cdot93''$;
      (iii) $AB=2\cdot52''$, $AC=3\cdot92''$;   (iv) $\widehat{A}=56°\ 19'$, $AC=3\cdot61''$;
      (v) $\widehat{A}=48°\ 35'$ $AB=2\cdot65''$.

5.1 (i) $3y = x + 3$; (ii) $y = 2x + 5$; (iii) $y = 4x - 3$;
  (iv) $2y = -4x - 1$; (v) $3y = 6 - 2x$; (vi) $8y = -6x - 7$.

5.2 (1) (i) $6y = 5x + 13$; (ii) $8y = 29 - 5x$; (iii) $6y = 29 - 11x$;
  (iv) $8y = 11x + 13$; (v) $8y = 5x + 29$; (vi) $6y = 11x - 29$;
  (vii) $8y = 11x - 13$; (viii) $6y = 13 - 5x$; (ix) $8y = -11x - 13$;
  (x) $6y = -11x - 29$; (xi) $8y = 13 - 11x$; (xii) $6y = -5x - 13$;
  (xiii) $8y = -5x - 29$; (xiv) $6y = 11x + 29$; (xv) $8y = 5x - 29$;
  (xvi) $6y = 5x - 13$.

(2) Specimen answers are (i) $y = \frac{5}{6}x + \frac{13}{6}$, $5x - 6y + 13 = 0$,
  (viii) $y = -\frac{5}{6}x + \frac{13}{6}$, $5x + 6y - 13 = 0$, and

  (xiii) $y = -\frac{5}{6}x - \frac{29}{8}$, $5x + 8y + 29 = 0$.

(3) (i) $-\frac{3}{2}, 2\frac{1}{3}, 3\frac{1}{2}$; (ii) $-\frac{3}{2}, -\frac{7}{3}, -\frac{7}{2}$; (iii) $\frac{3}{2}, \frac{7}{3}, -\frac{7}{2}$
  (iv) $\frac{3}{2}, -\frac{7}{3}, \frac{7}{2}$; (v) $\frac{3}{2}, -\frac{7}{3}, \frac{7}{2}$; (vi) $\frac{3}{2}, \frac{7}{3}, -\frac{7}{2}$;
  (vii) $-\frac{3}{2}, -\frac{7}{3}, -\frac{7}{2}$; (viii) $-\frac{3}{2}, \frac{7}{3}, \frac{7}{2}$.

(4) $y = \frac{4}{5}x + 570$ where $y = $ cost and $x = $ quantity; £890.

(1) (i) $\sqrt{2}$; (ii) 4 ; (iii) $2\sqrt{2}$; (iv) $5\sqrt{2}$; (v) 1.97 ;
(2) 14.8 ; (3) peri $= 20$; area $= 0$—points lie in straight line.

5.4 (1) $-0.5, 3.0$; (2) $-0.514, 3.014$; (3) 1.185 ;
  (4) 0.647 ; (5) $x = \frac{10}{19}, y = 11\frac{11}{19}$;
  (6) $x = -0.185$ and 16.185 ; (7) $x = -0.19$ and -15.81 ;
  (8) $x = -0.878$ and 0.734 ; (9) $x = -1.11$ and 1.
  (10) 0 ; (11) $x = \pm 1.896$ radians.

6.1 (1) ellipse ; (2) hyperbola ; (3) hyperbola ;
  (4) rectangular hyperbola ; (5) parabola ;
  (6) parallel straight lines ; (7) intersecting straight lines ;
  (8) circle ; (9) point circle ; (10) imaginary circle.

6.2 Put $2/(1 - \frac{1}{2} \cos \theta) = 1 \Big/ \cos\left(\theta - \frac{\pi}{4}\right)$. Using $\cos\left(\theta - \frac{\pi}{4}\right) = \frac{1}{\sqrt{2}}(\cos \theta + $

$\sin \theta)$ obtain equation in form $(2\sqrt{2} + 1)x + 2\sqrt{2} \sqrt{1 - x^2} = 2$ where
$x = \cos \theta$. Solve for x, obtaining 0.8771 and -0.2013. Whence inter-
sections at $(3.56, -28° 42')$ and $(1.82, 101° 35')$.

9.1 (2) (i) $6x$; (ii) $35x^4$; (iii) $15x^2$; (iv) $16x^3$; (v) $4/x^2$;
  (vi) $-10/x^3$; (vii) $-49/x^8$; (viii) $3x^2/5$; (ix) $\frac{3}{2}\sqrt{x}$;

  (x) $\frac{3}{2}\sqrt{x}$; (xi) $\frac{2}{3}x^{-1/3}$; (xii) $\frac{2}{3}\sqrt[3]{\frac{2}{x}}$

9.2 (1) (a) 3.4 ; (b) 3.0 ;
  (2) (a) $(32 - 9\sqrt{3})/3\sqrt{3}$; (b) 2.5 ; (c) 2.5.

9.3 (1) (i) $12x^2 + 6x$; (ii) $12x^2 - 6x$; (iii) $12x^2 + 6x + 2$;
  (iv) $40x^7 - 7$; (v) $27x^2 + 1$; (vi) $14 - 27x^2$;

  (vii) $\dfrac{18}{x^3} - \dfrac{14}{x^2}$; (viii) $6x + \dfrac{3\sqrt{x}}{2}$.

(2) (a) 0·24, (b) 0·50 ;   (3) (a) −0·085, (b) −0·285 ;
(4) (a) 8·05d., (b) 0·19, (c) −0·06.

9.4 (1)   (i) $24x^2 + 20x + 3$;   (ii) $3x^2 + 20x + 22$;
(iii) $66x^2 - 100x^4 - 10x - 6$;   (iv) $96x^{11} + 28x^6$;
(v) $(8x + 3)(2x^3 - x)(x^5 + x + 17) + (6x^2 - 1)(4x^2 + 3x)(x^5 + x + 17)$
$$+ (5x^4 + 1)(4x^2 + 3x)(2x^3 - x).$$

(2) $14\tfrac{2}{7}$d., 0·81, 0·42.

9.5 (1)   (i) $(8x^2 + 8x + 3)/(2x + 1)^2$;
(ii) $-(x^2 + 6x + 20)/(x^2 + 7x + 1)^2$;
(iii) $(26x^2 - 20x^4 + 10x - 6)/(3 - 5x^2)^2$;
(iv) $(16x^{11} + 28x^6)/(8x^5 + 4)^2$.
(2) Equilibrium price is $p = 2·438$.
Supply: $\eta = p(-2000 + 8p + 40p^2)/(1 + 10p)^2 x = -0·748$ at   equilibrium.
Demand: $\eta = p(-2501 - 8p + 120p^2)/(1 + 10p)^2 x = -0·776$.

10.1 (1)   (i) $y' - 20x^4 - 6x + 2$; $y'' = 80x^3 - 6$; convex ;
(ii) $20x^4 - 6x + 8$; $80x^3 - 6$; convex ;
(iii) $(x^2 - 4x - 3)/(x - 2)^2$; $14/(x - 2)^3$; concave ;
(iv) $(2x^4 - 27x^2 - 21)/(2x^2 - 7)^2$; $26x(2x^2 + 21)/(2x^2 - 7)^3$; concave.
(2) $40 + 8t + 0·6t^2$; $8 + 1·2t$; 180 f.p.s. ; 20 f.p.s.p.s.

10.2 (1)   (i) $4(4x^3 + 2x^2 - 3x + 8)^3(12x^2 + 4x - 3)$;
(ii) $3(12x^2 - 3)(4x^3 - 3x + 8)[7(4x^3 - 3x + 8) + 2]$;
(iii) $(3x + 1)/\sqrt{3x^2 + 2x}$;
(iv) $(12x^2 + 4x - 1)/2\sqrt{4x^3 + 2x^2 - x + 1}$;
(v) $(1 - 4x - 12x^2)/2(4x^3 + 2x^2 - x + 1)^{3/2}$;
(vi) $(x^3 + 3x^2 + 7)^2(x^2 + 2x)^3[4(2x + 2)(x^3 + 3x^2 + 7)$
$$+ 3(3x^2 + 6x)(x^2 + 2x)]$$;
(vii) $(x^2 + 2x)^3[4(2x + 2)(x^3 + 3x^2 + 7)$
$$- 3(3x^2 + 6x)(x^2 + 2x)]/(x^3 + 3x^2 + 7)^4$$;
(viii) $(x^3 + 3x^2 + 7)^2(x^2 + 2x)^3(11x^7 + 38x^6 + 24x^5 - 120x^4 - 572x^3 - 480x^2$
$$- 448x - 448)/(x^3 - 8)^3.$$ (Use result of (vi) above.)

10.4   (i) $2x + \cos x$;   (ii) $2x - \cos x$;
(iii) $(x^2 + 2) \cos x + 2x \sin x$;
(iv) $(3x^2 + 1) \cos x - (x^3 + x) \sin x$;
(v) $2x(\sin x - \cos x) + x^2(\cos x + \sin x)$;
(vi) $2x(\cos x - \sin x) - (x^2 + 2)(\cos x + \sin x)$;
(vii) $\tan x + x \sec^2 x$;
(viii) $\cos^2 x + \sin x + x(\cos x - 2 \cos x \sin x)$;
(ix) $(3 \cos 3x \cos x + \sin 3x \sin x)/\cos^2 x$;
(x) $(3 \cos 3x \cos 2x + 2 \sin 3x \sin 2x)/\cos^2 2x$;
(xi) $[(x^2 + x) \cos x + \sin x]/(x + 1)^2$;
(xii) $-2 \cos x \sin x = -\sin 2x$;
(xiii) $4 \sin^3 x \cos x$;   (xiv) $\cos x/2\sqrt{\sin x}$;   (xv) $\sec^2 x/2\sqrt{\tan x}$;
(xvi) $\sqrt{\sin x} + x \cos x/2\sqrt{\sin x}$.

11.1 (1) (i) max. when $x = \dfrac{1}{\sqrt{2}}$, min. when $x = -\dfrac{1}{\sqrt{2}}$, P.I. when $x = 0$;

(ii) min. when $x = \frac{1}{3}$;

(iii) max. when $x = 1$, min. when $x = 3$, P.I. when $x = 2$;

(iv) max. when $x = -\frac{1}{9}$, min. when $x = 1$, P.I. when $x = \frac{4}{9}$;

(v) max. when $x = \dfrac{\pi}{2}, \dfrac{5\pi}{2}, \dfrac{9\pi}{2}$, etc., min. when $x = \dfrac{3\pi}{2}, \dfrac{7\pi}{2}, \dfrac{11\pi}{2}$, etc.,

PI. when $x = \pi, 2\pi, 3\pi$, etc.

(vi) max. when $x = \dfrac{\pi}{4}, \dfrac{5\pi}{4}, \dfrac{9\pi}{4}$, etc., min. when $x = \dfrac{3\pi}{4}, \dfrac{7\pi}{4}, \dfrac{11\pi}{4}$,

etc., P.I. when $x = \dfrac{\pi}{2}, \pi, \dfrac{3\pi}{2}$, etc.

(vii) max. when $x = 0, 2\pi, 4\pi$, etc., min. when $x = \pi, 3\pi, 5\pi$, etc.

P.I. when $x = \dfrac{\pi}{2}, \dfrac{3\pi}{2}, \dfrac{5\pi}{2}$, etc.

(viii) No max. or min. P.I. when $x = 0, 2\pi, 4\pi$, etc.

11.2 $\dfrac{dx}{dp} = -60p + 3p^2$ showing an upward slope for $3p^2 > 60p$, i.e., $p > 20$;

(2) Upward sloping for $p > \frac{50}{3}$

(3) £28·2 per ton.

12.1 (1) $1\frac{1}{3}$; (2) 6·4 ; (3) 1/128.

12.3 (1) $1\frac{1}{4}$; (2) 31/5 ; (3) 63/128.

12.4 (1) $110\frac{5}{6}$; (2) $12\frac{2}{3}$; (3) $18\frac{3}{5}$.

12.5 (1) (i) $-48, 73$; (ii) $144, 165\frac{1}{3}$; (iii) 2, 2 ; (iv) 4, 0 ;
(v) 0, 8 ; (vi) $-5\frac{1}{3}$, 8.

15.1 (1) (i) $(6x^2 - 2x)/(2x^3 - x^2 + 3)$; (ii) $-6 \sin x/2 \cos x$;
(iii) $8(35x^4 - 9x^2)(7x^5 - 3x^3)^7$;

(2) $[(6x - 2) \sin x + (3x^2 - 2x + 1) \cos x]/(x^5 + 3x^2)^3$
$\qquad - [3(5x^4 + 6x)(3x^2 - 2x + 1) \sin x]/(x^5 + 3x^2)^4$;

(3) $(10x^4 + 3)/(2x^5 + 3x) \log_e 10$;

(4) (i) $(6x^2 - 3)e^{2x^3 - 3x} + e^x$; (ii) $(4x + 1)(1 + 4^{2x^2 + x} \log_e 4)$;

(5) (i) $(2x + 3)(x^3 - 4x^2 + x) \sin x + (3x^2 - 8x + 1)(x^2 + 3x) \sin x$
$\qquad\qquad + (x^2 + 3x)(x^3 - 4x^2 + x) \cos x$;

(ii) $(2x + 3)(x^3 - 4x^2 + x)^2 \sqrt{x^2 + 2} \sin x + 2(x^2 + 3x)(3x^2 - 8x + 1)$
$(x^3 - 4x^2 + x) \sqrt{x^2 + 2} \sin x + (x^2 + 3x)(x^3 - 4x^2 + x)^2 \sqrt{x^2 + 2} \cos x$
$\qquad\qquad + [(x^2 + 3x)(x^3 - 4x^2 + x)^2 \sin x]/2\sqrt{x^2 + 2}$;

(6) (i) $(2x + 4)/(x^2 + 4x + 2) \log_e 10$; (ii) $(\cot x)/\log_e 10$;
(iii) $(\cot x)/\log_e 4$;

(7) (i) $e^{x^2 - \sin x}(2x - \cos x)(2x^2 - 2 \sin x + 1)/2\sqrt{x^2 - \sin x}$; (ii) $e^x e^{e^x}$.

(8) (i) max. when $x = \frac{1}{3}$; (ii) min. when $x = 2$, P.I. when $x = 2 \pm \dfrac{1}{\sqrt{2}}$;

(iii) max. when $x = -2$, min. when $x = 0$, P.I. when $x = -2 \pm \sqrt{2}$;
(iv) min. when $x = 0$.

16.1 (i) $(x^2 + 3x + 5)^5/5 + C$ where C is an arbitrary constant.

(ii) $\log (x^3 + 7x^2 + 2) + C = \log A(x^3 + 7x^2 + 2)$;

(iii) $e^{x^3} + c$; (iv) $4^{x^2+5x+1}/\log 4 + C$; (v) $-e \cos x + C$;

(vi) $\frac{1}{3} \log A(x^3 + 3x^2 + 6x)$; (vii) $\frac{1}{8}(x^2 + 8x + 3)^4 + C$;

(viii) $\frac{1}{4}e^{x^4 + 4x^2 + 3} + C$; (ix) $4 \log Ax$;

(x) $\log A \tan x$; (xi) $\log A \tan x$; (xii) $a^{\tan x}/\log a + C$;

(xiii) $\frac{1}{3} \sin^3 x + C$; (xiv) $\frac{1}{5}(x^2 + 5x + 1)^5 + C$;

(xv) $3^{x^2 - \cos x}/\log 3 + C$. (xvi) $e \sin x + C$.

16.2 The arbitrary constant is henceforward omitted to save printing.

(i) $\frac{1}{15} (3x + 5)^5$; (ii) $\sin^{-1} \dfrac{x}{3}$; (iii) $\log (\sec x + \tan x)$;

(iv) $\frac{1}{2} \sin^{-1} \dfrac{2x}{3}$; (v) $-\frac{1}{2} \cos 2x$; (vi) $\tan^{-1} \dfrac{x}{3}$;

(vii) $\tan^{-1} \dfrac{x+1}{3}$; (viii) $\tan^{-1} \dfrac{x+2}{4}$.

16.3 (i) $-(x+1)e^{-x}$; (ii) $-(x^2 + 2x + 2)e^{-x}$; (iii) $a(x - a)e^{x/a}$;

(iv) $\dfrac{x^2}{4} (2 \log x - 1)$; (v) $-x \cos x + \sin x$;

(vi) $\frac{1}{8}(\sin 2x - 2x \cos 2x)$; (vii) $x \tan x + \log \cos x$;

(viii) $x \log 2x - x$; (ix) $\frac{1}{3} \tan^3 x + \tan x$.

17.1 (2) The first three converge for all values of x ; the fourth only for $-1 < x \leqslant 1$;

(3) $y = 1 + x + x^2 + x^3 + \ldots + x^n + \ldots$ converges for $|x| < 1$;

(4) $1 - x + x^2 - x^3 \ldots + (-1)^n x^n + \ldots$ convergent for $|x| < 1$;

$$1 - \frac{x}{2} + \frac{1.3}{2!} \left(\frac{x}{2}\right)^2 - \frac{1.3.5}{3!} \left(\frac{x}{2}\right)^3 + \ldots$$
$$+ (-1)^n \frac{1.3.5 \ldots (2n-1)}{n!} \left(\frac{x}{2}\right)^n + \ldots, |x| < 1;$$

$$1 + \frac{x}{2} + \frac{1.3}{2!} \left(\frac{x}{2}\right)^2 + \frac{1.3.5}{3!} \left(\frac{x}{2}\right)^3 + \ldots$$
$$+ \frac{1.3.5 \ldots (2n-1)}{n!} \left(\frac{x}{2}\right)^n + \ldots, |x| < 1;$$

(5) $\dfrac{1}{\sqrt{2}} \left[-1 + x + \dfrac{x^2}{2!} - \dfrac{x^3}{3!} - \dfrac{x^4}{4!} + \dfrac{x^5}{5!} + \dfrac{x^6}{6!} - \ldots \right]$

$\dfrac{1}{\sqrt{2}} \left[1 - x - \dfrac{x^3}{2!} + \dfrac{x^3}{3!} + \dfrac{x^4}{4!} + \dfrac{x^5}{5!} - \dfrac{x^6}{6!} + \ldots \right]$

$\dfrac{1}{\sqrt{2}} \left[1 + x - \dfrac{x^2}{2!} - \dfrac{x^3}{3!} + \dfrac{x^4}{4!} + \dfrac{x^5}{5!} - \dfrac{x^6}{6!} - \dfrac{x^7}{7!} + \ldots \right]$

18.1 (1) (i) $z_x = 3$, $z_y = 4$; (ii) $z_x = 2x + y$, $z_y = x + 4y$;

(iii) $z_x = x(5x^3 + 3xy^2 - 2y^3)$, $z_y = y(2x^3 - 3x^2y - 5y^3)$;

(iv) $z_x = -x(x^3 + 3xy^2 + 2y^3)/(x^3 - y^3)^2$, $z_y = y(2x^3 + 3x^2y + y^3)/(x^3 - y^3)^2$;

(v) $z_x = 2ze^{x^2+3y^2} = 2xz$, $z_y = 6yz$;

(vi) $z_x = \dfrac{2x}{x^2 + 3y^2}$, $z_y = \dfrac{6y}{x^2 + 3y^2}$.

18.2 (i) $z_x = 2x + y$, $z_y = x + 2y$, $z_{xx} = z_{yy} = 2$, $z_{xy} = z_{yx} = 1$;

(ii) $z_x = 4x^3 + 9x^2y + 4xy^2 + 3y^3$, $z_y = 3x^3 + 4x^2y + 9xy^2 + 4y^3$,
$z_{xx} = 12x^2 + 18xy + 4y^2$, $z_{yy} = 4x^2 + 18xy + 12y^2$,
$z_{xy} = z_{yx} = 9x^2 + 8xy + 9y^2$;

(iii) $z_x = 3y^2(y + 4x^2)$, $z_y = xy(8x^2 + 9y)$, $z_{xx} = 24xy^2$, $z_{yy} = 2x(4x^2 + 9y)$,
$z_{xy} = z_{yx} = 3y(3y + 8x^2)$;

(iv) $z_x = x(5x^3 + 3xy^2 + 10y^3)$, $z_y = y(2x^3 + 15x^2y + 25y^3)$,
$z_{xx} = 2(10x^3 + 3xy^2 + 5y^3)$, $z_{yy} = 2(x^3 + 15x^2y + 25y^3)$,
$z_{xy} = 6xy(x + 5y) = z_{yx}$;

(v) $z_x = 2x \sin y$, $x_y = x^2 \cos y$, $z_{xx} = 2 \sin y$, $z_{yy} = - x^2 \sin y$,
$z_{xy} = z_{yx} = 2x \cos y$;

(vi) $z_x = 2x(\cos x + \cos y) - \sin x(x^2 + y^3)$, $z_y = 3y^2(\cos x + \cos y)$
$\qquad - \sin y(x^2 + y^3)$, $z_{xx} = \cos x(2 - x^2 - y^3) - 4x \sin x + 2 \cos y$,
$z_{yy} = \cos y(8y - x^2 - y^3) - 6y^2 \sin y + 6y \cos x$,
$z_{xy} = - 3y^2 \sin x - 2x \sin y = z_{yx}$;

(vii) $z_x = 2xz$, $z_y = 2yz$, $z_{xx} = 2(1 + 2x^2)z$, $z_{yy} = 2(1 + 2y^2)z$,
$z_{xy} = z_{yx} = 4xyz$;

(viii) $z_x = 2(x + y)z$, $z_y = 2(x + 3y)z$, $z_{yy} = 2[1 + 2(x + y)^2]z$,
$z_{yy} = 2[3 + 2(x + 3y)^2]z$, $z_{xy} = 2[1 + 2(x + y)(x + 3y)]z = z_{yx}$;

(ix) $z_x = 2x/(x^2 + y^2)$, $z_y = 2y/(x^2 + y^2)$, $z_{xx} = 2(y^2 - x^2)/(x^2 + y^2)^2 = - z_{yy}$,
$z_{xy} = z_{yx} = - 4xy/(x^2 + y^2)^2$;

(x) $z_x = 3x^2/(x^3 + y^4)$, $z_y = 4y^3/(x^3 + y^4)$, $z_{xx} = 3x(2y^4 - x^3)/(x^3 + y^4)^2$,
$z_{yy} = 4y^2(3x^3 - y^4)/(x^3 + y^4)^2$, $z_{xy} = (3x^2 - 4y^3)/(x^3 + y^4)^2 = z_{yx}$;

(xi) $z_x = \cos x/(\sin x + \cos y)$, $z_y = - \sin y/(\cos x + \sin y)$,
$z_{xx} = z_{yy} = - (1 + \sin x \cos y)/(\cos x + \sin y)^2$,
$z_{xy} = z_{yx} = (\cos x \sin y)/(\sin x + \cos y)^2$;

(xii) $z_x = \cot x$, $z_y = \tan y$, $z_{xx} = - \mathrm{cosec}^2 x$, $z_{yy} = \sec^2 y$, $z_{xy} = z_{yx} = 0$;

(xiii) $z_x = [3x^2 - (x^3 + 4y^2) \cot x]/y \sin x$, $z_y = [4y^2 - x^3]/y^2 \sin x$,
$z_{xy} = z_{yx} = - [3x^2 + (4y^2 - x^3) \cot x]/y^2 \sin x$,
$z_{xx} = 6x/y \sin x - \cos x[6x^2 \sin x - (x^3 + 4y^2)(1 + \cos x)]/y \sin^3 x$,
$z_{yy} = 2x^3/y^3 \sin x$;

(xiv) $z_x = (2xy^3 - 3x^2y^2 - x^4) \sin x/(x^3 + y^3)^2 \sin y$
$\qquad + (x^2 + y^2) \cos x/(x^3 + y^3)^2 \sin y = [2x/(x^2 + y^2) - 3x^2/(x^3 + y^3)$
$\qquad + \cot x]z$,
$z_y = [2y/(x^2 + y^2) - 3y^2/(x^3 + y^3) - \cot y]z$,
$z_{xx} = [2(y^2 - x^2)/(x^2 + y^2)^2 + (3x^4 - 6y^3x)/(x^3 + y^3)^2 - \mathrm{cosec}^2 x]z$
$\qquad\qquad + [2x/(x^2 + y^2) - 3x^2/(x^3 + y^3) + \cot x]^2z$
$z_{yy} = [2(x^2 - y^2)/(x^2 + y^2)^2 + (3y^4 - 6yx^3)/(x^3 + y^3)^2 + \mathrm{cosec}^2 y]z$
$\qquad\qquad + [2y/(x^2 + y^2) - 3y^2/(x^3 + y^3) - \cot y]^2z$,
$z_{xy} = z_{yx} = [9x^2y^2/(x^3 + y^3)^2 - 4xy/(x^2 + y^2)^2]z + [2x/(x^2 + y^2)$
$\qquad - 3x^2/(x^3 + y^3) + \cot x][2y/(x^2 + y^2) - 3y^2/(x^3 + y^3) - \cot y]z$.

19.1 (1) $2e^{2t}(\cos 2t - \sin 2t) = 2x(x - y)$;

(2) $e^{3t}(1 + t)^3(7 + 3t) = x^3y(3y + 4\sqrt{y})$;

(3) $2e^{2t}(1+t)^5(4+t) = 2x^2y^2(3\sqrt{x}+x)$;

(4) $e^t[\cos(1+t^2) - 2t\sin[(1+t^2)] = x\cos y - 2t\sin y)$;

(5) $te^{2t}\sin^2 t[2\sin t(1+t) + 3t\cos t] = x^2y^2[2y(t+w) + 3w\cos t]$.

19.2 (1) (i) $2x - 3x\sqrt{1-x^2}$; (ii) $2x\sin y - x^3\cos y/\sqrt{1-x^2}$;

(iii) $3x^2\sin y + x^3\cos y(\cos x - x\sin x)$; (iv) $2xe^{x^2+y^3}(1+2y^2)$;

(2) $(3x+2)/(x+2)^3 - 0.02x - 0.1$.

20.1 (i) $(0, 0)$. Doubtful case—saddle point.

(ii) $(0, 0)$. Doubtful case, $(+\sqrt{2}, -\sqrt{2})$, $(-\sqrt{2}, +\sqrt{2})$ minima;

(iii) $(0, 0)$. Doubtful case—saddle point;

(iv) $(0, 0)$. Doubtful case—saddle point;

(v) Minima at $(-\pi, \pm\pi/2)$, $(0, \pm\pi/2)$, $(\pi, \pm\pi/2)$;

Maxima at $(\pm\pi/2, -\pi)$, $(\pm\pi/2, 0)$, $(\pm\pi/2, \pi)$.

20.2 (3) £1·30 and £1·37 (approx.).

21.1 (1) $(0, \pm3)$ minima, $(\pm4, 0)$ maxima;

(2) $(3, 0)$ minimum, $(-3, 0)$ maximum, saddle points at the four points $(\pm\sqrt{3}, \pm\sqrt{6})$;

(3) $\left(\dfrac{a}{3}, \dfrac{a}{3}, \dfrac{a}{3}\right)$ maximum, $(0, 0, a)$, $(0, a, 0)$, $(a, 0, 0)$ saddle points.

22.1 (ii), (vii) and (x) are not homogeneous. The others are homogeneous of the following degrees (i) 3, (iii) 2, (iv) 2, (v) 2, (vi) 0, (viii) 0, (ix) 1.

24.1 (i) $(3x^2 - 8x + 3)dx$; (ii) $(5x^4 + 9x^2 - 4)dx$;

(iii) $[(3x^2 + 1)\cos x - (x^3 + x)\sin x]dx$;

(iv) $[(x^3 + 6x)\cos x - (x^3 - 6x^2 + 6x)\sin x]dx$.

24.2 (i) $(2x+y)dx + (x+2y)dy = 0$;

(ii) $(3x^2 + 2xy + y^2)dx + (x^2 + 2xy + 3y^2)dy = 0$;

(iii) $(2x\sin y + y^2\cos x)dx + (x^2\cos y + 2y\sin x)dy = 0$;

(iv) $\cos x \sin y\, dx + \sin x \cos y\, dy = 0$;

(v) $(x^2 - y)dx + (y^2 - x)dy = 0$;

(vi) $(y\sec^2 x + \tan y)dx + (x\sec^2 y + \tan x)dy = 0$;

(vii) $y(x\log y + y)dx + x(y\log x + x)dy = 0$;

(viii) $(\cosh y + y\sinh x)dx + (\cosh x + x\sinh y)dy = 0$;

(ix) $(\sinh y + y\operatorname{sech}^2 x)dx + (x\cosh y + \tanh x)dy = 0$;

(x) $(2x\tanh y + y^2\sinh x)dx + (x^2\operatorname{sech}^2 y + 2y\cosh x)dy = 0$.

24.3 (1) $d^2z =$

(i) $2dx^2 + 2dx\,dy + 2dy^2$; (ii) $2(y+3x)dx^2 + 4x\,dx\,dy + 6y\,dy^2$;

(iii) $2\sin y\,dx^2 + 4x\cos y\,dx\,dy - x^2\sin y\,dy^2$;

(iv) $(2\sin y - y^2\sin x)dx^2 + 4(x\cos y + y\cos x)dx\,dy$
$+ (2\sin x - x^2\sin y)dy^2$;

(v) $(2\sin y - y^2\cos x)dx + 4(x\cos y - y\sin x)dx\,dy$
$+ (2\cos x - x^2\sin y)dy^2$;

(vi) $2y\sec^2 x \tan x\,dx^2 + 2(\sec^2 y + \sec^2 x)dx\,dy + 2x(\sec^2 y \tan y)dy^2$

(vii) $-\dfrac{y}{x^2}dx^2 + 2(1/y + 1/x)dx\,dy - \dfrac{x}{y^2}dy^2$;

(viii) $6x\sin^2 y\,dx^2 + 6x^2\sin 2y\,dx\,dy + 2x^3\cos 2y\,dy^2$.

(2) $d^2z =$

(i) $2dx^2 + 2dx\, dy + 2dy^2 + (2x+y)d^2x + (x+2y)d^2y$;

(ii) $2(y+3x)dx^2 + 4x\, dx\, dy + 6y\, dy^2 + (3x^2+2xy)d^2x + (2x^2+3y^2)d^2y$;

(iii) $2\sin y\, dx^2 + 4x\cos y\, dx\, dy - x^2\sin y\, dy^2 + 2x\sin y\, d^2x$
$$+ x^2\cos y\, d^2y$$

(iv) $(2\sin y - y^2\sin x)dx^2 + 4(x\cos y\cos x)dx\, dy$
$$+ (2\sin x - x^2\sin y)dy^2 + (2x\sin y + y^2\cos x)d^2x$$
$$+ (2y\sin x - x^2\cos y)d^2y.$$

24.4 (i) $y = \tfrac{1}{4}$ at $(-\tfrac{1}{2}, 0, 0)$; (ii) None ; (iii) $y = -\tfrac{1}{4}$ at $(\tfrac{1}{2}, -\tfrac{1}{2}, \tfrac{1}{2})$;
(iv) $y = 0$ at $(0, 0, 0)$ and $y = 1/48$ at $(1/6, 1/12, 0)$.

25.1 (1)   (i) $z = -1 \pm 2i$;   (ii) $-1/2 \pm 3/2i$;   (iii) $1/6 \pm \sqrt{35}/6i$;
(iv) $-3/25 \pm 4/25i$.

(2)   (i) $|z| = \sqrt{5}$, arg. $z = \theta$ where $\cos\theta = -\dfrac{1}{\sqrt{5}}$, $\sin\theta = \pm\dfrac{2}{\sqrt{5}}$;

(ii) $|z| = \sqrt{\tfrac{5}{2}}$, arg. $z = \theta$ where $\cos\theta = -\dfrac{1}{\sqrt{10}}$, $\sin\theta = \pm 3/\sqrt{10}$;

(iii) $|z| = 1$, arg. $z = \theta$ where $\cos\theta = \tfrac{1}{6}$, $\sin\theta = \pm\dfrac{\sqrt{35}}{6}$;

(iv) $|z| = \tfrac{1}{5}$, arg. $z = \theta$ where $\cos\theta = -\tfrac{3}{5}$, $\sin\theta = \pm\tfrac{4}{5}$;

(3)   (i) $x = 1, y = 2$;   (ii) $x = 2, y = 1$;   (iii) $x = 3, y = -1$.

25.2 Define θ as the *acute* angle such that $\tan\theta = \tfrac{4}{3}$. Then the solutions are;
(i) $5(\cos\theta + i\sin\theta)$;   (ii) $5(\cos\theta - i\sin\theta)$;

(iii) $5\left[\cos\left(\dfrac{\pi}{2}-\theta\right) + i\sin\left(\dfrac{\pi}{2}-\theta\right)\right]$;

(iv) $5\left[\cos\left(\dfrac{\pi}{2}-\theta\right) - i\sin\left(\dfrac{\pi}{2}-\theta\right)\right]$.

(v) $5\left[\cos\left(\pi-\theta\right) + i\sin\left(\pi-\theta\right)\right]$.

(vi) $5\left[\cos\left(\pi-\theta\right) - i\sin\left(\pi-\theta\right)\right]$;

25.3 (1)   (i) $\theta = n\pi$;   (ii) $\theta = \dfrac{4n+1}{4}\pi$;   (iii) $\theta = \dfrac{4n+3}{4}\pi$;

(iv) $\theta = n\pi + \dfrac{\pi}{6}$;   (v) $\theta = (2n\pi/3 - \pi/9)$;   (vi) $\theta = (2n\pi/3 + 2\pi/9)$.

25.4 (1)   (i) $-14 + 22i$;   (ii) $26 - 2i$;   (iii) $-26 + 2i$;
(iv) $14 - 22i$;   (v) $-14 - 22i$;   (vi) $-26 - 2i$;
(vii) $-14 + 22i$;   (viii) 34 ;   (ix) $-16 + 30i$;
(x) $-94 + 2i$;   (xi) $45 - 93i$;   (xii) $-103 - 81i$.

(2)   (i) $6[\cos 7\pi/12 + i\sin 7\pi/12]$;   (ii) $6[\cos\pi/12 + i\sin\pi/12]$;
(iii) $12[\cos\pi/2 + i\sin\pi/2]$;   (iv) $12[\cos 7\pi/12 - i\sin 7\pi/12]$.

25.5 (1)   (i) $(13 - i)/10$;   (ii) $(-7 + 11i)/10$;   (iii) $(13 + i)/10$;
(iv) $(43 + 49i)/10$;   (v) $(331 - 117i)/290$;   (vi) $(-253 - 204i)/169$.

(2)   (i) $x = 13/20$, $y = -\tfrac{1}{30}$;   (ii) $x = 13/30$, $y = -1/20$.

26.1 (i) $y = Ae^{x^2/6} - 2$; (ii) $3y^2 + 24y + 6\log(y+2) + 2x^3 = A$;

(iii) $y^2 = x^2 - Ax$; (iv) $3y^3 \log Ay = x^3$;

(v) $\sqrt{13}\log(Y^2 - XY - 3X^2) + 3\log\left[\dfrac{2Y - (1+\sqrt{13})X}{2Y - (1-\sqrt{13})X}\right] = A$;

where $Y = y+1$ and $X = x - 1$;

(vi) $x \sin y = A$.

26.2 (i) $Ae^{-6x} + Be^{2x}$; (ii) $e^{-2x}(A + Bx)$;

(iii) $Ae^{3x} + e^{-2x}(B + Cx)$; (iv) $e^{-2x}(A \cos \sqrt{5}x + B \sin \sqrt{5}x)$.

26.3 (i) $e^{3x}/19$; (ii) $-e^{2x}/5$; (iii) $xe^{2x}/24$; (iv) $-x^2e^x/2$;

(v) $-(18x^2 + 27x + 13)/81$; (vi) $-(x^2+1)/2$;

(vii) $-(\sin 5x)/21$; (viii) $(\sin 3x - \cos 3x)/12$.

26.4 The Complete Solutions are given by adding the P.I.'s of Exercise 25.3 to the following :

(i) $e^{-2x}(Ae^{\sqrt{6}x} + Be^{-\sqrt{6}x})$; (ii) $Ae^{3x} + B \sin x + C \cos x$;

(iii) $Ae^{2x} + e^{-5x/2}(Be^{\sqrt{15}x/2} + Ce^{-\sqrt{15}x/2})$;

(iv) $e^x(A + Bx) + Ce^{2x}$; (v) $Ae^{3x} + Be^{-3x}$;

(vi) $Ae^{3x} + Be^{-2x}$; (vii) $A \cos 2x + B \sin 2x$;

(viii) $e^{-x}(A \cos\sqrt{2}x + B \sin \sqrt{2}x)$.

27.1 (1) (i) $3Y_t + 3\Delta Y_t + \Delta^2 Y_t - 6t = 0$; (ii) $4Y_t + 3\Delta Y_t + \Delta^2 Y_t - 3t + 4 = 0$;

(iii) $4Y_t + 3\Delta Y_t + \Delta^2 Y_t - 3t - 8 = 0$; (iv) $3\Delta Y_t + 2\Delta^2 Y_t - 8t - 17 = 0$;

(2) (i) $2Y_{t+1} - Y_t = 3t$; (ii) $2Y_t - 2Y_{t+1} + Y_{t+2} = 3t$;

(iii) $3Y_t - 3Y_{t+1} + Y_{t+2} = 4t$; (iv) $5Y_t - 4Y_{t+1} + Y_{t+2} = 4t$.

27.2 (1) (i) $Y_t = \frac{7}{10}$; (ii) $Y_t = \frac{8t}{3}$; (iii) $Y_t = \frac{2t}{5}$; (iv) $Y_t = 7$;

(v) $Y_t = 7^{t+2}/172$;

(2) (i) $Y_t = A6^t + B(-1)^t$; (ii) $Y_t = (A + Bt)3^t$;

(iii) $Y_t = A(-2)^t + B(-3)^t$;

(iv) $Y_t = A(3i)^t + B(-3i)^t = C3^t \cos\left(\dfrac{\pi t}{2} - \phi\right)$;

(v) $Y_t = A + B3^t \cos\left(\dfrac{\pi t}{2} - \phi\right)$.

27.3 Note that the four equations in this exercise have been taken from Exercise 26.2. The solutions are given in a general form, with the values of the arbitrary constants that satisfy the initial conditions.

(i) $Y_t = \frac{7}{2} + A6^t + B(-1)^t$, $A = 1$, $B = \frac{1}{2}$;

(ii) $Y_t = \frac{3}{4} + (A + Bt)3^t$, $A = \frac{1}{4}$, $B = \frac{1}{2}$;

(iii) $Y_t = A + B(\frac{1}{4})^t + \dfrac{8t}{3}$, $A = \frac{4}{9}$, $B = \frac{32}{9}$;

(iv) $Y_t = 5^{t+2}/34 + A3^t \cos\left(\dfrac{\pi t}{2} - \phi\right)$, $A = \sqrt{2}$, $\phi = \dfrac{\pi}{4}$.

APPENDIX ANSWERS

A1   (i) $(2x+3)(x+1)$;   (ii) $(2x+1)(x+3)$;   (iii) $(2x+3)(x+5)$;
        (iv) $(2x+3)(2x+5)$;   (v) $(4x+3)(x+5)$;   (vi) $(8x+5)(x+2)$;
        (vii) $2(4x+5)(x+1)$;   (viii) $2(4x+1)(x+5)$;   (ix) $(4x-1)(2x+3)$;
        (x) $(4x-3)(2x+1)$;   (xi) $(8x-3)(x+1)$;   (xii) $(5x-4)(4x-3)$;
        (xiii) $2(5x-6)(2x-1)$; (xiv) $2(5x+6)(2x-1)$.

A6.1 (1)   (i) 10;   (ii) -19;   (iii) 0;   (iv) -14;   (v) 14;
          (vi) 10;   (vii) 10;   (viii) -5;   (ix) x^2-1;   (x) $-x^2-1$.
     (2)   (i) $x=5/7$, $y=17/7$; (ii) $x=57/31$, $y=-2/31$;   (iii) $x=11$, $y=6$;
          (iv) $x=7/12$, $y=35/36$; (v) $x=2/11$, $y=3/11$;
          (vi) $x=18/7$, $y=8/7$.

A6.2   (i) 40;   (ii) -40; (iii) 0; (iv) 0; (v) $x^3-10x+10$;   (vi) $9x$;
     (vii) 18; (viii) 0;   (ix) $abc-af^2-bg^2-ch^2+2fgh$;
     (x) 143;   (xi) 2;   (xii) 0.

A6.4 (1)   (i) 40; (ii) 12.
     (2)   (i) 0;   (ii) 3;   (iii) -32;
          (iv) $(a-b)(a-c)(a-d)(b-c)(b-d)(c-d)$ [Hint : if $a=b$, 2 rows the
             same];
          (v) $4(1-x)(2-x)(3-x)$ [Hint : If $x=1$ in original then 2 rows the
             same];
          (vi) $-(a+b+c+d)(a-b+c-d)[(a-c)^2+(b-d)^2]$.

A6.5 (i) $x=1/2$, $y=1$, $z=-\frac{1}{2}$;   (ii) $x=0$, $y=3$, $z=1$;
     (iii) $x=\frac{55}{6}$, $y=\frac{29}{12}$, $z=-\frac{11}{12}$, $w=-\frac{20}{3}$;
     (iv) $a=3/2$, $b=2$, $c=5/2$, $d=-7/2$;   (v) inconsistent;
     (vi) $a=-1$, $b=-7$, $c=5$, $d=-1$.

A7.1 (i), (ii) and (viii) are positive definite; the remainder are indefinite.

TABLES

	0	1	2	3	4	5	6	7	8	9	1 2 3	4 5 6	7 8 9
10	0000	0043	0086	0128	0170	0212	0253	0294	0334	0374	5 9 13	17 21 26	30 34 38
											4 8 12	16 20 24	28 32 36
11	0414	0453	0492	0531	0569	0607	0645	0682	0719	0755	4 8 12	16 20 23	27 31 35
											4 7 11	15 18 22	26 29 33
12	0792	0828	0864	0899	0934	0969	1004	1038	1072	1106	3 7 11	14 18 21	25 28 32
											3 7 10	14 17 20	24 27 31
13	1139	1173	1206	1239	1271	1303	1335	1367	1399	1430	3 6 10	13 16 19	23 26 29
											3 7 10	13 16 19	22 25 29
14	1461	1492	1523	1553	1584	1614	1644	1673	1703	1732	3 6 9	12 15 19	22 25 28
											3 6 9	12 14 17	20 23 26
15	1761	1790	1818	1847	1875	1903	1931	1959	1987	2014	3 6 9	11 14 17	20 23 26
											3 6 8	11 14 17	19 22 25
16	2041	2068	2095	2122	2148	2175	2201	2227	2253	2279	3 6 8	11 14 16	19 22 24
											3 5 8	10 13 16	18 21 23
17	2304	2330	2355	2380	2405	2430	2455	2480	2504	2529	3 5 8	10 13 15	18 20 23
											3 5 8	10 12 15	17 20 22
18	2553	2577	2601	2625	2648	2672	2695	2718	2742	2765	2 5 7	9 12 14	17 19 21
											2 4 7	9 11 14	16 18 21
19	2788	2810	2833	2856	2878	2900	2923	2945	2967	2989	2 4 7	9 11 13	16 18 20
											2 4 6	8 11 13	15 17 19
20	3010	3032	3054	3075	3096	3118	3139	3160	3181	3201	2 4 6	8 11 13	15 17 19
21	3222	3243	3263	3284	3304	3324	3345	3365	3385	3404	2 4 6	8 10 12	14 16 18
22	3424	3444	3464	3483	3502	3522	3541	3560	3579	3598	2 4 6	8 10 12	14 15 17
23	3617	3636	3655	3674	3692	3711	3729	3747	3766	3784	2 4 6	7 9 11	13 15 17
24	3802	3820	3838	3856	3874	3892	3909	3927	3945	3962	2 4 5	7 9 11	12 14 16
25	3979	3997	4014	4031	4048	4065	4082	4099	4116	4133	2 3 5	7 9 10	12 14 15
26	4150	4166	4183	4200	4216	4232	4249	4265	4281	4298	2 3 5	7 8 10	11 13 15
27	4314	4330	4346	4362	4378	4393	4409	4425	4440	4456	2 3 5	6 8 9	11 13 14
28	4472	4487	4502	4518	4533	4548	4564	4579	4594	4609	2 3 5	6 8 9	11 12 14
29	4624	4639	4654	4669	4683	4698	4713	4728	4742	4757	1 3 4	6 7 9	10 12 13
30	4771	4786	4800	4814	4829	4843	4857	4871	4886	4900	1 3 4	6 7 9	10 11 13
31	4914	4928	4942	4955	4969	4983	4997	5011	5024	5038	1 3 4	6 7 8	10 11 12
32	5051	5065	5079	5092	5105	5119	5132	5145	5159	5172	1 3 4	5 7 8	9 11 12
33	5185	5198	5211	5224	5237	5250	5263	5276	5289	5302	1 3 4	5 6 8	9 10 12
34	5315	5328	5340	5353	5366	5378	5391	5403	5416	5428	1 3 4	5 6 8	9 10 11
35	5441	5453	5465	5478	5490	5502	5514	5527	5539	5551	1 2 4	5 6 7	9 10 11
36	5563	5575	5587	5599	5611	5623	5635	5647	5658	5670	1 2 4	5 6 7	8 10 11
37	5682	5694	5705	5717	5729	5740	5752	5763	5775	5786	1 2 3	5 6 7	8 9 10
38	5798	5809	5821	5832	5843	5855	5866	5877	5888	5899	1 2 3	5 6 7	8 9 10
39	5911	5922	5933	5944	5955	5966	5977	5988	5999	6010	1 2 3	4 5 7	8 9 10
40	6021	6031	6042	6053	6064	6075	6085	6096	6107	6117	1 2 3	4 5 6	8 9 10
41	6128	6138	6149	6160	6170	6180	6191	6201	6212	6222	1 2 3	4 5 6	7 8 9
42	6232	6243	6253	6263	6274	6284	6294	6304	6314	6325	1 2 3	4 5 6	7 8 9
43	6335	6345	6355	6365	6375	6385	6395	6405	6415	6425	1 2 3	4 5 6	7 8 9
44	6435	6444	6454	6464	6474	6484	6493	6503	6513	6522	1 2 3	4 5 6	7 8 9
45	6532	6542	6551	6561	6571	6580	6590	6599	6609	6618	1 2 3	4 5 6	7 8 9
46	6628	6637	6646	6656	6665	6675	6684	6693	6702	6712	1 2 3	4 5 6	7 7 8
47	6721	6730	6739	6749	6758	6767	6776	6785	6794	6803	1 2 3	4 5 5	6 7 8
48	6812	6821	6830	6839	6848	6857	6866	6875	6884	6893	1 2 3	4 4 5	6 7 8
49	6902	6911	6920	6928	6937	6946	6955	6964	6972	6981	1 2 3	4 4 5	6 7 8

LOGARITHMS

	0	1	2	3	4	5	6	7	8	9	123	456	789
50	6990	6998	7007	7016	7024	7033	7042	7050	7059	7067	1 2 3	3 4 5	6 7 8
51	7076	7084	7093	7101	7110	7118	7126	7135	7143	7152	1 2 3	3 4 5	6 7 8
52	7160	7168	7177	7185	7193	7202	7210	7218	7226	7235	1 2 2	3 4 5	6 7 7
53	7243	7251	7259	7267	7275	7284	7292	7300	7308	7316	1 2 2	3 4 5	6 6 7
54	7324	7332	7340	7348	7356	7364	7372	7380	7388	7396	1 2 2	3 4 5	6 6 7
55	7404	7412	7419	7427	7435	7443	7451	7459	7466	7474	1 2 2	3 4 5	5 6 7
56	7482	7490	7497	7505	7513	7520	7528	7536	7543	7551	1 2 2	3 4 5	5 6 7
57	7559	7566	7574	7582	7589	7597	7604	7612	7619	7627	1 2 2	3 4 5	5 6 7
58	7634	7642	7649	7657	7664	7672	7679	7686	7694	7701	1 1 2	3 4 4	5 6 7
59	7709	7716	7723	7731	7738	7745	7752	7760	7767	7774	1 1 2	3 4 4	5 6 7
60	7782	7789	7796	7803	7810	7818	7825	7832	7839	7846	1 1 2	3 4 4	5 6 6
61	7853	7860	7868	7875	7882	7889	7896	7903	7910	7917	1 1 2	3 4 4	5 6 6
62	7924	7931	7938	7945	7952	7959	7966	7973	7980	7987	1 1 2	3 3 4	5 6 6
63	7993	8000	8007	8014	8021	8028	8035	8041	8048	8055	1 1 2	3 3 4	5 5 6
64	8062	8069	8075	8082	8089	8096	8102	8109	8116	8122	1 1 2	3 3 4	5 5 6
65	8129	8136	8142	8149	8156	8162	8169	8176	8182	8189	1 1 2	3 3 4	5 5 6
66	8195	8202	8209	8215	8222	8228	8235	8241	8248	8254	1 1 2	3 3 4	5 5 6
67	8261	8267	8274	8280	8287	8293	8299	8306	8312	8319	1 1 2	3 3 4	5 5 6
68	8325	8331	8338	8344	8351	8357	8363	8370	8376	8382	1 1 2	3 3 4	4 5 6
69	8388	8395	8401	8407	8414	8420	8426	8432	8439	8445	1 1 2	2 3 4	4 5 6
70	8451	8457	8463	8470	8476	8482	8488	8494	8500	8506	1 1 2	2 3 4	4 5 6
71	8513	8519	8525	8531	8537	8543	8549	8555	8561	8567	1 1 2	2 3 4	4 5 5
72	8573	8579	8585	8591	8597	8603	8609	8615	8621	8627	1 1 2	2 3 4	4 5 5
73	8633	8639	8645	8651	8657	8663	8669	8675	8681	8686	1 1 2	2 3 4	4 5 5
74	8692	8698	8704	8710	8716	8722	8727	8733	8739	8745	1 1 2	2 3 4	4 5 5
75	8751	8756	8762	8768	8774	8779	8785	8791	8797	8802	1 1 2	2 3 3	4 5 5
76	8808	8814	8820	8825	8831	8837	8842	8848	8854	8859	1 1 2	2 3 3	4 5 5
77	8865	8871	8876	8882	8887	8893	8899	8904	8910	8915	1 1 2	2 3 3	4 4 5
78	8921	8927	8932	8938	8943	8949	8954	8960	8965	8971	1 1 2	2 3 3	4 4 5
79	8976	8982	8987	8993	8998	9004	9009	9015	9020	9025	1 1 2	2 3 3	4 4 5
80	9031	9036	9042	9047	9053	9058	9063	9069	9074	9079	1 1 2	2 3 3	4 4 5
81	9085	9090	9096	9101	9106	9112	9117	9122	9128	9133	1 1 2	2 3 3	4 4 5
82	9138	9143	9149	9154	9159	9165	9170	9175	9180	9186	1 1 2	2 3 3	4 4 5
83	9191	9196	9201	9206	9212	9217	9222	9227	9232	9238	1 1 2	2 3 3	4 4 5
84	9243	9248	9253	9258	9263	9269	9274	9279	9284	9289	1 1 2	2 3 3	4 4 5
85	9294	9299	9304	9309	9315	9320	9325	9330	9335	9340	1 1 2	2 3 3	4 4 5
86	9345	9350	9355	9360	9365	9370	9375	9380	9385	9390	1 1 2	2 3 3	4 4 5
87	9395	9400	9405	9410	9415	9420	9425	9430	9435	9440	0 1 1	2 2 3	3 4 4
88	9445	9450	9455	9460	9465	9469	9474	9479	9484	9489	0 1 1	2 2 3	3 4 4
89	9494	9499	9504	9509	9513	9518	9523	9528	9533	9538	0 1 1	2 2 3	3 4 4
90	9542	9547	9552	9557	9562	9566	9571	9576	9581	9586	0 1 1	2 2 3	3 4 4
91	9590	9595	9600	9605	9609	9614	9619	9624	9628	9633	0 1 1	2 2 3	3 4 4
92	9638	9643	9647	9652	9657	9661	9666	9671	9675	9680	0 1 1	2 2 3	3 4 4
93	9685	9689	9694	9699	9703	9708	9713	9717	9722	9727	0 1 1	2 2 3	3 4 4
94	9731	9736	9741	9745	9750	9754	9759	9763	9768	9773	0 1 1	2 2 3	3 4 4
95	9777	9782	9786	9791	9795	9800	9805	9809	9814	9818	0 1 1	2 2 3	3 4 4
96	9823	9827	9832	9836	9841	9845	9850	9854	9859	9863	0 1 1	2 2 3	3 4 4
97	9868	9872	9877	9881	9886	9890	9894	9899	9903	9908	0 1 1	2 2 3	3 4 4
98	9912	9917	9921	9926	9930	9934	9939	9943	9948	9952	0 1 1	2 2 3	3 4 4
99	9956	9961	9965	9969	9974	9978	9983	9987	9991	9996	0 1 1	2 2 3	3 3 4

ANTILOGARITHMS

	0	1	2	3	4	5	6	7	8	9	1 2 3	4 5 6	7 8 9
·00	1000	1002	1005	1007	1009	1012	1014	1016	1019	1021	0 0 1	1 1 1	2 2 2
·01	1023	1026	1028	1030	1033	1035	1038	1040	1042	1045	0 0 1	1 1 1	2 2 2
·02	1047	1050	1052	1054	1057	1059	1062	1064	1067	1069	0 0 1	1 1 1	2 2 2
·03	1072	1074	1076	1079	1081	1084	1086	1089	1091	1094	0 0 1	1 1 1	2 2 2
·04	1096	1099	1102	1104	1107	1109	1112	1114	1117	1119	0 1 1	1 1 2	2 2 2
·05	1122	1125	1127	1130	1132	1135	1138	1140	1143	1146	0 1 1	1 1 2	2 2 2
·06	1148	1151	1153	1156	1159	1161	1164	1167	1169	1172	0 1 1	1 1 2	2 2 2
·07	1175	1178	1180	1183	1186	1189	1191	1194	1197	1199	0 1 1	1 1 2	2 2 2
·08	1202	1205	1208	1211	1213	1216	1219	1222	1225	1227	0 1 1	1 1 2	2 2 3
·09	1230	1233	1236	1239	1242	1245	1247	1250	1253	1256	0 1 1	1 1 2	2 2 3
·10	1259	1262	1265	1268	1271	1274	1276	1279	1282	1285	0 1 1	1 1 2	2 2 3
·11	1288	1291	1294	1297	1300	1303	1306	1309	1312	1315	0 1 1	1 2 2	2 2 3
·12	1318	1321	1324	1327	1330	1334	1337	1340	1343	1346	0 1 1	1 2 2	2 2 3
·13	1349	1352	1355	1358	1361	1365	1368	1371	1374	1377	0 1 1	1 2 2	2 3 3
·14	1380	1384	1387	1390	1393	1396	1400	1403	1406	1409	0 1 1	1 2 2	2 3 3
·15	1413	1416	1419	1422	1426	1429	1432	1435	1439	1442	0 1 1	1 2 2	2 3 3
·16	1445	1449	1452	1455	1459	1462	1466	1469	1472	1476	0 1 1	1 2 2	2 3 3
·17	1479	1483	1486	1489	1493	1496	1500	1503	1507	1510	0 1 1	1 2 2	2 3 3
·18	1514	1517	1521	1524	1528	1531	1535	1538	1542	1545	0 1 1	1 2 2	2 3 3
·19	1549	1552	1556	1560	1563	1567	1570	1574	1578	1581	0 1 1	1 2 2	3 3 3
·20	1585	1589	1592	1596	1600	1603	1607	1611	1614	1618	0 1 1	1 2 2	3 3 3
·21	1622	1626	1629	1633	1637	1641	1644	1648	1652	1656	0 1 1	2 2 2	3 3 3
·22	1660	1663	1667	1671	1675	1679	1683	1687	1690	1694	0 1 1	2 2 2	3 3 3
·23	1698	1702	1706	1710	1714	1718	1722	1726	1730	1734	0 1 1	2 2 2	3 3 4
·24	1738	1742	1746	1750	1754	1758	1762	1766	1770	1774	0 1 1	2 2 2	3 3 4
·25	1778	1782	1786	1791	1795	1799	1803	1807	1811	1816	0 1 1	2 2 2	3 3 4
·26	1820	1824	1828	1832	1837	1841	1845	1849	1854	1858	0 1 1	2 2 3	3 3 4
·27	1862	1866	1871	1875	1879	1884	1888	1892	1897	1901	0 1 1	2 2 3	3 3 4
·28	1905	1910	1914	1919	1923	1928	1932	1936	1941	1945	0 1 1	2 2 3	3 4 4
·29	1950	1954	1959	1963	1968	1972	1977	1982	1986	1991	0 1 1	2 2 3	3 4 4
·30	1995	2000	2004	2009	2014	2018	2023	2028	2032	2037	0 1 1	2 2 3	3 4 4
·31	2042	2046	2051	2056	2061	2065	2070	2075	2080	2084	0 1 1	2 2 3	3 4 4
·32	2089	2094	2099	2104	2109	2113	2118	2123	2128	2133	0 1 1	2 2 3	3 4 4
·33	2138	2143	2148	2153	2158	2163	2168	2173	2178	2183	0 1 1	2 2 3	3 4 4
·34	2188	2193	2198	2203	2208	2213	2218	2223	2228	2234	1 1 2	2 3 3	4 4 5
·35	2239	2244	2249	2254	2259	2265	2270	2275	2280	2286	1 1 2	2 3 3	4 4 5
·36	2291	2296	2301	2307	2312	2317	2323	2328	2333	2339	1 1 2	2 3 3	4 4 5
·37	2344	2350	2355	2360	2366	2371	2377	2382	2388	2393	1 1 2	2 3 3	4 4 5
·38	2399	2404	2410	2415	2421	2427	2432	2438	2443	2449	1 1 2	2 3 3	4 4 5
·39	2455	2460	2466	2472	2477	2483	2489	2495	2500	2506	1 1 2	2 3 3	4 5 5
·40	2512	2518	2523	2529	2535	2541	2547	2553	2559	2564	1 1 2	2 3 4	4 5 5
·41	2570	2576	2582	2588	2594	2600	2606	2612	2618	2624	1 1 2	2 3 4	4 5 5
·42	2630	2636	2642	2649	2655	2661	2667	2673	2679	2685	1 1 2	2 3 4	4 5 6
·43	2692	2698	2704	2710	2716	2723	2729	2735	2742	2748	1 1 2	3 3 4	4 5 6
·44	2754	2761	2767	2773	2780	2786	2793	2799	2805	2812	1 1 2	3 3 4	4 5 6
·45	2818	2825	2831	2838	2844	2851	2858	2864	2871	2877	1 1 2	3 3 4	5 5 6
·46	2884	2891	2897	2904	2911	2917	2924	2931	2938	2944	1 1 2	3 3 4	5 5 6
·47	2951	2958	2965	2972	2979	2985	2992	2999	3006	3013	1 1 2	3 3 4	5 5 6
·48	3020	3027	3034	3041	3048	3055	3062	3069	3076	3083	1 1 2	3 4 4	5 6 6
·49	3090	3097	3105	3112	3119	3126	3133	3141	3148	3155	1 1 2	3 4 4	5 6 6

	0	1	2	3	4	5	6	7	8	9	1 2 3	4 5 6	7 8 9
·50	3162	3170	3177	3184	3192	3199	3206	3214	3221	3228	1 1 2	3 4 4	5 6 7
·51	3236	3243	3251	3258	3266	3273	3281	3289	3296	3304	1 2 2	3 4 5	5 6 7
·52	3311	3319	3327	3334	3342	3350	3357	3365	3373	3381	1 2 2	3 4 5	5 6 7
·53	3388	3396	3404	3412	3420	3428	3436	3443	3451	3459	1 2 2	3 4 5	6 6 7
·54	3467	3475	3483	3491	3499	3508	3516	3524	3532	3540	1 2 2	3 4 5	6 6 7
·55	3548	3556	3565	3573	3581	3589	3597	3606	3614	3622	1 2 2	3 4 5	6 7 7
·56	3631	3639	3648	3656	3664	3673	3681	3690	3698	3707	1 2 3	3 4 5	6 7 8
·57	3715	3724	3733	3741	3750	3758	3767	3776	3784	3793	1 2 3	3 4 5	6 7 8
·58	3802	3811	3819	3828	3837	3846	3855	3864	3873	3882	1 2 3	4 4 5	6 7 8
·59	3890	3899	3908	3917	3926	3936	3945	3954	3963	3972	1 2 3	4 5 5	6 7 8
·60	3981	3990	3999	4009	4018	4027	4036	4046	4055	4064	1 2 3	4 5 6	6 7 8
·61	4074	4083	4093	4102	4111	4121	4130	4140	4150	4159	1 2 3	4 5 6	7 8 9
·62	4169	4178	4188	4198	4207	4217	4227	4236	4246	4256	1 2 3	4 5 6	7 8 9
·63	4266	4276	4285	4295	4305	4315	4325	4335	4345	4355	1 2 3	4 5 6	7 8 9
·64	4365	4375	4385	4395	4406	4416	4426	4436	4446	4457	1 2 3	4 5 6	7 8 9
·65	4467	4477	4487	4498	4508	4519	4529	4539	4550	4560	1 2 3	4 5 6	7 8 9
·66	4571	4581	4592	4603	4613	4624	4634	4645	4656	4667	1 2 3	4 5 6	7 9 10
·67	4677	4688	4699	4710	4721	4732	4742	4753	4764	4775	1 2 3	4 5 7	8 9 10
·68	4786	4797	4808	4819	4831	4842	4853	4864	4875	4887	1 2 3	4 6 7	8 9 10
·69	4898	4909	4920	4932	4943	4955	4966	4977	4989	5000	1 2 3	5 6 7	8 9 10
·70	5012	5023	5035	5047	5058	5070	5082	5093	5105	5117	1 2 4	5 6 7	8 9 11
·71	5129	5140	5152	5164	5176	5188	5200	5212	5224	5236	1 2 4	5 6 7	8 10 11
·72	5248	5260	5272	5284	5297	5309	5321	5333	5346	5358	1 2 4	5 6 7	9 10 11
·73	5370	5383	5395	5408	5420	5433	5445	5458	5470	5483	1 3 4	5 6 8	9 10 11
·74	5495	5508	5521	5534	5546	5559	5572	5585	5598	5610	1 3 4	5 6 8	9 10 12
·75	5623	5636	5649	5662	5675	5689	5702	5715	5728	5741	1 3 4	5 7 8	9 10 12
·76	5754	5768	5781	5794	5808	5821	5834	5848	5861	5875	1 3 4	5 7 8	9 11 12
·77	5888	5902	5916	5929	5943	5957	5970	5984	5998	6012	1 3 4	5 7 8	10 11 12
·78	6026	6039	6053	6067	6081	6095	6109	6124	6138	6152	1 3 4	6 7 8	10 11 13
·79	6166	6180	6194	6209	6223	6237	6252	6266	6281	6295	1 3 4	6 7 9	10 11 13
·80	6310	6324	6339	6353	6368	6383	6397	6412	6427	6442	1 3 4	6 7 9	10 12 13
·81	6457	6471	6486	6501	6516	6531	6546	6561	6577	6592	2 3 5	6 8 9	11 12 14
·82	6607	6622	6637	6653	6668	6683	6699	6714	6730	6745	2 3 5	6 8 9	11 12 14
·83	6761	6776	6792	6808	6823	6839	6855	6871	6887	6902	2 3 5	6 8 9	11 13 14
·84	6918	6934	6950	6966	6982	6998	7015	7031	7047	7063	2 3 5	6 8 10	11 13 15
·85	7079	7096	7112	7129	7145	7161	7178	7194	7211	7228	2 3 5	7 8 10	12 13 15
·86	7244	7261	7278	7295	7311	7328	7345	7362	7379	7396	2 3 5	7 8 10	12 13 15
·87	7413	7430	7447	7464	7482	7499	7516	7534	7551	7568	2 3 5	7 9 10	12 14 16
·88	7586	7603	7621	7638	7656	7674	7691	7709	7727	7745	2 4 5	7 9 11	12 14 16
·89	7762	7780	7798	7816	7834	7852	7870	7889	7907	7925	2 4 5	7 9 11	13 14 16
·90	7943	7962	7980	7998	8017	8035	8054	8072	8091	8110	2 4 6	7 9 11	13 15 17
·91	8128	8147	8166	8185	8204	8222	8241	8260	8279	8299	2 4 6	8 9 11	13 15 17
·92	8318	8337	8356	8375	8395	8414	8433	8453	8472	8492	2 4 6	8 10 12	14 15 17
·93	8511	8531	8551	8570	8590	8610	8630	8650	8670	8690	2 4 6	8 10 12	14 16 18
·94	8710	8730	8750	8770	8790	8810	8831	8851	8872	8892	2 4 6	8 10 12	14 16 18
·95	8913	8933	8954	8974	8995	9016	9036	9057	9078	9099	2 4 6	8 10 12	15 17 19
·96	9120	9141	9162	9183	9204	9226	9247	9268	9290	9311	2 4 6	8 11 13	15 17 19
·97	9333	9354	9376	9397	9419	9441	9462	9484	9506	9528	2 4 7	9 11 13	15 17 20
·98	9550	9572	9594	9616	9638	9661	9683	9705	9727	9750	2 4 7	9 11 13	16 18 20
·99	9772	9795	9817	9840	9863	9886	9908	9931	9954	9977	2 5 7	9 11 14	16 18 20

NATURAL SINES

Degrees	0' 0°.0	6' 0°.1	12' 0°.2	18' 0°.3	24' 0°.4	30' 0°.5	36' 0°.6	42' 0°.7	48' 0°.8	54' 0°.9	Mean Differences				
											1	2	3	4	5
0	·0000	0017	0035	0052	0070	0087	0105	0122	0140	0157	3	6	9	12	15
1	·0175	0192	0209	0227	0244	0262	0279	0297	0314	0332	3	6	9	12	15
2	·0349	0366	0384	0401	0419	0436	0454	0471	0488	0506	3	6	9	12	15
3	·0523	0541	0558	0576	0593	0610	0628	0645	0663	0680	3	6	9	12	15
4	·0698	0715	0732	0750	0767	0785	0802	0819	0837	0854	3	6	9	12	15
5	·0872	0889	0906	0924	0941	0958	0976	0993	1011	1028	3	6	9	12	14
6	·1045	1063	1080	1097	1115	1132	1149	1167	1184	1201	3	6	9	12	14
7	·1219	1236	1253	1271	1288	1305	1323	1340	1357	1374	3	6	9	12	14
8	·1392	1409	1426	1444	1461	1478	1495	1513	1530	1547	3	6	9	12	14
9	·1564	1582	1599	1616	1633	1650	1668	1685	1702	1719	3	6	9	12	14
10	·1736	1754	1771	1788	1805	1822	1840	1857	1874	1891	3	6	9	12	14
11	·1908	1925	1942	1959	1977	1994	2011	2028	2045	2062	3	6	9	11	14
12	·2079	2096	2113	2130	2147	2164	2181	2198	2215	2232	3	6	9	11	14
13	·2250	2267	2284	2300	2317	2334	2351	2368	2385	2402	3	6	8	11	14
14	·2419	2436	2453	2470	2487	2504	2521	2538	2554	2571	3	6	8	11	14
15	·2588	2605	2622	2639	2656	2672	2689	2706	2723	2740	3	6	8	11	14
16	·2756	2773	2790	2807	2823	2840	2857	2874	2890	2907	3	6	8	11	14
17	·2924	2940	2957	2974	2990	3007	3024	3040	3057	3074	3	6	8	11	14
18	·3090	3107	3123	3140	3156	3173	3190	3206	3223	3239	3	6	8	11	14
19	·3256	3272	3289	3305	3322	3338	3355	3371	3387	3404	3	5	8	11	14
20	·3420	3437	3453	3469	3486	3502	3518	3535	3551	3567	3	5	8	11	14
21	·3584	3600	3616	3633	3649	3665	3681	3697	3714	3730	3	5	8	11	14
22	·3746	3762	3778	3795	3811	3827	3843	3859	3875	3891	3	5	8	11	14
23	·3907	3923	3939	3955	3971	3987	4003	4019	4035	4051	3	5	8	11	14
24	·4067	4083	4099	4115	4131	4147	4163	4179	4195	4210	3	5	8	11	13
25	·4226	4242	4258	4274	4289	4305	4321	4337	4352	4368	3	5	8	11	13
26	·4384	4399	4415	4431	4446	4462	4478	4493	4509	4524	3	5	8	10	13
27	·4540	4555	4571	4586	4602	4617	4633	4648	4664	4679	3	5	8	10	13
28	·4695	4710	4726	4741	4756	4772	4787	4802	4818	4833	3	5	8	10	13
29	·4848	4863	4879	4894	4909	4924	4939	4955	4970	4985	3	5	8	10	13
30	·5000	5015	5030	5045	5060	5075	5090	5105	5120	5135	3	5	8	10	13
31	·5150	5165	5180	5195	5210	5225	5240	5255	5270	5284	2	5	7	10	12
32	·5299	5314	5329	5344	5358	5373	5388	5402	5417	5432	2	5	7	10	12
33	·5446	5461	5476	5490	5505	5519	5534	5548	5563	5577	2	5	7	10	12
34	·5592	5606	5621	5635	5650	5664	5678	5693	5707	5721	2	5	7	10	12
35	·5736	5750	5764	5779	5793	5807	5821	5835	5850	5864	2	5	7	10	12
36	·5878	5892	5906	5920	5934	5948	5962	5976	5990	6004	2	5	7	9	12
37	·6018	6032	6046	6060	6074	6088	6101	6115	6129	6143	2	5	7	9	12
38	·6157	6170	6184	6198	6211	6225	6239	6252	6266	6280	2	5	7	9	11
39	·6293	6307	6320	6334	6347	6361	6374	6388	6401	6414	2	4	7	9	11
40	·6428	6441	6455	6468	6481	6494	6508	6521	6534	6547	2	4	7	9	11
41	·6561	6574	6587	6600	6613	6626	6639	6652	6665	6678	2	4	7	9	11
42	·6691	6704	6717	6730	6743	6756	6769	6782	6794	6807	2	4	6	9	11
43	·6820	6833	6845	6858	6871	6884	6896	6909	6921	6934	2	4	6	8	11
44	·6947	6959	6972	6984	6997	7009	7022	7034	7046	7059	2	4	6	8	10

NATURAL SINES

Degrees	0′ 0°·0	6′ 0°·1	12′ 0°·2	18′ 0°·3	24′ 0°·4	30′ 0°·5	36′ 0°·6	42′ 0°·7	48′ 0°·8	54′ 0°·9	Mean Differences 1 2 3	4 5
45	·7071	7083	7096	7108	7120	7133	7145	7157	7169	7181	2 4 6	8 10
46	·7193	7206	7218	7230	7242	7254	7266	7278	7290	7302	2 4 6	8 10
47	·7314	7325	7337	7349	7361	7373	7385	7396	7408	7420	2 4 6	8 10
48	·7431	7443	7455	7466	7478	7490	7501	7513	7524	7536	2 4 6	8 10
49	·7547	7558	7570	7581	7593	7604	7615	7627	7638	7649	2 4 6	8 9
50	·7660	7672	7683	7694	7705	7716	7727	7738	7749	7760	2 4 6	7 9
51	·7771	7782	7793	7804	7815	7826	7837	7848	7859	7869	2 4 5	7 9
52	·7880	7891	7902	7912	7923	7934	7944	7955	7965	7976	2 4 5	7 9
53	·7986	7997	8007	8018	8028	8039	8049	8059	8070	8080	2 3 5	7 9
54	·8090	8100	8111	8121	8131	8141	8151	8161	8171	8181	2 3 5	7 8
55	·8192	8202	8211	8221	8231	8241	8251	8261	8271	8281	2 3 5	7 8
56	·8290	8300	8310	8320	8329	8339	8348	8358	8368	8377	2 3 5	6 8
57	·8387	8396	8406	8415	8425	8434	8443	8453	8462	8471	2 3 5	6 8
58	·8480	8490	8499	8508	8517	8526	8536	8545	8554	8563	2 3 5	6 8
59	·8572	8581	8590	8599	8607	8616	8625	8634	8643	8652	1 3 4	6 7
60	·8660	8669	8678	8686	8695	8704	8712	8721	8729	8738	1 3 4	6 7
61	·8746	8755	8763	8771	8780	8788	8796	8805	8813	8821	1 3 4	6 7
62	·8829	8838	8846	8854	8862	8870	8878	8886	8894	8902	1 3 4	5 7
63	·8910	8918	8926	8934	8942	8949	8957	8965	8973	8980	1 3 4	5 6
64	·8988	8996	9003	9011	9018	9026	9033	9041	9048	9056	1 3 4	5 6
65	·9063	9070	9078	9085	9092	9100	9107	9114	9121	9128	1 2 4	5 6
66	·9135	9143	9150	9157	9164	9171	9178	9184	9191	9198	1 2 3	5 6
67	·9205	9212	9219	9225	9232	9239	9245	9252	9259	9265	1 2 3	4 6
68	·9272	9278	9285	9291	9298	9304	9311	9317	9323	9330	1 2 3	4 5
69	·9336	9342	9348	9354	9361	9367	9373	9379	9385	9391	1 2 3	4 5
70	·9397	9403	9409	9415	9421	9426	9432	9438	9444	9449	1 2 3	4 5
71	·9455	9461	9466	9472	9478	9483	9489	9494	9500	9505	1 2 3	4 5
72	·9511	9516	9521	9527	9532	9537	9542	9548	9553	9558	1 2 3	3 4
73	·9563	9568	9573	9578	9583	9588	9593	9598	9603	9608	1 2 2	3 4
74	·9613	9617	9622	9627	9632	9636	9641	9646	9650	9655	1 2 2	3 4
75	·9659	9664	9668	9673	9677	9681	9686	9690	9694	9699	1 1 2	3 4
76	·9703	9707	9711	9715	9720	9724	9728	9732	9736	9740	1 1 2	3 3
77	·9744	9748	9751	9755	9759	9763	9767	9770	9774	9778	1 1 2	3 3
78	·9781	9785	9789	9792	9796	9799	9803	9806	9810	9813	1 1 2	3 3
79	·9816	9820	9823	9826	9829	9833	9836	9839	9842	9845	1 1 2	2 3
80	·9848	9851	9854	9857	9860	9863	9866	9869	9871	9874	0 1 1	2 2
81	·9877	9880	9882	9885	9888	9890	9893	9895	9898	9900	0 1 1	2 2
82	·9903	9905	9907	9910	9912	9914	9917	9919	9921	9923	0 1 1	2 2
83	·9925	9928	9930	9932	9934	9936	9938	9940	9942	9943	0 1 1	1 2
84	·9945	9947	9949	9951	9952	9954	9956	9957	9959	9960	0 1 1	1 2
85	·9962	9963	9965	9966	9968	9969	9971	9972	9973	9974	0 0 1	1 1
86	·9976	9977	9978	9979	9980	9981	9982	9983	9984	9985	0 0 1	1 1
87	·9986	9987	9988	9989	9990	9990	9991	9992	9993	9993	0 0 0	1 1
88	·9994	9995	9995	9996	9996	9997	9997	9997	9998	9998	0 0 0	0 0
89	·9998	9999	9999	9999	9999	1·000	1·000	1·000	1·000	1·000	0 0 0	0 0
90	1·000											

NATURAL COSINES

[Numbers in difference columns to be subtracted, not added.]

Degrees	0' 0°·0	6' 0°·1	12' 0°·2	18' 0°·3	24' 0°·4	30' 0°·5	36' 0°·6	42' 0°·7	48' 0°·8	54' 0°·9	Mean Differences				
											1	2	3	4	5
0	1·000	1·000	1·000	1·000	1·000	1·000	·9999	9999	9999	9999	0	0	0	0	0
1	·9998	9998	9998	9997	·9997	9997	9996	9996	9995	9995	0	0	0	0	0
2	·9994	9993	9993	9992	9991	9990	9990	9989	9988	9987	0	0	0	1	1
3	·9986	9985	9984	9983	9982	9981	9980	9979	9978	9977	0	0	1	1	1
4	·9976	9974	9973	9972	9971	9969	9968	9966	9965	9963	0	0	1	1	1
5	·9962	9960	9959	9957	9956	9954	9952	9951	9949	9947	0	1	1	1	2
6	·9945	9943	9942	9940	9938	9936	9934	9932	9930	9928	0	1	1	1	2
7	·9925	9923	9921	9919	9917	9914	9912	9910	9907	9905	0	1	1	2	2
8	·9903	9900	9898	9895	9893	9890	9888	9885	9882	9880	0	1	1	2	2
9	·9877	9874	9871	9869	9866	9863	9860	9857	9854	9851	0	1	1	2	2
10	·9848	9845	9842	9839	9836	9833	9829	9826	9823	9820	1	1	2	2	3
11	·9816	9813	9810	9806	9803	9799	9796	9792	9789	9785	1	1	2	2	3
12	·9781	9778	9774	9770	9767	9763	9759	9755	9751	9748	1	1	2	3	3
13	·9744	9740	9736	9732	9728	9724	9720	9715	9711	9707	1	1	2	3	3
14	·9703	9699	9694	9690	9686	9681	9677	9673	9668	9664	1	1	2	3	4
15	·9659	9655	9650	9646	9641	9636	9632	9627	9622	9617	1	2	2	3	4
16	·9613	9608	9603	9598	9593	9588	9583	9578	9573	9568	1	2	2	3	4
17	·9563	9558	9553	9548	9542	9537	9532	9527	9521	9516	1	2	3	3	4
18	·9511	9505	9500	9494	9489	9483	9478	9472	9466	9461	1	2	3	4	5
19	·9455	9449	9444	9438	9432	9426	9421	9415	9409	9403	1	2	3	4	5
20	·9397	9391	9385	9379	9373	9367	9361	9354	9348	9342	1	2	3	4	5
21	·9336	9330	9323	9317	9311	9304	9298	9291	9285	9278	1	2	3	4	5
22	·9272	9265	9259	9252	9245	9239	9232	9225	9219	9212	1	2	3	4	6
23	·9205	9198	9191	9184	9178	9171	9164	9157	9150	9143	1	2	3	5	6
24	·9135	9128	9121	9114	9107	9100	9092	9085	9078	9070	1	2	4	5	6
25	·9063	9056	9048	9041	9033	9026	9018	9011	9003	8996	1	3	4	5	6
26	·8988	8980	8973	8965	8957	8949	8942	8934	8926	8918	1	3	4	5	6
27	·8910	8902	8894	8886	8878	8870	8862	8854	8846	8838	1	3	4	5	7
28	·8829	8821	8813	8805	8796	8788	8780	8771	8763	8755	1	3	4	6	7
29	·8746	8738	8729	8721	8712	8704	8695	8686	8678	8669	1	3	4	6	7
30	·8660	8652	8643	8634	8625	8616	8607	8599	8590	8581	1	3	4	6	7
31	·8572	8563	8554	8545	8536	8526	8517	8508	8499	8490	2	3	5	6	8
32	·8480	8471	8462	8453	8443	8434	8425	8415	8406	8396	2	3	5	6	8
33	·8387	8377	8368	8358	8348	8339	8329	8320	8310	8300	2	3	5	6	8
34	·8290	8281	8271	8261	8251	8241	8231	8221	8211	8202	2	3	5	7	8
35	·8192	8181	8171	8161	8151	8141	8131	8121	8111	8100	2	3	5	7	8
36	·8090	8080	8070	8059	8049	8039	8028	8018	8007	7997	2	3	5	7	9
37	·7986	7976	7965	7955	7944	7934	7923	7912	7902	7891	2	4	5	7	9
38	·7880	7869	7859	7848	7837	7826	7815	7804	7793	7782	2	4	5	7	9
39	·7771	7760	7749	7738	7727	7716	7705	7694	7683	7672	2	4	6	7	9
40	·7660	7649	7638	7627	7615	7604	7593	7581	7570	7559	2	4	6	8	9
41	·7547	7536	7524	7513	7501	7490	7478	7466	7455	7443	2	4	6	8	10
42	·7431	7420	7408	7396	7385	7373	7361	7349	7337	7325	2	4	6	8	10
43	·7314	7302	7290	7278	7266	7254	7242	7230	7218	7206	2	4	6	8	10
44	·7193	7181	7169	7157	7145	7133	7120	7108	7096	7083	2	4	6	8	10

NATURAL COSINES

[Numbers in difference columns to be subtracted, not added.]

Degrees	0′ 0°·0	6′ 0°·1	12′ 0°·2	18′ 0°·3	24′ 0°·4	30′ 0°·5	36′ 0°·6	42′ 0°·7	48′ 0°·8	54′ 0°·9	Mean Differences 1 2 3	4 5
45	·7071	7059	7046	7034	7022	7009	6997	6984	6972	6959	2 4 6	8 10
46	·6947	6934	6921	6909	6896	6884	6871	6858	6845	6833	2 4 6	8 11
47	·6820	6807	6794	6782	6769	6756	6743	6730	6717	6704	2 4 6	9 11
48	·6691	6678	6665	6652	6639	6626	6613	6600	6587	6574	2 4 7	9 11
49	·6561	6547	6534	6521	6508	6494	6481	6468	6455	6441	2 4 7	9 11
50	·6428	6414	6401	6388	6374	6361	6347	6334	6320	6307	2 4 7	9 11
51	·6293	6280	6266	6252	6239	6225	6211	6198	6184	6170	2 5 7	9 11
52	·6157	6143	6129	6115	6101	6088	6074	6060	6046	6032	2 5 7	9 12
53	·6018	6004	5990	5976	5962	5948	5934	5920	5906	5892	2 5 7	9 12
54	·5878	5864	5850	5835	5821	5807	5793	5779	5764	5750	2 5 7	9 12
55	·5736	5721	5707	5693	5678	5664	5650	5635	5621	5606	2 5 7	10 12
56	·5592	5577	5563	5548	5534	5519	5505	5490	5476	5461	2 5 7	10 12
57	·5446	5432	5417	5402	5388	5373	5358	5344	5329	5314	2 5 7	10 12
58	·5299	5284	5270	5255	5240	5225	5210	5195	5180	5165	2 5 7	10 13
59	·5150	5135	5120	5105	5090	5075	5060	5045	5030	5015	3 5 8	10 13
60	·5000	4985	4970	4955	4939	4924	4909	4894	4879	4863	3 5 8	10 13
61	·4848	4833	4818	4802	4787	4772	4756	4741	4726	4710	3 5 8	10 13
62	·4695	4679	4664	4648	4633	4617	4602	4586	4571	4555	3 5 8	10 13
63	·4540	4524	4509	4493	4478	4462	4446	4431	4415	4399	3 5 8	10 13
64	·4384	4368	4352	4337	4321	4305	4289	4274	4258	4242	3 5 8	11 13
65	·4226	4210	4195	4179	4163	4147	4131	4115	4099	4083	3 5 8	11 13
66	·4067	4051	4035	4019	4003	3987	3971	3955	3939	3923	3 5 8	11 14
67	·3907	3891	3875	3859	3843	3827	3811	3795	3778	3762	3 5 8	11 14
68	·3746	3730	3714	3697	3681	3665	3649	3633	3616	3600	3 5 8	11 14
69	·3584	3567	3551	3535	3518	3502	3486	3469	3453	3437	3 5 8	11 14
70	·3420	3404	3387	3371	3355	3338	3322	3305	3289	3272	3 5 8	11 14
71	·3256	3239	3223	3206	3190	3173	3156	3140	3123	3107	3 6 8	11 14
72	·3090	3074	3057	3040	3024	3007	2990	2974	2957	2940	3 6 8	11 14
73	·2924	2907	2890	2874	2857	2840	2823	2807	2790	2773	3 6 8	11 14
74	·2756	2740	2723	2706	2689	2672	2656	2639	2622	2605	3 6 8	11 14
75	·2588	2571	2554	2538	2521	2504	2487	2470	2453	2436	3 6 8	11 14
76	·2419	2402	2385	2368	2351	2334	2317	2300	2284	2267	3 6 8	11 14
77	·2250	2233	2215	2198	2181	2164	2147	2130	2113	2096	3 6 9	11 14
78	·2079	2062	2045	2028	2011	1994	1977	1959	1942	1925	3 6 9	11 14
79	·1908	1891	1874	1857	1840	1822	1805	1788	1771	1754	3 6 9	11 14
80	·1736	1719	1702	1685	1668	1650	1633	1616	1599	1582	3 6 9	12 14
81	·1564	1547	1530	1513	1495	1478	1461	1444	1426	1409	3 6 9	12 14
82	·1392	1374	1357	1340	1323	1305	1288	1271	1253	1236	3 6 9	12 14
83	·1219	1201	1184	1167	1149	1132	1115	1097	1080	1063	3 6 9	12 14
84	·1045	1028	1011	0993	0976	0958	0941	0924	0906	0889	3 6 9	12 14
85	·0872	0854	0837	0819	0802	0785	0767	0750	0732	0715	3 6 9	12 15
86	·0698	0680	0663	0645	0628	0610	0593	0576	0558	0541	3 6 9	12 15
87	·0523	0506	0488	0471	0454	0436	0419	0401	0384	0366	3 6 9	12 15
88	·0349	0332	0314	0297	0279	0262	0244	0227	0209	0192	3 6 9	12 15
89	·0175	0157	0140	0122	0105	0087	0070	0052	0035	0017	3 6 9	12 15
90	·0000											

NATURAL TANGENTS

Degrees	0' 0°·0	6' 0°·1	12' 0°·2	18' 0°·3	24' 0°·4	30' 0°·5	36' 0°·6	42' 0°·7	48' 0°·8	54' 0°·9	Mean Differences				
											1	2	3	4	5
0	·0000	0017	0035	0052	0070	0087	0105	0122	0140	0157	3	6	9	12	15
1	·0175	0192	0209	0227	0244	0262	0279	0297	0314	0332	3	6	9	12	15
2	·0349	0367	0384	0402	0419	0437	0454	0472	0489	0507	3	6	9	12	15
3	·0524	0542	0559	0577	0594	0612	0629	0647	0664	0682	3	6	9	12	15
4	·0699	0717	0734	0752	0769	0787	0805	0822	0840	0857	3	6	9	12	15
5	·0875	0892	0910	0928	0945	0963	0981	0998	1016	1033	3	6	9	12	15
6	·1051	1069	1086	1104	1122	1139	1157	1175	1192	1210	3	6	9	12	15
7	·1228	1246	1263	1281	1299	1317	1334	1352	1370	1388	3	6	9	12	15
8	·1405	1423	1441	1459	1477	1495	1512	1530	1548	1566	3	6	9	12	15
9	·1584	1602	1620	1638	1655	1673	1691	1709	1727	1745	3	6	9	12	15
10	·1763	1781	1799	1817	1835	1853	1871	1890	1908	1926	3	6	9	12	15
11	·1944	1962	1980	1998	2016	2035	2053	2071	2089	2107	3	6	9	12	15
12	·2126	2144	2162	2180	2199	2217	2235	2254	2272	2290	3	6	9	12	15
13	·2309	2327	2345	2364	2382	2401	2419	2438	2456	2475	3	6	9	12	15
14	·2493	2512	2530	2549	2568	2586	2605	2623	2642	2661	3	6	9	12	16
15	·2679	2698	2717	2736	2754	2773	2792	2811	2830	2849	3	6	9	13	16
16	·2867	2886	2905	2924	2943	2962	2981	3000	3019	3038	3	6	9	13	16
17	·3057	3076	3096	3115	3134	3153	3172	3191	3211	3230	3	6	10	13	16
18	·3249	3269	3288	3307	3327	3346	3365	3385	3404	3424	3	6	10	13	16
19	·3443	3463	3482	3502	3522	3541	3561	3581	3600	3620	3	7	10	13	16
20	·3640	3659	3679	3699	3719	3739	3759	3779	3799	3819	3	7	10	13	17
21	·3839	3859	3879	3899	3919	3939	3959	3979	4000	4020	3	7	10	13	17
22	·4040	4061	4081	4101	4122	4142	4163	4183	4204	4224	3	7	10	14	17
23	·4245	4265	4286	4307	4327	4348	4369	4390	4411	4431	3	7	10	14	17
24	·4452	4473	4494	4515	4536	4557	4578	4599	4621	4642	4	7	11	14	18
25	·4663	4684	4706	4727	4748	4770	4791	4813	4834	4856	4	7	11	14	18
26	·4877	4899	4921	4942	4964	4986	5008	5029	5051	5073	4	7	11	15	18
27	·5095	5117	5139	5161	5184	5206	5228	5250	5272	5295	4	7	11	15	18
28	·5317	5340	5362	5384	5407	5430	5452	5475	5498	5520	4	8	11	15	19
29	·5543	5566	5589	5612	5635	5658	5681	5704	5727	5750	4	8	12	15	19
30	·5774	5797	5820	5844	5867	5890	5914	5938	5961	5985	4	8	12	16	20
31	·6009	6032	6056	6080	6104	6128	6152	6176	6200	6224	4	8	12	16	20
32	·6249	6273	6297	6322	6346	6371	6395	6420	6445	6469	4	8	12	16	20
33	·6494	6519	6544	6569	6594	6619	6644	6669	6694	6720	4	8	13	17	21
34	·6745	6771	6796	6822	6847	6873	6899	6924	6950	6976	4	9	13	17	21
35	·7002	7028	7054	7080	7107	7133	7159	7186	7212	7239	4	9	13	18	22
36	·7265	7292	7319	7346	7373	7400	7427	7454	7481	7508	5	9	14	18	23
37	·7536	7563	7590	7618	7646	7673	7701	7729	7757	7785	5	9	14	18	23
38	·7813	7841	7869	7898	7926	7954	7983	8012	8040	8069	5	9	14	19	24
39	·8098	8127	8156	8185	8214	8243	8273	8302	8332	8361	5	10	15	20	24
40	·8391	8421	8451	8481	8511	8541	8571	8601	8632	8662	5	10	15	20	25
41	·8693	8724	8754	8785	8816	8847	8878	8910	8941	8972	5	10	16	21	26
42	·9004	9036	9067	9099	9131	9163	9195	9228	9260	9293	5	11	16	21	27
43	·9325	9358	9391	9424	9457	9490	9523	9556	9590	9623	6	11	17	22	28
44	·9657	9691	9725	9759	9793	9827	9861	9896	9930	9965	6	11	17	23	29

NATURAL TANGENTS

Degrees	0' 0°·0	6' 0°·1	12' 0°·2	18' 0°·3	24' 0°·4	30' 0°·5	36' 0°·6	42' 0°·7	48' 0°·8	54' 0°·9	Mean Differences 1	2	3	4	5
45	1·0000	0035	0070	0105	0141	0176	0212	0247	0283	0319	6	12	18	24	30
46	1·0355	0392	0428	0464	0501	0538	0575	0612	0649	0686	6	12	18	25	31
47	1·0724	0761	0799	0837	0875	0913	0951	0990	1028	1067	6	13	19	25	32
48	1·1106	1145	1184	1224	1263	1303	1343	1383	1423	1463	7	13	20	27	33
49	1·1504	1544	1585	1626	1667	1708	1750	1792	1833	1875	7	14	21	28	34
50	1·1918	1960	2002	2045	2088	2131	2174	2218	2261	2305	7	14	22	29	36
51	1·2349	2393	2437	2482	2527	2572	2617	2662	2708	2753	8	15	23	30	38
52	1·2799	2846	2892	2938	2985	3032	3079	3127	3175	3222	8	16	24	31	39
53	1·3270	3319	3367	3416	3465	3514	3564	3613	3663	3713	8	16	25	33	41
54	1·3764	3814	3865	3916	3968	4019	4071	4124	4176	4229	9	17	26	34	43
55	1·4281	4335	4388	4442	4496	4550	4605	4659	4715	4770	9	18	27	36	45
56	1·4826	4882	4938	4994	5051	5108	5166	5224	5282	5340	10	19	29	38	48
57	1·5399	5458	5517	5577	5637	5697	5757	5818	5880	5941	10	20	30	40	50
58	1·6003	6066	6128	6191	6255	6319	6383	6447	6512	6577	11	21	32	43	53
59	1·6643	6709	6775	6842	6909	6977	7045	7113	7182	7251	11	23	34	45	56
60	1·7321	7391	7461	7532	7603	7675	7747	7820	7893	7966	12	24	36	48	60
61	1·8040	8115	8190	8265	8341	8418	8495	8572	8650	8728	13	26	38	51	64
62	1·8807	8887	8967	9047	9128	9210	9292	9375	9458	9542	14	27	41	55	68
63	1·9626	9711	9797	9883	9970	2·0057	2·0145	2·0233	2·0323	2·0413	15	29	44	58	73
64	2·0503	0594	0686	0778	0872	0965	1060	1155	1251	1348	16	31	47	63	78
65	2·1445	1543	1642	1742	1842	1943	2045	2148	2251	2355	17	34	51	68	85
66	2·2460	2566	2673	2781	2889	2998	3109	3220	3332	3445	18	37	55	73	92
67	2·3559	3673	3789	3906	4023	4142	4262	4383	4504	4627	20	40	60	79	99
68	2·4751	4876	5002	5129	5257	5386	5517	5649	5782	5916	22	43	65	87	108
69	2·6051	6187	6325	6464	6605	6746	6889	7034	7179	7326	24	47	71	95	119
70	2·7475	7625	7776	7929	8083	8239	8397	8556	8716	8878	26	52	78	104	131
71	2·9042	9208	9375	9544	9714	9887	3·0061	3·0237	3·0415	3·0595	29	58	87	116	145
72	3·0777	0961	1146	1334	1524	1716	1910	2106	2305	2506	32	64	96	129	161
73	3·2709	2914	3122	3332	3544	3759	3977	4197	4420	4646	36	72	108	144	180
74	3·4874	5105	5339	5576	5816	6059	6305	6554	6806	7062	41	81	122	163	204
75	3·7321	7583	7848	8118	8391	8667	8947	9232	9520	9812	46	93	139	186	232
76	4·0108	0408	0713	1022	1335	1653	1976	2303	2635	2972	53	107	160	213	267
77	4·3315	3662	4015	4374	4737	5107	5483	5864	6252	6646					
78	4·7046	7453	7867	8288	8716	9152	9594	5·0045	5·0504	5·0970	Mean differences cease				
79	5·1446	1929	2422	2924	3435	3955	4486	5026	5578	6140	to be sufficiently accurate.				
80	5·6713	7297	7894	8502	9124	9758	6·0405	6·1066	6·1742	6·2432					
81	6·3138	3859	4596	5350	6122	6912	7720	8548	9395	7·0264					
82	7·1154	2066	3002	3962	4947	5958	6996	8062	9158	8·0285					
83	8·1443	2636	3863	5126	6427	7769	9152	9·0579	9·2052	9·3572					
84	9·5144	9·677	9·845	10·02	10·20	10·39	10·58	10·78	10·99	11·20					
85	11·43	11·66	11·91	12·16	12·43	12·71	13·00	13·30	13·62	13·95					
86	14·30	14·67	15·06	15·46	15·89	16·35	16·83	17·34	17·89	18·46					
87	19·08	19·74	20·45	21·20	22·02	22·90	23·86	24·90	26·03	27·27					
88	28·64	30·14	31·82	33·69	35·80	38·19	40·92	44·07	47·74	52·08					
89	57·29	63·66	71·62	81·85	95·49	114·6	143·2	191·0	286·5	573·0					
90	∞														

LOGARITHMS
$\log_e X$

(1) $1 \cdot 0 \leqslant X \leqslant 9 \cdot 9$

	·0	·1	·2	·3	·4	·5	·6	·7	·8	9
1	0·0000	0·0953	0·1823	0·2624	0·3365	0·4055	0·4700	0·5306	0·5878	0·6419
2	0·6931	0·7419	0·7885	0·8329	0·8755	0·9163	0·9555	0·9933	1·0296	1·0647
3	1·0986	1·1314	1·1632	1·1939	1·2238	1·2528	1·2809	1·3083	1·3350	1·3610
4	1·3863	1·4110	1·4351	1·4586	1·4816	1·5041	1·5261	1·5476	1·5686	1·5892
5	1·6094	1·6292	1·6487	1·6677	1·6864	1·7047	1·7228	1·7405	1·7579	1·7750
6	1·7918	1·8083	1·8245	1·8405	1·8563	1·8718	1·8871	1·9021	1·9169	1·9315
7	1·9459	1·9601	1·9741	1·9879	2·0015	2·0149	2·0281	2·0412	2·0541	2·0669
8	2·0794	2·0919	2·1041	2·1163	2·1282	2·1401	2·1518	2·1633	2·1748	2·1861
9	2·1972	2·2083	2·2192	2·2300	2·2407	2·2513	2·2618	2·2721	2·2824	2·2925

(2) $X \geqslant 10$

Proceed as follows, using the table below.

Since $290 = 2 \cdot 90 \times 100$

$$\log_e 290 = \log_e 2 \cdot 90 + \log_e 100$$
$$= 1 \cdot 0647 \text{ (from table above)} + 4 \cdot 6052 \text{ (from table below)}$$
$$= 5 \cdot 6699.$$

X	10	100	1,000	10,000	100,000	1,000,000
$\log_e X$	2·3026	4·6052	6·9078	9·2103	11·5129	13·8155

(3) $0 < X \leqslant 0 \cdot 99$

Proceed as follows, using the table below.

Since $0 \cdot 29 = 2 \cdot 9 \times 10^{-1}$

$$\log_e 0 \cdot 29 = \log_e 2 \cdot 9 + \log_e 10^{-1}$$
$$= 1 \cdot 0647 + \bar{3} \cdot 6974 = 1 \cdot 0647 + 0 \cdot 6974 + \bar{3}$$
$$= 1 \cdot 7621 + \bar{3}$$
$$= \bar{2} \cdot 7621.$$

X	10^{-1}	10^{-2}	10^{-3}	10^{-4}	10^{-5}	10^{-6}
$\log_e X$	$\bar{3} \cdot 6974$	$\bar{5} \cdot 3948$	$\bar{7} \cdot 0922$	$\bar{10} \cdot 7897$	$\bar{12} \cdot 4871$	$\bar{14} \cdot 1845$

INDEX

This index is to enable the reader to find definitions, the more important results and the pages on which main discussions start. It does not try to list every reference.